Environmental Division

Presentations at the 2009 AIChE Annual Meeting

Nashville, Tennessee, USA
8-13 November 2009

ISBN: 978-1-61567-933-1

Printed from e-media with permission by:

Curran Associates, Inc.
57 Morehouse Lane
Red Hook, NY 12571

Some format issues inherent in the e-media version may also appear in this print version.

Printed by Curran Associates, Inc. (2010)

For permission requests, please contact AIChE
at the address below.

AIChE
3 Park Avenue
New York, NY 10016-5991

Phone: (203) 702-7660
Fax: (203) 775-5177

www.aiche.org

Additional copies of this publication are available from:

Curran Associates, Inc.
57 Morehouse Lane
Red Hook, NY 12571 USA
Phone: 845-758-0400
Fax: 845-758-2634
Email: curran@proceedings.com
Web: www.proceedings.com

TABLE OF CONTENTS

Fate and Transport of Metal Oxide Nano-Particles in Aquatic System

E. Sahle-Demessie[1], Z. Li,[2], Q. Zhao[1], G. Sorial[2]

[1]U.S. Environmental Protection Agency, Office of Research and Development, NRMRL, 26 W. Martin Luther King Drive (MS 443), Cincinnati, Ohio 45268.

[2]Department of Civil and Environmental Engineering, University of Cincinnati P.O. Box 210071, Cincinnati, OH 45221-0071

Abstract

Physical and chemical properties of nanoscale CeO_2 particles, including size, composition and crystallinity were measured. Size measurements using DLS methods after sonication differs from values reported by vendors due to aggregation during storage. The effect of water chemistry on long term aquatic stability of dispersed nanoparticle was evaluated. The zeta potential of nanoparticles were measured at selected pH and ionic strengths based on levels of NaCl concentration (1 mM and 10 mM). Nano-CeO_2 showed negative surface charges as the pH and NaCl concentration increase. The stability results were correlated with the experimental data on their transport through sand-packed columns. Transport study of nano-CeO_2 and the influence of pH and ionic concentration showed that the mobility of CeO_2 nanoparticles is greatly affected by the water chemistry. The favorable condition for nanoparticles to transport through sand-packed column is lower ionic concentration (<10 mM) and higher pH (≥6). A mathematical model of the transport and retention of the nanoparticles in environmental aqueous systems was developed.

Keywords CeO_2 nano-particles aggregation, stability and transport, octano-water partitioning

1. Introduction

The increased production and use of manufactured nanoparticles presents concerns for human health and the environmental [1]. When nano-scale particles are released to the environment, their stability, transport and interaction with other pollutants and natural organic matter play significant role in determining risk. The transport and fate of commercial available nanoparticles has greater practical importance in determining how to respond to accidental releases, and handling of end-of-life and disposal issues. The objective of the study is to investigate the transport and retention of commercial available nano-size CeO_2 particles in water saturated porous media. Experimental studies are used to determine the effects of water chemistry on the transport of the particles in packed bed columns, and to develop mathematical models that predicts the transport and retention of nano-materials in environmental aquatic systems.

2. Characterization of CeO_2 Nano Particles

CeO_2 nanoparticles were received from Alfa Aesar Co. in a suspension form, 18 wt % in water with dispersants and polishing additives. The suspension was then diluted to 50mg/L with Milli-Q water. The pH of the suspensions were adjusted by adding appropriate amounts from 1N HCl or 1 M NaOH standard solutions.

Solid phase characterization of nano-particles includes particle shape and size (transmission electron microscopy, TEM), crystalinity (X-Ray Diffraction, SEM-EDAX), surface area (BET) and UV-Vis absorbance. Particles were sonicated for 20 min before deposited onto a copper mesh grid to obtain the images with the Transmission Electron Microscopy (Joul). TEM images of CeO_2, shown in Figure 1, indicated that particles are poly dispersed. Powder X-Ray microanalysis were performed to determine the chemical composition and crystallographic structure of the nanoparticles, and the data were collected with Cu Kα radiation (λ= 0.15059 nm) with a 0.04 2θ step size at the speed of 2 θ /min (Figure 2). Nanoparticle suspensions will be characterized with dilute water for turbidity, zeta-potential and electron mobilities. The size distribution of the CeO_2 nanoparticles was determined by Laser diffraction Particle Size Analyzer (Beckman Coulter LS230) and the hydrodynamic diameter was determined using dynamic light scattering system (DLS, Malvern ZetaSizer Nano ZS) (Figure 3).. Scanning Electron Microscopy (SEM) and Energy Dispersive Spectroscopy (EDS) were also used to determine surface composition of the particles to measure the mass and atomic ratios.

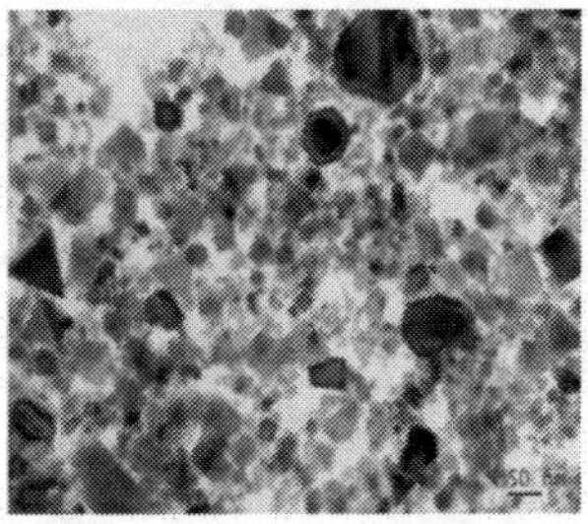

Figure 1. Transmission Electron microscopic image of CeO_2 particles

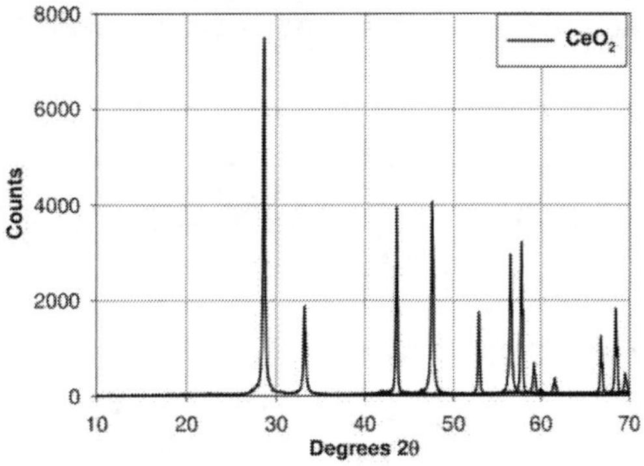

Figure 2 X-Ray diffraction pattern of nano-cerium oxide

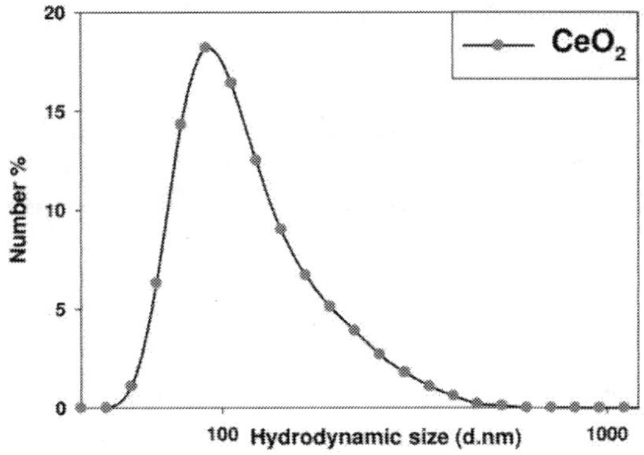

Figure 3. Hydrodynamic size distribution of CeO2 obtained from DLS is based on the intensity of scattered light

3. Stability of CeO_2 nanoparticles in Aqueous System

Much colloidal dispersions have kinetic stability, even though they are unstable thermodynamically. Stability of nanoparticles in aqueous system can have significant influence in the risk analysis: exposure, bioavailability, mobility and reactivity. In this task we study kinetic stability of two-phase hydrophobic nanoparticle colloids that include coalescence and flocculation. With coalescence the small particles are fused and surface area reduced, whereas in flocculation the small nanoparticles clump together like a bunch of grapes. In this research we examine the characteristics, dispersion and stability of nano-particles in water as well as the effects of water chemistry and the influence of natural organic matter.

The ζ-potential is a very good index of the magnitude of repulsive interaction between colloidal particles and measurement of ζ-potential is used to assess the stability of CeO_2 colloidal. This includes the condition in which the particles resist flocculation or aggregation and exhibits a long shelf-life. This will depend on the balance of the repulsive and attractive forces that exist between particles as they approach one another. If all the particles have a mutual repulsion then the dispersion will remain stable. However, if the particles have little or no repulsive force then some instability mechanism will eventually take place e.g. flocculation, aggregation etc. Figure 4 shows zeta-potential as a function of pH for CeO_2 in different solution chemistry. Study indicated that long term stability of colloids, and the influence of natural organic matter depended on electrolytic composition, ionic strength, and pH on– humic acids on stability and aggregation on nanoparticles.

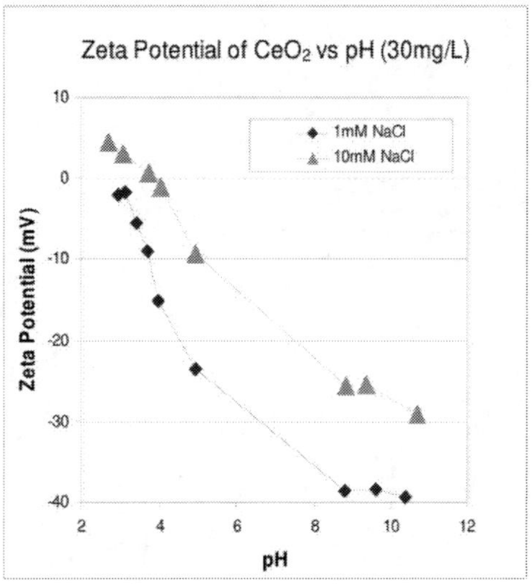

Figure 4. ζ-Potential of CeO_2 at to different ion concentrations

4. Transport study

A multi-column experimental system, shown on Figure 1, has been used in studying nano-particle the transport. In order to experimentally evaluate the role of water chemistry on the deposition and entrainment of nano-CeO_2 the flow composition was changed in three steps following procedures in previous studies [2,3]. At first Glass columns (25.4 mm ID., 450 mm length) were thoroughly cleaned with Milli-Q water, and were saturated with ion free MilliQ water. The columns were then saturated with a particle-free solution which has the same ionic concentration and pH with the nanoparticle suspension. Concentrations of NaCl in the effluent stream were monitored using conductivity prob. During the deposition phase, Phase 1, 50 mg/L CeO_2 nanoparticle suspension was continuously fed to the column at a rate of 10 ml/min. The pH and ionic concentration of the nanoparticle suspensions was maintained constant for each column but varied for different columns. During the rinse step, Phase 2, the particle-free solution was fed to rinse the attached particles at the same ionic strength as the deposition stage. During Phase 3, the feed flow was switched to Milli-Q water to lower the ionic concentration and the fraction of particles that were deposited at

Phase I are now re-retrained.

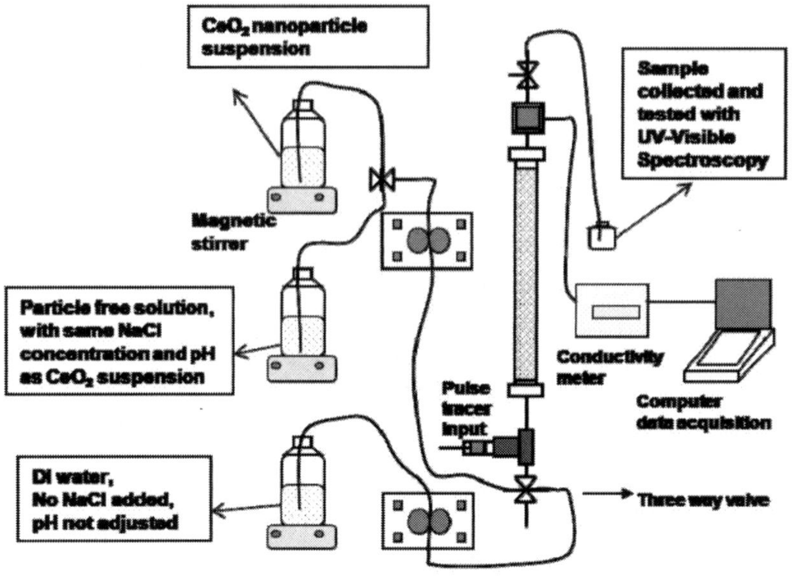

Figure 5. Transport experimental setup, Multi-column experimental system

4.1 Porous media

The sand used in this study was obtained from AGSCO Co. composed mainly quartz and trace quantities of feldspar. The columns were packed with sieve fraction 20 x 30 mesh, ranging from 0.595 mm to 0.851 mm, was used as packed bed material. Prior to use, sand was filled soaked in 1M nitrate acid for 1 hr, followed by rinsing with the sand with Milli-Q water till the rinse solution achieved pH 6~7. The water was drained and the sand was kept in an oven at 550 °C for 12h to dry and remove all the attached organics.

4.2 Analytical Procedures

Concentrations of CeO_2 particles in effluent stream were monitored by collecting ml fractions every 2 min. and analyzing with UV-vis spectrometer at the wavelength of $\lambda = 309$

nm. Multi-concentration standard samples of nano-CeO_2 were prepared for calibration and quantitative analysis. Effluent samples were collected and measured by the UV-vis spectroscopy

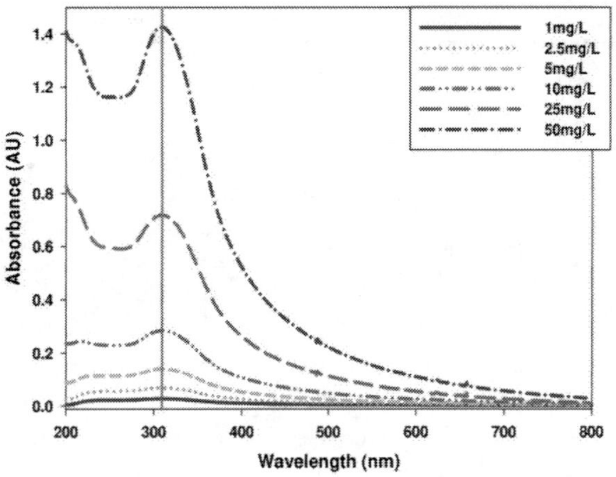

Figure 6. Uv-Vis spectroscopy of CeO_2 nano-particles at increased concentrations

5. Results and Discussions

Breakthrough study

A representative breakthrough profile is presented in Figure 2. The breakthrough curves at the presence of various ionic concentrations are shown by the normalized effluent nanoparticle concentration (C/C_0) as a function of cumulative volume normalized to pore volumes. The pH was maintained at 6.

The results show that the water chemistry has a great impact on the transport of CeO_2 nanoparticles. In Phase1, for ionic concentration lower than 10mM, the C/C_0 begin to increase after the first pore volume, and keep increasing until the feed was switched to the particle free solution. A fast decrease of C/C_0 was observed in Phase 2. During Phase 3, particles which were previously deposited in Phase 2 was partly detached due to a sudden decrease of ionic concentration. Sharp peaks were observed during this phase.

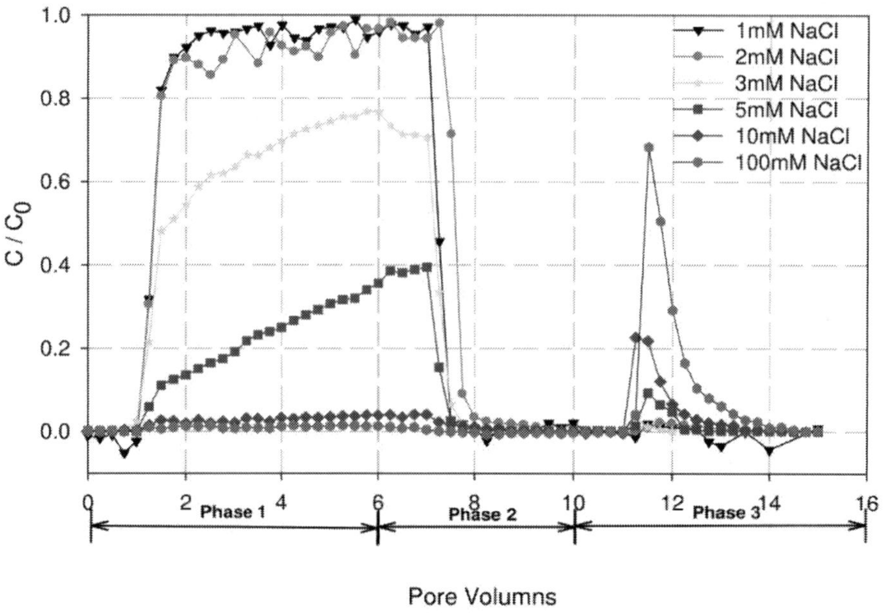

Figure 7. The effect of NaCl concentration on the deposition and transport of nano-CeO_2 at pH 6, NaCl concentration 1mM~100 mM.

Collector Efficiency

$$\eta = \frac{1}{\pi a_c^2 U C_o} \qquad 1$$

Where a_c is the radius of sand particle, U is the superficial velocity and Co bulk particle concentration. The accumulation of particles in the volume element

$$\frac{3}{4}\eta (1 - \varepsilon) \frac{U}{a_c} C A_c dz = - Q dC \qquad 2$$

Using boundary conditions at Z = 0, is C = Co and at Z = L, C = Ce where L is the bed length noting U = Q/Ao

$$\int_{Co}^{Ce} \frac{dc}{C} = - \frac{3}{4 a_c} \eta (1 - \varepsilon) \int_0^L dZ \qquad 3$$

The actual (experimental) single collector efficiency, ηexp, is derived from this equation and the experimental attachment efficiency, which can characterize the full be is estimated using:

$$\alpha_{exp} = -\frac{4}{3}\frac{a_c}{(1-\varepsilon)\,L\,\eta_o}\ln\left(\frac{C_e}{C_o}\right) \qquad 4$$

Where Ce/Co = value is taken from the break through curve in the initial single stage of deposition, since at the early stage of the deposition process, , can be viewed as the "clean-bed" single collector removal efficiency, η_o.

The fraction of particles recovered during the rinse phase, Phase 2 and re-entrainment phase, Phase 3, are expressed as FRC_2 and FRE_3, respectively. To calculated the fraction of nanoparticles recovered during phase 2 and 3 the following the mathematical expressions were followed.

$$FRC_2 = \frac{\sum \text{Mass of NP recovered during Phase 2}}{\sum \text{Mass Deposited during Phase 1}}$$

$$FRE_3 = \frac{\sum \text{Mass of NP recovered during Phase 3}}{\sum \text{Mass Deposited during Phase 1 - Mass recovered phase 2}}$$

Here the d_c is the diameter of the bed material, ε is the bed porosity, L is the bed length and η_0 is the single collector efficiency which was calculated by the similar method [4]. Monitoring the effluent concentration of nanoparticles showed that the break through times from one pore volume flow to practically no breakthrough. The effect of ionic concentration on the particle deposition and retention are demonstrated in Figure 3.

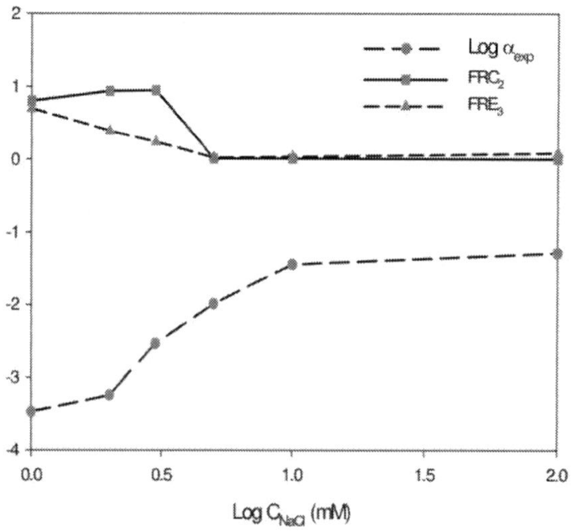

Figure 8. Effects of ionic concentration on the deposition and re-entrainment of nano-CeO_2 particles (pH =6, ε=0.37, dp= 69.7 nm, dc=0.707 mm, η_0 =0.0835)

Pollutant adsorption on Nanoparticles

The fate of organic compound released into the environment is primarily controlled by three factors: 1) the prevailing environmental conditions at the point(s) of discharge, transport and subsequent residence, 2) the physicochemical properties of the compound, and 3) patterns of use (that is, locus and timing of introduction). The physico-chemical properties relating to the environmental behavior of hydrophobic organic compounds are mainly affected by the aqueous solubility and octanol-water partition coefficient. Aqueous solubility (S_w) is the equilibrium distribution of a solute between water and solute phases. In other words it is the maximum solute concentration possible at equilibrium, and it can function as a limiting factor in concentration dependent (for example, kinetic) processes The octanol-water partition coefficient is the ratio of the concentration of a chemical in octanol and in water at equilibrium. Octanol is an organic solvent that is used as a surrogate for natural organic matter. This parameter is used in many environmental studies to help determine the fate of chemicals in the environment, such as to predict the extent a contaminant will bioaccumulate in fish. The octanol-water partition coefficient has been correlated to water solubility; therefore, the water solubility of a substance can be used to estimate its octanol-water partition coefficient. The presence of nano-particles can influence the octanol-water partitioning of hydrophobic organic water contaminants. Figure 9 shows thermogravimetric analysis of nano_TiO_2 used in determining octanol-water partitioning of phenatherene. The effects of water chemistry, pH and exposure to light are shown in this study.

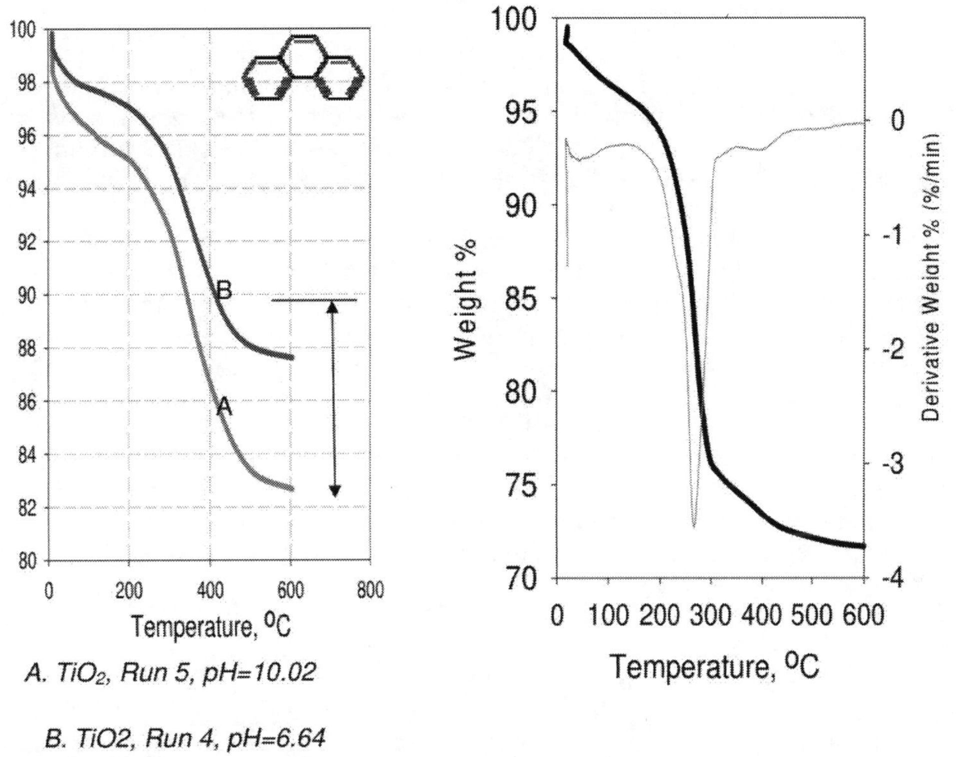

Figure 9 Thermogravimetric analysis of phenatherene adsorption on nano-TiO_2 showing th a) effects of pH and b) light exposure

References

(1) A. Baun N, B.Hartmanm, K.Grieger, K.O.Kusk, *Ecotoxicology*, 2008, 17, 387-395

(2) McDowell-Boyer, L.M. Environmental Science and Technology, 1992, 26, 586-593

(3) Franchi, A. and O'Melia, C.R. *Environ. Sci & Techn.*, 2003, 37, 1122-1229

(4) Yao, KM, Haboboam, M.T. and O'Melia, C.R. *Environ. Sci & Techn*, 1971, 5, 1105-1112.

ADSORPTION OF ARSENATE ON ALGINATE ENCAPSULATED MAGNETIC SORBENT WITH THE PRESENCE OF HUMIC SUBSTANCES: A SPECTROSCOPY ANALYSIS

Soh Fong Lim, Department of Chemical Engineering and Energy Sustainability, University Malaysia Sarawak, Kota Samarahan, Malaysia
J. Paul Chen, Division of Environmental Science and Engineering, National University of Singapore, Singapore, Singapore

Abstract

The adsorption of arsenate onto alginate encapsulated magnetic sorbent with the presence of humic substances is studied in this paper. The purpose of this investigation is to explore the adsorption chemistry of arsenate with the presence of humic substances through the surface morphology, microstructure, and element constitution of the sorbent, as well as the binding characteristics between arsenate and magnetic sorbent. The scanning electron microscopy (SEM) analysis for elemental compositions was used to study the morphology of the sorbent. X-ray Photoelectron Spectroscopy (XPS) was employed to investigate the interactions between the sorbent and arsenate. Batch kinetic experiments of arsenate using the magnetic sorbent with the presence of humic substances show that equilibrium sorption capacity is reduced. This finding is in parallel with the XPS analysis results where the metal oxide bond decreases with the presence of humic substances. The arsenate adsorption equilibrium time is not affected by the presence of humic substances. The SEM analysis indicates that the precipitates of humic substances create irregular homogeneous structure on the sorbent surface which consequently reduces the sorbent pores. The XPS analysis reveals that humic substances and arsenate simultaneously adsorb onto the magnetic sorbent which creates a competition phenomenon. The XPS results also show that the solid state reduction of arsenate to arsenite is promoted by the presence of humic substances.

Introduction

Humic substances contribute a major fraction of dissolved organic matter in surface water and groundwater (Afcharian et al., 1997; Nagao et al., 2007). They usually consist of complex mixtures of natural organic elements such as humic acid and fulvic acids. They have various chemically reactive functional groups, such as carboxylic, phenolic, alchoholic hydrolytic, and quinine groups (Chen et al., 2007). The existence of humic substances in water environment has created many problems to the implementation of water and waste water treatment processes such as membrane fouling, competition with adsorbate, and bacteria regrowth (Gustafsson, 2006; Jermann et al., 2007). Numerous literatures have shown that humic substances are able to adsorb onto various materials such as soils, polymers, and bio-ploymers (Zhang and Bai, 2003; Chen and Wu, 2004; Gustafsson, 2006). This adsorptive characteristic has indirectly created a problem in adsorption process for water treatment. The problem addressed is the competition of humic substances with other anions such as arsenate, phosphate, and sulfate (Gustafsson, 2006).

Arsenate is a type of inorganic dissolved arsenic element which has attracted much attention in water research environment. Drinking water that is rich in arsenic over a long period of time can lead to arsenic poisoning, an issue often described as arsenicosis. By considering the adverse health effects of arsenic in humans, United States Environmental Protection Agency (EPA) and World Health Organization (WHO) have implemented the reduction of arsenic MCL in drinking water from 50 μg/L to 10 μg/L. This implementation has prompted global attention to study and research in order to obtain a better removal and treatment technologies.

In this article, adsorption behavior of arsenate onto alginate encapsulated magnetic sorbent with the presence of humic substances was studied through spectroscopic analysis. This magnetic sorbent was developed in our recent study (Lim and Chen, 2007; Lim et. al, 2008). The SEM was used to study the morphology and components of the sorbent. The XPS was employed to investigate the interactions between the sorbent and arsenic.

Experimental

Materials

Sodium arsenate ($Na_2HAsO_4{\cdot}7H_2O$) from Fluka (Switzerland) and nitrate acid and sodium hydroxide from Merck (Germany) were used. Humic acid sodium salt (humic substances) was purchased from Aldrich (Milwaukee, WI). The calcium-alginate magnetic sorbent used is composed of complexation between iron oxide particles and alginate with the existence of calcium ions, which was prepared by a similar approach illustrated in our previous paper and had been proven to have positive attraction towards magnetic fields (Lim and Chen, 2007). Nitrate acid and sodium hydroxide solutions in various concentrations were used for pH adjustments. All the chemicals used were of reagent grades.

Concentration of metal ions was determined via an Inductively Coupled Plasma Emission Spectroscopy (ICP-OES) (Perkin-Elmer Optima 3000, U.S.A). All the samples were filtered through a 0.45 μm Whatman Autovial filter (U.S.A) before the analysis. The initial and final solution pHs were measured using a pH meter (Orion 525A, Boston, M.A.).

Adsorption experiments

In order to observe the effect of humic substances on arsenate removal, two sets of batch kinetic experiments were conducted; one set in the presence of humic substances and the other set in the absence of humic substances. Both of the batch kinetic experiments were carried out with a fixed arsenate initial concentration.

In this study, arsenate stock solutions from sodium arsenate were used to represent the metal-contaminated stream. The stock arsenate solutions were added with humic substances accordingly. Subsequently, solution pH was adjusted to 5 before the magnetic sorbent was added into the solution. All the adsorption experiments were conducted by adding a dosage of 0.5 g/L sorbent. The solutions were then rotated on a rotary shaker with the rotation speed of

200 rpm at 293 K. Samples were taken at appropriate time intervals and filtered with 0.45μm filters before analysis for metal ions concentrations. The concentrations of the metal ions were measured by the ICP-OES. All the kinetic experiments were repeated twice for precision and accuracy.

Spectroscopic analyses

The SEM (JEOL, JSM-5600V, Japan) was performed on the magnetic sorbent samples which were after arsenate sorption both in the presence and in the absence of humic substances. The SEM analyses enable direct visualization on the sample surface morphologies through observing the surface microstructures.

The XPS analyses were performed by a Kratos AXIS His-165 Ultra (Shimadzu, Japan), with monochromatized Al Kα X-ray source (1486.6 eV) working at 150 watt, 15 kV and 10 mA and base pressure of 3 x 10^{-8} Torr in the analytical chamber. The XPS analyses provide information on elemental composition and chemical states of the samples. For wide scan spectra, an energy range of 0 to 1100 eV was used with pass energy of 80 eV and step size of 1 eV. The high resolution scans were conducted according to the element being examinated with pass energy of 40 eV and step size of 0.05 eV. The XPS spectra data were collected in binding energy forms. The spectral interpretation was performed using a nonlinear least-square curve fitting decoding program with the subtraction of background (linear type background for non-metal elements and Shirley type background for transition metal elements) and the ratio of Lorentzian (0%)-Gaussian (100%).

Results and Discussion

Experimental

Batch kinetic experiments were conducted to study the effect of humic substances on the arsenate adsorption performance. Results for the arsenate adsorption kinetic experiments with different humic concentrations at pH 5 are shown in Figure 1. It is found that arsenate ion adsorption capacity increases with increasing contact time for all the cases. The kinetic experimental data can be analyzed as a two-step reaction process. Initially, it is a rapid adsorption reaction phase and then, followed by a slower to steady phase. When the steady phase is observed, it shows that equilibrium phase has been attained. The equilibrium phase can be attained within 25 h for both conditions. This would suggest that the presence of humic substances does not significantly influence the kinetic of the adsorption reaction. In other words, the presence of humic substances does not interrupt the transport of arsenate ion from bulk solution to the magnetic sorbent's surface and also attachment to the surface.

One of the most important findings in this study is that the presence of humic substances has reduced the equilibrium adsorption capacity of arsenate by the magnetic sorbent. It is very obvious in the second reaction phase (slow adsorption phase) that the adsorption capacities are lower when humic substances present in the solution (Figure 1). The adsorption capacity of arsenate with the presence of humic substances remains lower than the other until equilibrium phase. The decrease of metal adsorption capacity with the presence of humic substances is also observed in adsorption of copper ion by activated carbon (Chen and Wu, 2004). In that study, it was found that similar copper adsorption capacity can still be attained with a longer reaction time. However, this was not observed in this study. One possible explanation for this is through the difference of ion charges between copper and arsenate ions. Naturally, free-arsenate ion has a negative charge, whereas, free-copper ion has a positive charge. Humic substances are also negatively charged as reported in literature (Zhang and Bai, 2003). The similarity of charges between arsenate and humic substances might create competitive phenomena during the attachment reaction to the magnetic sorbent's surface. This finding raises a number of questions especially regarding the mechanisms of the competition between humic substances and arsenate ions which will be discussed in next section.

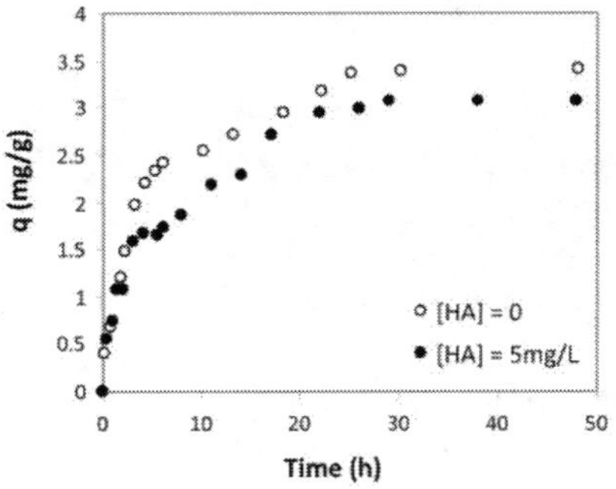

Figure 1. Effect of humic acid, HA on arsenic sorption kinetics; $[As (V)]_o$= 3 mg/L, m= 0.5 g/L, T= 293 K, pH= 5.

SEM analyses

The SEM analyses were used to observe the surface microstructure of the alginate encapsulated magnetic sorbent after arsenate adsorption for both in the presence and absence of humic substances. Figure 2 shows SEM images of the magnetic sorbent after arsenate adsorption for two different conditions; in the absence of humic substances and in the presence of humic substances.

Generally, it is observed that the surface of the magnetic sorbent is uneven structure with irregular and eccentric pores. The SEM image of the magnetic sorbent after arsenate sorption in the presence of humic substances (Figure 2) shows lesser pore compared to the SEM image of the sorbent in the absence of humic substances (Figure 2). This indicates that the precipitates of humic substances may create irregular homogeneous structure on the surface of the magnetic sorbent which subsequently has clogged some of the pores.

These SEM results are consistent with the findings reported in the literature. Zhang and Bai (2003) found that significant amount of humic substances adsorbed onto chitosan-coated granules. Chen and Wu (2004) also reported that humic substances in solution would have simultaneous adsorption with copper ions onto activated carbon. Therefore, it is possible that the precipitated particles which attached on the surface of the magnetic sorbent are humic substances.

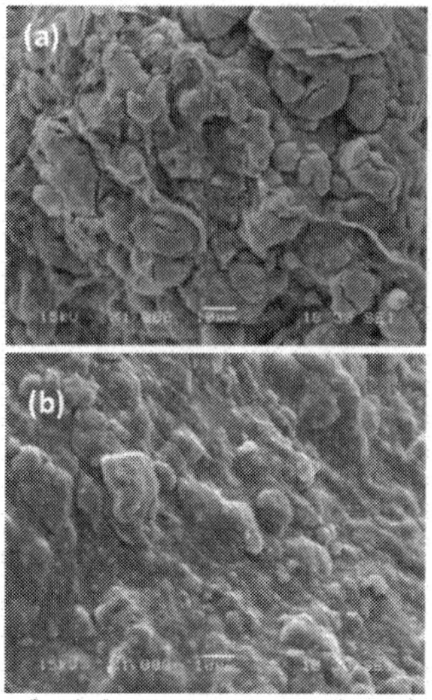

Figure 2. SEM images (x1000) of alginate encapsulated magnetic sorbent after arsenate adsorption: (a) in the absence of humic substances and (b) in the presence of humic substances.

XPS analyses

Variations in binding energies and chemical intensity shift on XPS spectra depend on the concentration of compound's functional elements on the surface of the sorbent. These variations are reflected from the difference in chemical potential and element's polarity. The results of the XPS spectra for arsenic loaded magnetic sorbent in the absence of humic substances and in the presence of humic substances are tabulated in Table 1.

Oxygen is observed predominantly in the form of C=O and also in forms of As-O, Fe-O, C-OH, and C-O-C. Metal oxides binding fingerprints are detected as As-O and Fe-O. No binding energy shift is observed for the metal oxides peak (529.4 eV) in the presence of humic substances during arsenate removal. These findings indicate that similar chemical bonding occurs between those atoms. One interesting finding is that the metal oxide bond decreases with the existence of humic substances. This is observed by the decrease in relative content of the metal oxide atomic ratio from 37.3 to 21.7 %. The decrease in the metal oxide content might be attributed to the lesser formation of As-O. This finding is in parallel with the batch kinetic experimental results obtained where sorption capacity is slightly lower when humic acid is present in the solution. Another interesting finding is that the relative contents of C-O-C and C-OH increase from 25.1 to 27.5 % with the presence of humic substance during arsenate adsorption. This change might be a consequence of surface reaction between the humic substances and the magnetic sorbent. Thus, this suggests chemical bonding of humic substances on the magnetic sorbent.

Changes in peak positions also occur in the C1s spectrum after adsorption of arsenate ions, both with and without humic substances. The carbon atoms can be found in three chemically different positions in the adsorbed-magnetic sorbent, leading to three differing C1s binding energies; C-C, C-O, and C=O. A very important point is the C-C peak with humic substance during arsenate adsorption has significantly higher atomic ratio than that without the presence of humic substance as summarized in Table 1. The results indicate that there is chemical bonding between humic substances and the magnetic sorbent which contributes to the additional C-C bonding.

The appearance of a weak N1s peak in the samples for arsenate sorption in the presence and absence of humic substances is observed in high resolution scans. No nitrogen peak is found on fresh alginate encapsulated magnetic sorbent. The appearance of N1s peak in the sample for arsenate sorption in the absence of humic substances is mainly attributed by the addition of nitric acid (HNO_3) which was used for pH adjustment. The deconvoluted N1s peaks are attributed to the =N- and –NH- at 398.2 eV and 399.8 eV, respectively (Zhang and Bai, 2003). Thus, this explains why the N1s bond is observed during the high resolution scan.

Examination of the high resolution N1s peaks for the magnetic sorbents after arsenate sorption in the presence of humic substances shows minor changes in the relative areas of the fitted peaks than the peaks in the absence of humic substances (Table 1). After adsorption of arsenate in the presence of humic substances, the results in Table 1 show that the –NH- relative content is increased to about 32% with about 16% increment. This is most likely to be the spectroscopically evidence that the humic substances do attach on the magnetic sorbent's sites as the –NH- group is usually found on humic substance according to the study reported by Brunetti et al. (2007). These results are in agreement with the results obtained from SEM analysis where a change of surface morphology was observed. The initial indication that humic substances exist on the surface of the magnetic sorbent has therefore been further clarified.

Another closer examination reveals that the As3d also can be detected in the high resolution scan. The arsenic element is observed as an adsorbate on the magnetic sorbent. Two As3d photoelectron peaks occur at 45.2 and 46.7 eV, which correspond to the characteristic peak position of As(III) and As(V) (Wagner et al., 1979). The result in Table 1 indicates 80 % and 64 % of inorganic arsenate ions are reduced to arsenite after their adsorption onto the magnetic sorbent for in the presence and in the absence of humic substances, respectively. However, no arsenite is detected in the solution. Thus, this result gives a strong indication that the reduction of arsenate to arsenite is promoted by the presence of humic substances. This finding is parallel with the work reported by Palmer et al. (2006). They found that inorganic arsenates were reduced to arsenite by homogeneous aqueous solution of several humic and fulvic acids.

Table 1. Binding Energy and Relative Content of Elements in Different Forms.

Valence state	Sample	Assignment	Binding energy (eV)	Intensity (counts·s^{-1})	Relative content (%)
O1s	**MS-As**	metal oxide	529.4	805.7	37.3
		C=O	531.0	813.1	37.6
		C-OH,C-O-C	532.1	543.4	25.1
	MS-As+HA	metal oxides	529.4	420.0	21.7
		C=O	531.1	985.9	50.9
		C-OH,C-O-C	532.1	532.1	27.5
C1s	**MS-As**	C-C	284.5	487.4	34.8
		C-O (ether or alcohol)	286.2	628.3	44.8
		C=O (carboxyl or quinone)	287.9	286.5	20.4
	MS-As+HA	C-C	284.5	2041.8	78.6
		C-O (ether or alcohol)	286.2	268.7	10.3
		C=O (carboxyl or quinone)	287.9	287.8	11.1
As3d	**MS-As**	As(III)	45.2	46.2	64.0
		As(VI)	46.7	26.0	36.0
	MS-As+HA	As(III)	45.2	61.6	80.0
		As(VI)	46.7	15.4	20.0
N1s	**MS-As**	=N-	398.2	136.6	83.5
		-NH-	399.8	26.9	16.5
	MS-As+HA	=N-	398.2	137.2	67.6
		-NH-	399.8	65.9	32.4

MS-As: arsenate loaded magnetic sorbent without humic substance; MS-As+HA: arsenate loaded magnetic sorbent with humic substance

Conclusions

Batch kinetic experiments of arsenate using the alginate encapsulated magnetic sorbent with the presence of humic substances show that the equilibrium sorption capacity is reduced. The arsenate adsorption equilibrium time is not affected by the presence of humic substances. The SEM analysis indicates that the precipitates of humic substances create an irregular homogeneous structure on the surface of the sorbent which consequently reduces the sorbent pores. The XPS analysis reveals that humic substances and arsenate simultaneously adsorb onto the magnetic sorbent which creates a competition phenomena.

Acknowledgements

The authors would like to thank National University of Singapore for the use of laboratory facilities and equipments. This project is registered with University Malaysia Sarawak and the financial supports from the University Malaysia Sarawak are highly appreciated.

References

Afcharian, A., Levi, Y., Kiene, L., Scribe, P. (1997), "Fractionation of dissolved organic matter from surface waters using macroporous resins," Water Research, Vol. 31, pp. 2989-2996.

Nagao, S., Kodama, H., Aramaki, T., Fujitake, N., Yonebayashi, K. (2007), "Variations in $\Delta^{14}C$ of humic substances in the Lake Biwa waters," Nuclear Instruments and Methods in Physics Research Section B: Beam Interactions with Materials and Atoms, Vol. 259, pp. 552-557.

Chen, C., Wang, X., Jiang, H., Hu, W. (2007),"Direct observation of macromolecular structures of humic acid by AFM and SEM," Colloids and Surfaces A: Physicochemical and Engineering Aspects, Vol. 302, pp. 121-125.

Gustafsson, J. P. (2006), "Arsenate adsorption to soils: Modelling the competition from humic substaces," Geoderma, Vol. 136, pp. 320-330.

Jermann, D., Pronk, W., Meylan, S., Boller, M. (2007), "Interplay of different NOM fouling mechanisms during ultrafiltration for drinking water production," Water Research, Vol. 41, pp. 1713-1722.

Zhang, X., Bai, R. (2003), " Mechanisms and kinetics of humic acid adsorption onto chitosan-coated granules," Journal of Colloid and Interface Science. Vol. 264, pp. 30-38.

Chen, J.P., Wu, S. (2004), "Simultaneous adsorption of copper ions and humic acid onto an activated carbon," Journal of Colloid and Interface Science, Vol. 280, pp. 334–342.

Lim, S. F., and Chen, J. P. (2007) "Synthesis of an innovative calcium-alginate magnetic sorbent for removal of multiple contaminants," *Applied Surface Science*, Vol. 253, pp. 5772–5775.

Lim, S.F., Zheng, Y.M., Zou, S.W., Chen, J.P. (2008), "Investigation of copper adsorption onto an alginate encapsulated magnetic sorbent by FTIR, XPS, and modeling simulation," *Environmental Science & Technology*, Vol. 42, *pp.* 2551–2556.

Brunetti, G. Plaza, C., Clapp, C.E., Senesi, N. (2007), "Compositional and functional features of humic acids from organic amendments and amended soils in Minnesota, USA," Soil Biology & Biochemistry, Vol. 39, pp. 1355–1365.

Wagner, C. D., Riggs, W. M., Davis, L. E., Moulder, J. F., Muilenberg, G. E. (1979), "Handbook of X-ray photoelectron spectroscopy," Physical Electronics Division, Perkin-Elmer Corp.: Eden Prairie, MN.

Palmer, N. E., Freudenthal, J. H., and von Wandruszka, R. (2006), "Reduction of arsenates by humic materials," Environ. Chem., Vol. 3, pp. 131-136.

Chemotaxis Enhances Apparent Microbial Dispersion in Systems That Are Chemically and Physically Heterogeneous

Meng Wang, Department of Chemical Engineering, University of Virginia, Charlottesville, VA

Roseanne M. Ford, Chemical Engineering, University of Virginia, Charlottesville, VA

A substantial research effort has been devoted to describe and predict bacterial migration in natural subsurface environments. Previous mathematical models, in which bacteria were usually considered as immobile colloids, have been used extensively to describe bacterial transport behaviors in both laboratory tests and field-scale studies. However, neglect of bacterial own swimming properties in the transport models, such as motility and chemotaxis, may wrongly estimate the productivity of microbial populations in conditions when those biological attributes are significant, such as bioremediation and microbial oil enhanced recovery.

A 2-dimensional mathematical model was developed to simulate transport phenomena of chemotactic bacteria in a sand-packed column designed with structured physical heterogeneity in the presence of a localized chemical source. First, the hydrodynamic conditions were quantified by fitting the model output to the experimental observations of a conservative tracer (nitrate). Next, model simulations fit to data for nonchemotactic bacteria provided the essential microbial transport parameters. Finally, parameter values were input into the mathematical model to predict bacterial concentration in the column effluent under the influence of a chemoattractant gradient.

Consistency between experimental observation and model prediction supported the assertions that (1) dispersion-induced microbial transfer between adjacent conductive zones occurred at the interface and had little influence on bacterial transport in the bulk flow of the permeable layers and (2) the enhanced transverse bacterial migration in chemotactic experiments relative to nonchemotactic controls were mainly due to directed migration toward the chemical source zone. Based on parameter sensitivity analysis, chemotactic parameters determined in bulk aqueous fluid were adequate to predict the microbial transport in our intermediate-scale porous media. Further analysis of bacterial transport over a range of flow rates revealed that bacterial chemotaxis might be impeded at high fluid velocity and high shear stress, as reported previously in the literature. Comparison of motile bacteria with passive microspheres revealed the importance of accounting for motility when evaluating the microbial filtration rate. By quantitatively describing bacterial fate and transport in a heterogeneous system, this mathematical model serves to advance our understanding of chemotaxis and motility in porous media. Moreover, it provides insights for improving the design of in-situ microbial processes.

See more of this Session: Colloidal and Interfacial Phenomena in Aquatic Systems
See more of this Group/Topical: Environmental Division

Transformation of Gold Nanoparticles in Presence of Humic Substances

Vasanta Pallem, Department of Chemical Engineering and Center for the Management, Utilization and Protection of Water Resources, Tennessee Technological University, Cookeville, TN

Holly A. Stretz, Department of Chemical Engineering, Tennessee Technological University, Cookeville, TN

Martha J. M. Wells, Center for the Management, Utilization and Protection of Water Resources and Department of Chemistry, Tennessee Technological University, Cookeville, TN

Bethany Wigington, Ecosystems Research Division, USEPA National Exposure Research Laboratory, Athens, GA

The interaction of biomedical imaging gold nanoparticles with environmental entities such as humic substances was investigated applying Dynamic light scattering (DLS). The increasing applications of gold nanoparticles in biomedical imaging and cancer therapy indicate potential for their subsequent release into surface waters. Therefore, it is primarily important to understand the interactions of gold nanoparticles with natural organic matter (humic substances), which will play a major role in the fate and transport of these particles in aquatic systems. The current study investigates the size changes taking place on citrate coated gold nanoparticles (core sizes 5 and 10 nm) in the presence of commercial humic acid having concentrations of 2 and 8 ppm. Humic acid controls at the same concentration exhibited micron sized aggregates based on volume- average size distributions, however the humic acid aggregates were absent when mixed with the citrate-coated gold nanoparticles. The size analysis data suggests that humic acid stabilizes gold nanoparticles and acts as a carrier.

See more of this Session: Colloidal and Interfacial Phenomena in Aquatic Systems
See more of this Group/Topical: Environmental Division

How Size and Crystallinity of Nanoparticles (NPs) Affect Their Chemical Activity in Water? Acid-Base Properties of Metal Oxide Nps and Adsorption of Oxy-Anions and Fatty Acids

Irina Chernyshova, Earth and Environment Engineering, Columbia University, New York, NY

Sathish Ponnurangam, Earth and Environment Engineering, Columbia University, New York, NY

Ponisseril Somasundaran, Columbia University, New York, NY

Particle size effects on (bio)chemical properties of metal (hydr)oxide nanoparticles (NPs) are of fundamental scientific and technological interest. We present results of the first multi-pronged study of the effect of nanoparticle size and crystallinity on acid-base and adsorption properties of hematite (ƒÑ-Fe2O3). Using measurements of point of zero charge, electrokinetic mobility, in situ FTIR spectroscopy, and DFT modeling, we discovered that particle size controls adsorption properties of hematite significantly. In particular, inner-sphere mononuclear monodentate carbonate dominates on 7 nm NPs where as an increase in the particle size to 40 nm (or the lattice order) results in an increase in the relative concentration of the inner-sphere bidentate binuclear complexes. Interestingly, the surface concentration of adsorbed carbonate also increases with such increase in size. No or very little adsorption of carbonate was detected on 60 nm and larger NPs. In the case of sodium laurate, a decrease in either particle size or crystallinity lowers 1) the molecular form of the adsorbed surfactants vs. the ionic form; 2) packing order of the hydrocarbon chains and hence hydrophobicity of the surfactant-coated surface. It was shown that laurate adsorbs as both inner-sphere monodentate complex and water-surface-hydration separated ion pair (outer-sphere bridging bidentate complex). The latter is protonated with increasing particle size, which explains the observed increase in the packing order of the laurate chains. These effects are explained by a decrease in basicity of hematite with increasing particle size due to the quantum size effects, size-induced changes in structure of the NP surface and interior, surface delocalization of electronic charge, and the synthesis route.

See more of this Session: Colloidal and Interfacial Phenomena in Aquatic Systems
See more of this Group/Topical: Environmental Division

Reduction of Immobilized Metal Ions to Zero-Valent Nanoparticles On Functionalized Silica for Dechlorination

Noah D. Meeks, Dept of Chemical and Materials Engineering, University of Kentucky, Lexington, KY

D. Bhattacharyya, Chemical and Materials Engineering, University of Kentucky, Lexington, KY

Dr. Leonidas Bachas, Chemistry, University of Kentucky, Lexington, KY

Zero-valent iron nanoparticles (or bi-metallic iron/palladium particles) used for the dechlorination of hazardous organic groundwater pollutants are produced by reduction of iron ions immobilized on nanoscale platforms. These platforms include silica nanoparticles functionalized with sulfonate or poly(aspartic acid), as well as functionalized silica with both hydrophobic and hydrophilic functional groups. The latter platform is useful for creating hydrophobic nanodomains near the active site of dechlorination reaction. Extent of silylation for the various platforms is determined by IR, thermogravimetric analysis, total organic carbon analysis, total organic sulfur analysis, and calcium sorption. Extent of functionalization for sulfonate is about 1 molecule per square nm of surface area. The functionalized silica platforms are then used to immobilize ferrous ions. Sulfonate functionalized particles have the advantage of being useful at lower pH, and about 0.1 mmol ferrous ions are immobilized on these particles at pH = 3.5. Reducing agents (including borohydride and ascorbic acid) are used to form zero-valent iron from the immobilized ions. These particles are found (by SEM and TEM) to have possibly experienced directed growth as many of the particles are rod-shaped; the particle length is about 50 nm, and particles are also characterized by X-ray diffraction and other techniques. Palladium is preferentially deposited on these iron particles over their functionalized silica platforms, and these bi-metallic particles are characterized through electron microscopy, energy dispersive X-ray spectroscopy, X-ray diffraction, and X-ray photoelectron spectroscopy. The bi-metallic nanoparticles are shown to be reactive for the dechlorination of trichloroethylene and other chlorinated organic pollutants. This research is supported by the NIEHS-SBRP program.

See more of this Session: Environmental Applications of Nanotechnology and Nanomaterials
See more of this Group/Topical: Environmental Division

Multifunctional Nanoparticulate Systems for in-Situ Remediation of Chlorinated Hydrocarbons

Bhanukiran Sunkara, Chemical and Biomolecular Engineering, Tulane University, New Orleans, LA

Jingjing Zhan, Department of Chemical and Biomolecular Engineering, Tulane University, New Orleans, LA

Jibao He, Central Instrumental Facilities, Tulane University, New Orleans, LA

Gary McPherson, Department of Chemistry, Tulane University, New Orleans, LA

Gerhard Piringer, Department of Earth and Environmental Science, Tulane University, New Orleans, LA

Vijay T. John, Department of Chemical and Biomolecular Engineering, Tulane University, New Orleans, LA

The widespread occurrence of dense non-aqueous phase liquids (DNAPLs) in groundwater and in soil is of serious environmental concern. Nanoscale zerovalent iron particles (NZVI) are a preferred option for reductive dehalogenation of TCE due to their environmentally benign nature, high efficiency and low cost. However, it is difficult to transport these particles to the source of contamination due to aggregation. This study describes a novel approach to the preparation of ZVI containing particulate systems that are effectively transported to contaminant sites and serve as targeted delivery agents for remediation. We describe the synthesis of highly uniform colloidal carbon microspheres embedded with zerovalent iron nanoparticles. These materials have multiple functionalities (a) they are reactive and function effectively in reductive dehalogenation (b) they are highly adsorptive thereby bringing the chlorinated compound to the proximity of the reactive sites and also serving as adsorption materials for decontamination (c) they are of the optimal size for transport through sediments (d) they have amphiphilic chemical functionalities that help stabilize them when they reach the DNAPL target zones. These multiple functionalities can be designed at low cost and the materials are environmentally innocuous. The detailed characterization of these multifunctional colloids through high resolution electron microscopy, and their functional properties will be described. The presentation will focus on the fundamental aspects of adsorption, transport and reaction using these materials.

See more of this Session: Environmental Applications of Nanotechnology and Nanomaterials
See more of this Group/Topical: Environmental Division

Reactivity of Singlet Oxygen Generated Using Nanoporous Silicon for Transformation of Acrylonitrile

Veera Boddu, Environmental Processes Branch, US Army Engineer Research and Development Center, Champaign, IL

Deereje G. Worku, Environmental Process Branch, US Army Engineer Research and Development Center, Champaign, IL

Donald M. Cropek, Construction Engineering Research Laboratory, US Army Engineer Research and Development Center, Champaign, IL

Crystalline nanoporous silicon (PSi) mediated generation of singlet oxygen is by far one of the most efficient photoexcitation processes with up to 80% quantum efficiency. The singlet oxygen is effective in breaking double bonds (-ene type) between two carbon atoms and also bonds between heteroatoms. This singlet oxygen, however, is short lived (~few milliseconds). Depending on the reaction media and other quenching reactions, it leaves many unknown parameters related to reaction chemistry, kinetics, and electron transfer processes. Our research focuses on testing the hypothesis that the excitons generated on PSi are sufficiently reactive to break the double and triple bonds (C=N and N=N; C"kN and N"kN) present in an organic molecule. Thus, the singlet oxygen potentially offers selective transformation of functional groups in complex molecules. Also, singlet oxygen based chemistry is one of the less-exploited strands of green chemical technologies. Only a small number of photochemical or photocatalytic processes have been commercialized on a large scale due to limited understanding of the reaction chemistry and other mass transfer limiting parameters. Understanding the degradation and transformation mechanisms and pathways will foster development of efficient processes for mitigation of Army waste streams prior to introduction of munitions compounds into service. The presentation will focus on preliminary results of our experimental studies on transformations of acrylonitrile (C=N and C"kN) using singlet oxygen generated on PSi.

See more of this Session: Environmental Applications of Nanotechnology and Nanomaterials
See more of this Group/Topical: Environmental Division

Photocatalytic Activity of Multi-Doped TiO_2 Nanoparticles for Degradation of Rhodamine B

Xiangxin Yang, Chundi Cao, Larry Erickson, Keith Hohn, Ronaldo Maghirang, Kenneth Klabunde*

Department of Chemical Engineering, Kansas State University, Manhattan, KS 66506, USA
Department of Biological and Agricultural Engineering, Kansas State University, Manhattan, KS 66506, USA
Department of Chemistry, Kansas State University, Manhattan, KS 66506, USA

Abstract

Titanium dioxide (TiO_2), as a chemically stable, nontoxic, highly efficient, and relatively inexpensive photocatalyst, has been widely used for water and air purification since many environmental pollutants can be degraded by oxidation and reduction processes on TiO_2 surface. However, the application of TiO_2 is limited by its UV activation requirement because of its large band gap. The current presents a facile sol-gel method for the synthesis of visible-light-active TiO2-based catalysts with multiple dopants. The catalysts were characterized by X-ray diffraction, UV-vis spectroscopy, and X-ray photoelectron spectroscopy. Results revealed that TiO_2's structure was slightly distorted by the substitution of carbon, nitrogen, and sulfur dopants for some oxygen atoms and Fe^{3+} for Ti^{4+} in the lattice of TiO_2. These dopants were responsible for narrowing the band gap and shifting the optical response shift of TiO_2 from ultraviolet (UV) to the visible light region. Photocatalytic reaction showed that the reactivity of multi-doped TiO_2 was significantly enhanced.

Introduction

Titanium dioxide (TiO_2) has been widely used for environmental cleaning since many environmental pollutants can be degraded by oxidation and reduction processes on TiO_2 surface.[1,2,3] Under irradiation with light of sufficient energy to span TiO_2's band gap, the electron (e^-) can be promoted into the conduction band (CB), leaving a hole (h^+) in the valence band (VB). Then, electrons and holes can initiate redox reactions on the surface of TiO_2. Finally, pollutants can be degraded into CO_2 and H_2O.[1] However, the application of TiO_2 is limited by its UV activation requirement because of its large band gap (3.2 or 3.0 eV in the anatase or rutile crystalline phase, respectively). Therefore, efforts have been directed towards shifting the optical sensitivity of TiO_2 from UV to the visible light region.

Transition metal dopants have been widely used to extend TiO_2's light absorption into the visible light region. Choi et al.[4] studied the effects of 21 different dopants on the photoreactivity of TiO_2. Fe^{3+} dopant proved to be better than Ru^{3+}, V^{4+}, Mo^{5+}, Os^{3+}, and Re^{5+}. It is generally accepted that the Fe^{3+} dopant formed shallow charge traps within TiO_2 crystal lattice through the substitution of Ti^{4+}, which reduced the electron-hole recombination and improved the photocatalytic efficiency.[5,6]

Although transition metal modified TiO_2 photocatalysts are active under visible light illumination, the efficiency is still low for practical use. Recently, doping nonmetal atoms, such as C, N, and S, has received much attention. Wang et al.[7] reported the shift of photo response of TiO_2 from UV to the infrared region by carbon dopant. S-doped TiO_2 showed high activity for degradation of MB in water under irradiation at wavelengths longer than 440 nm.[8] Visible light-active N-doped TiO_2 photocatalyst has been prepared by thermal nitridation of TiO_2, amination of TiO_2, sputtering, and hydrolyzing titanium precursors in ammonia.[9,10,11,12]

It is of great interest to investigate the synergetic effects of multiple dopants on the optical shift, crystallinity, surface areas, and activity of TiO_2. It was reported that N and F co-doped TiO_2 had higher visible light activity than N-doped or F-doped TiO_2 since N dopant improved the visible light absorption and the doped F atom enhanced the surface acidity and the adsorption of agents.[13] Our previous research showed two dopants had more beneficial effects than single one for enhancing the absorption in the visible light region and improving the photocatalytic activity of TiO_2.[14,15] In this work, multiple dopants modified TiO_2 was developed via a facile sol-gel process, which showed an enhanced activity compared to TiO_2.

Experimental

Catalyst synthesis

Thiourea,iron chloride, and titanium isopropoxide (97%) were purchased from Aldrich without further treatment. The general procedure for the preparation was as follows: first, 3.8 g thiourea was added to a beaker containing 150 mL deionized water, which was kept in a water-ice bath. While under vigorous stirring, a certain amount of iron chloride, according to the desired Fe^{3+} doping content, was added to the solution. Then 11 mL of titanium isopropoxide was added dropwise to aqueous solution to suppress the active hydrolysis reaction. Water was removed by drying in air at 80 °C. The dried powder was ground and calcined at 500 °C in air for 2 h. Fe doping content (x) was chosen as 0, 0.1, 0.3, and 0.5 wt%, which was the mass percentage of Fe^{3+} in the titanium oxide powder (denoted as CNS-xFe-TiO_2).

Catalyst characterization

X-ray diffracton (XRD) patterns were obtained with a Bruker D8 diffractometer with Cu Kα radiation (1.5406) at 40 kV and 40 mA. Measurements were recorded in steps of 0.025° with a count time of 2 s in the 2θ range of 20-65°. The phases were identified with the aid of Joint Committee on Powder Diffraction Standards (JCPDS) files. UV-vis absorbance spectra of catalysts were recorded with a Cary 500 Scan UV-vis NIR spectrophotometer. All spectra were referenced to poly (tetrafluoroethylene). X-ray photoelectron spectroscopy (XPS) data were recorded using a Perkin-Elmer PHI 5400 electron spectrometer. The spectrometer utilizes acrochromatic Al Kα radiation (1486.6 eV). The XPS binding energies were measured with a precision of 0.1 eV.

Results and discussion

UV-Vis spectroscopy

Figure 1 shows UV-vis diffuse reflectance spectra of the pure and doped TiO_2. Pure TiO_2 showed absorption only in the UV region (less than 400 nm). The onset of the absorption spectrum of CNS-xFe-TiO_2 was shifted towards visible light, a result of carbon, nitrogen, sulfur, and iron dopants. It has been reported that when O atom of the TiO_2 lattice was substituted by C, N, or S, the p-orbital of these dopants overlapped with the valence band O-2p orbital[16,17]. Therefore, the band gap of TiO_2 became narrower and the absorption edge shifted to the lower energy. With the addition of Fe^{3+}, the absorbance in the visible light range was enhanced, which can be attributed to the interaction between Fe^{3+} d electrons and the TiO_2 conduction or valence band since Fe dopant could introduce impurity levels within the band gap of TiO_2.[18] The electronic transitions between the impurity level and the valence or conduction band effectively red shifted the band edge absorption threshold.[19]

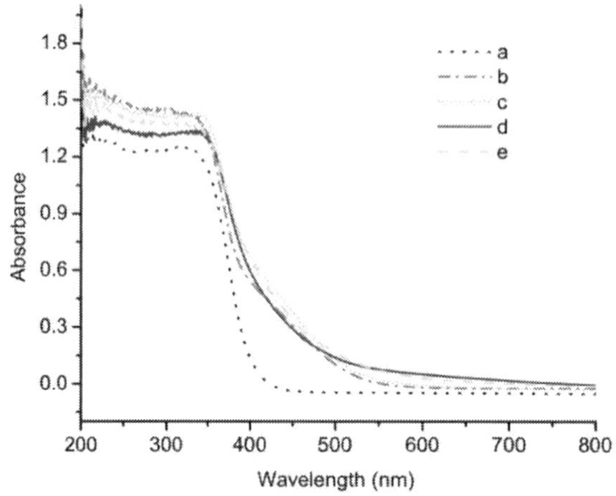

Figure 1 The UV diffuse reflectance spectra of TiO_2 and CNS-xFe-TiO_2 catalysts (a: TiO_2, b:x=0, c: x=0.1, d: x=0.3, e: x=0.5)

X-ray diffraction

Figure 2 exhibits the XRD patterns of pure and doped TiO_2. The characteristic (101) plane diffraction peak ($2\theta = 25.3°$) of anatase phase was clearly observed, no other phase feature peaks (rutile or brookite) can be detected. For CNS-0Fe-TiO_2, the width of this peak was broadened and its intensity was significantly decreased. With the addition of iron, this phenomenon was further enhanced, although not as pronounced as previously. It has been reported that Fe^{3+} has retarding effect on TiO_2 anatase crystallites growth, which favored the formation of smaller TiO_2 crystallites. Since the radii of Fe^{3+} and Ti^{4+} for coordination number of 6 were 0.79 and 0.75 Å, respectively, the substitution of Ti^{4+} site by Fe^{3+} ion should be easy without a significant crystalline distortion.[4,20] The replacement of Ti^{4+} by Fe^{3+} was proven by the further slight shift of the (101) peak with iron introduction.

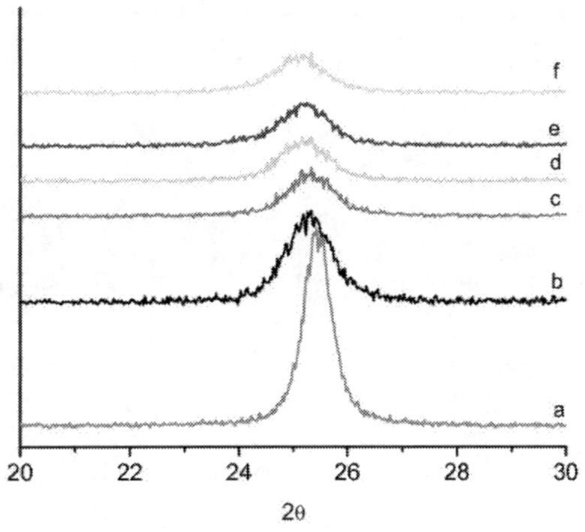

Figure 2 XRD patterns of TiO_2 and CNS-xFe-TiO_2 catalysts (a: TiO_2, b:x=0, c: x=0.1, d: x=0.3, e: x=0.5)

X-ray photoelectron spectroscopy

As shown in Figure 3, the C 1s XPS spectra showed a strong peak at 285 eV and a shoulder at around 288 eV. After sputtering, a new peak appeared at 281 eV. Three forms of carbon species have been previously detected: elemental carbon with binding energy of 285 eV; surface adsorbed carbonate species (288 V); Ti-C bond resulting from substituting the lattice oxygen atoms by carbon (281 eV).[7,21] Therefore, multiple carbon species, namely, substitutional and interstitial carbon atoms and carbonate species, existed in the lattice of TiO_2. The S 2p XPS spectrum showed a peak at 168 eV. The assignment of this peak was controversial. A broad peak attributed to S 2p3/2 at around 168 eV suggested several oxidation states of S atoms: the peak of the S^{6+} state appeared at 168.2 eV and the peak at 167.5 eV was assigned to S^{4+}.[22] This peak was attributed to S^{6+} state since the sample was treated in air at 500°C for 2 h. After sputtering, a new peak appeared at 160.2 eV, indicating that sulfur entered the lattice of TiO_2 and substituted for oxygen atoms. The N 1s XPS spectrum showed a very broad peak between 398 and 402 eV. After sputtering, the broad peak disappeared and a new one was observed at around 396 eV, which can generally be assigned to N^{2-} anions resulting from substitution for oxygen sites by nitrogen atoms in the TiO_2 lattice.[23] The Fe 2P XPS spectrum showed a peak at 709.1 eV. The peak can be assigned to Fe^{3+}.[24,25] The spectrum was very similar with the one after sputtering, indicating Fe^{3+} existed both on the surface and inside the bulk of TiO_2 powder.

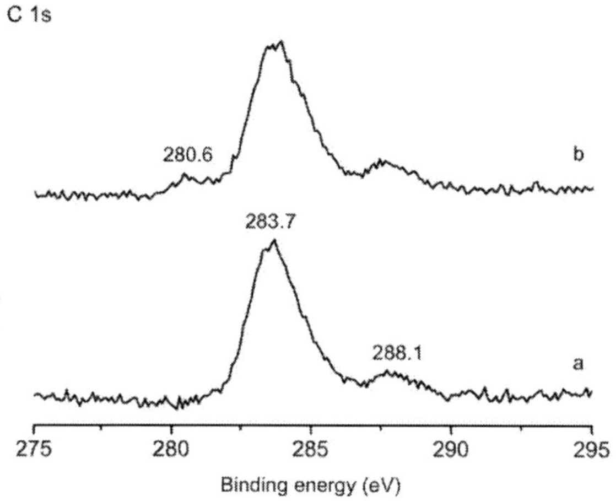

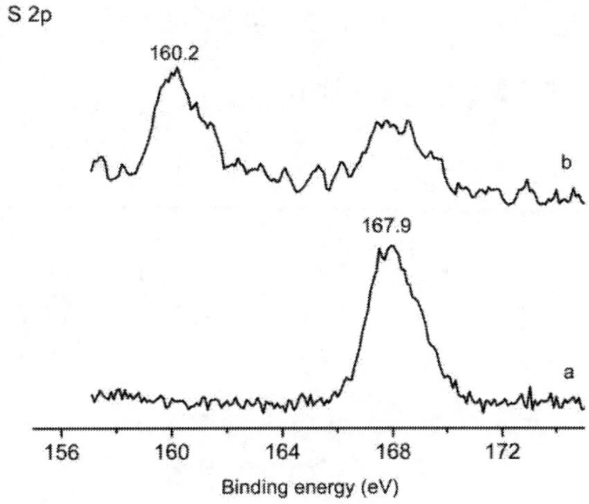
S 2p
160.2
167.9
b
a
156
160
164
168
172
Binding energy (eV)

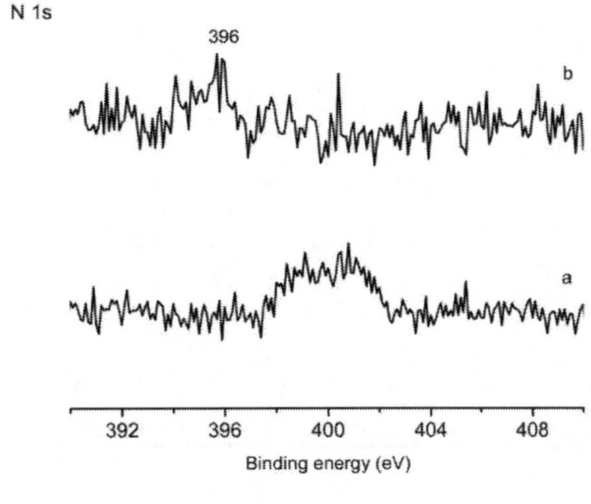
N 1s
396
b
a
392
396
400
404
408
Binding energy (eV)

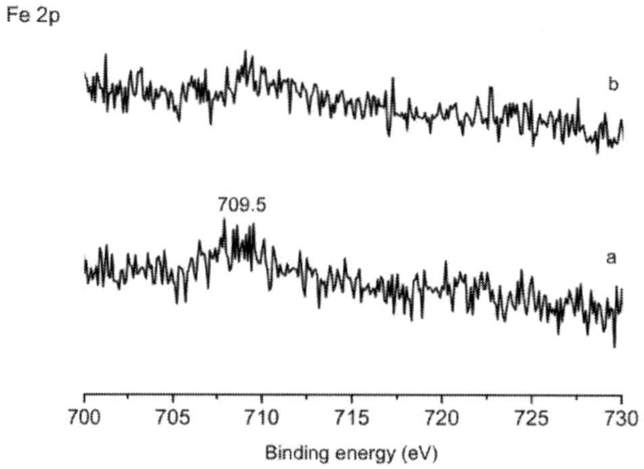

Figure 3 XPS spectra for the C 1s, N 1s, S 2p core levels of CNS-0Fe-TiO_2 and Fe 2p of CNS-0.3Fe-TiO_2

Photocatalytic activity measurements

The photocatalytic activity of catalysts was evaluated by the degradation of RB in an aqueous solution. The light source was a 1000-W high-pressure Hg arc lamp (Oriel Corporation). The combination of a Vis-NIR long pass filter (400 nm) and colored glass filter (>420 nm) was used to eliminate UV radiation during visible light experiments. A general procedure was carried out as follows: First, 100 mL aqueous RB solution (1×10^{-5} mol/L) was placed in a water-jacketed reactor, which was maintained at a temperature of 25 °C. Then, 30 mg catalyst was suspended in the solution. The suspension was stirred vigorously for 60 min in the dark to establish adsorption-desorption equilibrium of RB. The suspension was then irradiated under visible light. Samples were withdrawn periodically from the reactor and centrifuged and analyzed by recording variations in the absorption in the UV-vis spectra of RB using a Cary 500 UV-vis spectrophotometer at its characteristic wavelength (λ = 553 nm).

For the comparison of reaction rate among different catalysts, pseudo-first-order reaction equation was introduced. Figure 4 shows the degradation of RB as a function of reaction time. The rate constants k were 0.0016, 0.0019, 0.0029, and 0.0010 min^{-1} for CNS-xFe-TiO_2 with x equals 0, 0.1, 0.3, and 0.5, respectively, while k for pure TiO_2 was 0.0002 min^{-1}. Therefore, the visible-light reactivity of TiO_2 was significantly enhanced with these dopants. It was assumed that the substitution for oxygen atoms by carbon, nitrogen, and sulfur atoms and Ti^{4+} by Fe^{3+} in the crystal structure of TiO_2, the electrons can be promoted to the conduction band from the newly formed impurity levels. The electrons can also be promoted from the valence band to the impurity level or from the lower to the higher impurity levels. Therefore, doped CNS-xFe-TiO_2 photocatalysts had narrower band gap than pure TiO_2 and then increased the absorption in the visible light region. Subsequently, the departed electrons and holes reacted with surface adsorbed O_2 and H_2O, respectively, forming $O^{2\bullet-}$ and $OH^{\bullet}$. It should be pointed out that the degradation of RB did not occur clearly in the absence of catalysts or visible light under the experimental conditions.

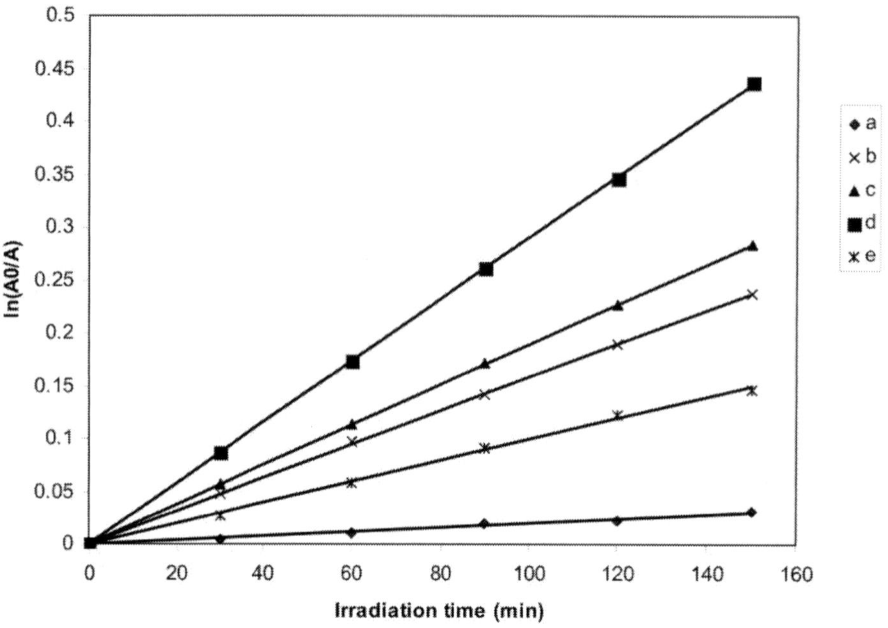

Figure 5: The degradation of RB using different catalysts under visible light irradiation (a: TiO_2, b:x=0.1, c: x=0.5, d: x=0.3, e: x=0)

Conclusions

Visible-light-active multi-doped TiO_2 photocatalysts were successfully synthesized. These multi-dopants retarded the growth and aggregation of TiO_2 particles. Multiple carbon species, substitutional and interstitial carbon atoms and carbonate species, existed in the lattice of TiO_2. Nitrogen and sulfur atoms were incorporated into the lattice of TiO_2. The incorporated iron dopant existed as Fe^{3+}. The diffuse reflectance spectra suggested the photo absorption of multi-doped TiO_2 was extended to the visible light region. It was assumed that these dopants introduced impurity levels between the conduction and valence band of TiO_2, leading to narrower band gap and enhancing the visible light absorption. Compared with TiO_2, these catalysts showed much higher activity for the degradation of RB under visible light irradiation.

References

[1] X. Chen, S. Mao, Chem. Rev. 107 (2007) 2891

[2] M. Hoffmann, S. Martin, W. Chen, D. Bahnemannt, Chem. Rev. 95 (1995) 69

[3] A. Linsebigler, G. Lu, J. Yates Jr, Chem. Rev. 95 (1995) 735

[4] W. Choi, A. Termin, M. R. Hoffmann, J. Phys. Chem. 98 (1994) 13669

[5] C. Wang, Q. Li, R. Wang, J. Mater. Sci. 39 (2004) 1899

[6] Z. Zhang, C. Wang, R. Zakaria, J. Ying, J. Phys. Chem. B 102 (1998) 10871

[7] X. Wang, S. Meng, X. Zhang, H. Wang, W. Zhang, Q. Du, Chem. Phys. Lett. 444 (2007) 292

[8] T. Ohno, T. Mitsui, M. Matsumura, Chem. Lett. 32 (2003) 364

[9] H. Chen, A. Nambu, W. Wen, J. Graciani, Z. Zhong, J. Hanson, E. Fujita, J. Rodriguez, J. Phys. Chem. C 111 (2007) 1366

[10] C. Burda, Y. Lou, X. Chen, A. C. S. Samia, J. Stout, J. L. Gole, Nano Lett. 3(2003) 1049

[11] R. Asahi, T. Morikawa, T. Ohwaki, K. Aoki, Y. Taga, Science 293 (2001) 269

[12] T. Ihara, M. Miyoshi, Y. Iriyama, O. Matsumoto, S. Sugihara, Appl. Catal. B: Environ. 42 (2003) 403

[13] D. Li, N. Ohashi, S. Hishita, T. Kolodiazhnyi, H. Haneda, J. Solid State Chem. 178 (2005) 3293

[14] X. Yang, C. Cao, K. Hohn, L. Erickson, R. Maghirang, K. Klabunde, J. Catal. 252 (2007) 296

[15] X. Yang, C. Cao, L. Erickson, K. Hohn, R. Maghirang, K. Klabunde, J. Catal. 260 (2008) 128

[16] R. Asahi, T. Morikawa, T. Ohwaki, K. Aoki, Y. Taga, Science 293 (2001) 269

[17] C. Enache, J. Schoonman, R. Krol, J. Electroceram. 13 (2004) 177

[18] Y. Cong, J. Zhang, F. Chen, M. Anpo, D. He, J. Phys. Chem. C 111 (2007) 10618

[19] M. Litter, J. Navio, J. Photochem. Photobiol. A 98 (1996) 171

[20] M. S. Nahar, K. Hasegwa, S. Kagaya, Chemosphere 65 (2006) 1976

[21] Z. Shi, X. Ye, K. Liang, S. Gu, F. Pan, J. Mater. Sci. 22 (2003) 1255

[22] T. Ohno, T. Tsubota, K. Nishijima, Z. Miyamoto, Chem. Lett. 33 (2004) 750

[23] H. Irie, Y. Watanabe, K. Hashimoto, J. Phys. Chem. B 107 (2003) 5483

[24] X. Shen, J. Guo, Z. Liu, S. Xie, Applied Surface Science 254 (2008) 4726

[25] M. Zhou, J. Yu, B. Cheng, J. Hazard. Mater. B 137 (2006) 1838

Photocatalytic Oxidation Systems for Indoor Air Cleaning Applications

Lynette Vera Bayless, Chemical Engineering, Kansas State University, Manhattan, KS

Xiangxin Yang, Chemical Engineering, Kansas State University, Manhattan, KS

Larry Erickson, Chemical Engineering, Kansas State University, Manhattan, KS

Ronaldo G. Maghirang, Biological and Agricultural Engineering, Kansas State University, Manhattan, KS

The use of photocatalytic oxidation (PCO) is a promising component of an effective indoor air cleaning system. Numerous PCO studies have focused on the evaluation and improvement of the system for a typical indoor environment. Development of energy-efficient photocatalysts is one of the vital aspects that can facilitate the commercialization of PCO air cleaning devices. One of the recent advances on photocatalyst development is the synthesis of 2% C- and V-doped TiO_2. The catalyst contains some carbon and 2% vanadium. Experimental studies on a laboratory scale have confirmed the catalyst's comparable activity both under visible light and dark conditions. This prominent characteristic of the catalyst combined with the potential utilization of the visible light spectrum from fluorescent lamps and solar radiation make the catalyst an attractive element of energy-efficient PCO air cleaning systems. This study is one of the first attempts to investigate the potential application in a typical indoor air environment of the two promising PCO design options that incorporates the C and V co-doped TiO_2. Mathematical modeling and simulation techniques were utilized to assess the use of packed bed and thin film systems in the photocatalytic oxidation of VOCs that are commonly found in offices, residential buildings, and other facilities. Using various operating conditions, the pressure drop through the packed bed and the influence of mass transfer limitations on the removal rate of the contaminants indoors for the two systems were investigated.

Keywords: air cleaning, energy-efficient, indoor air, mass transfer, mathematical model, packed bed, photocatalytic oxidation, simulation, thin films, visible light, VOCs

See more of this Session: Environmental Applications of Nanotechnology and Nanomaterials
See more of this Group/Topical: Environmental Division

Towards Predicting Impacts of Nanomaterials in the Environment

Leslie M. Shor[1,2]

[1]Department of Chemical, Materials and Biomolecular Engineering, and [2]Center for Environmental Science and Engineering (CESE), University of Connecticut, Storrs CT 06269

Abstract

Substantial and ongoing research investment has launched a (projected) trillion-dollar nanotech industry. As the sector matures from research into large-scale production, sub-100 nm-sized particles are becoming ubiquitous in textiles, personal care products, medical diagnostics, and consumer goods. Meanwhile, a growing body of nanotoxicology research is providing evidence that many nanoparticle forms including fullerene and metal oxides cause a variety of toxic endpoints. Over time, with lipid accumulation amplifying effective concentrations many times over those in environmental media, even dilute environmental concentrations of nanoparticles may cause serious harmful effects to predatory fish, birds, and humans.

From the literature on environmental fate and transport of hydrophobic organic contaminants, risk drivers for nanoparticles are likely to include: (1) the ability to be readily transported long distances and to accumulate and persist in soil and sediment ecosystems, (2) the potential to bioaccumulate up the food chain, and (3) toxicity to a variety of endpoints. The goal of this research is to develop a conceptual framework that shows how a limited number of physiochemical parameters can be used to predict regional-scale transport modeling and benthic accumulation of nanoparticles in rivers and estuaries. Together with an industrial ecology approach to quantify spatially distributed non-point source emissions, this approach may help policy makers identify key nanoparticle properties most likely to influence long-term environmental impacts.

Motivation

Emerging technologies using nano-scale materials will make possible dramatic changes in medical diagnostics [1], quantum improvements in sensors and computing power [2], and enormous gains in the energy efficiency of lighting and electronics [3]. However, a steadily growing literature in nanotoxicology demonstrates several nanomaterial forms cause a variety of toxic effects [4], including C_{60} [5,6,7], T_iO_2 nanoparticles [8], quantum dots [9], and carbon nanotubes [10]. For example, C_{60} has been demonstrated to cause cytotoxic effects in human cell lines, oxidative stress in fish, and reproductive effects and death in *Daphnia* [11]. Over time, even small-scale releases may accumulate in sensitive environmental compartments, become absorbed and accumulate fatty tissues, and cause harmful effects to predatory fish, birds, and humans [12], just as novel hydrophobic organic chemicals did half a century ago [13]. Society may be repeating old mistakes by allowing the pace of innovation to outstrip a comprehensive understanding of the environmental impacts [14]. Foresight into long-term environmental health and safety risks could steer production away from the most harmful forms, and in the process, support the sustainability of the nanotech industry.

Approach

Key risk drivers for anthropogenic contaminants in the environment include: (1) Persistence: the potential to accumulate and persist in the environment, such that concentrations will continue to build over time; (2) Bioaccumulation: the combined effect of selective uptake and/or trophic transfer that effectively magnify contaminant concentrations in biota over that in environmental media; and (3) Toxicity: demonstrated harmful effects for nanoparticulate forms.

Nanoparticle Occurrence and Persistence

Several different nanoparticle forms have received special attention in the environmental impacts literature. Some forms occur naturally in the environment. C_{60} can result from natural combustion, and colloid-sized minerals including nano-TiO_2, nano-SiO_2, and nano-ZnO can be formed by wear of larger minerals [12]. Other commonly studied nanomaterials are not found in nature, such as CdS Quantum Dots [20], and metal-doped or surface-functionalized forms of more conventional nanoparticles [21]. Novel materials are believed to a pose greater chance of environmental impacts than naturally-occurring materials for a given exposure dose [14], although many novel materials are so expensive to produce they are unlikely to be released in appreciable quantities. Rare earth oxides including Eu_2O_3 constitute an important industry [22] with applications including backlighting for LCD display devices, plasma display phosphors, solid state fluorescent lighting, and electric discharge lighting. There is also the potential for significant releases of rare earth oxides due to growing stockpiles of used television screens, computer monitors, and high-strength magnets.

A critical factor influencing nanoparticle behavior under environmental conditions is aggregation [12]. There have been several recent reports on the effect of natural conditions on the aggregation state of C_{60} and nano-TiO_2. The solubility of C_{60} in aqueous solution increases with divalent cation concentration, and decreases with increasing ionic strength [23]. Extended mixing forms aggregated C_{60}, with aggregate size depending on C_{60} concentration and NOM concentration [24]. Other authors found 4-5 nm nano-TiO_2 particles readily form stable aggregates with 50-60 nm diameter at pH 4.5 and ionic strength of 0.0045 M NaCl [25]. Microbial activities can alter the surface properties and thereby the aggregation of nanoparticles, principally through the release of extracellular enzymes that oxidize surfaces of C_{60} and promote aqueous solubility [12].

Nanoparticle Bioaccumulation

For hydrophobic organic contaminants, the key parameter determining bioaccumulation potential is the octanol-water partition coefficient of the contaminant and the lipid content of the biota [26]. Specific information on the bioaccumulative potential of nanoparticles is scarce, although one recent report found evidence for accumulation of carboxylated and biotintylated quantum dots by *Tetrahymena* (a freshwater protozoan), and trophic transfer to rotifers [27].

Nanoparticle Toxicity

There have been approximately 50 published studies to date demonstrating ecotoxicity of various nanoparticles. Most freshwater studies employed well-established toxicity protocols with specific model organisms, especially *Daphnia magnia* [4]. Other common indicator species are bacteria including *E. coli* and *B. subtilis*, mussels, and fish. The mechanism for toxicity for both nano-TiO_2 and C_{60} may involve the formation of reactive oxygen species (ROS) [4]. In addition, C_{60} is able to readily penetrate the cell membranes, perhaps resulting in cytoplasm leakage, or enabling other harmful materials to enter. Oxidation of C_{60} reduces its toxicity to *E. coli* [28] and *B. subtilis* [6]. In molecular dynamics simulations, functionalization of C_{60} to $C_{60}(OH)_{20}$ reduces translocation rates through a model PPC membrane by eight orders of magnitude [29].

Particle and aggregate size are also important parameters impacting toxicity. In general, smaller particles and aggregates are reported to be more toxic [7]. If surface reactivity is the mechanism for toxicity, then the surface area not the mass concentration should be determined experimentally. Nanoparticles may also act as carriers for other toxic compounds, especially polycyclic aromatic hydrocarbons (PAHs) [4].

Risk Assessment Framework

Adverse environmental impacts can be objectively quantified and evaluated using ecological risk assessment. The approach has been used to establish sediment quality criteria [16] and to prioritize environmental cleanup projects [17]. Together with emission and fate and transport modeling, risk assessment tools have the potential to forecast future impacts of nanoparticles in the environment [18]. Key parameters for performing quantitative risk assessments include (i) environmental flow coefficients, especially for aqueous-sediment partitioning; (ii) transformation, degradation, and bioaccumulation; (iii) toxicity data using standardized toxicity tests (no observed effects concentrations [NOECs] reported for nano-TiO_2 span a factor of 1000); and (iv) studies on the impact of size, shape, modification, and purity of nanoparticles on risk [18].

In conducting a risk assessment, a predicted or measured environmental concentration is **compared with a known dose-response function to determine the probability of adverse effects.** Adverse effects may include chronic effects such as changes in mobility, grazing rates or reproduction rates as well as lethal acute toxicity, and must be defined for a specific exposure scenario for a particular organism and endpoint. In past work, concentrations of environmental contaminants were predicted in air and surface water using an integrated approach that encompassed industrial ecology to determine activity-dependent primary emission throughout a region home to 10M people, then modeled transmission from point of primary release to the harbor boundary. Fate and transport modeling was based on literature values of key parameters, including interfacial partitioning (evaporation, dissolution, sorption, etc), particle transport, and abiotic and biological PAH degradation rates [31]. Comparing predicted environmental concentrations with measured environmental concentrations for individual compounds provides an external validation on the process. The advantage of modeling concentrations rather than measuring them is that environmental impacts can be described as a function of consumer use of alternative products, watershed management, or other policy decisions.

Evaluating tradeoffs is intuitively how people make decisions, and placing the environmental impacts of nanoparticles in a similar context will allow a more public policy.

References

1. Fortina P, Kricka L, Surrey S, Grodzinski P (2005) Nanobiotechnology: the promise and reality of new approaches to molecular recognition. Trends Biotechnol 23: 168-173.
2. Yu B, Meyyappan M (2006) Nanotechnology: Role in emerging nanoelectronics. Solid State Electronics 50: 536-544.
3. Chau R, Datta S, Doczy M, Doyle B, Jin B, et al. (2005) Benchmarking nanotechnology for high-performance and low-power logic transistor applications. IEEE Transactions on Nanotechnology 4: 153-158.
4. Farré M, Gajda-Schrantz K, Kantiani L, Barceló D (2009) Ecotoxicity and analysis of nanomaterials in the aquatic environment Analytical and Bioanalytical Chemistry 393: 81-95.
5. Dhawan A, Taurozzi JS, Pandey AK, Shan W, Miller SM, et al. (2006) Stable Colloidal Dispersions of C60 Fullerenes in Water: Evidence for Genotoxicity. Envionmental Science & Technology 40: 7394–7401.
6. Fortner JD, Lyon DY, Sayes CM, Boyd AM, Falkner JC, et al. (2005) C60 in Water: Nanocrystal Formation and Microbial Response. Environmental Science & Technology 39: 4307–4316.
7. Lyon DY, Adams LK, Falkner JC, Alvarez PJJ (2006) Antibacterial Activity of Fullerene Water Suspensions: Effects of Preparation Method and Particle Size. Environmental Science & Technology 40: 4360–4366.
8. Lovern SB, Strickler JR, Klaper R (2007) Behavioral and Physiological Changes in Daphnia magna when Exposed to Nanoparticle Suspensions (Titanium Dioxide, Nano-C60, and C60HxC70Hx). Environmental Science & Technology 41: 4465–4470.
9. Hoshino A, Fujioka K, Oku T, Suga M, Sasaki YF, et al. (2004) Physicochemical Properties and Cellular Toxicity of Nanocrystal Quantum Dots Depend on Their Surface Modification. Nano Letters 4: 2163 -2169.
10. Ghafari P, St-Denis CH, Power ME, Jin X, Tsou V, et al. (2008) Impact of carbon nanotubes on the ingestion and digestion of bacteria by ciliated protozoa. Nature Nanotechnology 3: 347 - 351.
11. Baun A, B. Hartmann N, Grieger K, O. Kusk K (2008) Ecotoxicity of engineered nanoparticles to aquatic invertebrates: a brief review and recommendations for future toxicity testing. Ecotoxicology and Environmental Safety 17: 387-395.
12. Klaine SJ, Alvarez PJJ, Batley GE, Fernandes TF, Handy RD, et al. (2008) Nanomaterials in the Environment: Behavior, Fate, Bioavailability, and Effects Environmental Toxicology and Chemistry 27: 1825–1851.
13. Carson R (1962) Silent Spring: Houghton Mifflin.
14. Hansen SF, Maynard A, Baun A, Tickner JA (2008) Late lessons from early warnings for nanotechnology. Nature Nanotechnology 3: 444-447.
15. US Environmental Protection Agency (2005) Persistent Bioaccumulative and Toxic (PBT) Chemical Program.
16. DiTorro DM, Zarba CS, Hansen DJ, Berry WJ, Swartz RC, et al. (1991) Technical basis for establishing sediment quality criteria for nonionic organic chemicals using equilibrium partitioning. Environmental Toxicology and Chemistry 10: 1541-1583.
17. Shor LM, Kosson DS, Rockne KJ, Young LY, Taghon GL (2004) Combined Effects of Contaminant Desorption and Toxicity on Risk from PAH Contaminated Sediments. Risk Analysis 24: 1109-1120.

18. Mueller NC, Nowack B (2008) Exposure Modeling of Engineered Nanoparticles in the Environment. Environmental Science & Technology 42: 4447–4453.
19. U.S. Environmental Protection Agency (1999) Risk assessment guidance for superfund. Volume I: Human health evaluation manual. Part E. Supplementary guidance for dermal risk assessment. Washington, D.C.: Office of Emergency and Remedial Response.
20. Striolo A, Ward J, Prausnitz JM, Parak WJ, Zanchet D, et al. (2002) Molecular weight, osmotic second virial coefficient, and extinction coefficient of colloidal CdSe nanocrystals. Journal of Physical Chemistry B 106: 5500-5505.
21. Mahajan SV, Dickerson JH (2007) Synthesis of monodisperse sub-3 nm RE2O3 and Gd2O3 : RE3+ nanocrystals. Nanotechnology 18.
22. Mueller A (2008) Of markets, patents, recycling and networking at Phosphor Global Summit 2008. Intertech Pira Conference Blog.
23. Li Q, Xie B, Hwang YS, Xu Y (2009) Kinetics of C60 Fullerene Dispersion in Water Enhanced by Natural Organic Matter and Sunlight. Environmental Science & Technology 43: 3574–3579.
24. Duncan LK, Jinschek JR, Vikesland PJ (2008) C-60 colloid formation in aqueous systems: Effects of preparation method on size, structure, and surface, charge. Environmental Science & Technology 42: 173-178.
25. French RA, Jacobson AR, Kim B, Isley SL, Penn RL, et al. (2009) Influence of Ionic Strength, pH, and Cation Valence on Aggregation Kinetics of Titanium Dioxide Nanoparticles. Environmental Science & Technology 43: 1354–1359.
26. Mackay D, Shiu WY, Ma KC (1992) Illustrated Handbook of Physical-Chemical Properties and Environmental Fate for Organic Chemicals: Polynuclear Aromatic Hydrocarbons, Polychlorinated Dioxins, and Dibenzofurans. Chelsea, MI: Lewis Publishers.
27. Holbrook RD, Murphy KE, Morrow JB, Cole KD (2008) Trophic transfer of nanoparticles in a simplified invertebrate food web. Nature Nanotechnology 3: 352 - 355.
28. Lee J, Cho M, Fortner JD, Hughes JB, Kim J-H (2009) Transformation of Aggregated C60 in the Aqueous Phase by UV Irradiation. Environmental Science & Technology 43: 4878–4883.
29. Qiao R, Roberts AP, Mount AS, Klaine SJ, Ke PC (2007) Translocation of C60 and Its Derivatives Across a Lipid Bilayer. Nano Letters 7: 614–619.
30. Lovern SB, Klaper R (2006) *Daphnia magna* mortality when exposed to titanium dioxide and fullerene (C_{60}) nanoparticles Environmental Toxicology and Chemistry 25: 1132–1137.
31. Rodenburg LA, Valle S, Panero M, Shor L (2009) Mass Balances on Selected Polycyclic Aromatic Hydrocarbons (PAHs) in the NY/NJ Harbor. *In Review.*

Synthesis of Traceable Nanoparticles to Enable the Study of the Fate and Transport of Engineered Nanomaterials

Alex Yokochi, School of Chemical, Biological and Environmental Engineering, Oregon State University, Corvallis, OR

Jeffrey A. Nason, School of Chemical, Biological and Environmental Engineering, Oregon State University, Corvallis, OR

Kevin Caple, School of Chemical, Biological and Environmental Engineering, Oregon State University, Corvallis, OR

Dylan Stankus, School of Chemical, Biological and Environmental Engineering, Oregon State University, Corvallis, OR

Robert Kimmell, School of Chemical, Biological and Environmental Engineering, Oregon State University, Corvallis, OR

Nicholas AuYeung, School of Chemical, Biological and Environmental Engineering, Oregon State University, Corvallis, OR

Engineered nanomaterials hold great promise for technological innovation due to unique properties that emerge at the nanoscale. As a result, nanomaterials are increasingly being incorporated into consumer products ranging from electronics to cosmetics, and are being evaluated for novel drug delivery and treatment of contaminated water. Unfortunately, the boom in nanoscience and nanotechnology research has not been accompanied by an equal effort investigating the environmental implications of nanomaterials and little is known about the prevalence, behavior and risks of nanomaterials in the environment. Concerns about the environmental health and safety aspects of nanomaterials are global: production, distribution, use and disposal of nanoparticles will undoubtedly result in their wide release into the environment. It is therefore imperative that methodology to track specific nanomaterials through the environment be developed.

A major difficulty is the fact that at present, very few analytical techniques are capable of unambiguously identifying and characterizing nanomaterials in environmental matrices, especially when nanoparticles composed entirely of elements with high environmental background concentrations are concerned. We have worked to develop methodology to prepare nanoparticles labeled with elements present in low background concentrations in the environment (e.g., lanthanides, gold, etc) where the concentration of the tracer element can therefore be used as a proxy identifier for the presence of the nanoparticles.

We will present our initial results and observations in this work.

See more of this Session: Environmental Applications of Nanotechnology and Nanomaterials
See more of this Group/Topical: Environmental Division

Characterization of Microbes in a Regenerative Biofilter System for VOC Removal

Wen-Hsuan Huang, Department of Biomedical and Chemical Engineering, Syracuse University, Syracuse, NY

Zhiqiang Wang, Mechanical and Aerospace Engineering, Syracuse University, Syracuse, NY

Beverly Guo, Mechanical and Aerospace Engineering, Syracuse University, Syracuse, NY

Jianshun Zhang, Mechanical and Aerospace Engineering, Syracuse University, Syracuse, NY

Dacheng Ren, Biomedical and Chemical Engineering, Syracuse University, Syracuse, NY

Effective removal of volatile organic compounds (VOCs) is critical for indoor air quality (IAQ) control. The performance of existing technologies of VOC removal are limited by inadequate selection of filter media, poor airflow management inside the cleaning devices, insufficient catalytic reaction surface area, and poor distribution of UV light irradiation. A new technology based on the Wolverton air filtration system uses a plant root bed of activated carbon, porous shale pebbles, microbes and a wet scrubber to remove VOCs and radon from the air in tightly sealed buildings. To better understand this system and optimize its performance, we characterized the microbes in this system and their activities in formaldehyde removal. We chose golden pathos as the plant and isolated the microbes from the plant roots and pebbles. By using universal primers to amplify 16S rRNA gene, 6 different species of bacteria were identified including Arthrobacter aurescens, Leifsonia xyli subsp.xyli, Bacillus sp.NRRL B-14911, Bacillus cereus, Arthrobacter oxydans, and Pseudomonas putida. In addition, a strain of Pseudomonas putida isolated from this system was found to reduce the concentration of formaldehyde by 70% after incubation for 24 hours. The distribution of the microbes in the system will also be discussed.

See more of this Session: Fundamentals of Environmental Biotechnology
See more of this Group/Topical: Environmental Division

Evaluation of a Model Biological Fuel Cell System for Reduction Efficiency and Bioremediation Potential

Lewis Hsu, Sonny Astani Department of Civil and Environmental Engineering, University of Southern California, Los Angeles, CA

Shelly Ann Masuda, Sonny Astani Department of Civil and Environmental Engineering, University of Southern California, Los Angeles, CA

Kenneth H. Nealson, Department of Earth Sciences, University of Southern California, Los Angeles, CA

Massoud Pirbazari, Sonny Astani Department of Civil and Environmental Engineering, University of Southern California, Los Angeles, CA

Bioelectrochemical systems and in particular microbial fuel cells (MFCs) have generated a large amount of interest and research in the past few years. These systems are largely focused on the oxidation of an inexpensive substrate, e.g. wastewater, to generate electrical energy. Such a process has the potential to eliminate several costs associated with treating wastewater, if operated in lieu of conventional treatment involving activated sludge and anaerobic digestion processes. Large uncertainties exist on the fundamental process governing the operation of these reactor systems. These include transport mechanisms for various species, mechanisms of biocatalysis, dominant members of a mixed community biocatalysis, etc. For these reasons, specific electrochemically active bacteria, *Shewanella* spp. and *Geobacter* spp., have been used as model organisms due to the large amount of information already available for these species.

One aspect of MFC systems currently under investigation is utilizing the reducing potential at the cathode to perform treatment as well. For example, denitrification or sulfate reduction may be employed at the cathode as part of a treatment system in tandem to the oxidation of wastewater. Of more interest is the potential of utilizing the cathode to reduce polluting heavy metals such as uranium or chromium. Oxidized forms of these metals are soluble and highly mobile. Upon reduction, these metals generally form insoluble precipitates, facilitating their immobilization and preventing their exposure to the environment.

In this study, the focus of the MFC systems is aimed at catalyzing the reduction reaction occurring at the cathode. More specifically, several strains of *Shewanella* (*S oneidensis* MR-1, *S. amazonensis* SB2B, *S. loihica* PV-4, *S. species* W3-18-1, *S. species* ANA-3, and *S. species* MR-4) were evaluated as a cathodic biocatalyst for the reduction of hexavalent chromium, a common soil and groundwater pollutant. The aim of these experiments was to evaluate the efficiency and extent of chromium removal. While all species were able to provide some reduction of chromium, the speed and extent of reduction were different among species. For example, *S. species* W3-18-1 was able to reduce hexavalent chromium levels to less than 1 ppb, a level unattainable by the other species.

Electron microscopy was also employed to evaluate morphological characteristics and biofilm characteristics. Several different biofilm types were observed. Spotty, clumped coverage was seen for MR-1, SB2B, ANA-3, and MR-4. A high degree of even, monolayer coverage was seen for W3-18-1. One species, PV-4, did not show an appreciable amount of biofilm development. Moreover, very little attachment to the electrode surface was observed. Selected area elemental analysis was also conducted by energy dispersive spectroscopy to determine the spatial distribution of chromium precipitates located at the electrode and biofilm surfaces.

The differences observed with these species of *Shewanella* operating as cathodic biocatalysts give some insight as to governing design aspects (i.e. biomass attachment, areas of chromium deposition, rate of reduction, etc.) associated with MFC performance as a hexavalent chromium bioremediation tool.

Key words: chromium, bioremediation, microbial fuel cell, Shewanella, cathodic biocatalyst

See more of this Session: Fundamentals of Environmental Biotechnology
See more of this Group/Topical: Environmental Division

Characterizing the Abundance and Activity of Soil Microbes Using Single-Strand Conformational Polymorphism by Capillary Electrophoresis

Alice C. Jernigan, Chemical Engineering, University of Arkansas, Fayetteville, AR

Duane C. Wolf, Dept. of Crop, Soil and Environmental Sciences, University of Arkansas, Fayetteville, AR

Greg Thoma, Dept. of Chemical Engineering, University of Arkansas, Fayetteville, AR

Christa Hestekin, Dept of Chemical Engineering, University of Arkansas, Fayetteville, AR

Bioremediation is the process by which microbes are used to break down soil contaminants. The traditional method for looking at bioremediation is by viable plate counts or microscopic examination. These methods are both time consuming and can underestimate the numbers due to the inability to culture many soil organisms. The increased speed and automation of capillary electrophoresis (CE) allows for more rapid and reliable results. The identification and assessment of activity of microbial soil communities can be accomplished by isolating DNA and RNA from soil samples. These DNA/RNA samples can then be characterized by a technique called single-strand conformational polymorphism (SSCP) analysis of the 16S gene and its variable regions. These variable regions have been amplified by polymerase chain reaction (PCR) giving species-dependent fragments of the gene sequence. SSCP, a method that is normally used to detect mutations, is the heat denaturing of these fragments resulting in single strand DNA (ssDNA). Snap cooling of the ssDNA causes it to refold into different secondary structures (conformations) depending on the nucleotide sequence. These different conformations can be separated due to their different electrophoretic mobility, even if they are the same size. CE-SSCP has been used to identify mixed populations from DNA extracted from soil samples and to identify activity of the microorganisms by examining the products of reverse transcription (RT)-PCR which relates to the sample's RNA. By isolating both DNA and RNA from the same soil sample, at the same time, a more accurate picture of the microbial community in those samples has been made available. This is an important development in the study of contaminated soils and the monitoring of the progress of bioremediation.

See more of this Session: Fundamentals of Environmental Biotechnology
See more of this Group/Topical: Environmental Division

Electron Transfer and Local pH Variations in Anodic Biofilms

Alim Dewan, Chemical Engineering, Washington State University, Pullman, WA

Hung Duc Nguyen, School of Chemical Engineering and Bioengineering, Washington State University, Pullman, WA

Haluk Beyenal, School of Chemical Engineering and Bioengineering, Washington State Univeristy, Pullman, WA

Proton transport out of the anodic biofilms is a limiting factor of current in microbial fuel cells. It has been demonstrated that the limitation of proton transport causes local pH changes inside biofilms. However, the correlation between local pH changes and current is not quantified. In this study, we have quantified local pH changes in biofilms with current generated from a microbial fuel cell. A two compartment microbial fuel cell was used where biofilms of *Shewanella oneidensis* MR - 1 were grown on the graphite plate anode. In the cathodic, oxygen was used as the electron acceptor. The local pH changes inside the biofilms were measured using pH microelectrode. We found that there is a linear correlation between current and local pH. Our results also showed that the pH variation was higher near the electrode surface compare to that in the bulk solution and that the direction of pH change depended on the polarization potential relative to the equilibrium potential of the anode.

See more of this Session: Fundamentals of Environmental Biotechnology
See more of this Group/Topical: Environmental Division

Biological Sulfate Reduction in Fluidized Bed Reactors: Byproduct Control and Recovery

Atosa Vahdati, Civil and Environmental Engineering, University of Southern California, Los Angeles, CA

Varadarajan Ravindran, University of Southern California, Los Angeles, CA

Massoud Pirbazari, University of Southern California, Los Angeles, CA

Industrial wastewaters with high sulfate concentrations ranging from 1,500 to 20,000 mg/L are produced by a large spectrum of industries of economic significance, namely, paper and pulp, sea-food processing, fermentation for starch production, fertilizer production, leather tanning, textile processing, electroplating, and mineral processing. These wastewaters with high sulfate concentrations cannot be directly discharged into water bodies under the present environmental regulations of the National Pollution Discharge Elimination Systems (NPDES) program because of the limits on the chemical oxygen demand (COD) for surface water discharges. Industrial wastewaters rich in sulfate pose a major problem for industries because they increase the total dissolved solids content, and interfere with methanogenic processes during biological treatment. The interference of sulfate with methanogenic processes results in poor production of methane in anaerobic processes, a major setback from an energy recovery standpoint.

Microbial sulfate reduction has certain advantages over other technologies besides high efficiency and potential for energy recovery. The process is effective in handling industrial wastewaters containing heavy metals because it facilitates the precipitation of metals as sulfides. Furthermore, sulfate reducing bacteria have adaptability and can acclimate themselves to different environmental conditions such as pH, temperature and toxic metals. Nevertheless, the process has certain disadvantages that must be overcome before it can be used on a variety of industrial applications. It produces hydrogen sulfide that is toxic and is inhibitive to microbial sulfate reduction and reduces the process efficiency. Additionally, hydrogen sulfide is undesirable owing to its toxicity to humans and animals, and its bad odor.

Our previous study investigated the increase in the effectiveness of microbial sulfate reduction in anaerobic fluidized bed reactor (FBR) systems for the treatment of industrial wastewaters of high sulfate content. The FBR system employed granular activated carbon (GAC) as the packing medium for microbial support and its choice was attributed to its ability to promote the growth of microbial biofilms, and to remove biologically toxic and inhibitive constituents in wastewaters. Lactate and acetate were identified as favorable electron donors for microbial sulfate reduction from thermodynamic considerations. The FBR studies were conducted for these electron donors to evaluate the biological sulfate reduction efficiency.

The present study was directed at for recovery of sulfur, methane and hydrogen as by-products of the FBR system. Furthermore, the study was intended to examine methods for hydrogen sulfide reduction in the liquid phase and the gaseous phase. An anaerobic biofiltration system (ABS) was employed as an approach for hydrogen sulfide control within the FBR system, and simultaneous recovery of elemental sulfur. Two groups of anoxygenic photolithotrophic autotrophs were identified as the major bacterial strains responsible for the transformation of hydrogen sulfide to elemental sulfur.

The ABS was evaluated for hydrogen sulfide conversion with the idea of increasing methanogenic activity of microorganisms for the production of methane or hydrogen in the FBR system. The study also examined the mass-transfer of hydrogen sulfide produced in the FBR from the liquid phase to the gaseous phase, and subsequent conversion of hydrogen sulfide to elemental sulfur. The increase in methanogenic activity potentially improved methane or hydrogen production for energy recovery. A sulfur mass balance analysis scheme was employed for the FBR system using lactate and acetate as electron donors so as to measure the quantity of hydrogen sulfide, and the amount of elemental sulfur recoverable. This approach of using an ABS conferred on FBR microbial sulfate reduction process a dual advantage of sulfur recovery and energy production.

See more of this Session: Fundamentals of Environmental Biotechnology
See more of this Group/Topical: Environmental Division

96-well culture microplates as a microreactor platform for microalgae research

Ileana Hernández-Mireles, Cecilia García-Martínez, Adriana Pacheco, Mario M. Álvarez, Tecnológico de Monterrey, Monterrey, N.L., México.

Abstract

Platforms for high-throughput microalgae screening are not yet commercially available. We analyzed the feasibility of utilizing 96-well microplates as microreactor platforms for microalgae research. Particularly, the use of the system as a screening tool for systematic optimization of microalgae culture media was demonstrated. Six microalgae strains were grown at three different culture medium water dilutions in a randomized experimental design in the central subset of the microplate. At each well, microalgae growth was evaluated by light absorbance at 680 nm and the specific growth rate was calculated from the exponential phase of the growth curve. A statistical analysis of a one species microplate experiment showed that the system was robust as a microreactor platform for microalgae kinetic studies, i.e. the data exhibited a normal distribution with no significant effect of well position. Based on our analysis, recommendations are given for the adequate design of growth screening experiments of microalgae in 96-well microplates.

Introduction

Microalgae are considered photoautotrophic microorganisms, as light and carbon dioxide serve as their sole source of energy and carbon, respectively. For decades, microalgae have been culture for aquaculture nutrition and as key participants in wastewater treatment and bioremediation systems (1, 2, 3). Furthermore, their potential for a wide range of applications has been suggested in the cosmetic, pharmaceutical, and food industries (4, 5, 6). Nowadays, microalgae biomass is receiving great attention as a potential energy source; a system that captures carbon dioxide and does not contribute to global warming (7, 8). In despite of all these interest, microalgae are practically an unexploited resource, of 25 000 known species only fifteen are used commercially (6).

Microalgae screening techniques have important throughput limitations. Research groups have worked with small to medium scale cultures, mostly flasks and column bioreactors, ranging from 250 ml to 75 L (9, 10, 11). However, screening experiments at these scales demand considerable time and space. In addition, collection of data for kinetic characterization of microalgae cultures is a time-consuming, cumbersome, and not necessarily trivial task. Techniques to do it effectively and massively are not well developed, as there are not commercially available validated bioreactors for microalgae screening. Relative high-throughput screening protocols for selection and kinetic characterization of microalgae would be highly desirable to expedite research oriented to fully exploit their potential.

In this document, we present and discuss two experiments to validate the use of commercial 96-well microplates for microalgae screening. Microplates are mainly used to conduct immunoassay tests and culture mammalian cells. Nevertheless, they have been used in other applications such as colorimetric determination of pollutants (12), ecotoxicology assays utilizing bacteria (13) and even employing microalgae (14, 15). Although their application for microalgae culture has been suggested (16), there are no reports on their statistical validation. Our intention is to demonstrate that 96-well microplates can be a robust, reliable, cost-effective, and practical platform for relatively high-throughput microalgae screening applications and kinetic characterization.

Materials and methods

Experimental system

Microalgae were cultured in polystyrene flat bottom clear 96-well cell culture microplates (Corning®, New York, USA). Evaporation effects in border wells, evident in preliminary experiments (data not shown), were minimized by adding culture medium to border columns and rows. These wells were not inoculated. Therefore, microplate effective area was reduced to 6x10 wells. Microplates were incubated in an anaerobic chamber A143 (VWR International Inc, USA) under an atmosphere of 20% (v/v) CO2/air, 20°C, and 24 h light period provided by cool white fluorescent lamps with an intensity of 2400 lux.

Strains and culture conditions

Four of the microalgae strains used in this study were kindly donated by UTEX The Culture Collection of Algae (University of Texas in Austin, USA): *Chlorococcum a*ureum (UTEX 1768), *Scenedesmus* sp. (UTEX 1589), *Chlorella kessleri* (UTEX 397), and *Botryococcus sudeticus* (UTEX 2629). Two additional strains were isolated from environmental samples; from water of a natural spring (Strain I) and from an aquarium (Strain II). Currently, these strains are part of the CEMEX-ITESM Microalgae Collection (Centro de Biotecnologia del Tecnologico de Monterrey, Mexico). All collection strains were cultured in 125 ml flasks accordingly to the recommended culture media by UTEX: modified bold 3N medium (MB3N) for *Chlorococcum aureum* and *Scenedesmus* sp., proteose medium for *Chlorella kessleri*, and soil extract medium for *Botryococcus sudeticus*. Strains isolated from environmental samples were cultured in BG11 medium (16). Microalgae were cultured until they reached ¾ exponential growth before inoculating the microplates. Each microplate well was filled with 200 μl of MB3N medium, used as a general-purpose minimal medium. To prepare the inoculum, 2 ml of each microalgae culture were centrifuged at 3000 rpm for 1.5 min. The cell pellet was suspended in 2 ml of MB3N medium and centrifuged again. The procedure was repeated twice to eliminate any traces of the previous medium and each well was inoculated with 20 μl of the final 2 ml cell suspension.

Estimation of specific growth rates

Microalgae growth was measured by absorbance at 680 nm in a Multi-Detection Microplate Reader Synergy HT (BioTek, USA). Absorbance readings were taken at the time of inoculation and every 24 h for 8 days. Readings were taken after 30 s of microplate agitation at the highest level. To kinetically characterize microalgae growth at each well, the specific growth rate (μ) was calculated. A first order kinetic growth behavior with respect to biomass concentration (expressed as absorbance at 680 nm) was assumed: $d[X]/dt = \mu[X]$, where [X] is the biomass concentration, t is time, and μ (day-1) is the specific growth rate. The integrated form of this equation is $\ln[X] = \mu t + \ln[X]_o$, where $[X]_o$ is the initial biomass concentration. Therefore, μ can be calculated from the slope of the straight line resulting from plotting $\ln([X]/[X]_o)$ versus t, corresponding to the exponential growth phase of each culture.

Statistical analysis

Standard statistical methodologies were used to analyze normal probability plots, one-sample t-tests, analysis of variance (ANOVA), and Tukey´s test. Minitab 14®, JMP (SAS Institute Inc, North Carolina, USA), and Excel® data analysis tools were employed.

Results and discussion

We present and discuss two different experiments to validate the use of commercial 96-well cell culture microplates for microalgae screening applications: (a) a randomized experiment that demonstrates a strategy for culture media optimization for six microalgae strains in a single microplate, and (b) an experiment to statistically validate that once discarding the border rows and columns, the analyzed system can be considered as a set of 60 independent and statistically equivalent microreactors for microalgae culture and screening, eliminating the need of randomization of experiments.

An example of a randomized screening experiment

In a first experiment, we illustrated the use of the system to evaluate the effect of culture media concentration (100%, 70%, and 50% (v/v) of reference medium concentration to sterile deionized water, no effect on pH was observed) on growth of six microalgae strains by utilizing a randomized experimental design. Therefore, we required a 3x6 experiment with three repetitions of each treatment; a total of 54 experiments (3x6x3). Discarding the border rows and columns, the microplate system can still accommodate the necessary experiments in a single plate. The individual experiments were randomly distributed within 54 of the 60 central wells of the microplate, as shown in Figure 1A.

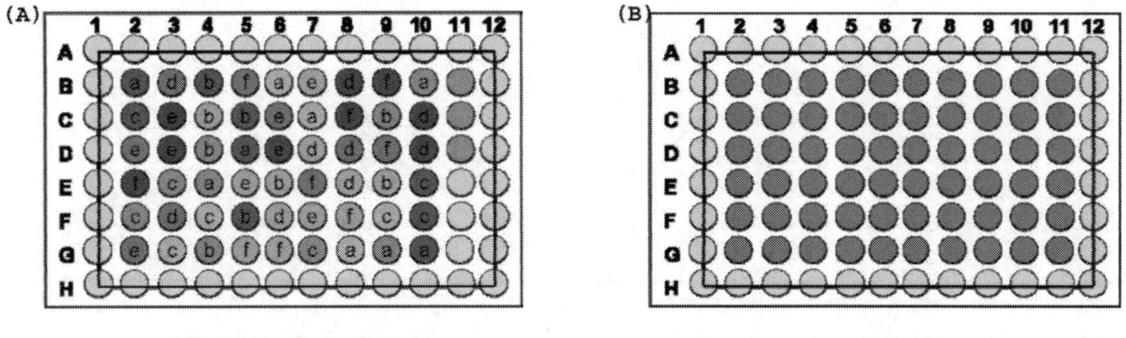

Figure 1. Experimental microplate layouts. (**A**) Randomized experiment to evaluate six microalgae strains growing at three different MB3N medium concentrations. Different letters indicate different strains. Gradient of grays indicate different medium concentration (100, 70 and 50%). The experiment was run in triplicate. (**B**) Experiment layout to evaluate an environmental microalgae isolate (Strain I). Microplate borders were not inoculated (wells marked with a line).

The optical density of each culture was monitored throughout the experiment and the specific growth rate was calculated from the exponential phase of each growth profile accordingly to the algorithm described in materials and methods. A plot of ln[X] versus t rendered a straight line for the exponential growth phase of each culture, corroborating the assumption of a first order differential growth model with respect to biomass concentration expressed as optical density at 680 nm. Figure 2 shows the specific growth rates (μ) for each microalgae strain at the three different culture medium concentrations. Standard error bars indicate the experimental dispersion of the data: 0.3 to 3%, 1 to 5%, and 1 to 12% for the experiments at 50%, 70%, and 100% of culture medium concentration, respectively. Meaningful kinetic information can be derived from this simple experimental array. For example, Tukey´s test ($P<0.05$) corroborates that only *Chlorella kessleri* differed in their growth rate means between treatments. However, no statistical differences in growth rate means were observed for the rest of the strains evaluated. In a biotechnology application, these strains could be grown with a significant decrease in nutrient concentrations (by diluting the culture medium) with the assurance of no detrimental effects on the specific growth rates during exponential growth. For proof of concept, we selected as the output variable the specific growth rate in exponential growth phase. However, the statistical analysis of other relevant variables such as the length of the lag phase or the final biomass concentration can also be informative and easily obtain from the 54 individual growth curves generated.

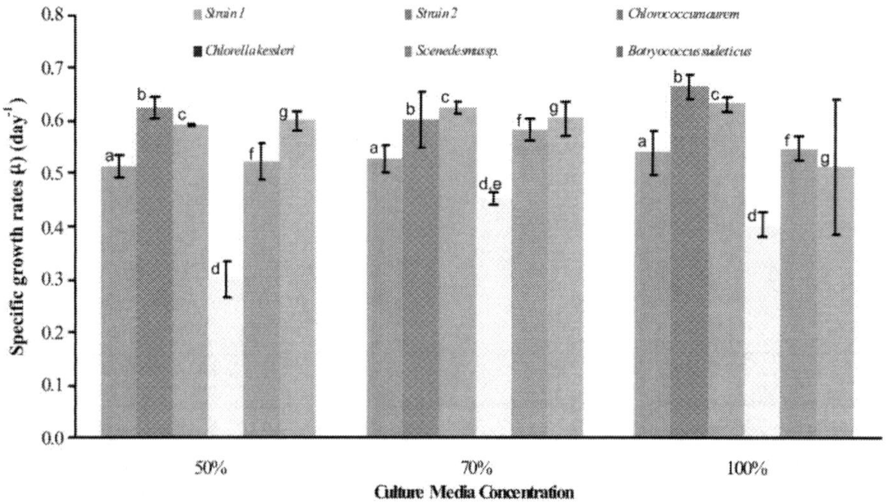

Figure 2. Specific growth rates (mean ± SE, n = 3) of six microalgae strains cultured in a 96-well microplate at three different MB3N culture medium concentrations. Means with the same letter are not significantly different (Tukey´s test, $P<0.05$).

A set of 60 statistically equivalent bioreactors

We conducted a second experiment to validate if all 60 wells (central subset of the microplate) were statistically equivalent. If they were, the requirement of randomized experiments could be relaxed, further simplifying the experimental design. We cultured one of the environmental microalgae isolates, Strain I, in each of the wells of the microplate (Fig. 1B). The specific growth rate (μ) mean and standard error for the set was 0.864 ± 0.011 (n = 60). Figure 3a shows the histogram of the μ values. To accept the hypothesis of equivalence, a normal distribution of the output parameter (μ) should be observed, implying a normal distribution of the experimental errors throughout the system. The frequency bars shown in Figure 3A fitted a Gaussian curve, evidencing a normal distribution of the data. The probability plot corroborated that the data adjusted to a normal distribution; the data felt into a straight line (Fig. 3B). Consequently, each well can function as an independent microreactor.

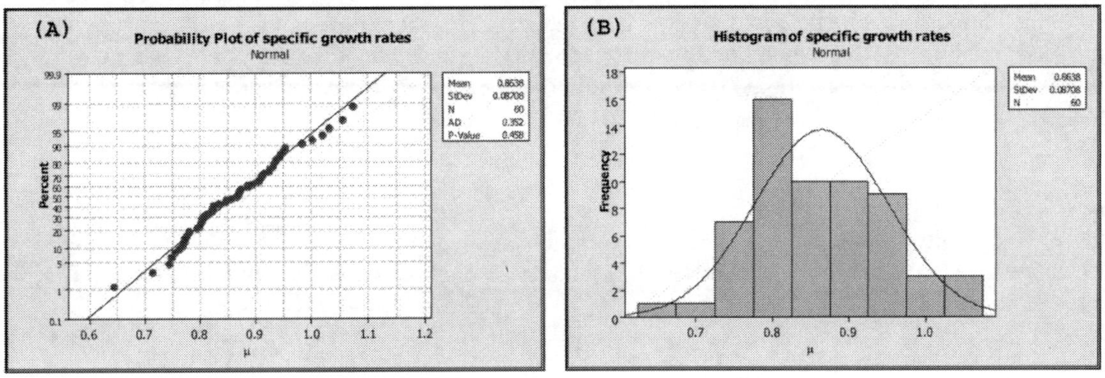

Figure 3. Distribution of calculated values of μ in an experiment where Strain I was cultured in the central region of a 96-well microplate. (A) Histogram of the calculated μ values. (B) Normal probability plot of the calculated μ values.

In addition, an analysis of variance of the specific growth rates values (μ) for rows and columns was conducted for this set of data (same microalgae strain in all wells, Fig. 1B). The ANOVA results, summarized in Table 1, further confirmed the normality behavior of the specific growth rates throughout the microplate. When comparing average means from different rows and columns independently, the null hypothesis, means are not different, was accepted (P>0.05). This implies that well location within the central subset of the microplate had no effect over the data and that the effects of external not accounted factors were randomly distributed throughout the microplate. It is not our intention to suggest this result as a general conclusion. A similar analysis is recommended for each particular experimental setting, since position of the illumination source, gas circulation patterns of different incubators, among other relevant variables, could originate a position dependence effect.

Table 1. ANOVA of specific growth rates (μ) for microplate columns and rows.

Source	DF	SS	MS	F	*P*-value
Columns	9	0.108	0.012	1.79	0.094
Error	50	0.338	0.007		
Total	59	0.447			
Rows	5	0.024	0.005	0.62	0.685
Error	54	0.423	0.008		
Total	59	0.447			

We have demonstrated the feasibility of commercially available 96-well microplates as microbioreactor platforms for microalgae related research as: optimization of culture medium, high-throughput screening for fast growing microalgae under particular culture conditions (i.e. 20% CO2), and kinetic characterization of microalgae growth. In addition, it was proven that the system could be used as a platform of 60 independent and statistically equivalent microreactors for microalgae kinetic studies. Overall, the experimental platform for microalgae culture and the analysis algorithm presented here provide high quality kinetic data for up to sixty independent microalgae cultures. Moreover, the worldwide availability of microplates and their convenience (i.e. easy handling, substrate and space savings, and fast absorbance readings) makes this micro-scale culture technique very practical for microalgae screening.

Acknowledgements

We thankfully recognize the funding and technical support provided by CEMEX and acknowledge the support of the Tecnologico de Monterrey through its Research Funding Program of Catedras de Investigacion.

References

1. Muñoz, R., and Guieysse, B. (2006) Algal-bacterial processes for the treatment of hazardous contaminants: A review. Water Research 40: 2799-2815.
2. Muñoz, R., Köllner, C., and Guieysse, B. (2008) Biofilm photobioreactors for the treatment of industrial wastewaters Journal of Hazardous Materials 161: 29-34.
3. Silva-Aciares, F.R., and Riquelme, C.E. (2008) Comparisons of the growth of six diatom species between two configurations of photobioreactors Aquacultural Engineering 38: 26-35.
4. Skjanes, K., Lindbald, P., and Muller, J. (2007) BioCO2 - A multidisciplinary, biological approach using solar energy to capture CO2 while producing H2 and high value products. Biomolecular Engineering 24: 405-413.
5. Vairappan, C.S., Nie, G.P., and Yen, A.M. (2007) Antibacterial Activity of Fine Chemicals from *Scenedesmus acuminatus* (Scenedesmaceae, Chlorococcales, Chlorophyta). Malaysian Journal of Science 26: 65-72.
6. Raja, R., Hemaiswarya, S., Kumar, N.A., Sridhar, S., and Rengasamy, R. (2008) A perspective on the biotechnological potential of microalgae. Critical Reviews in Microbiology 34: 77-88.
7. Huntley, M., and Redalje, D. (2007) CO2 mitigation and renewable oil from photosynthetic microbes: a new appraisal. Mitigation and Adaptation Strategies for Global Change 12: 573-608.
8. Chisti, Y. (2008) Biodiesel from microalgae beats bioethanol. Trends in Biotechnology 26: 126-131.
9. Stepan, D.J., Shockey, R.E., Moe, T.A., and Dorn, R. (2002) Carbon dioxide sequestering using microalgal systems. *In* 2.3. Pittsburgh: U.S. Department of Energy, pp. 1-27.
10. de Morais, M.G., and Costa, J.A.V. (2007) Isolation and selection of microalgae from coal fired thermoelectric power plant for biofixation of carbon dioxide. Energy Conversion and Management 48: 2169-2173.
11. Ono, E., and Cuello, J.L. (2007) Carbon Dioxide Mitigation using Thermophilic Cyanobacteria. Biosystems Engineering 96: 129-134.
12. Zuck, P., O'Donnell, G.T., Cassaday, J., Chase, P., Hodder, P., Strulovici, B., and Ferrer, M. (2005) Miniaturization of absorbance assays using the fluorescent properties of white microplates. Analytical Biochemistry 342: 254-259.

13. Bitton G., Junk K., and Koopman B. (1994). Evaluation of a microplate assay specific for heavy metal toxicity. Archives of Environmental Contamination and Toxicology 27(1):25-28.
14. Eisentraeger A., Dott W., Klein J., and Hahn S. (2003). Comparative studies on algal toxicity testing using fluorometric microplate and Erlenmeyer flask growth-inhibition assays. Ecotoxicology and Environmental Safety 54: 346-354.
15. Gregor J., Jancula D., and Marsalek B. (2008). Growth assay with with mixed cultures of cyanobacteria and algae aaessed by in vivo fluorescence: one step closer to real ecosystems? Chemosphere 70: 1873-1878.
16. Andersen, R.A. (2005) Algal Culturing Techniques: Elsevier Academic Press.

Sustainability of Products and Processes: a Review

Subhas Sikdar, National Risk Management Research Lab, US EPA, Cincinnati, OH

For measuring sustainability performance of products and processes, it is generally believed that we need to identify metrics that neatly fall into economic, environmental, and societal categories. The task of sustainability analysis then is some integration scheme that provides an overall assessment. Many corporations acknowledge the useful fact, however, that the three dimensions of sustainability are connected with one another: an action in one category changes factors in others. Nevertheless, these metrics are used to conform to the so-called triple bottom line characterization of sustainable development. At the process and product level, many have argued that the societal impacts have been ignored. Except for a very few uniquely societal metrics, societal impacts are already included in extant metrics. This talk presents an overview of research on sustainability metrics. In particular, here is discussed a metrics classification that incorporates the interconnections among measurable metrics, and a method of analysis that depends on identifying the metrics with their dimensionality, prioritizing them according to their extent of sensitivity to overall impacts, and finally aggregating them in a single indicator. This aggregate indicator can be conveniently used for comparative analyses for sustainability. Several examples will be used to illustrate the methodology.

See more of this Session: Fundamentals of Environmental Sustainability
See more of this Group/Topical: Environmental Division

Understanding Sustainability Based On the Thermodynamics of Industrial and Ecological Systems

Bhavik R. Bakshi, Lowrie Department of Chemical and Biomolecular Engineering, The Ohio State University, Columbus, OH

The overall goal of manufacturing systems is to produce low entropy products from higher entropy resources. The second law implies that any decrease in entropy in one system must result in a larger increase in the entropy of the surroundings. This entropy increase often occurs in the environment, as is common in industrial systems, in the form of environmental harm. Ecosystems are also subject to the same laws of thermodynamics, but seem capable of avoiding such harm due to entropy increase. In this talk, I will introduce the thermodynamics of ecological systems and compare it with that of industrial systems via several examples. The resulting insight is useful for determining what kind of technologies or technological systems may be sustainable.

See more of this Session: Fundamentals of Environmental Sustainability
See more of this Group/Topical: Environmental Division

Sustainability and Life Cycle Assessments

Raymond L. Smith, US EPA, Cincinnati, OH

Sustainability can be defined as the ongoing achievement of desired states, for instance through meeting "the needs of the present without compromising the ability of future generations to meet their own needs" (Brundtland Commission, 1987). While the desired (social, economic, and environmental) states of individuals and cultures can conflict with each other there are some environmental impacts upon which people generally agree. These environmental impacts include acidification, smog formation, climate change, toxicity, etc. To study these impacts in systems requires tools that are capable of panoramic views. One such tool is life cycle assessment. Aspects of LCA will be described in this presentation with a view towards differences in hot spot versus consequential analyses, whether inventory and impact assessment methods are sufficient for analyses, and how system boundaries, allocation and value choices can influence results. LCA helps bridge a gap between our perceived worries (perhaps the above listed environmental impacts) and a systems analysis of one or more issues. While LCA doesn't resolve the normative issues associated with conflicting desires, it can present a scientific representation of such issues.

See more of this Session: Fundamentals of Environmental Sustainability
See more of this Group/Topical: Environmental Division

A Fuzzy Logic Global Optimization Approach to Sustainability Assessment

Jeremy A. Conner, Chemical Engineering Department, University of California at Los Angeles, Los Angeles, CA

Yannis A. Phillis, Department of Production Engineering and Management, Technical University of Crete, Chania, Greece

Vasilios I. Manousiouthakis, Chemical & Biomolecular Engineering, UCLA, Los Angeles, CA

Fuzzy logic has been used extensively to evaluate the sustainability of an entity based on a number of basic indicators (air, land, economy, health, etc.). Prior models have taken specific values of basic indicator data, created a fuzzy set hierarchy, and computed an exact value for an overall sustainability index of the entity. The latter step required a "defuzzification" process that allowed the exact calculation of that index.

In reality, exact values for basic indicators are either difficult to come by or, even when known, contain significant levels of uncertainty. To address this shortcoming, in this work we propose a novel approach that allows basic indicator data to be provided in the form of intervals which are guaranteed to contain all possible basic indicator values. As a result, all possible values of the overall sustainability index are guaranteed to belong to an interval. This work quantifies the tightest possible interval range for the overall sustainability index, whose computation is achieved through global optimization techniques. Properties of the underlying global optimization problems are first established, and then utilized in the development of an algorithm that exactly quantifies the range of the overall sustainability index, in a finite number of iterations.

See more of this Session: Fundamentals of Environmental Sustainability
See more of this Group/Topical: Environmental Division

Development of a New Recycling Process for Concrete Sludge Generated From Concrete Pole Plant

Atsushi Iizuka, Environmental Systems, The University of Tokyo, Kashiwa, Chiba, Japan

Akihiro Yamasaki, Department of Materials and Life Science, Seikei University, Musashino, Japan

Yukio Yanagisawa, Environmental Systems, The University of Tokyo, Kashiwa, Chiba, Japan

A new recycling process for concrete sludge using carbon dioxide (CO2) was investigated. Concrete sludge is a slurry waste mainly consists of water and cement particles. An actual concrete sludge sample generated from a centrifugal-reinforced concrete pole plant in Japan was collected and used in this study. An alkali component, mainly calcium hydroxide (Ca(OH)2), is repeatedly extracted from cement particles by repeating water addition and subsequent solid-water separation. Calcium carbonate (CaCO3) is precipitated by introducing gaseous CO2 into the obtained Ca(OH)2 solution. In this process, concrete sludge is neutralized by gaseous CO2, and CaCO3 particles can be obtained. The calcium extraction and CaCO3 precipitation reaction rates were measured, and feasibility of the recycling process was discussed.

See more of this Session: Fundamentals of Environmental Sustainability
See more of this Group/Topical: Environmental Division

EVALUATION OF THE OPERATION AND EFFICIENCY OF A SOLAR PANEL LOCATED IN THE BOGOTA CITY, COLOMBIA [1]

Castillo Lorena[2]
Olaya Claudia
Camargo Gabriel[3]

ABSTRACT

This investigation presents the assessment of the operation and efficiency of the solar panel located in the Bogotá city, Colombia this investigation covers the characterization of current conditions of the system of solar panels which is used to maintain the temperature of a lake between 30 ° C and 38 ° C to ensure the survival of the plant Victoria Regia. Establish the technical, economic and environmental viability panel solar and analysis of compliance with the standards in the design of the Colombian solar panels.

The initial phase of the project was the collection of information and data in both field and in the offices of IDEAM (Institute of Hydrology, Meteorology and Environmental Studies of Colombia), Secretary of Environment of Bogota and ICONTEC (Colombian Institute of Technical Standards and Certification).

Data were obtained as temperature, slope of solar panels, piping drawings, flow, environment of the panel, panel materials and components, etc..

With data generated daily radiation at the station IDRD (Institute for Sport and Recreation District), started a collection and analysis of the radiation statistics annual, monthly and hourly through graphics. With these data and some data obtained in practice proceeded to evaluate the thermal efficiency of the collector, η that is to say the total incident radiation (direct + diffuse) on the area of the collector.

Then proceeded to identify nonconformities that the system presented according to the NTC (Colombian Technical Standards).

According to the analysis were generated strategies and recommendations for improve the system working and consequently increase the efficiency of the collector.

KEY WORDS

Solar termal energy, radiation, efficiency, solar panel, Victoria Regia.

---

[1] This investigation is in the draft grade "EVALUATION OF THE OPERATION AND EFFICIENCY SOLAR PANEL LOCATED IN THE BOTANICAL GARDEN JOSE CELESTINO MUTIS" Universidad Libre

[2] Students X semestre de Ingeniería Ambiental. Universidad Libre. Bogota, Colombia

[3] Teacher Universidad Libre. Bogota, Colombia

INTRODUCTION

The energy consumption has taken as an indicator of industrial development: more consumption, more development. The availability of energy resources, distribution, use and environmental effects is one of the largest occupations in the contemporary world: In developing countries, it lacks the economic and technological means to provide and meet the demand for energy especially in their areas rural, by conventional technologies. This is where the so-called renewable or alternative energy supply provide an interesting option especially with decentralized supply schemes that allow the use of local resources to supply power at lower costs (Child, 2006).

The International Energy Agency (IEA) said that the foundation of modern life of the world depends on oil to 80% and that as countries industrialize and their populations increase, so does energy consumption.

Solar thermal systems are installed in Colombia has generated a level of acceptable satisfaction among users, but with some reservations that result in technical errors of the companies that have manufactured and marketed. Technical errors, coupled with ignorance, generate distrust among potential users.

In Colombia, the primary energy[4] production comes from hydropower, for the abundance of water in most parts of the country, and a second place of fossil fuels (oil, gas and coal), whose reserves are already depleted[5].

So the Government in recent years has invested in the development and implementation of alternative technologies for energy production, operating with renewable resources, to solve the problem of global energy crisis and contribute to a cleaner environment.

According to National Planning Unit Energy Mining (UPME), renewables now cover about 20% of global electricity consumption.

Among renewable energy are solar, geothermal, biomass, etc. Since solar energy the most important of these.

[6] Solar energy is free, generates no emissions and is silent

MATERIALS AND EQUIPMENT

The research was conducted between January 2008 and January 2009, conducting field visits to the Botanical Gardens of Bogotá José Celestino Mutis, obtaining temperature data in the system copper piping, recirculating water temperature in the

---

[4] Primary energy is that contained in raw fuels and other energy forms received by a system as an input to the system.

[5] http://www.mineducacion.gov.co/cvn/1665/article-117028.html

[6] http://www.mineducacion.gov.co/cvn/1665/article-117028.html

pipe and the temperature of the lake where is the Victoria Regia; for obtaining these temperatures was used a gun infrared and thermometer BRAND. Was measured the inclination of solar panel with a level topography. In addition was measured the distance of nearby obstacles to solar panel. In observation visits were assessed the conditions presented by the system, obtaining data for the map location of pipelines and the overall system, it was observed the design of solar panel components, and its operation and its conditions. Radiation data were collected in the IDEAM and District Environment Secretariat. According to the data, the study required an annual collection, monthly and hourly data supplied by the District IDEAM and Secretary of Environment, which was performed with a graphical analysis where the variation was observed radiation between 1990 and 2008.
Since the graphical analysis of the radiation, we calculated the theoretical efficiency of the collector. Conditions were compared with standards system and the NTC 4368[7] GTC 108.

Finally made an analysis of technical feasibility, economic and environmental project.

[7] Eficiencia Energética. Sistema de Calentamiento de Agua con Energia Solar y Componentes. 1997

VICTORIA REGIA

Scientific name or Latino: Victoria amazonica

Common or usual name: Victoria regia, Abatiyu, Agoape, Aguape, leaf of sun, Iguape, Irupe, Corn water Maruru, real Ninfa, water dish, giant lotus.

Family: Nymphaeaceae (Ninfaceas).

Origin: South America, including the Amazon River. Your tropical home acuatico extends to Brazil, Colombia, Peru and Guyana.

Etymology: the specific name of Victoria regia is dedicated to the Queen
Victoria of England, as it was described by the British botanical s. XIX, but then to Victoria amazonica exchange alluding to his place of discovery.

Aquatic plants of enormous dimensions.

Perpendicular edges of 10-20 cm, green and reticulate over the top, red and the veins arranged in a spiral in the face that is in contact with water.

The venation of these sheets has a peculiar architecture.

This flange prevents the water penetration which allows heavy weights to hold up to 20 kilos, so it is common to see birds and small mammals resting on them.

The blade is supported by a long stalk that binds to a submerged rhizome which is sinking into the mud.

A solitary flower is born in the dark of night, perfuming the air in the summer moon with a delicate and sweet fragrance.

Large flowers are 20 to 40 centimeters, with cotton white or pink petals, which are becoming ruby red.

The flower born at night, remains open during the day and the next night closes submerged in water.

RESULTS AND CONCLUSIONS

Theoretical and ideal calculations for Solar Energy Utilization.

The hourly variation, diurnal and annual solar radiation intensity is determined by astronomical parameters, geographical, weather conditions and human activities that alter the natural conditions of the atmosphere, as the weakening of the ozone layer due to higher use of CFS. A level of troposphere the presence of clouds, dust, aerosols and ozone concentration are absorbers of photons of energy or solar radiation.

Figure 1. Global Radiation IDRD Meteorological Station and El Dorado, Bogota

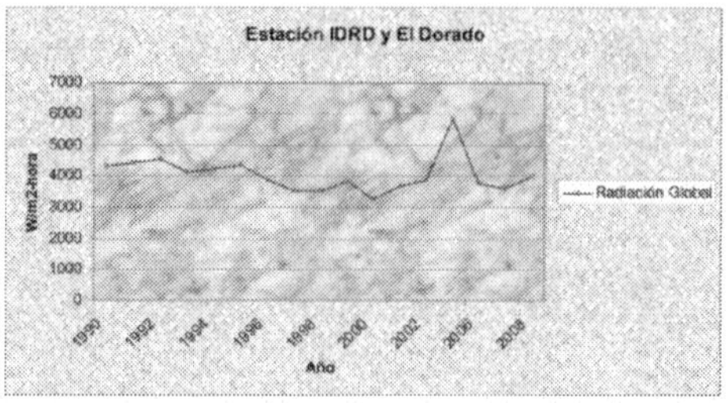

Source: Authors

The more radiation that has been presented in recent years was in 2004, so seeing this in filing an average of 4,000 W/m2-hour.

Figure 2. Global Radiation Weather Station IDRD Year 2005

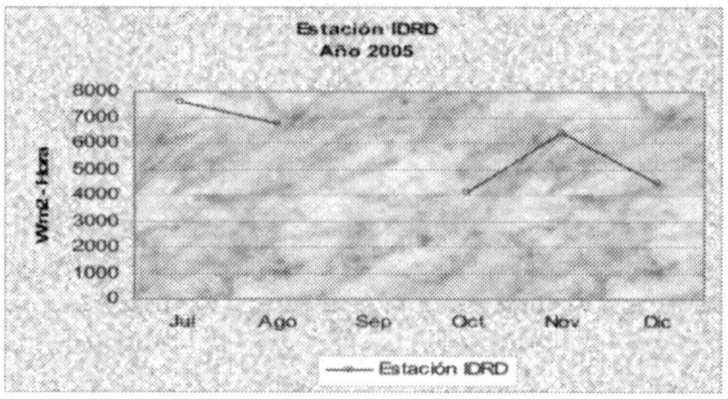

Source: Authors

The figure shows empty spaces that represent missing data.

There is a higher radiation in July and less radiation for the month of October.

Figura 3. Global Radiation Weather Station IDRD Year 2006

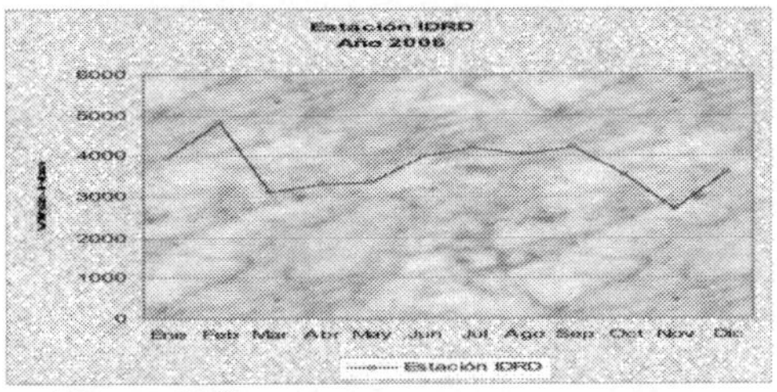

Source: Authors

In 2006, more radiation is presented in February and less radiation for the month of November.

Figure 4. Global Radiation Graph Weather Station IDRD Year 2007.

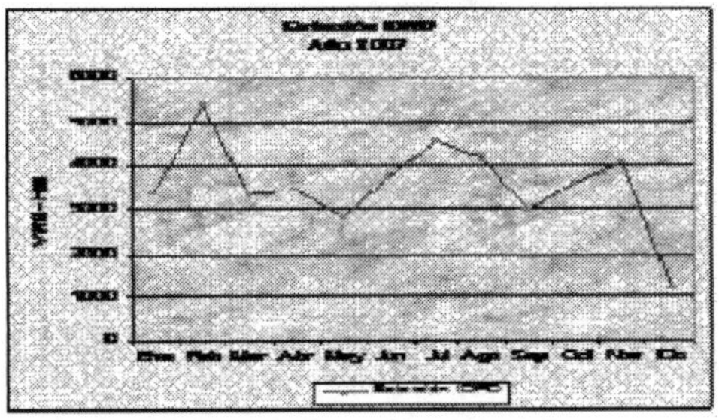

Source: Authors

In 2007, more radiation is presented in February and less radiation for the month of December.

Figure 5. Global Radiation Graph Weather Station IDRD Year 2008.

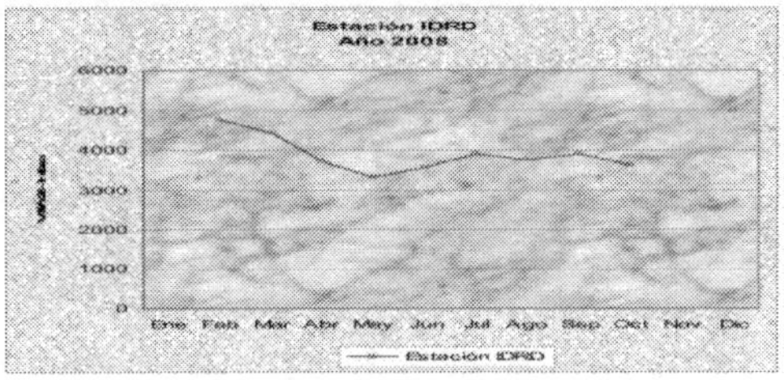

Source: Authors

The graph shows the month of November emptied, for lack of data.

In 2008, more radiation is presented in December and less radiation for the month of May.

After analyzing the graph of monthly global radiation in the years 2005 - 2008, concludes that most radiation occurs in the months of January and February and the lowest in the months of May, June and November.

Giving the months of highest radiation efficiency in the system.

GLOBAL RADIATION ANALYSIS TAKING time stats.

The intensity of solar radiation varies with time of day. During the early morning and late afternoon sunlight falls almost horizontally on the point of incidence. During its course of radiation can be absorbed and scattered by gas molecules, aerosol particles or water, in what is called the diffuse component of radiation. When the sun pass directly through the atmosphere without being absorbed, are the direct component of radiation.

Analyzing the February and March 2008, as they were in which temperature data were obtained in the system, February is also one of the months which has a higher radiation.

Figure 6. Global Radiation Grafica Weather Station IDRD February Year 2008.

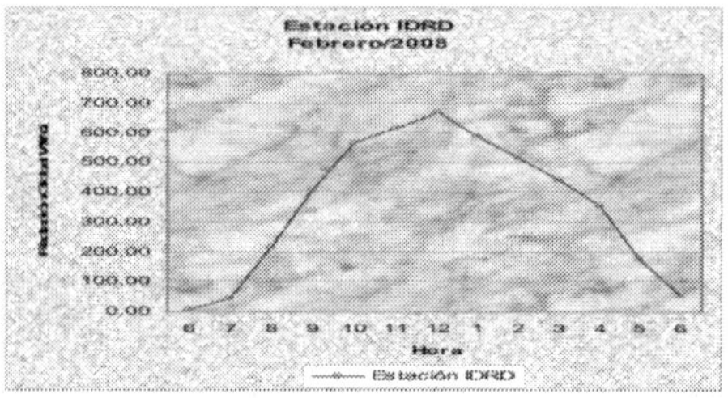

Source: Authors

In February 2008, more radiation is present between 12 m and 1 pm.

Figure 7. Global Radiation Graph Weather Station IDRD March Year 2008.

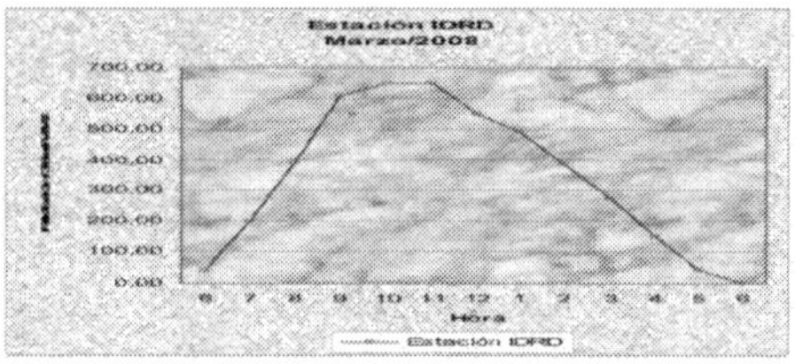

Source: Authors

In March 2008, more radiation is present between 11 m and 12 p.m.

Other factors that influence the intensity of solar radiation are the walls and asphalt refraction, and reflection from the windows of buildings. Grass reflects at least 10% of the incident radiation. Solar radiation also varies with altitude, the more radiation at higher altitudes.

Figure 8. Global Radiation Graph IDRD Meteorological Station in April Year 2008

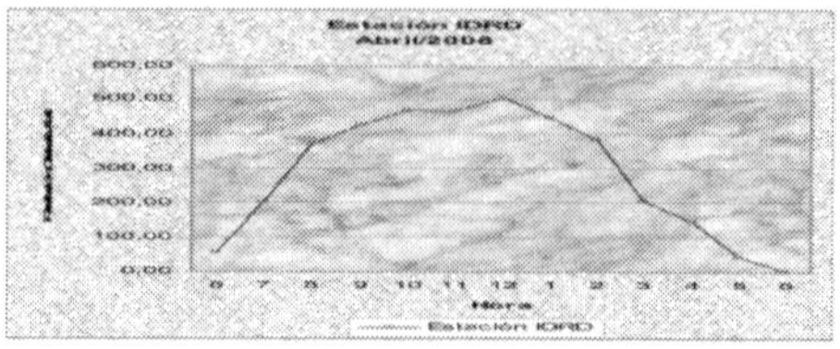

Source: Authors

In April 2008, more radiation is present between 11 m and 1 pm

Calculation of the Efficiency of Solar Collector.

Basic equations of the collector. The thermal efficiency of the collector, N, is the relationship heat gained by the flow passing through the collector, Q_u, total incident radiation (direct + diffuse) on the collector area.

$$\eta = Q_u / G.Aa$$

The useful energy (Q_u) transferred to the flow in quasi-stationary regime; the collector plane is expressed, with very good accuracy by the expression:

$$Q_u = m\, C_p\, (T_e - T_s)$$

[8]C_p = 4.169 J/Kg/K

[8] It is the energy required to raise a kelvin at that temperature in unit mass of the substance. http://www2.udec.cl/~dfiguero/curso/capacidadfases/capacidadfases.htm

Mass flow rate of flow of transfer:

The recirculation system carries water from the solar panel cleared with constant flow temperature 42 ° C and density 992.2 kg/m3, the pipe is 61 m long and ¾ inch diameter copper pipe:

$m = \rho v$

$v = \pi$ (radius) 2 (length) $= \pi$

(0.01905m) 2 (61m)

v = 0.069 m3

Therefore, the mass is:

$m = \rho v$ = (992.2 kg/m3) (0.069m3)

m = 68.46Kg

The temperature of the recirculating water through the pipe was measured with infrared gun resulting in 28.91 ° C at the inlet manifold and 49.53 ° C at the collector output (time just before coming to recirculate in the lake).

The useful energy (Q_u) transferred to the flow.

$$Q_u = m\,C_p\,(T_e - T_s)$$

$$Q_u = (68.46\text{Kg})\ (4.169\ \text{J/Kg/K})\ (49.53 - 28.91)$$

$$Q_u = 5885.15\text{W}$$

Thermal Efficiency of the collector, averaging the data obtained in the year 2005 – 2008.

$$\eta = Q_u / G.Aa$$

$$\eta = 5885.15\text{W} / G.Aa$$

$$\eta = (5885.15\text{W}) / (577.83\text{W/m}^2)\ (12\text{m}^2)$$

$$\eta = 84.87\ \%$$

CONCLUSIONS

The calculations obtained result in an overall efficiency of the collector of 85.54%, which can be interpreted as a high efficiency, taking into account that the solar panel does not meet some of the recommendations outlined in the 4368 NTC and one of the collectors not working properly, however despite this the solar panel is meeting our goal of having the desired temperature of the lake and thus keep alive the Victoria Regia.

Average radiation was evaluated in the months of February, March and April (see discussion taking time stats global radiation) of 2008, because the analysis was obtained during the first months of the year have increased radiation. Similarly measurements taken were made during peak hours of radiation, (see Figures 6,7 and 8). Concluding that the system works smoothly, but this efficiency could be improved by modifying some components of the panel and made to perform regular maintenance.

The load from low efficiency resulting in the Series 3 and 5 of the solar panel, which is due to the presence of a tree (see figure 11) located less than 70cm away from the series 5, the location of this tree influences the reception in this last series solar panel. We did not identify the causes of the decrease in the efficiency of the 3 series.

The use of solar panels in Colombia is limited; this is because the initial investment is made, unlike other energy sources. In Colombia, the electric heaters were replaced by natural gas, as this is a more economical energetic in terms of initial investment and billing. Despite this, if we compare the solar heating with natural gas is more economically and environmentally sustainable use of the former, although the initial investment is higher, as solar water heaters do not increase the cost of the bill energy or gas, and no environmental impact as might be the release of pollutants into the atmosphere.

The physico - chemical lake water shows where the Victoria Regia is suitable for the survival of it.

There must be a system-wide maintenance, taking into account the recommendations made.

It is recommended that the tree is now at a distance of 70cm Series 5 of the solar panel is moved to a different place or are taking any measure which does not affect the efficiency of the panel by this.

It is recommended to keep clear of vegetation the bottom of the collector.

We recommend draining the collector at least once a year, unless the water supply conditions involving more frequent cleaning.

REFERENCES

Stu Campbell, "Construya su Propio Calentador Solar", Impreso en México: Gustavo Gili 1985, Ed. Exlibris, 117 páginas.

Ley 629 Diciembre 27 de 2000

Ley 697 del 3 de noviembre del 2001

Ministerio de Minas y Energía, División de Fuentes no Convencionales, Instituto de Ciencias Nucleares y Energías Alternativas, HIMAT, Atlas Radiación Solar en Colombia. Pág. 60.

Ministerio de Ambiente, Vivienda y Desarrollo Territorial. Instituto de Hidrología, Meteorología y Estudios Ambientales IDEAM. Atlas Climatológico de Colombia.

NTC 2631 ENERGIA SOLAR. CALCULO DE TRANSMITANCIA Y REFLECTANCIA FOTOMETRICAS EN MATERIALES SOMETIDOS A RADIACION SOLAR.

NTC 4368 EFICIENCIA ENERGETICA. SISTEMAS DE CALENTAMIENTO DE AGUA CON ENERGIA SOLAR Y COMPONENTES.

NTC 1736 ENERGIA SOLAR. DEFINICIONES Y NOMENCLATURA.

ORTEGA RODRIGUEZ, Mario. Energías Renovables, Publicaciones Thomson Paraninfo. 328 páginas.

GUILLEAN T. Prance. Victoria amazónica ou Victoria Regia

FRANZINI, Joseph B. y FINNEMORE, E. John. Mecánica de Fluidos con aplicaciones en ingeniería. Novena Edición. 1999.

Gasification of Hybrid Feedstock Using Animal Manures and Hays

Kyoung S. Ro, Ph.D., P.E.
USDA-ARS
Coastal Plains Soil, Water & Plant Research Center
2611 W. Lucas Street
Florence, SC 29501

Introduction

Currently, animal production annually provides 35 million dry tons of sustainable biomass/manure feedstock that can be converted to various forms of renewable biomass energy (Perlack et al., 2005). Using a simple arithmetic average HHV of different animal manures (13.4 MJ $kgTS^{-1}$), the annual energy content of the 35 million dry tons of manure is estimated to be approximately 0.43 EJ, about 14% of the total renewable biomass energy consumed in the U.S. (Ro et al., 2009). There are two pathways of extracting renewable energy from animal manure – biochemical and thermochemical pathways. Although biochemical technologies such as anaerobic digestion technology are well-established and robust, these technologies are typically very slow with processing times of days and weeks. In addition, biochemical processes still leave substantial amounts of unconverted sludge and supernatant that require further treatment and disposal. In contrast, emerging thermochemical conversion (TCC) technologies such as gasification and pyrolysis only require treatment times in the span of minutes to hours (Cantrell et al., 2007 and 2008; He et al., 2000; Priyadarsan et al., 2004; Ro et al., 2007). Furthermore, they destroy pathogens and convert most organic matter into energy-rich and value-added end products such as combustible gases, liquids, and charcoals (hereafter called biochar). The distribution of these TCC end products depends on the operating temperature, pressure, heating rate, and residence time (Bridgwater, 2003; Huber et al., 2006). These TCC end products can be used as energy intermediates for combined heat and power generation (CHP) or feedstocks for downstream catalytic conversion processes to produce higher value products such as liquid transportation fuels. Although there is an abundance of literature on gasification and pyrolysis of biomass and coals, the literature on TCC technologies for animal manures is very limited. Only few researchers published the study on the gasification of chicken litter, swine and feedlot manures (Priyadarsan et al., 2004; Sheth and Turner, 2002).

The objective of this study is to present the up-to-date results of on-going research effort to evaluate the efficiency of a skid-mounted gasification/pyrolysis system in producing combustible gas from animal manures: swine solids; chicken litter, and swine solids blended with rye grass.

Materials and Methods

USIG Skid-Mounted Gasification/Pyrolysis System

The skid-mounted pyrolysis system by the US Innovation Group, Inc. (USIG, Indianapolis, IN) is a proprietary thermo-chemical conversion system that can produce combustible gases from various carbon wastes such as auto shredder residuals (ASM), sewage sludge, municipal solid wastes (MWS), etc. The process flow diagram, a picture, and detailed description of each units of the skid mounted system are shown in Figure 1 and Table 1.

Eight to 19 L of dried swine solids, chicken litter, and blended swine solids (29% rye and 79% swine solids, w/w) were pyrolyzed in the reactor at 620°C (893K) for two hours. Devolatilized gas was scrubbed by water with a venturi scrubber, compressed, and fed into the gas storage tank. The oil from the scrubbing water was separated and collected for later analysis. The scrubbing water retuned back to the recirculation tank for continual scrubbing. Gas, liquid, and solid end products from the USIG pyrolysis reactor system were collected and subsequently analyzed for their chemical and thermal properties.

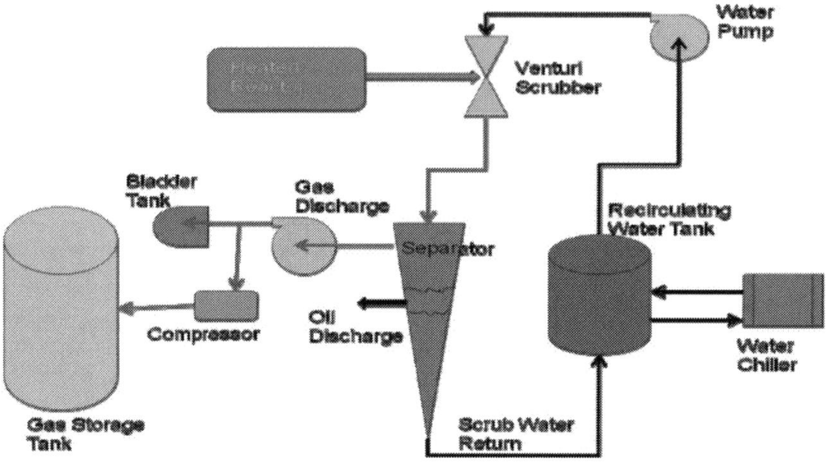

Figure 1 – Flow diagram of the USIG skid-mounted pyrolysis system

Table 1 – USIG Skid-Mounted Pyrolysis System Components

System Component	Description
Reactor	Electrically heated 0.06 m^3 cylinder (304 stainless steel)
Water recirculation tank	80 gal tank
Gas storage tank	80 gal tank
Chiller and scrubber pump	0.5 HP MP Pump Inc. (Model 29905), Flow: proprietary
Blower	0.6 HP Gardner Denver (Type SAP501024210143), Pressure/Flow: Proprietary
Venturi	Penberthy (Model D1752), proprietary modification for scrubbing use
Chiller	Thermal Transfer Products (Model UPA-50-1), Flow: Proprietary

Feedstock Characteristics.

The swine solid was obtained from a solid-liquid separation system treating flushed manure from a 5600-head fishing swine operation in North Carolina (Vanotti et al., 2009). Chicken litter was obtained from a 52,000-bird broiler farm in South Carolina. Rye grass sample was obtained from the Clemson University Pee Dee Research and Education Center, Florence, SC. Characteristics of these feedstock samples are shown in Table 2. The hybrid feedstock was produced by blending 3.0/7.2 (m/m) ratio of rye grass and swine solids.

Table 2 – Characteristics of feedstock samples used in this study

Parameters	*Chicken Litter*	*Swine Solids (dried)*	*Rye Grass*
	Proximate Analyses[2]		
Moisture (%)	10.2 ± 0.07	12.8 ± 0.3	6.2 ± 0.5
Volatile Matter ($\%_{db}$)[1]	51.1 ± 0.5	60.6 ± 1.1	76.0 ± 1.1
Fixed C ($\%_{db}$)[‡]	8.2 ± 0.5	8.1 ± 0.6	14.8 ± 0.8
Ash ($\%_{db}$)[2]	30.6 ± 0.5	18.5 ± 0.2	3.2 ± 0.2
HHV (MJ/kg)	13.0 ± 0.3	19.5 ± 0.2	18.6 ± 0.1
	Ultimate Analyses[3]		
H ($\%_{db}$)	3.7 ± 0.07	5.2 ± 0.1	5.3 ± 0.1
C ($\%_{db}$)	30.9 ± 0.4	41.3 ± 0.4	46.7 ± 0.3
O ($\%_{db}$)	21.1 ± 0.1	17.5 ± 0.4	31.1 ± 0.3
N ($\%_{db}$)	2.9 ± 0.02	4.0 ± 0.1	0.4 ± 0.0
S ($\%_{db}$)	0.7 ± 0.1	0.8 ± 0.0	0.0 ± 0.0

[1]ASTM D3175-07; [2]ASTM D3172; [3]ASTM D3176-02; [‡] calculated as 100-Volatile Matter – Ash

Thermogravimetric analysis.

Dried manure samples with an average sample size of 9- 21 mg were subjected to pyrolysis using a thermogravimetric analyzer (Mettler-Toledo TGA SDTA851e). The mass loss and temperature changes were recorded simultaneously. This unit operated under a three-point calibration using Indium, Aluminum, and Gold. All samples were placed in an aluminum oxide (Al_2O_3) 70 μL crucible and pyrolyzed using nitrogen carrier gas at a flow rate of 60 mL/min within a temperature range from 323 to 1163K at a constant temperature ramping rate of 10K/min.

Analytical Methods

Producer gases from gasifying animal manure were analyzed for major hydrocarbons and sulfur containing gases (ASTM D1945/D1946, D3588-98, D6228-98). Proximate and ultimate analyses were performed for the biochars. Surface functional groups and adsorption capacity for ammonia and hydrogen sulfide were analyzed with FT-IR, 13C-NMR, and photoacoustic gas analyzer. Liquid oil/tar were separated from the scrubbing water and subsequently analyzed for calorific values. Scrubbing water was analyzed for water quality parameters such as chemical oxygen demand (COD), biochemical oxygen demand (BOD), and pH.

Results and Discussion

Thermal decomposition characteristics

Figure 2 shows the thermogravimetric diagrams (TGs) of the three different feedstocks. Thermal decomposition characteristics of these feedstocks were very similar although slightly more rapid thermal decomposition rate was observed around 600K for rye grass. At the USIG reactor operation temperature of 893K, 47, 38, and 38% of initial masses remained for chicken litter, swine solids, and rye grass, respectively.

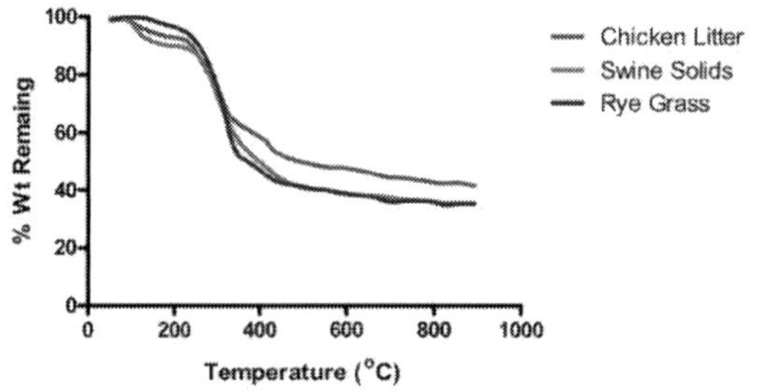

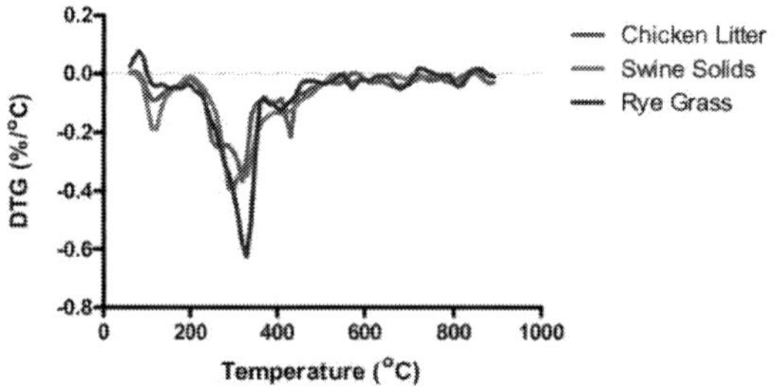

Figure 2 – TG and DTG diagrams of the three feedstocks

Producer Gas Characteristics

Gasification/pyrolysis of swine solid produced gas with the highest heating value slightly lower than that of natural gas as shown in Figure 3. Major gaseous compounds were H_2, CH_4, CO, and CO_2 (Figure 4). Although H_2S was low, surprisingly high concentrations of carbonyl sulfide were detected as shown in Figure 5. Additional gas cleaning steps may be required for subsequent combustion to reduce SO_x emission.

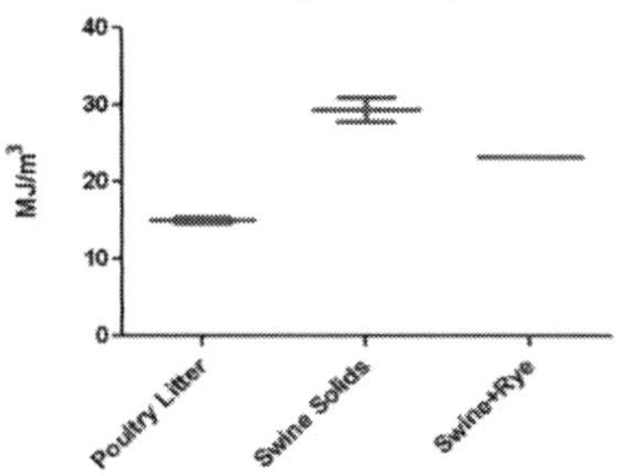

Figure 3 – Heating values (HHV) of produced gas

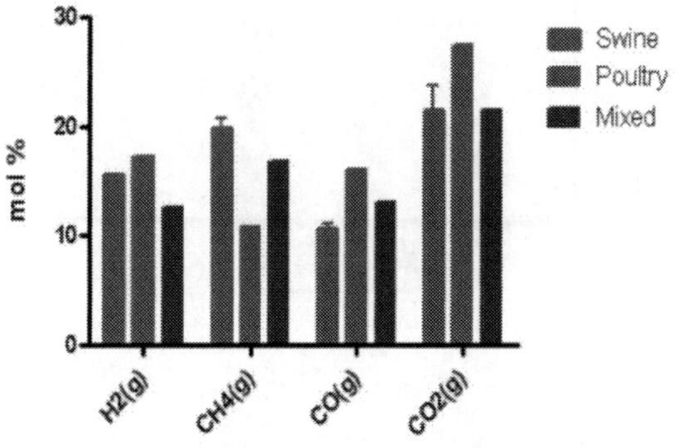

Figure 4 – Produced gas compositions

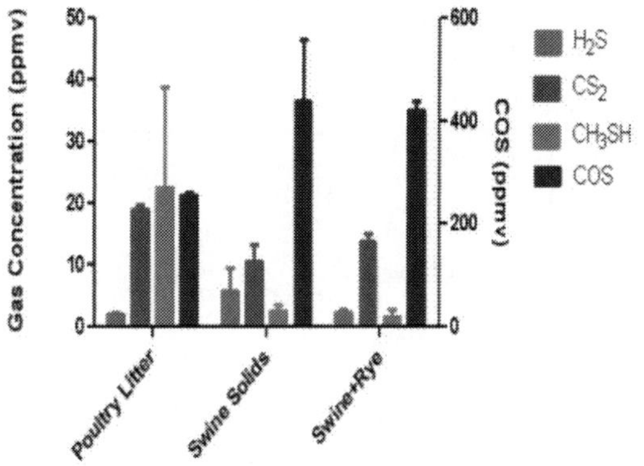

Figure 5 – Sulfur containing gases

Biochar Characteristics

Fixed carbon and volatile matter of biochars were presented in Figure 6. The hybrid feedstock (i.e., swine solid mixed with rye grass) has the highest fixed carbon followed by swine solids and poultry litter. Approximately the same volatile matter was measured for all three feedstocks. The heating values of the biochar made from the hybrid feedstock and swine solids were in between high and low rank coals as shown in Figure 7. The heating value of the poultry litter biochar was slightly below that of low rank lignite. Approximately 50% of the feedstock energy was retained in biochar, 25% in produced gase, and 25% in oil/tar as shown in Figure 8.

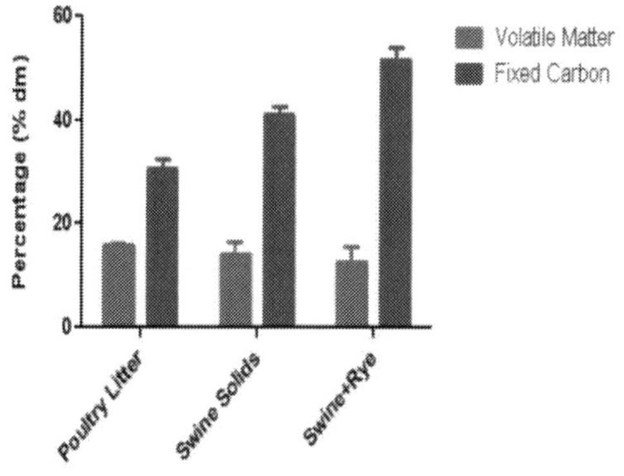

Figure 6 – volatile matter and fixed carbon of biochar

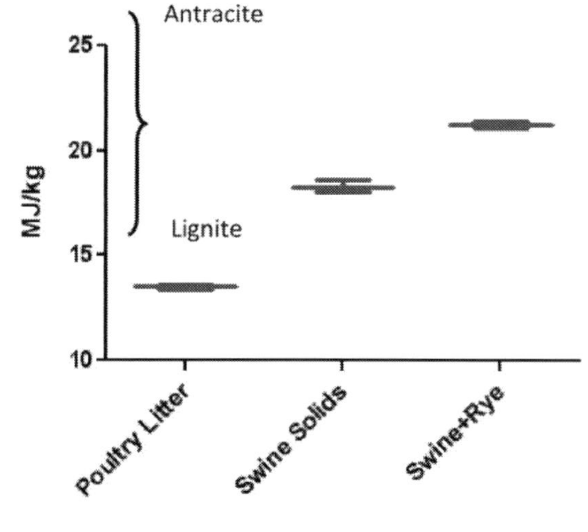

Figure 7 – Heating values of biochar

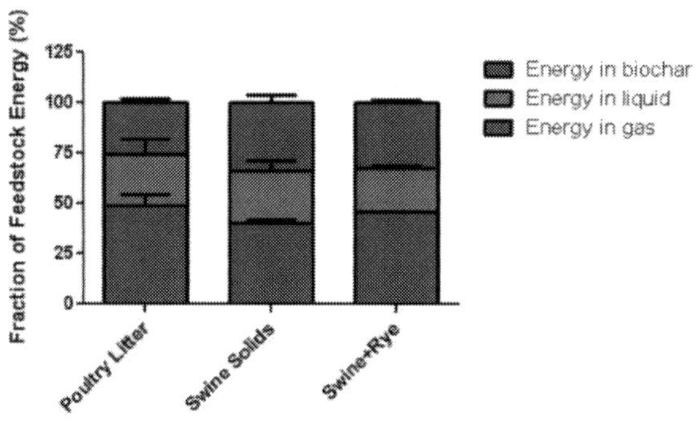

Figure 8 – Distribution of original feedstock energy

Acknowledgement.

This research was conducted under the USDA-ARS NFCA 58-6657-9-235. The authors thank Dr. Joel Keller of R&A Energy Solutions, LLC and Mr. Brian Rayles of US Innovation Group for their cooperation. The authors are grateful to Mr. Mel Johnson and Jerry Martin II for aid in sample collection and analyses. The authors are also grateful to personnel from Super Soil Systems USA, Clinton, NC, and Mitch Tyner chicken farm in SC for providing samples. Mention of trade name, proprietary product, or vendor is for information only and does not guarantee or warrant the product by the USDA and does not imply its approval to the exclusion of other products or vendors that may also be suitable.

References

Cantrell, K. Ro, K., Mahajan, D., Anjom, M., Hunt, P.G. 2007. Role of thermochemical conversion in livestock waste-to-energy treatments: obstacles and opportunities. *Ind. Eng. Chem. Res.* 46: 8918-8927.

Cantrell, K., Ducey, T, Ro, K., Hunt, P.G. 2008. Livestock waste-to-bioenergy generation opportunities. *Bioresour. Technol.* 99: 7941-7953.

He, B.J., Zhang, Y., Funk, T.L., Riskowski, G.L., Yin, Y., 2000. Thermochemical conversion of swine manure: An alternative process for waste treatment and renewable energy production. *Trans. ASAE*, 43, 1827-1833.

Perlack, R.D., Wright, L.L., Turhallow, A.F., Graham, R.L., Stokes, B.J., Erbach, D.C., 2005. Biomass as a feedstock for a bioenergy and bioproducts industry: The technical feasibility of a billion-ton annual supply. DOE/GO-102995-2135. in: U.D.o. Energy (Ed.).

Priyadarsan, S., Annamalai, K., Sweeten, J.M., Mukhtar, S., Holtzapple, M.T., 2004. Fixed-bed gasification of feedlot manure and poultry litter biomass. *Trans. ASAE*, 47, 1689-1696.

Ro, K.S., Cantrell, K., Elliott, D.C., Hunt, P.G., 2007. Catalytic wet gasification of municipal and animal wastes. Ind. Eng. Chem. Res. 46: 8839-8845.

Ro, K.S., Cantrell, K.B., Hunt, P.G., Ducey, T.F., Vanotti, M.B., Szogi, A.A. 2009. Thermochemical conversion of livestock wastes: carbonization of swine solids. *Bioresource Technol.* 100 (22): 5466-5471.

Sheth, A.C. and Turner, A.D. 2002. Kinetics and economics of catalytic steam gasification of broiler litter. *Tr. ASAE* 45(4): 1111-1121.

Vanotti, M.B., A.A. Szogi, P.D. Millner, and J.H. Loughrin (2009) Development of a second-generation environmentally superior technology for treatment of swine manure in the USA. Bioresour. Technol. 2009, doi:10.1016/j.biortech.2009.02.019

Development and Testing of a Proposed Protocol for Tar Sampling for a Biomass Gasification System

R. Mark Bricka, Chemical Engineering, Mississippi State University, Mississippi State, MS

Prashanth Buchrieddy, Chemical Engineering, Mississippi State University, Mississippi State, MS

Biomass gasification produces tars as a byproduct. These tars consist of a variety of hydrocarbons but are predominated by polycyclic aromatic hydrocarbons (PAHs). Typically these tars proved to be quite troublesome, fouling equipment and downstream catalysts. Many studies have reported the results of tars generation from gasification operations but tar analysis (and subsequent quantitation) is highly dependent on the sampling technique. As a result comparison and analysis of different studies discussing tar generation are very difficult if not impossible.

This is a pressing need to develop a standardized method for tar sample collection and analysis. The European community has develop and accepted a standard tar sampling protocol and this method is being considered for standardization in the United States. This method is loosely based off of the US Environmental Protection Agency method 5 sampling protocol.

This study presents the details of the sampling protocol and the results of using this protocol to sample a pilot downdraft gasification unit at Mississippi State University. The reproducibility of the method as well as it eases of use, and pitfalls will be presented.

See more of this Session: Technology Advances and Applications of Thermal Chemical Conversion of Various Carbon Sources
See more of this Group/Topical: Environmental Division

Biomass Gasification: Steam Reforming of Tars Using Nickel Supported Montmorillonite

Prashanth R. Buchireddy, Mississippi State University, Starkville, MS

R. Mark Bricka, Chemical Engineering, Mississippi State University, Mississippi State, MS

Biomass is considered as a potential feedstock for sustainable energy production that can be converted to chemicals, fuels, electricity etc., either by thermochemical or biological processes. Gasification is one of the thermochemical methods of converting biomass to synthetic gas (CO and H2), which could potentially be used to produce different chemicals such as acetic acid and ethanol via chemical or biological processes and/or generate electricity by using engines, turbines, fuel cells etc. However, the synthetic gas produced may contain different organic (Tars) and inorganic impurities, which have the potential to damage the process equipment used in different conversion processes. Hence, the synthetic gas has to be cleaned to remove these impurities, and the extent to which the gas has to be cleaned is dictated by the end use application.

Laboratory scale tests have been performed to evaluate catalytic tar removal. Nickel supported clays have been the focus of this investigation, since clay is relatively cheap and abundantly available. Nickel impregnated clay catalyst was prepared by wet impregnation technique and catalyst characterization was performed using XRD, BET, FTIR, SEM, and ICP. This catalyst has been tested on a laboratory scale flow reactor using naphthalene as a model compound to evaluate the removal efficiencies. Tests have shown that naphthalene removal efficiencies of greater than 95% were achieved with 7.5% Nickel Montmorillonite catalyst under the conditions tested (S/C - 5, GHSV- 9450 h-1, Napthalene Feed - 10.5 g/m3, Temperature - 750oC, Particle Size - 20-40 mesh, Test Duration - 100 Hours). Tests were also carried out to evaluate the effect of S/C ratio, gas hourly space velocity, varying nickel content, naphthalene feed, and temperature; the results of which will be discussed.

See more of this Session: Technology Advances.and Applications of Thermal Chemical Conversion of Various Carbon Sources
See more of this Group/Topical: Environmental Division

Coal Power Generation with High Efficiency and near-Zero Emissions

Thomas A. Adams II, Chemical Engineering, Massachusetts Institute of Technology, Cambridge, MA

Paul I. Barton, Chemical Engineering, Massachusetts Institute of Technology, Cambridge, MA

Recent advances in coal gasification technologies have opened the door for new, environmentally friendly electricity generation processes. For example, integrated gasification combined cycle (IGCC) processes typically achieve higher electrical efficiencies than traditional pulverized coal power plants. Using absorption techniques, IGCC processes can be enabled to achieve 90% carbon capture at the expense of a significant energy penalty [1]. Alternative techniques such as oxycombustion [2] and chemical looping [3] can be employed to perform combustion using an oxygen source undiluted by nitrogen. This permits up to 100% post-combustion capture of CO_2 without the need for an absorption-based process, but at a significantly higher energy penalty.

A new design paradigm is required to achieve both high efficiency and 100% CO_2 capture. In this paper, a novel electricity generation process using integrated gasification and solid oxide fuel cells is presented. By replacing the combustion step with SOFC stacks, electricity can be generated without the need for high purity oxygen while preventing nitrogen from entering the fuel exhaust. This enables 100% downstream capture of CO_2 and avoids the need for energy-intensive pressure-swing absorption-essentially eliminating the energy penalty for CO_2 capture. As a result, the thermal efficiency of the SOFC-based process is higher than IGCC processes by about 10 absolute percentage points (%HHV).

In addition to 100% CO_2 capture, the process has zero atmospheric emissions. Any H_2S, SO_X, unreacted hydrocarbons, or any other pollutants are sequestered along with the CO_2. For a 750 MW power plant, this prevents up to 400 tonne/yr of SO_2 emissions. Fresh water consumption is 25% lower than IGCC and 50% lower than traditional pulverized coal. If waterless cooling systems are employed (such as air-cooling), a surplus of high-purity water is generated by the process, rather than consumed. Thus, the fuel-cell enabled process has significant environmental and efficiency advantages over the current alternatives.

References

[1] Woods MC, Capicotto PJ, Haslbeck JL, Kuehn NJ, Matuszewski M, Pinkerton LL, Rutkowski MD, Schoff RL, Vaysman V. Cost and Performance Baseline for Fossil Energy Plants. Volume 1: Bituminous Coal and Natural Gas to Electricity Final Report. DOE/NETL 2007/1281, Revision 1, August 2007.

[2] Duan L, Lin R, Deng S, Jin H, Cai R. A novel IGCC system with steam injected H_2/O_2 cycle and CO_2 recovery. *Energy Convers Manage* 45:797 809 (2004).

[3] Li F, Fan L-S. Clean coal conversion processes - progress and challenges. *Eng Environ Sci* 1:248-267 (2008).

See more of this Session: Technology Advances and Applications of Thermal Chemical Conversion of Various Carbon Sources
See more of this Group/Topical: Environmental Division

A High Temperature Solid Oxide Fuel Cell Operating On Phosphine Contaminated Coal Syngas

K. Channa R. De Silva, Chemical and Biomolecular Engineering, Ohio University, Athens, OH

Brian J. Kaseman, Mechanical Engineering, Ohio University, Athens, OH

David J. Bayless, Mechanical Engineering, Ohio University, Athens, OH

The current fossil fuel based power generation is problematic due to high emissions of greenhouse gases and low efficiencies of conversion, while the instability in price of oil highlights the need to focus on alternative energy sources. In striking a balance between efficient production of electricity and more environmentally benign technologies one cannot overlook research on solid oxide fuel cells (SOFCs). These solid state devices can generate electricity and have the advantage of being able to use a variety of gaseous fuels like hydrogen, coal syngas and natural gas. Coal is the most abundant fossil fuel in the United States with estimated reserves of two hundred years. Due to the low cost of coal there is great scope for research on SOFCs operating on coal syngas. One of the main problems associated with coal syngas is contaminants, such as hydrogen sulfide (H2S), hydrogen chloride (HCl), arsine (ASH3) and phosphine (PH3), which have proven to be detrimental to the operation of a solid oxide fuel cell anode. SOFCs have been tested with H2S, HCl and AsH3, and results have been published for the effects on anode operation. However the effects of phosphine are still being investigated by researchers from different institutions. This research intended to be performed using a new generation of SOFCs, may provide data that can be useful incorporating phosphine tolerant properties or may reveal the feasibility of developing such material.

See more of this Session: Technology Advances,and Applications of Thermal Chemical Conversion of Various Carbon Sources
See more of this Group/Topical: Environmental Division

Effects of Coal Interaction with Supercritical CO2: Chemical Structure

Wei-Yin Chen, Chemical Engineering, University of Mississippi, University, MS

Guang Shi, Chemical Engineering, University of Mississippi, University, MS

Benson Baari Gathitu, Chemical Engineering, University of Mississippi, University, MS

The study of the interactions between CO2 and coal has been gaining attention due to the potential application of CO2 sequestration in coal mines. Coal bed sequestration is an attractive approach because not only coal adsorbs CO2, but also CO2 replaces methane originally adsorbed on coal's surface. In a recent work, we reported the physical changes of coal structures after treatment [1]. The objective of this study is to enhance knowledge of solvent effects on coal chemical structure.

Lignite and bituminous coal are exposed to supercritical CO2 in a batch reactor at 80 and 130 "aC, 2000 to 8100 psig, and with 48 h residence time. The treated coals are characterized by temperature-programmed pyrolysis (TPP) up to 1650 "aC followed by online analysis by gas chromatography / mass spectrometer. Raw and treated samples are also characterized by Fourier-transform infrared spectrometer. Drying before treatment showed some effects on the physical structure of coal [1], and it is therefore included in the study for its chemical effects.

Results show that drying has no effect on CO emission during TPP for both bituminous coal and lignite. Interaction with supercritical CO2 drastically changes CO emission for treated bituminous coal above 1200 "aC, but almost no notable effect on treated lignite. Treatment pressure has a significant effect on CO emission for bituminous coal. When coals are treated at higher pressures, higher CO emissions are observed during TPP.

1. Gathitu, B.B., W.Y. Chen, and M.C. MaClure "Effects of Coal Interaction with Supercritical CO2: Physical Structure," Industrial and Engineering Chemistry Research, in press, DOI: 10.1021/ie9000162, 2009.

See more of this Session: Technology Advances and Applications of Thermal Chemical Conversion of Various Carbon Sources
See more of this Group/Topical: Environmental Division

Calcium Looping Process for Clean Coal Conversion

Shwetha Ramkumar, Chemical and Biomolecular Engineering, The Ohio State University, Columbus, OH

Liang-Shih Fan, Chemical and Biomolecular Engineering, The Ohio State University, Columbus, OH

The rising energy demand coupled with the depleting global oil reserves and the environmental degradation due to emissions has led to extensive research in the field of clean energy production. The Calcium Looping Process (CLP) is one of the clean coal technologies being developed for the production of hydrogen (H2), electricity and liquid fuels and is based on the syngas obtained from coal gasification. It integrates the water gas shift (WGS) reaction with in-situ carbon dioxide (CO2), sulfur and halide removal at high temperatures in a single reactor while eliminating the need for a WGS catalyst and reducing the overall foot print of the hydrogen production process. The CLP comprises of two reactors; the carbonation reactor where the thermodynamic constraint of the WGS reaction is overcome by the incessant removal of the CO2 product and high-purity H2 is produced with contaminant removal and the calciner where the calcium sorbent is regenerated and a sequestration-ready CO2 stream is produced. The purity of H2 is increased by a large extent when the carbonation reaction is integrated with the WGS reaction. The steam addition for the WGS reaction is also reduced to stoichiometric quantities which aids in reducing the parasitic energy consumption of the process. The integration of the calcium looping process results in an increase in the energy efficiency by about 7% when compared to the state of the art coal to hydrogen process. In this paper, the reaction chemistry occurring in the calcium looping process, thermodynamics, lab and bench scale experimental studies, process intensification strategies and energy conversion efficiencies will be discussed.

See more of this Session: Technology Advances,and Applications of Thermal Chemical Conversion of Various Carbon Sources
See more of this Group/Topical: Environmental Division

Modeling pyrite behavior in gasification environments

Esam Jassim, Steven A. Benson, Wayne S. Seames, and Michael D. Mann, Department of Chemical Engineering, 241 Centennial Drive Stop 7101, Grand Forks, ND 58202-7101

Abstract

The behavior of pyrite and other sulfide minerals in gasification processes are of fundamental importance to efficient and clean coal utilization technologies. Previous work has established that the physical and chemical transformations of pyrite and other impurities depend on mineral composition, mineral physical properties, the mineral association with coal particles, and the presence of organically associated inorganic elements. Further, the studies pyrite transformations found that under gasification conditions the existence of sticky reduced forms iron phases is prolonged and sulfur is released producing hydrogen sulfide. The properties of the iron rich pyrite derived phases impact the release of other elements and their ability to interact with other particles.

This paper is aimed at developing a more quantitative model of pyrite particle transformations during combustion and gasification processes. Modeling studies are being conducted to determine the impact of residence time and temperature on particle fragmentation, coalescence and release of volatile elements. The detailed particle model provides an accurate characterization of particle temperature, sulfur release rates, sulfur reactions, and pyrite conversion rates to intermediate species over wide ranges of gas temperatures, pressures, and compositions. This information can be used more accurately permit the accurately predict of pyrite mass loss rates, pollutant formation, and iron rich intermediate ash particle physical characteristics.

Preliminary results show that sulfur release is greatest early in the oxidation process as a result of the conversion of pyrite to pyrrhotite and pyrrhotite decomposition. In addition, the process of pyrite particle fragmentation was found to enhance the release of sulfur during the devolatilization process and to increase particle temperature as a result of sulfur oxidization. However, when the amount of oxygen in the bulk environment is reduced, the formation of hydrogen sulfide increases and pyrite particle fragmentation decreased.

1. Introduction

Coal will continue to play a major role in meeting energy demands for the next several decades in United States, China and India (Mitchell et al., 2007). Future coal-fired power plants must be highly efficient and likely equipped with carbon dioxide separation and capture systems to reduce greenhouse gas emission. Such aims can only be achieved by improving the coal conversion technologies through a clear understanding of the chemical and physical processes of coal gasification and combustion combined with accurate computational modeling of key processes.

Raw coal contains various types of impurities such as minerals included in the carbon matrix, organically associated inorganic elements, and excluded minerals not associated with the coal matrix (Benson et al., 1993 and Seames, 2003). The included mineral particles and organically associated impurities tend to interact and coalesce inside a coal particle during combustion (Yan et al., 2002). The excluded minerals particles tend to fragment during combustion (Lawrence et al., 1995).

The most common mineral forms in coal include: quartz, alumino-silicate clays, kaolinite, illite, calcite, and pyrite (Srinivasachar S. et al, 1989). The pyrite mineral grains have long been recognized as a host site for trace elements (Finkelman, 1994) and a key contributor to wall slagging in combustion systems (Lawrence et al., 1995).

Experiments have been performed to describe the thermal history of pyrite, using drop tube furnace (Groves S.J. et al, 1987; Asaki et al, 1985). These studies have developed a better understanding of the progress of decomposition of individual pyrite particles and showed that pyrite initially decomposes to pyrrhotite, and then melts to form liquid iron sulfides at temperatures > 1373K (Srinivasachar and Boni, 1989).

However since recent studies suggest that pulverized coal fly ash particle distribution is best described as a tri-model particle size distribution including: a submicron fume region, a fine fragmentation region centered at approximately 2.0 μm diameter, and a bulk fragmentation region (Seames, 2003), it becomes essential to include the fragmentation process in the excluded minerals combustion progress, which is the main theme of this study.

Fragmentation of excluded minerals into a number of smaller particles is important as it influences the particle size distribution of fly ash, particularly for a coal with a significant fraction of excluded minerals. For instant, Helble et al (1990) reported North Dakota lignite with a significant fraction of excluded pyrite exhibited a much finer PSD of combustion ash compared to that of minerals in the parent coal.

Yan et al (2001) have examined the fragmentation behavior of pure pyrite and calcite during coal combustion at 1300 °C using Poisson distribution to simulate the particle size variation resulting from the fragmentation of minerals grains. Their study concluded that pyrite likely generated four pieces of fragments, whereas calcite expected to produce three segments during fragmentation.

2. Model of Excluded Mineral Transformation

The fragmentation mechanisms associated with excluded mineral grain transformations has not been examined in detail for coal conversion systems. Although numerous studies have been performed to address the thermal history of excluded minerals, particularly pyrite, there is no complete model for predicting the progress of oxidation of exclusion element in coal gasification and combustion environments. In most studies, a limited number of steps in the transformation process have been included and sometimes the model is not consistent with experimental observations. For example, the model developed by Asaki et al. (1985) that assumed the formation of sulfur dioxide instead of sulfur monoxide, as the gaseous oxidation during pyrrhotite oxidation (Srinivasachar and Boni, 1989)

Figure 1 illustrates pathway for the excluded pyrite particle showing the particle transformations during combustion. The process can be summarized as follows:

i. Decomposition of pyrite to pyrrhotite

ii. Fragmentation

iii. Oxidation to molten oxysulphide, then magnetite and finally hematite.

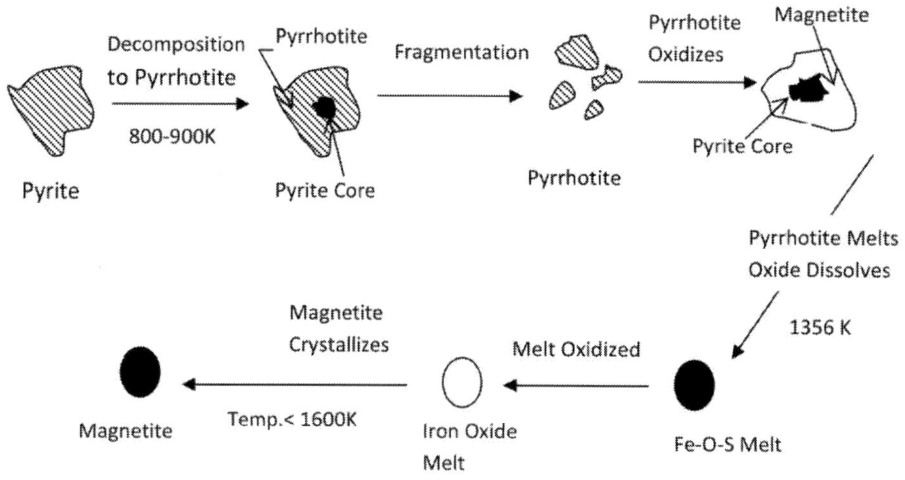

Figure 1. Reaction Pathway history for Excluded Pyrite (Sirnivasachar and Boni, 1989)

3. Mathematical formulas

The general transport equation used to track the particle temperature variation in terms of energy transferred to the particle by conduction and radiation, and the rate at which it is generated or consumed due to chemical reactions or phase changes is written as:

$$m_p\overline{C}_p\frac{dT_p}{dt}=\pi d_p^2\left[h(T_g-T_p)+\sigma\varepsilon(T_w^4-T_p^4)\right]+(-\Delta H)\dot{R} \quad (1)$$

where $\dot{R}$ is the reaction rate obtained by combining the intrinsic reaction and pore diffusion steps:

$$\dot{R}=\frac{(\pi d_p^2)(p_{O_2}/RT_g)(\phi)}{\frac{1}{k_i}+\frac{1}{k_{ext}}+\frac{1}{k_{\text{int}}}} \text{ where: } k_{ext}=\frac{ShD_m}{d_p} \text{ ;: } k_{\text{int}}=\frac{2D_{s2-pore}}{d_p} \quad (2)$$

and for small particle Reynolds number, Sh=2 and $Nu=\frac{hd_p}{k_m}=2$ where Nu is the Nusselt number and k_m is the thermal conductivity of the gas.

The stages of the pyrite transformation process starting form ambient conditions are:

Stage-1 Pyrite preheat

Solid pyrite is heated from ambient temperature to decomposition temperature (~870K as suggested by Srinivasachar, 1989). No reactions occur during this stage. Hence the particle mass and size is assumed to remain constant.

Stage-2 Pyrite conversion to pyrrhotite.

Once the pyrite particle has reached the decomposition temperature, it decomposes to form pyrrhotite.

$$FeS_{2(s)}\rightarrow 1.14Fe_{0.877}S_{(s)}+0.43S_{2(g)} \quad (3)$$

Sulfur is released until complete pyrite conversion to pyrrhotite (Fe1-xS) by 970K (Mitchell, 2002). Gaseous sulfur resulting from the dissociation of pyrite reacts with oxygen in the boundary layer. The heat derived from the exothermic reaction (ΔH=-109 kJ/mol O_2) is conducted back to the particle surface. Assuming the rate of sulfur leaving the particle equals the rate of diffusion of oxygen from bulk gas to the reaction front, the sulfur release rate is given by:

$$\dot{m}_s=\frac{\dot{Q}_{in}}{\Delta H_1} \text{ where} \quad (4)$$

$$\dot{Q}_{in}=\frac{4\pi k_{f1}(T_f-T_p)}{\frac{1}{r_p}-\frac{1}{r_f}}+4\pi r_p^2\sigma\varepsilon(T_f^4-T_p^4) \quad (5)$$

$$\text{with } r_f=\frac{\dot{m}_s}{4\pi D_m C_{O_2}} \quad (6)$$

The reaction front temperature (T_f) can be obtained by solving the equations for the sulfur release rate and the quasi-steady state heat balance at the reaction front simultaneously, using iteration scheme of Newton-Raphson. The heat balance is based on the fact that the energy released by the sulfur reaction is transferred to particle and bulk environment:

$$\dot{Q}_{in} + \dot{Q}_{out} = 4\pi.D_m.C_{O_2}.r_f.(-\Delta H_2) \quad (7)$$

$$\dot{Q}_{out} = 4\pi r_f k_{f2}(T_f - T_g) + 4\pi r_f^2 \sigma (T_f^4 - T_w^4) \quad (8)$$

Stage-3 Fragmentation

Fragmentation of pyrite grains was observed during in combustion experiments. The fragmentation of a mineral grain during combustion is stochastic since the number of offspring particles formed from parent mineral grain is random (Yan L., 2000).

Physically, fragmentation is based on Poisson distribution. Assuming that particles produced during fragmentation are in the same size and composition, the probability that the original mineral grain fragments into $b + 1$ segment is given by (Yan L., 2000):

$$P_b = \frac{e^{-\lambda}\lambda^b}{b!} \quad \text{(b= 0, 1, 2, ...)} \quad (9)$$

where the Poisson parameter λ is the average net growth number of particles per parent particle after fragmentation, which was found from fitting experiment data. Yan et al (2001) suggested that pyrite generated four fragments as an average, while calcite produced three fragments.

Stage-4 Pyrrhotite melting

When the particle temperature reaches the melting point of pyrrhotite (~1356K), the pyrrhotite portion of the particle starts to melt (ΔH=341.15 kJ/kg) whereas magnetite on the surface that formed during oxidation, dissolves in the pyrrhotite melt.

$$1.14Fe_{0.877}S_{(l)} + 1.2367O_2 \rightarrow 0.33Fe_3O_{4(l)} + 1.14SO_{(g)} \quad (10)$$

and ΔH= -190.1 kJ/mol O_2

Stage-5 Molten sulfide oxidation

During combustion process, the remaining sulfur in the Fe-S-O droplet is oxidized. Oxidization and particle temperature continue to increase in this stage until all the sulfide is consumed, leaving behind an oxide melt. Under gasification process, the remaining of the sulfur would be reacted with the hydrogen producing hydrogen sulfide.

4. Assumptions and Numerical Approach

For small pyrite particles, it is reasonable that the temperature distribution on the particle surface is uniform. Although appreciable loss of sulfur has been observed to begin at 700K and end by 970K (Srinivasachar and Boni, 1989), the decomposition processes are rapid enough to assume that it occurs at constant temperature (Lawrence et al, 1995).

The numerical solution involves solving the unsteady state heat balance for single particle using the Runge-Kutta-Fehlberg method in which the error in the solution can be estimated and controlled and an appropriate step size of time can be determined automatically.

After sulfur devolatilization, the pyrrhotite is exposed to the exothermic oxidation reaction that results in increasing the particle temperature to pyrrhotite melting point. As described model of fragmentation, the exothemic transformation stage causes the particle to breakup producing fragments.

4.1 CFD simulation

The pyrite particle is fixed in space in a freely convective air flow of high temperature and 1 atmosphere pressure environment. It is intended to study the process in both combustion and gasification perspectives. Accordingly, the mole concentration of O_2 in the study is chosen to be 21%, 5%, and 0.5% so that covers both combustion and gasification processes.

The CFD simulation is illustrated in Figure1a where a spherical particle placed in an open atmosphere, defined as length and width with about 10 particle diameters. The particle is exposed to standard atmospheric pressure (101.3 kPa) and temperature of 1400 K. The computational domain has to be selected in such a way so that it has minimal influence on the result. It was found from our studies that the far field effect becomes insignificant on the particle film temperature at a domain outer radius of $1.10r_p$ from the particle center. A Cartesian mixed grid of 7x4 in the boundary layer was employed to describe the temperature distribution in the film layer and compare it with the value estimated form the mathematical model. The thickness of the grid in the boundary layer was chosen to be $0.05r_p$ as concluded from Figure 1b.

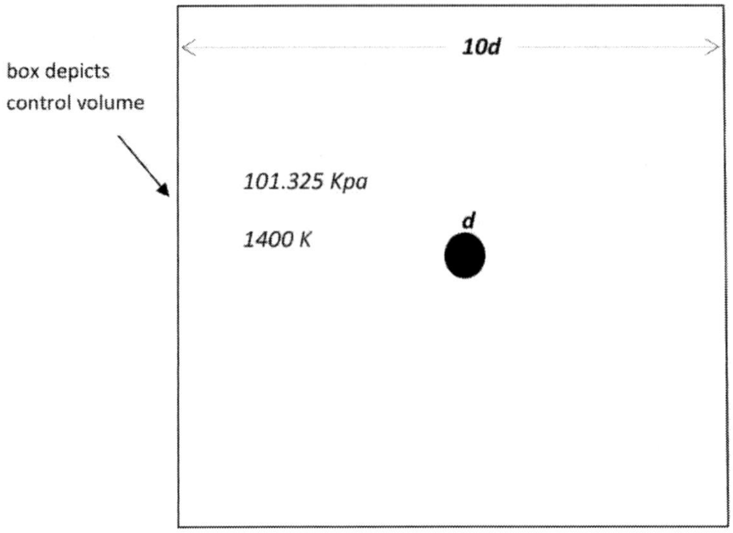

Figure 1a. Control volume shows the pyrite particle as injected into a hot gas environment

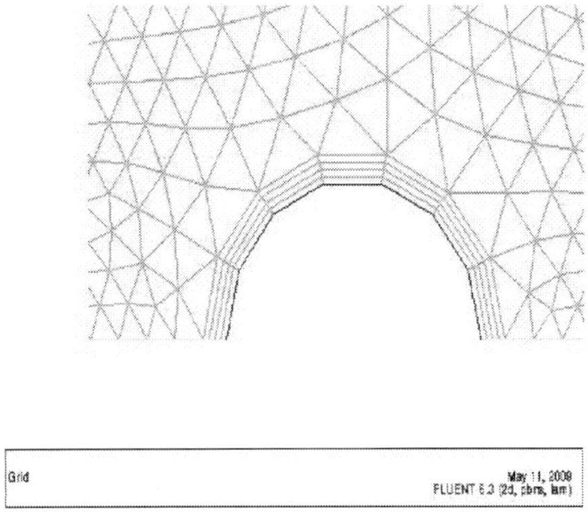

Figure 1b. Meshing scheme around axisymmetric particle surface

The control volume based technique consists of:

- The domain (as shown in Figure 1b) was divided into discrete control volumes using a computational grid;
- Integration of the governing equations over the individual control volumes was used to formulate non-linear algebraic equations for the discrete dependent variables of velocities, temperatures, species and reactions.

The solution algorithm for the non-linear algebraic equations involves:

- The linearization of the discretized equations and solution of the resultant linear equation system to yield updated values of the dependent variables starting with a given initial value and boundary conditions;
- The update continues until the scaled relative difference between two successive iterations has been reduced below a user-set threshold. In this study, the threshold value was set as

$$\frac{\left|\varphi^{k+1}-\varphi^{k}\right|}{\varphi^{k}} \leq 10^{-6}, \qquad (11)$$

where φ^k represents temperatures, species etc. of the kth iteration.

The computational domains and the partial differential equations describing the flow were discretized before being solved. Different meshing schemes were tested. It was found that mixed (quadrilateral and tri-geometrical) paving mesh converged quit fast although relatively larger number of control volumes is required.

Thus, an O-grid with a quadrilateral paving scheme was applied near the particle surface (boundary layer region) for the discretization of the computational domains, generating very fine resolution near the particle surface.

4.2 Grid dependency

The grid independence of the results, Figure 2, was achieved by changing the thickness of the boundary layer and comparing the average film temperature with the calculated value at selected time intervals. The influence of cell thickness and number of cells is shown in Figure 2. Three cases are compared namely, no boundary layer (r_f/r_p =1); boundary layer less than film thickness; and boundary layer greater than film thickness. The latter is investigated for the optimal number of discrete. It is clearly shown from Figure 2 that improper meshing in the boundary layer results in alteration in the prediction of the film temperature which in turn affect the prediction of the amount of sulfur released.

In regard of the number of mesh lines in the boundary layer region, the estimation of average film temperature initially increases and then remains constant after about 4 lines of cells with $0.05r_p$ thickness. This indicates that increasing the number of cells (control volumes) beyond a certain value will not have any significant effect on the film temperature (see Figure 2), but will unnecessarily increase the computational time.

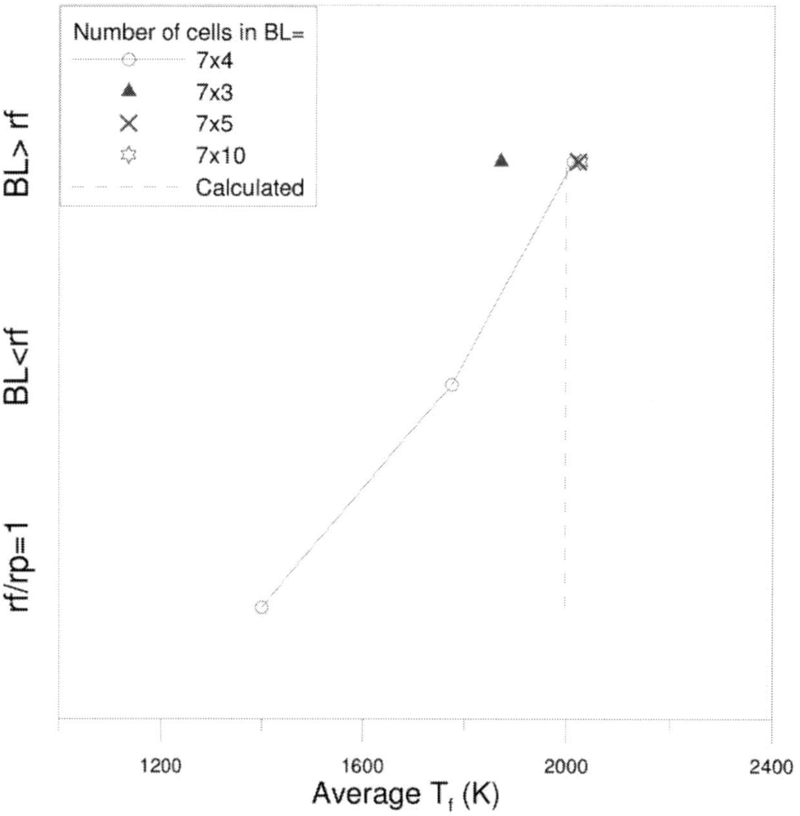

Figure 2. Grid dependency.

5. Results and Discussion

5.1 Temperature history

The computational fluid dynamics technique was developed to describe the transformation of a 45 μm diameter pyrite during combustion. Combustion conditions given in Table 1 are used as input parameters to the CFD software (FLUENT was used here). Figure 3 shows the profile of the particle temperature distribution for several time steps. The CFD simulation predictions clearly show a non-uniform temperature gradient at the particle surface that causes the particle to fragment. The average particle temperature of each time interval was estimated and compared with model prediction. Figure 4 illustrates the modeled thermal history of the pure pyrite particle initially injected in gas stream environment. The results indicate hat there is good agreement between the proposed models (Srinivasachar, 1989) and the CFD prediction, which gives an indication on the model reliability.

Table 1. Initial conditions for pyrite simulation.

Bulk temperature	1400 K
Wall temperature	1400 K
Pressure	101.325 Kpa
Oxygen mole fraction	0.21
Initial particle temperature	273
Particle emissivity	0.5

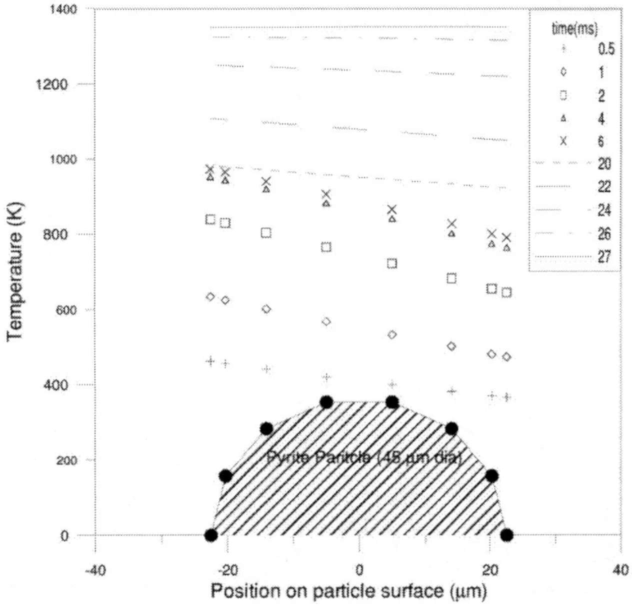

Figure 3. particle temperature distribution history predicted by CFD.

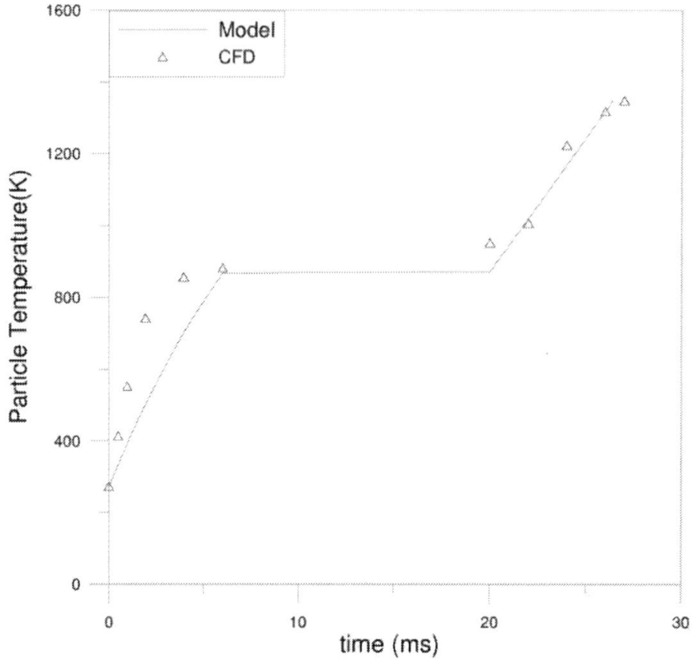

Figure 4. Comparison of particle temperature predicted by the proposed Model (Srinivasachar, 1989) and CFD.

5.2 Effect of fragmentation

The fragmentation behavior of pure pyrite during combustion is examined using the approach discussed earlier. The particle temperature with and without fragmentation is plotted in Figure 5. The figure shows that fragmentation results in an increase in the temperature of the fragmented particles decreases the duration of the melt phase during oxidation. Further, the fragmentation may increase the peak gas temperature perhaps to the level of melting point of magnetite and hematite causing sticking.

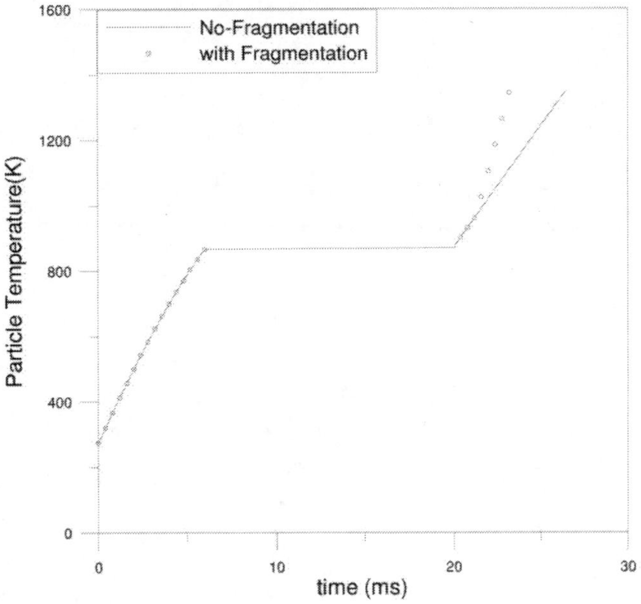

Figure 5. Particle thermal history with fragmentation.

When the particle temperature increases due to heat transfer from the bulk flow, sulfur begins to be released from the particle and form around the particle surface in the boundary layer region. Heat produced from the reaction of released sulfur with the bulk oxygen in the boundary layer region is fed back onto the particle surface, causing further increase in the particle temperature and devolatilization rate. The sulfur released is defined as the amount of sulfur removed from the particle, referenced by the original amount of sulfur contain in pyrite. The sulfur release rate is investigated and compared with CFD prediction. It was found that fragmentation increases the amount of sulfur released and that amount is greatest early in the oxidation process specifically at the formation and destruction of pyrrhotite. As shown by Figure 6, the data obtained from the model (Srinivasachar, 1989) is in excellent agreement with the CFD simulation.

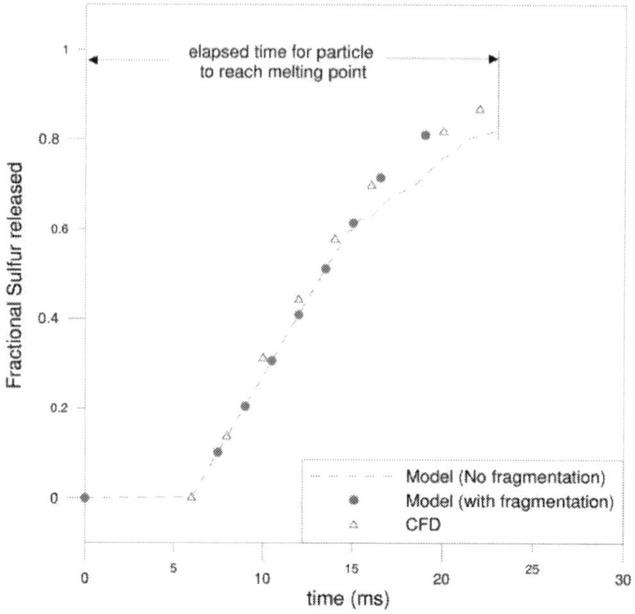

Figure 6. Percent of Sulfur released with fragmentation.

5.3 Effect of oxygen content

The concentration of oxygen in the bulk environment was examined to simulate gasification conditions. The oxygen level strongly influenced the amount of sulfur released from the pyrite particle. Figure 7 illustrates the influence of oxygen concentration on the releasing of sulfur. The time required for 60% sulfur release and almost completely released of sulfur for different concentration of oxygen namely 0.5%, 5%, and 21% are plotted and compared. As the bulk oxygen content decreases, the time required by the sulfur to completely release is significantly increases to 1000 ms as compared to 40 ms, for the 0.5 and 21% oxygen cases, respectively. This could be returned to two reasons; the first is low oxygen concentration means less penetration and diffusion of oxygen to the depth of particle to replace the sulfur released and the second reason refers to the quantitatively reduction in the exothermic sulfur/oxygen reaction due to low oxygen concentration results in decreasing the incremental rate in particle temperature. Hence it can be concluded that the time elapsed to remove sulfur is almost inversely proportion to the oxygen concentration in the bulk environment.

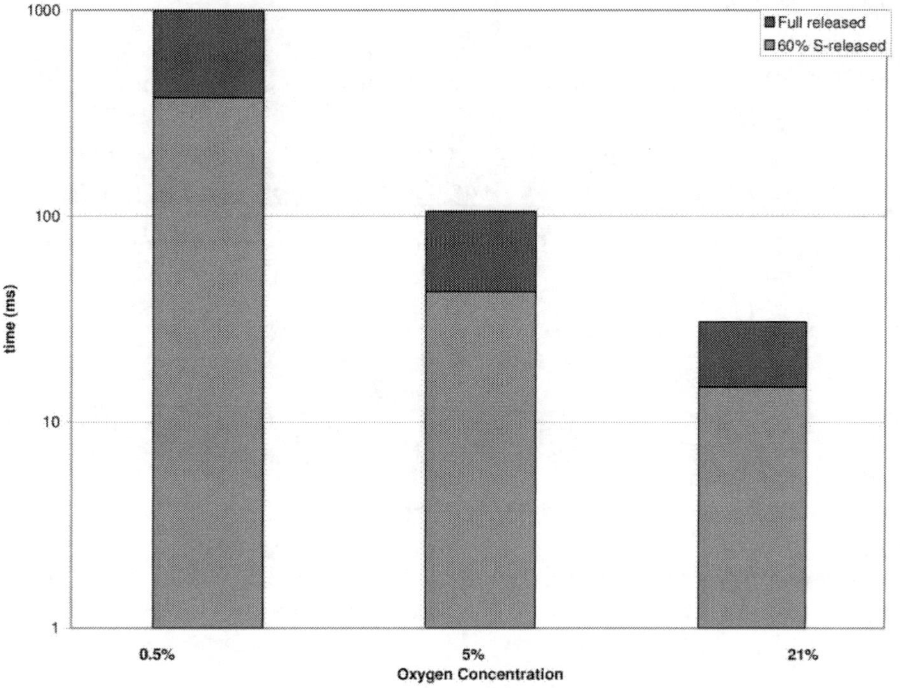

Figure 7. Time required to release sulfur as a function of oxygen content.

6. Conclusion

A mechanistic model of pyrite fragmentation during combustion and gasification processes was examined. The combustion and gasification processes influence the pyrite thermal history, fragmentation behavior, and rate of sulfur release. For example, an excluded pyrite particle exposed to combustion conditions results in particle fragmentation producing smaller size particles and releasing more sulfur. Under reducing conditions the rate of sulfur release is significantly reduced that will result in more low melting point sulfide and reduced iron phases. In addition, the degree of fragmentation was decreased.

The results of the CFD model were compared to other modeling efforts. The results of the comparisons showed good agreement.

Experimental data are needed to validate and enhance the model in the future work.

Acknowledgment

This material is based upon work supported by the Department of Energy office of Science under Award DE-FG02-06ER46292 (ND state grant, "coal combustion emission mechanisem").

References

Akan-Etuk, A. and Niksa, S., Paper No. 89-82 WSS/CI Fall Mtg., Livermore, CA, October 1989.

Akan-Etuk, A., "Pyrite Transformations During Pyrolysis," Report prepared under the Supervision of Prof. S. Niksa, Mechanical Engineering Department, Stanford University, 1990.

Asaki Z, Mori S, Ikeda M & Kondo Y "Oxidation of pyrrhotite particles falling through a vertical tube", Metallurgical Transaction B 16B: 627-638, 1985.

Benson, S.A.; Jones, M.L.; Harb, J.N. Ash Formation and Deposition. In Fundamentals of Coal Combustion for Clean and Efficient Use; Smoot, D.L., Ed.; Coal Science and Technology 20 Series; Elsevier: Amsterdam, 1993; Chapter 4, pp 299–373.

Finkelman, R.B., Modes of occurrence of potentially hazardous elements in coal: levels of confidence, Special Issue: Trace Elements Transformations in Coal-Fired Power Systems, Ed. S.A. Benson, E.N. Steadman, A.K. Mehta, and C.E. Schmidt, Fuel Processing Technology, 39, 21-24, 1994.

Groves S.J., Williamson J., and Sanyal A., "Decomposition of pyrite during pulverized coal combustion", Fuel 66, 461-466, 1987.

Helble J.J., Srinivasachar S., and Boni A., "factors influencing the transformation of minerals during pulverized coal combustion", Prog. Energy Combustion science, 16, 267-279, 1990.

Jorgensen F.R.A., "On Maximum Temperatures Attained During Single Particle Combustion of Pyrite," *Trans. Instn. Min. Metall.* C 90, pp. 10–16, 1981.

Lawrence E. Bool III, Thomas W., Jost O.L., "The partitioning of Iron during the combustion of pulverized coal", Combustion and Flame 100, 262-270, 1995.

Mitchell R.E., Ma L., Kim B., "On the behaviour of pulverized coal chars", Combustion and Flame 151, 426-436, 2007.

Mitchell R.E., "Mechanisms of Pyrite oxidation to Nonslagging Species", Report submitted to National Energy Technology Laboratory, The United States Department Of Energy,DE-FG22-94PC94205, Stanford University, 2002.

Rassak E., "Creation, Capture and Coalescence of Mineral Species in Coal Flame", Journal of the Institute of Energy, v57, 231-239, 1984.

Seames W.S., "An initial study of the fine fragmentation fly ash particle mode generated during pulverized coal combustion", Fuel process Technology, 81, 109-125, 2003.

Srinivasachar S., and Boni A., "A kinetic model for pyrite transformations in a combustion environment", Fuel 68, 829-836, 1989.

Srinivasachar S., Helble J.J., Boni A., Shah N., Huffman G., and Huggins F., "Mineral behaviour during coal combustion 2.Illite transformations", Progress Energy combustion Since 16, 293-302, 1990.

Ten Brink H.M., Eenkhoorn S., and Weeda M., "Flame transformations of coal-siderite", proceedings of engineering foundation conference on the impact of Ash Deposition on coal fired plants, Taylor & Francis, 373-383, 1993.

Ten Brink H.M., Eenkhoorn S., and Weeda M., "The behaviour of coal mineral carbonates in a simulated coal flame", Fuel Processing technology, v47, 233-243, 1996.

Tukey J.W., "Moments of Random Group Size Distribution", Annals of Mathematical Statistics, v20, 523-539, 1949.

Yan L., Gupta R.P., Wall T.F., "A mathematical model of ash formation during pulverized coal combustion", Fuel 81, 337-344, 2002.

Yan L., Gupta R.P., Wall T.F., "Fragmentation behaviour of pyrite and calcite during high-temperature processing and mathematical simulation", Energy &Fuels 15, 389-394, 2001.

Yan L. CCSEM analysis of minerals in pulverized coal and ash formation modeling, PhD thesis, The University of Newcastle, 2000.

Effects and Roles of Air Feed on the Noncatalytic Gasification of Jet Fuel in Supercritical Water

Jason Picou, Jonathan Wenzel, Michael Stever, Jared Bouquet, and Sunggyu Lee, Missouri University of Science and Technology, Rolla, MO

Abstract

The air oxidation of jet fuel can be beneficially incorporated in the novel supercritical water reformation. This paper experimentally and comparatively examines the roles of the concurrent oxidation reactions on the overall process chemistry, and specifically the effect of air feed position. The effect of air feed position, either at the inlet of the 48-inch long tubular reactor where the water/fuel mixture also enters, or a mid-reactor feed position 21.8" downstream from this point, on the noncatalytic reformation of jet fuel was evaluated using a tubular 0.4-L Haynes® Alloy 230 supercritical water reactor. The temperature, pressure and water-to-fuel ratio were kept nominally identical, while the air feed position and air flow rate were varied in order to elucidate its effect on reformation. Over the conditions examined, the mid-reactor air feed achieved higher hydrogen and carbon dioxide yields than the inlet air feed position for the same air feed rates, while the carbon conversion and methane yield for both feed positions was similar. The result indicates that the air feed position has little effect over the pyrolysis reaction while the mid-reactor air feed enhances the oxidation and water gas shift reactions more than the inlet air feed.

Introduction

The ability to efficiently produce hydrogen from a wide variety of hydrocarbon sources on demand from a small, portable apparatus is important to the future viability of the hydrogen economy. One of the more difficult to reform, but frequently used hydrocarbons, is jet fuel. The difficulty of reforming jet fuel is mostly due to the high concentration of sulfur, a variety of fuel additives used, the variety of hydrocarbon compounds and the longer chain and cyclic hydrocarbons that constitute jet fuel. Sulfur is a traditional poison for metallic catalysts, which are used in most reformation processes.

These inherent technological difficulties are overcome in a novel, non-catalytic approach using supercritical water partial oxidation for the reformation of jet fuel. In this process, supercritical water functions as a highly energized reforming agent, or a gasifying reactant, and also as a homogenizing reaction medium. Supercritical water is a non-polar solvent unlike its counterpart ambient water, allowing hydrocarbon molecules to be miscible with water molecules in all proportions, as well as oxygen.[1,2] The absence of a heterogeneous noble metal catalyst negates sulfur poisoning and coke fouling. The addition of oxygen in the form of air provides *in-situ* exothermic heat of combustion to supplement the endothermic reformation reaction. The purpose of these experiments is to determine if introducing air at a location farther inside the reactor, as opposed to the inlet of the reactor where the fuel and water mixture enters, will improve the process efficiency of the non-catalytic supercritical water reformation of jet fuel.

Apparatus & Chemicals

The water used in the reformer was deionized water, and the air used was Airgas Breathing Quality Grade D compressed air. The jet fuel used was both civilian jet fuel (Jet-A) and aviation logistic fuel (JP-8), both of which are an assortment of hydrocarbons including straight chain, branched and cyclic. An ASTM D2887 boiling range distribution analysis determined that the length of the carbon bonds varied from 7 to 17 carbons, with the average being 12 for both fuels. Therefore, the fuel was modeled as a single representative molecular species, n-dodecane, which has a chemical formula of $C_{12}H_{26}$. Both jet fuels were tested by Texas Oil Tech Laboratories for sulfur content by ASTM D4294 and found that the Jet-A contained 990 ppm sulfur and the JP-8 contained 810 ppm sulfur. Even though two different jet fuels were used due to difficulty in obtaining additional JP-8, the two samples were considered nominally the same from the standpoints of the fuel reformation and the specific process employed.

The supercritical water reformation system consists of a liquid feed system, feed preheater, reactor, reactor heaters, air feed system, sample collection system, and data acquisition and control system, of which a schematic process flow diagram is illustrated in Figure 1. The supercritical water reactor has a volume of 383 mL, and is constructed of Haynes® Alloy 230, which is an alloy of mostly nickel, chromium, tungsten, and molybdenum, among other elements.[3] The stress-strain properties, as well as creep and rupture properties, of Haynes® Alloy 230 allow the reactor to be operated at temperatures up to 800°C at a pressure of 36.2 MPa.

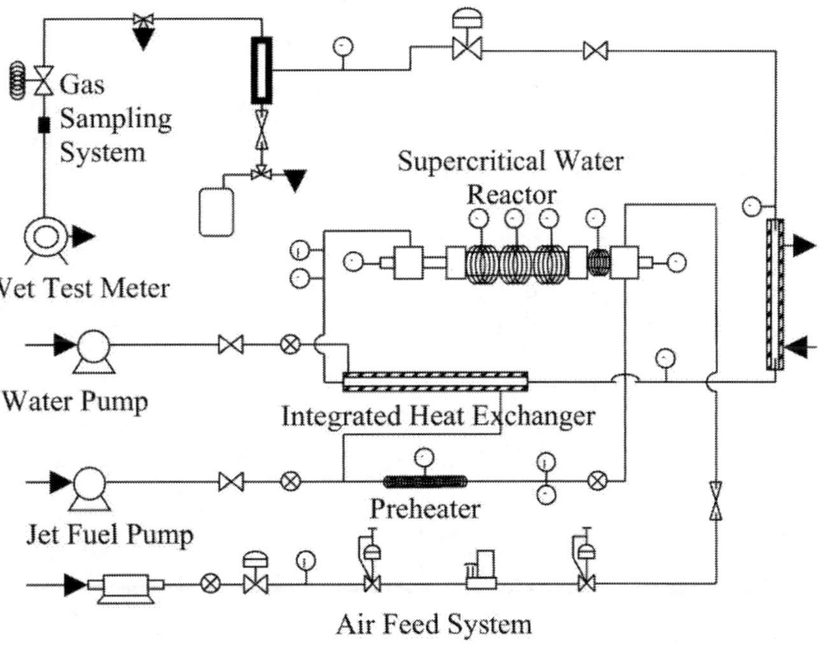

Figure 1. A schematic of the supercritical water reformation system at the Laboratory of Transportation Fuels and Polymer Processing, Missouri University of Science and Technology.

Three of the experiments were conducted with air fed into the inlet cross, directly across from where the fuel and water mixture enters, as shown in Figure 2. Three of the experiments were conducted with the air inlet inside the reactor 21.8" past the original feed spot. The first set initiates the partial oxidation of jet fuel at an entrance portion of the reactor, whereas the latter set brings about the partial oxidation of the reaction mixture in the mid-section of the reactor. Analysis of the gaseous effluent was performed using a HP 5890 Series A gas chromatograph with a thermal conductivity detector. The gas chromatograph is calibrated to detect hydrogen, nitrogen, carbon monoxide, methane, carbon dioxide, acetylene, ethylene and ethane. The carrier gas for the gas chromatograph was Airgas Ultra High Purity Argon with a purity of 99.9995%.

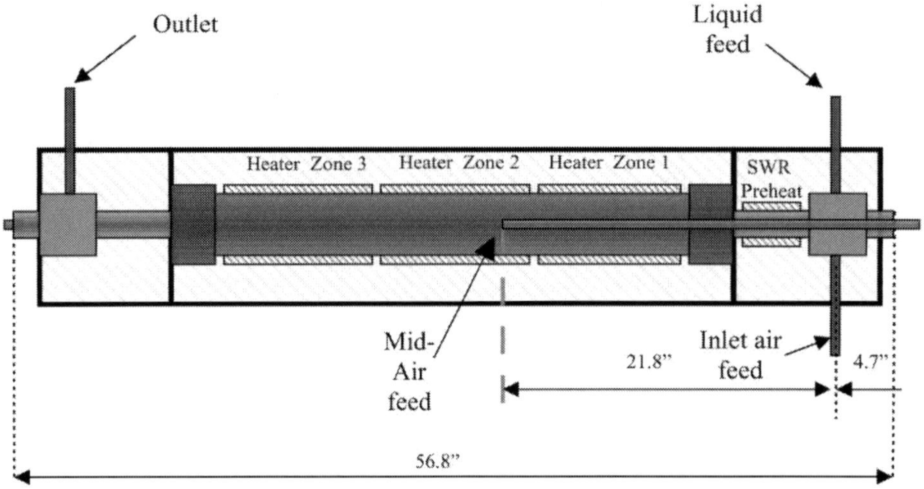

Figure 2. Air feed positions and lengths of the Haynes® Alloy 230 supercritical water reformation reactor at the Missouri University of Science and Technology.

Chemical Reactions

A variety of reactions are conceivable in the supercritical water reformation process, the most significant of which are listed below.[4] The overall reformation reaction of jet fuel may be written as:

$$C_{12}H_{26} + 12\ H_2O \rightarrow 12\ CO + 25\ H_2 \qquad \Delta H^\circ_{298} = 1866.2\ \text{kJ/mol} \qquad (1)$$

where jet fuel is represented by a model compound of n-dodecane for stoichiometric simplicity based on its average molecular weight. The above reaction is in competition and occurs in parallel with the pyrolysis reaction:

$$C_{12}H_{26} \rightarrow C_aH_b + C_xH_y + zH_2 \qquad (12 = a + x \text{ and } 26 = b + y + 2z) \qquad (2)$$

The endothermic pyrolysis reaction is thought to be responsible for any gaseous light hydrocarbons contained in the effluent gas. Repeated pyrolysis leaves hydrogen deficient fractions, which eventually become solid coke.[5]

In the presence of oxygen, another set of reactions occur. The first (reaction 3) is the partial oxidation of hydrocarbons, while the second (reaction 4) is the complete oxidation reaction:

$$C_{12}H_{26} + 6\ O_2 \rightarrow 12\ CO + 13\ H_2 \qquad \Delta H^\circ_{298} = -1035.6\ \text{kJ/mol} \qquad (3)$$

$$C_{12}H_{26} + 18\tfrac{1}{2}\ O_2 \rightarrow 12\ CO_2 + 13\ H_2O \qquad \Delta H^\circ_{298} = -7575.0\ \text{kJ/mol} \qquad (4)$$

In addition to these reactions, the water gas shift reaction also takes place. The forward reaction is a highly desirable reaction, since additional hydrogen is produced. The forward reaction is thermodynamically favored at temperatures of 815°C or below:[6]

$$CO + H_2O \leftrightarrow CO_2 + H_2 \qquad \Delta H^\circ_{298} = -41.2 \text{ kJ/mol} \qquad (5)$$

None of these reactions occur alone or independently as isolated events, nor are they mutually exclusive. For example, pyrolysis or oxidation could break down the original jet fuel hydrocarbons, then reformation occurs on the resulting products. Various other reactions, like methanation or the Boudouard reaction, are also possible, even though the thermodynamic equilibrium for the forward Boudouard reaction is not favorable for the process conditions of the current study. The discussion therefore is limited to the aforementioned reactions for simplicity and since they effectively describe all of the products observed.

Experimental

The air feed position and the feed rate of air was varied to investigate the effects upon the reformation of jet fuel in supercritical water. The temperature was held at 769.2 ± 0.9°C, while a constant pressure of 24.05 ± 0.02 MPa and a 15:1 water-to-fuel mass ratio was maintained. This corresponds to a water-to-carbon (H_2O/C) molar ratio of about 12, or a jet fuel concentration of 6.25 weight percent. Table 1 gives the experimental run number, temperature, pressure, fuel type, and water, fuel and air flow rates along with the air feed position for each experiment. The inlet air feed position is when the air feed is introduced directly across from the fuel-water mixture feed in the inlet cross, while mid or middle air feed is when the air is fed 21.8" downstream from this point. The experiments were conducted in the order given by the Run ID, from one to thirteen. Most of these experiments were performed without any air feed, in order to determine the effects of both normal and mid-reactor air feed on the reactions, as well as to determine the reproducibility of the experiments.

Table 1. Jet fuel reformation experiments in supercritical water. Water-to-fuel mass ratio is constant at 15:1, temperature at 769.3 ± 0.9°C, and pressure at 24.03 ± 0.02 MPa. The air flow rate and air feed position are varied.

Run No.	Temp (°C)	Pressure (MPa)	Water (g/min)	Fuel Type	Fuel Flow (g/min)	Air Flow (slpm)	Air feed position
13	768	23.9	14.5	JP-8	1.08	0.00	none
9	767	24.0	15.9	JP-8	1.01	0.00	none
8	769	24.0	14.1	JP-8	1.04	0.00	none
7	770	24.0	15.1	JP-8	0.93	0.00	none
3	770	24.1	15.4	JP-8	0.99	0.00	none
2	780	24.1	15.3	Jet-A	0.99	0.00	none
1	767	24.1	15.3	Jet-A	0.91	0.00	none
6	772	24.0	15.1	JP-8	0.95	0.50	inlet
12	766	24.0	15.1	JP-8	1.09	0.50	mid
5	772	24.1	15.1	JP-8	0.95	1.00	inlet
11	766	23.8	15.0	JP-8	1.08	0.99	mid
4	765	24.1	14.3	JP-8	1.00	3.01	inlet
10	769	24.1	15.7	JP-8	1.15	2.97	mid

Results and Discussion

The results of the experiments outlined in Table 1 are shown in Table 2, which lists the flow rate of gaseous components of the reactor effluent at ambient conditions (72°F and 1 atm) and the carbon gasification percentage. The carbon gasification percentage is defined as the amount of carbon in the gaseous reactor effluent divided by the carbon that entered as jet fuel.

Table 3. Gas effluent volumetric flow rates as measured at ambient conditions and the carbon gasification percentage for the experiments conducted.

Run #	Gas flow rate (L/min)							Carbon Gasification
	H_2	N_2	CO	CH_4	CO_2	C_2H_6	Total	
13	0.67	0.00	0.09	0.77	0.26	0.02	1.82	62.6%
9	0.59	0.00	0.13	0.71	0.19	0.01	1.64	60.5%
8	0.60	0.00	0.13	0.78	0.18	0.00	1.69	60.8%
7	0.57	0.00	0.12	0.76	0.21	0.02	1.68	71.0%
3	0.51	0.00	0.11	0.70	0.15	0.02	1.49	58.8%
2	0.57	0.00	0.10	0.70	0.21	0.01	1.60	60.8%
1	0.72	0.00	0.08	0.66	0.30	0.01	1.77	67.3%
6	0.58	0.45	0.15	0.69	0.25	0.02	2.13	68.2%
12	0.82	0.43	0.12	0.77	0.37	0.02	2.53	69.8%
5	0.56	0.88	0.18	0.66	0.31	0.01	2.60	71.4%
11	0.91	0.81	0.12	0.73	0.49	0.02	3.08	74.0%
4	0.70	2.44	0.25	0.60	0.65	0.00	4.64	87.2%
10	1.05	2.44	0.10	0.66	0.78	0.00	5.04	78.6%

The results of these experiments are shown in the following figures, for ease of comparison. Figure 3 is the carbon conversion as a function of increasing oxygen-to-carbon ratio. The oxygen-to-carbon ratio, since the fuel flow rate is kept constant, is a measure of air flow. Experiments with no air flow have a zero oxygen-to-carbon ratio.

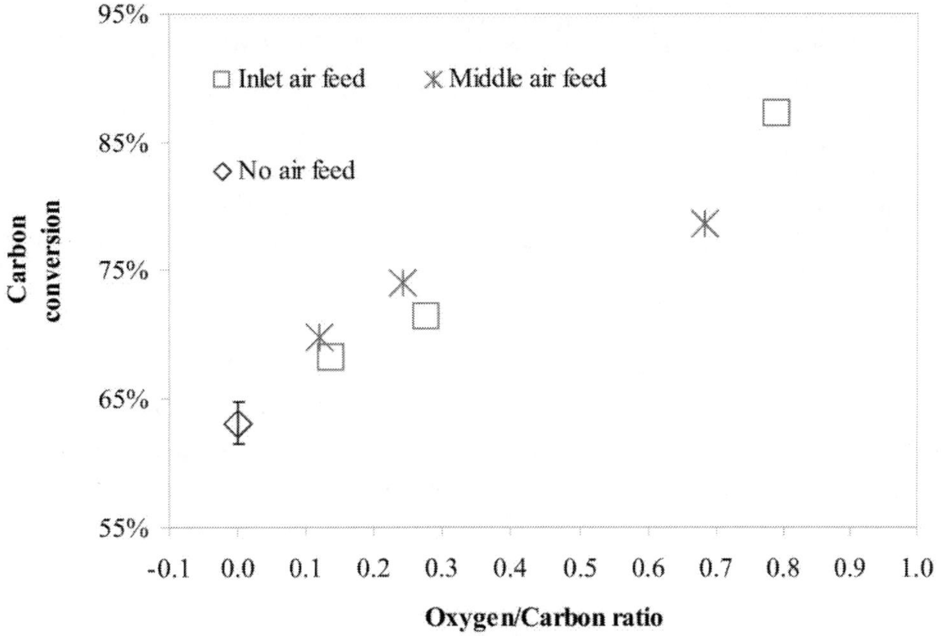

Figure 3. Carbon conversion as a function of oxygen-to-carbon ratio for both normal and middle reactor air feed.

As shown in Figure 3, the carbon conversion increases with increasing oxygen-to-carbon ratio for both air feed methods. The seven experiments performed without any air flow have a mean carbon conversion of 63.1%, with a standard error of 1.7%, which demonstrates the reproducibility of these experiments. When air is fed at the inlet position the carbon conversion is 68.2% at an oxygen-to-carbon ratio of 0.14, increases to 71.4% at 0.28, and reaches a maximum of 87.2% at a 0.79 oxygen/carbon ratio for the experiments conducted for this flow configuration. The middle air feed exhibits a similar progression in regards to carbon conversion as a function of oxygen feed. The carbon conversion is 69.8% at the lowest oxygen-to-carbon ratio, increases to 74.0% at 0.24, and is 78.6% at the highest oxygen-to-carbon ratio of 0.68 used for the mid-air feed experiments. This indicates that the air feed position has little effect on the carbon conversion for the conditions examined.

Figure 4 illustrates the hydrogen and carbon monoxide yields for the experiments conducted. The gas yield is calculated as the molar flow rate of that particular gaseous species out of the system divided by the molar flow rate of jet fuel into the system resulting in a dimensionless number that accounts for variations in the inlet fuel flow rate.

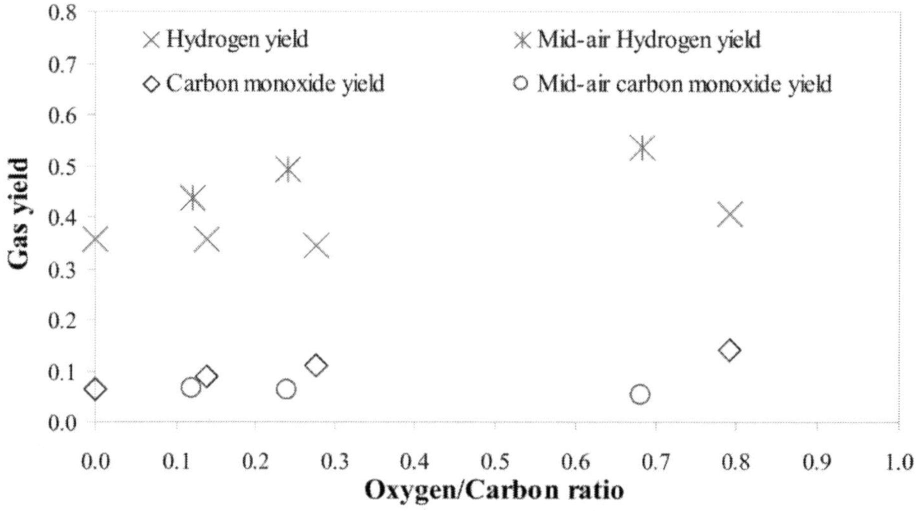

Figure 4. The hydrogen and carbon monoxide gas yield for changing oxygen to carbon ratios with normal and middle reactor air feed.

In Figure 4, the hydrogen and carbon monoxide yields for the normal air flow path and the experiments without air flow are combined for simplicity, the experiments with no air flow have a oxygen-to-carbon ratio of zero. The results for the air flow that began 21.8" from the inlet air entrance is denoted by the prefix "Mid air" while the inlet air feed has no prefix. Beginning with the baseline experiments with no air flow, the hydrogen yield was 0.36 ± 0.02 and 0.065 ± 0.004 for carbon monoxide, which is a 5% and 7% standard error respectively.

The hydrogen yield for inlet air flow is 0.36 at an oxygen-to-carbon ratio of 0.14 and 0.34 at a 0.27 O_2/C ratio. The hydrogen yield was 0.41 at the highest O_2/C ratio of 0.79, which is the highest hydrogen yield obtained at the inlet air feed position. When air is feed into the middle of the reactor, hydrogen yield increases with increasing oxygen-to-carbon ratio, with a yield of 0.44 at a 0.12 O_2/C ratio, 0.49 at a 0.24 O_2/C ratio, and a yield of 0.53 at a 0.68 O_2/C ratio. These mid-reactor hydrogen yields are 20% greater than the inlet hydrogen yields at similar oxygen-to-carbon ratios. While the feed location of air does not substantially affect the carbon conversion, as shown in Figure 3, it does affect hydrogen yield as shown in Figure 4.

The carbon monoxide yield, increases with increasing oxygen-to-carbon ratio for the normal air feed and decreases for the middle air feed. With normal air flow, the carbon monoxide yield goes from 0.09 at the lowest O_2/C ratio to a maximum at 0.14 at the highest O_2/C ratio. The middle air feed setup resulted in a carbon monoxide yield decrease from 0.07 at the lowest O_2/C ratio to 0.05 at the highest O_2/C ratio. These results indicate that either less carbon monoxide is produced yet more hydrocarbons are being converted to hydrogen in the mid-air feed system, or that the water gas shift reaction is more active, consuming carbon monoxide to produce more carbon dioxide and hydrogen gas.

The other major species present in the gaseous effluent are methane and carbon dioxide. Figure 5 illustrates the effects of the air feed position upon the methane and carbon dioxide yield for increasing O_2/C ratio, which complements the discussion of some of the effects presented in the Figure 4. It should be noted that the ethane yield for the different air feed positions is not compared in the figures due to its small yield as compared to the other gaseous species; however it is listed in Table 2.

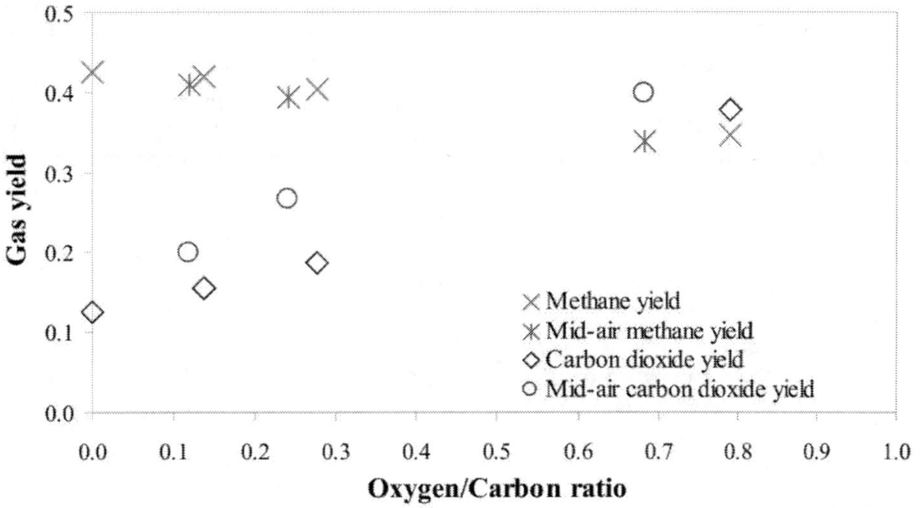

Figure 5. The methane and carbon dioxide gas yield for changing oxygen to carbon ratios with normal and middle reactor air feed.

As with Figure 4, the data points in Figure 5 for air fed into the middle of the reactor have "Mid-air" as a prefix while the data points for air fed at the inlet cross of the reactor do not have a prefix. The methane yield is 0.425 ± 0.009 and the carbon dioxide yield is 0.13 ± 0.01 with no air flow. Methane yields for both air feed systems decrease for increasing air feed, and the methane yield for any particular oxygen-to-carbon ratio are similar between the two feed systems. For the inlet air feed the methane yield decreases from 0.42 to 0.41 to 0.35 with increasing oxygen-to-carbon ratios. The middle reactor feed had a methane yield decreasing from 0.41 to 0.39 to 0.34 with increasing oxygen-to-carbon ratios. These results are nearly identical, and it may be concluded that the air feed position has limited impact on the production of methane.

In comparison to methane, carbon dioxide has a very different result. For both air feed positions the carbon dioxide yield increases with increasing oxygen-to-carbon ratio, which is expected since the oxidation reaction would increase with increasing oxygen feed. The normal air feed position increases from 0.16 to 0.19 to 0.38 for the highest O_2/C ratio, while the carbon dioxide yield for the mid-reactor air feed increases from 0.20 to 0.27 to 0.40 at the highest O_2/C ratio for this feed position. For the oxygen-to-carbon ratios of about 0.13 and 0.25, the middle reactor air feed had a carbon dioxide yield between 20% to 30% greater than the normal air feed. This indicates that either the oxidation reaction produces more carbon monoxide with air fed at the middle of the reactor, or that the forward-water gas shift reaction is converting more carbon monoxide into carbon dioxide and hydrogen, or both. The fact that the carbon monoxide yield remains relatively constant for air fed at the middle of the reactor, while hydrogen and carbon dioxide yields increase faster than normal air feed, indicates that the forward-water gas shift reaction contributes to the increased hydrogen yield.

Conclusion

The effect of air feed position, either at the point where the fuel and water mixture enters a tubular 0.4-L Haynes® Alloy 230 supercritical water reactor or 21.8" into the reactor from this point, on the supercritical water reformation and oxidation of jet fuel was studied using a 0.4-L Haynes® Alloy 230 supercritical water reactor. The temperature, pressure and water-to-fuel mass ratio were kept nominally identical at 769.3 ± 0.9°C, 24.03 ± 0.02 MPa and 15:1 respectively, while the air feed position and amount of air fed was varied. Seven duplicate experiments were conducted without air feed to determine the reproducibility of the experiments as well as to give a comparison to determine the effects of air flow. Carbon conversion, the amount of carbon that exits the reactor as gas divided by the amount of carbon in the fuel that entered, was not noticeably affected by air feed position, and for both feed positions carbon conversion increased with increasing air flow. The highest carbon conversion was 87%, which was achieved at a 0.79 oxygen-to-carbon ratio. The gas flow rate changed with increasing oxygen-to-carbon ratio, with hydrogen and carbon dioxide increasing and methane decreasing for both air feed positions. Hydrogen yield was 20% greater and carbon dioxide yield 30% greater when air was feed into the middle of the reactor when compared to inlet air flow. Carbon monoxide yield was less for mid-air feed, and methane yield in general was unchanged. The pyrolysis and methanation reactions are not affected by the air feed position, but the forward-water gas shift reaction and the oxidation reactions are affected and are more active with middle air feed in comparison to feeding air where the fuel and water enter the reactor.

Acknowledgements

This work was sponsored by the U.S. Army under PE Number 0602705A through a subcontract from DRS Technical Services, Inc. The authors are grateful for insightful technological guidance and supervision provided by Dr. Terry Dubois, CERDEC, U.S. Army, Ft. Belvoir, Virginia and Dr. H. Brian Lanterman of DRS Technical Services.

References

(1) Holliday, R.L.; Jong, B.Y,M.; Kolis, J.W. Organic synthesis in subcritical water: Oxidation of alkyl aromatics. *J. Supercrit. Fluids* **1998**, 12, 255-260.

(2) Klijn, J.E.; Engberts, J.B.F.N. Fast reactions 'on water'. *Nature* **2005**, 435, 746-747.

(3) *Haynes® 230 Alloy*, Haynes High-Temperature Alloys International, 2004 pp 2.

(4) Picou, J. W. Autothermal Non-Catalytic Reformation of Jet Fuel in a Supercritical Water Medium. M.S. Thesis, Missouri University of Science and Technology, Rolla, MO, June 2008.

(5) Lee, S., Lanterman, H., Wenzel, J., Edwards, N., Adams, A., Wootton, J., Garcia, A. Noncatalytic direct reformation of JP-8 fuel in supercritical water. *Prepr. Pap. Am. Chem. Soc., Div. Pet. Chem.* **2006**, *51*, 521-523

(6) Lee, S.; *Alternative Fuels*, Taylor & Francis, Philadelphia, 1996; pp.140-151.

A Framework for the Development of Dynamic Multi-Zonal Gasification Reactor Models

David C. Miller, Collaboratory for Process & Dynamic Systems Research, National Energy Technology Laboratory, U.S. Department of Energy, Morgantown, WV

Sara E. Zenner, Gasification Division, National Energy Technology Laboratory, U.S. Department of Energy, Morgantown, WV

Gasification of coal, coke and/or biomass offers a significant opportunity to cleanly produce energy, liquid fuels and/or chemicals. In order to understand how to best design gasification systems, accurate and computationally efficient models of the gasification reactor are required. Currently, most modeling approaches focus on either highly sophisticated computational fluid dynamics (CFD) models or overly simplified equilibrium reactors. While the former approach offers high accuracy, it suffers from long computation times that limit its ability to be used within a flowsheet process simulator. The latter approach improves calculation speed at the expense of detail and accuracy. Many simulations employing the equilibrium model approach make use of restricted equilibrium or approach temperature to better match data from experiments or more detailed models; however, these approaches typically limit the model applicability to a single operating point. Multi-zonal models offer a balance between the computational expense of a detailed CFD model and the speed of a simple equilibrium model. This paper will describe a framework for developing dynamic multi-zonal gasification reactor models. This approach can be implemented using CFD results, experimental data or a combination of the two. The effectiveness of the approach will be demonstrated using the equipment and process model development and simulation environment, Aspen Custom Modeler®.

See more of this Session: Advances in Gasification Research - II
See more of this Group/Topical: Energy and Transport Processes

Mathematical Modeling of a Transient 1D Entrained-Flow Texaco Gasifier Using Aspen Custom Modeler

Richard Turton, Chemical Engineering, West Virginia University, Morgantown, WV

Debangsu Bhattacharyya, Chemical Engineering, West Virginia University, Morgantown, WV

Stephen E. Zitney, Collaboratory for Process and Dynamic Systems Research, National Energy Technology Laboratory, U.S. Department of Energy, Morgantown, WV

Job Kasule, Chemical Engineering, West Virginia University, Morgantown, WV

Coal-fired integrated gasification combined cycle (IGCC) power plants produce synthesis gas that is subsequently converted into electricity through combustion in a gas turbine. Several mathematical models have been developed to study the processes taking place inside the gasifier. Such models range from simple one-dimensional steady-state models to sophisticated transient three dimensional CFD-based models that incorporate turbulence effects in the reactor. The practical operation of the gasifier is dynamic in nature but most one dimensional and some higher dimensional models are often steady state and although many higher order CFD based models are dynamic in nature, they are too complicated with many process parameters to be used directly in design and scale-up studies. Thus lower dimensional models are still useful in these types of studies.

In the current study, a one dimensional transient model for the entrained flow GE/Texaco type gasifier is developed using Aspen Custom Modeler ®, which is an equation-based industrial simulator for creating, editing and re-using models of process units.

The current model is based on mass, momentum and energy balances for the solid and gas phases that exist in the gasifier. The chemical reactions considered in the model are the devolatilization/pyrolysis, gasification and combustion, and homogeneous gas-gas reactions. This highly nonlinear system of equation is solved using the inbuilt Newton-based non-linear solvers.

The results to be presented are the transient and steady state solutions to our model equations giving the velocity, concentration (mass fractions), and temperature profiles in the gasifier. Parametric studies results for which the effect of the process parameters on the performance of the gasifier is investigated would be presented for different types of coals. Comparisons of the current model results to available Texaco pilot plant data and other published models will also be made.

See more of this Session: Advances in Gasification Research - II
See more of this Group/Topical: Energy and Transport Processes

Performance Prediction of Entrained Flow Coal Gasifier for IGCC System Using a Steady-State Process Model

Thanh D. B. Nguyen, Department of Chemical Engineering, Hankyong National University, Anseong-si, Gyonggi-do, 456-749, South Korea

Young-Il Lim, Department of Chemical Engineering, Hankyong National University, Anseong-si, Gyonggi-do, 456-749, South Korea

Byung-Ho Song, Department of Chemical Engineering, Kunsan National University, Gunsan, Jeonbuk 573 - 701, South Korea

Si-Moon Kim, IGCC Group, Korea Electric Power Research Institute, Daejeon, 305 - 380, South Korea

Yong-Jin Joo, IGCC Group, Korea Electric Power Research Institute, Daejeon, 305 - 380, South Korea

Dal-Hong Ahn, IGCC Group, Korea Electric Power Research Institute, Daejeon, 305 - 380, South Korea

In an Integrated-coal Gasification Combined-Cycle (IGCC) plant, the gasifier is one of the most important units, since it converts solid fossil fuels such as gaseous fuel, liquid fuel or chemical feedstocks into clean syngas. The entrained-flow coal gasifiers have been widely used in IGCC plants due to the high gasification rate achievement and the relatively clean gas production.

This study focuses on finding optimum operating conditions for the given entrained flow coal gasifiers using a steady-state process model to predict the process performance. The process model includes two-stage equilibrium submodels and a carbon burnout mechanism for which a kinetic model is used to calculate the amount of combusted coal and the heat released to provide the gasification process. The stoichiometric equilibrium model for char gasification and gas phase reactions is used to calculate the temperature and composition of the syngas at the gasifier exit. The parametric effects of operating pressure, oxygen/carbon ratio and steam/carbon ratio on the gasification performance are also studied.

See more of this Session: Advances in Gasification Research - II
See more of this Group/Topical: Energy and Transport Processes

Integrated Tool for IGCC Power Plants Design

Mar Perez-Fortes, Chemical Engineering Department, Universitat Politècnica de Catalunya - ETSEIB, Barcelona, Spain

Aaron D. Bojarski, Chemical Engineering Department, Universitat Politècnica de Catalunya - ETSEIB, Barcelona, Spain

Enric Velo, Universitat Politécnica de Catalunya, Barcelona, Spain

Jose M. Nougués, Chemical Engineering Department, Universitat Politècnica de Catalunya - ETSEIB, Barcelona, Spain

Luis Puigjaner, Chemical Engineering Department, Universitat Politècnica de Catalunya - ETSEIB, Barcelona, Spain

Abstract: Expanding economies and global warming are increasingly important and interrelated challenging issues which require cleaner and more efficient power plants designs to comply with sustainable energy demand are required. Integrated Gasification Combined Cycle (IGCC) power plants have an important role, because of their more efficient way of producing energy from fossil fuels. Actions are focused on clean power and H_2 from coal, and bioproducts. So, one of the future of coal-based power generation strategies should be CO_2 transport and storage in order to obtain a purer hydrogen stream. This paper proposes a methodology and supporting tool for estimating plant performance in terms of power effiency and environmental compliance and making an economic assessment for different scenarios. The technical performance has been modeled in Aspen Hysys and Aspen Plus, and models have been validated with real power plant data from ELCOGAS. Comparisons, in terms of power, emissions, efficiencies and costs between a wide variety of plant designs are presented: they enhance differences in raw material, purification units layout, and hydrogen obtention. Results obtained are examined and discussed towards future work.

Keywords: IGCC power plant, conceptual modelling, performance prediction Aspen Hysys modeling, plant layout.

1. MODELLING PERFORMANCE AND PREDICTIONS

In this paper, an integrated platform for different scenarios prediction has been done. On the one hand, Aspen Hysys is the software used as the main platform for building the conceptual model. On the other hand, Microsoft Excel is the tool used for the economic evaluation. The IGCC plant flowsheet can be changed (by the addition or the sustraction of units, or directly by changing the layout of the flowsheet). From the Aspen Hysys model, data such as power, emissions and efficiency can be obtained. From Microsoft Excel calculations it is possible to estimate preliminary costs: direct costs, the total capital requirement, operating costs and the cost of electricity (COE). The two platforms works in an integrated way.

1.1 Conceptual Modelling.

The model is implemented using two main chemical flowsheeting environments: Aspen Hysys and Aspen Plus, which are flexible tools for process simulation by providing thermodynamic models for the estimation of chemical properties and unit operation models for many processes. Aspen Hysys has been chosen as the platform for the overall process simulation because of its capability of accepting custom models as extensions, and its ease of coding. These models could represent complex chemical reactions (COS hydrolysis) or partial gasification steps (pyrolysis). The proposed platform also allows creating new chemical components not included in its database, such as non stoichiometric solids for raw material and char definition. Alternatively, Aspen Plus is used for calculations involving water systems and electrolytes. These ionic models are required to solve phase equilibrium problems for unit operation systems such as venturi scrubbers, SWS or MDEA absorbers. The aforementioned models have been integrated in Aspen Hysys by means of ANN extensions. Data required to train each of the ANN's comes from sensitivity analysis performed within Aspen Plus while the neural network training is carried out using the ANN package provided with Matlab 6.5. Matlab has been also used for gasifier model parameters estimation.

The global flowsheet studied in this work contemplates all components in a typical IGCC layout. It includes a PRENFLO gasifier and a series of purification units, i.e. a ceramic filter, a venturi scrubber, a COS hydrolysis reactor, a sour water stripper, a MDEA absorber and a Claus plant. An ASU is used to obtain oxygen at a purity of 85 wt%. Steam, oxygen and fuel raw materials enter the gasifier and they are converted into syngas, which is cooled down before entering the purification units. Heat is recovered by producing steam in the HRSG, which is used in a steam turbine. The ELCOGAS power plant is designed to produce around 335 MW of gross power with relatively low emissions. SO_2 emissions are controlled by removing sulfur species from the syngas before the combustion in the gas turbine cycle, and producing liquid sulphur. NO_x emissions are avoided by using preventive methods such as nitrogen saturation of the gas also before the combustion. It is worthwhile to mention that ashes are highly removed as slag in the gasification reactor due to high pressure and temperature conditions. Nevertheless, remaining particles are extracted from the syngas by means of filters. In relation to CO_2 and CO emissions, they are assumed to be produced in the gas turbine cycle. Other contaminants such as chlorine, mercury and heavy metals are distributed in the streams flowsheet by using experimental correlations, based on data developed within the AGAPUTE project (Advanced GAs Purification TEchnologies for co-gasification of coal, refinery by-products, biomass and waste, targeted to clean power produced from gas ans steam turbine-sets generator and fuel cells, supported by European Commission, project no. RFS-CR-04006, see Acknowledgements).

The model shows good agreement between simulated results and industrial data for outlet streams of purification gas units: venturi scrubber, sour water stripper, MDEA absorber, and Claus plant, as seen in Fig. 1. The error is a maximum of 18%, which is well within the accepted error margin at the preliminary design stage (25-40%).

Fig. 1. Purification models validation with real power plant data.

Also, in orther to assure the model confidence, the result of the whole plant has been evaluated by means of the clean gas composition (where again a maximum error of 18% has been observed). See also Fig. 1, right side.

1.2 Best plant design

Election of costs, efficiency and emissions as KPI is directly related to the conception of a profitable gasification plant. The three authors' main criteria for IGCC best design are:

- Economic evaluation: costs à Cost Of Energy (COE).

- Current challenges: environmental impact à Emissions, like CO2, NOx and SO2.

- Process synergy à Energy efficiency.

1.2.1 Results: Scenarios comparison

The scenarios comparison enhance sixteen different cases:

- Eight with different raw materials (coal , coke, a mistureof them, and a mixture of them plus a waste biomass, orujillo)

- Six more cases with different purification gas units layout: by avoiding some units, and also by adding a Selexol absorber or a Dolomite unit.

- Two more by adding a CO2 purification train (with water gas shift reactor, MEA absorber and PSA) and considering hydrogen to the market or to power production.

Pareto curves have been used to compare costs, environmental impact and efficiency of the different scenarios. A methodology to address the work main goal has been proposed: it consists of giving the tactic to find explicit and final solutions for concrete IGCC power plant flowsheet conditions. Costs, emissions and efficiency are considered as selection criteria. In this way, an exhaustive calculation of these parameters is developed in this work.

Acknowledgements: Data provided by partners (ELCOGAS, UPM and IJA-CSIC) of Agapute project (RFS-CR-04006,) and the support received from the European Commission are gratefully acknowledged.

REFERENCES

Christopher Frey H, Akunuri, N. (2001). Probabilistic Modeling and Evaluation of the Performance, Emissions, and Cost of Texaco Gasifier-Based Integrated Gasification Combined Cycle Systems Using ASPEN. Prepared by North Carolina State University for Carnegie Mellon University and U.S. Dept. of Energy, Pittsburgh.

See more of this Session: Advances in Gasification Research - II
See more of this Group/Topical: Energy and Transport Processes

Bioremediation

Dogga Singh, Chem, GEC, Jalandhar, India

Abe Singh, Chem, GEC, Jalandhar, India

See more of this Session: Contaminant Transport and Site Remediation
See more of this Group/Topical: Environmental Division

CHLOROFORM AEROBIC COMETABOLIC BIODEGARDATION IN A CONTINUOUS-FLOW REACTOR

Abstract ID#: 151306

Dario Frascari, Angelo Verboschi, Roberta Ciavarelli, Massimo Nocentini and Davide Pinelli. University of Bologna, Department of Chemical, Mining and Environmental Engineering, Bologna, Italy. email: dario.frascari@unibo.it

Introduction

Chlorinated aliphatic hydrocarbons (CAHs) are widespread subsurface contaminants [1,2]. The study of the CAH biodegradation began in the 1980s. CAH biodegradation can occur both under anaerobic (reductive dechlorination) and aerobic (oxidation) conditions. The anaerobic process is particularly interesting for its high potential of transformation of high-chlorinated CAHs [3,4]. On the other hand, aerobic biodegradation can lead to the complete and rapid dechlorination of low- and medium-chlorinated CAHs [5-16]. Besides, recent studies indicate that some aerobic strains can also perform the cometabolic degradation of CAHs with 4 chlorine atoms, such as 1,1,2,2-tetrachloroethane [13,17,18]. In most cases, aerobic CAH degradation occurs by cometabolism; therefore, it is necessary to provide a suitable electron donor (growth substrate) [5,6,10,12]. Indeed, the term "cometabolism" indicates the fortuitous degradation of a compound, not utilizable by the cell as a source of energy and carbon. This degradation is catalyzed by an enzyme produced by the cell in order to metabolize another compound, the primary substrate.

This study focuses on aerobic CAH biodegradation. Today, full-scale applications of CAH aerobic cometabolism are still rare, as several issues relative to the implementation of this technology to the remediation of CAH-contaminated sites still need to be addressed. Of these, two deserve particular mention: i) it is important to avoid the risk of a complete consumption of the supplied growth substrate within a short distance from the injection wells: in other words, the bioreactive zone must be sufficiently long to allow the complete biodegradation of the target contaminants; ii) aquifer clogging near the injection well can occur as a result of an excessive biomass growth on the growth substrate supplied. The supply of alternated pulses of growth substrate and oxygen represents an interesting solution, potentially effective both in the creation of a long bioreactive zone and in the control of aquifer clogging: as a result of hydrodynamic dispersion and substrate sorption, the over-lapping of substrate and oxygen occurs at low concentration, over a wide aquifer portion and, in each point, in a discontinuous way. This allows to control the biomass growth rate and to attain a wider bioreactive zone.

This study, focused on chloroform (CF) cometabolism by butane-grown bacteria, was conducted in slurry microcosms and in a 2-m continuous-flow column reactor simulating a portion of saturated aquifer. The main goals were: a) to evaluate the capacity of the indigenous biomass of a sandy aquifer to grow on butane and to degrade CF via aerobic cometabolism; b) to investigate the pulsed injection of growth substrate and oxygen as a tool to control clogging of the porous medium and to attain a wide bioreactive zone; and c) to identify a suitable kinetic and fluid-dynamic model to fit the experimental data of butane utilization and CF cometabolism.

Materials and Methods

I – Slurry microcosm tests. The first part of the study, conducted in 119-mL slurry microcosms, was aimed at evaluating the capacity of the indigenous biomass of a sandy aquifer to grow on butane and to degrade CF via aerobic cometabolism. Each microcosm contained 20 g (wet weight) of a non-polluted sandy soil, 50 mL of non-contaminated groundwater and 60 mL of headspace air. Microcosms were divided into two groups characterized by different experimental conditions. In the first group, the microcosms were initially spiked with only butane (0.1, 0.25, 0.50, 1 or 1.5 mg/L in the aq. phase), and CF (1 mg/L in the aq. phase) was added after 5-6 pulses of butane utilization (butane being supplied in each pulse at the same concentration of the initial pulse). In the second group, after a 32- or 98-day pre-exposition to CF (1, 5 or 10 mg/L) the bioreactors were spiked with butane (1.5 mg/L). Upon completion of the first butane pulse, butane (1.5 mg/L) was re-supplied in consecutive pulses, until the attainment of the complete degradation of the CF initially supplied. In each group, different sets of microcosms were operated at 15 °C (yearly average temperature of the aquifer where the groundwater was sampled), 25 °C or 30 °C (the optimal temperature for the metabolic activity of several mesophiles). Each condition was studied in triplicate microcosms. As an example, the butane and CF concentrations relative to the initial 35 days of operation of a microcosm of group 1 operated at 30 °C are shown in Fig. 1. The goal of groups 1 and 2 was to study the effect of different operational conditions on the lag time required by the indigenous biomass of the studied sandy aquifer to start utilizing butane and degrading CF. The experimental condition of group 2 aimed at simulating an in-situ treatment with growth substrate introduction in the contaminated zone some time after the contamination has occurred. The purpose of group 1 was to simulate a process where the specialized biomass develops before the introduction or the arrival of the contaminant, such as an on-site biological pump-and-treat system. All the experimental details relative to the set-up and operation of these microcosms are described in reference [10].

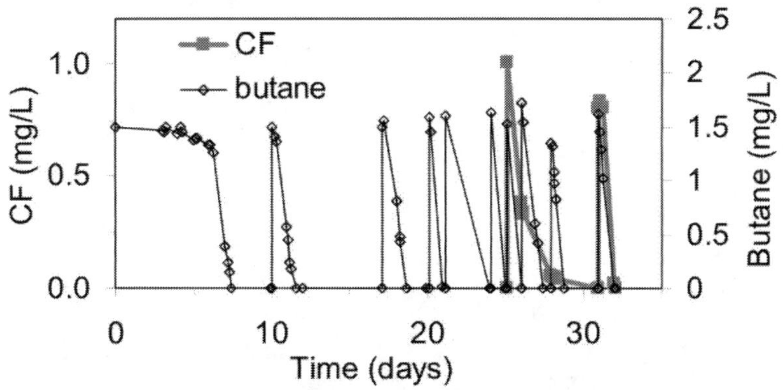

Fig. 1. CF and butane aqueous phase concentration vs. time during the initial 35 days of operation of a group 1 microcosm, operated at 30 °C.

II – Pure-culture tests in liquid-gas microcosms. The second part of the study, aimed at identifying a suitable kinetic model for the studied CF cometabolic biodegradation, was conducted with the single CF-degrading strain that resulted to be predominant among the cultivable strains found in the above-described slurry microcosms. The strain proved to be related to the genus *Rhodococcus* and to the species *R. aetherovorans*. It was named BCP1. The study, described in detail in reference [6], consisted in pure-culture, resting-cell degradation tests conducted with only butane, only CF or with both compounds. The experimental data were interpreted by means of a Monod-type model with competitive inhibition and transformation product toxicity, indicated by several authors as the most appropriate for the aerobic cometabolism of chlorinated solvents [14-16]. The model included, in the butane and CF kinetic equations, a second Monod-like term relative to the effect of the oxygen concentration. This term, which set equal to 1 in the pure-culture tests where oxygen limitations were avoided, was introduced to take into account the influence of the presence or absence of oxygen in the subsequent flow-reactor tests, fed with alternated butane and oxygen pulses. Accordingly, the following kinetic expressions were utilized respectively for butane utilization, CF degradation, cell growth and oxygen consumption:

$$R_B = -\frac{q_{\max,B} \cdot B}{K_{s,B} \cdot \left(1 + \frac{CF}{K_{i,CF}}\right) + B} \cdot \frac{O}{K_{s,O-B} + O} \cdot X \qquad (1)$$

$$R_{CF} = -\frac{q_{\max,CF} \cdot CF}{K_{s,CF} \cdot \left(1 + \frac{B}{K_{i,B}}\right) + CF} \cdot \frac{O}{K_{s,O-CF} + O} \cdot X \qquad (2)$$

$$R_X = -Y \cdot R_B - b \cdot X + \frac{R_{CF}}{T_{c,CF}} \qquad (3)$$

$$R_O = 4 \cdot R_B + 1.5 \cdot R_{CF} - k_{O,abio} \cdot O \qquad (4)$$

where $q_{max,B}$ (mg/(mg$_{protein}$ day)), B (mg/L) and $K_{s,B}$ (mg/L) refer respectively to the maximum specific degradation rate, aqueous phase concentration and half-saturation constant for butane, $q_{max,CF}$, CF and $K_{s,CF}$ to the corresponding parameters for chloroform, $K_{i,CF}$ (mg/L) to the constant of chloroform inhibition on butane utilization, $K_{i,B}$ (mg/L) to butane inhibition on chloroform utilization, C_O to the oxygen concentration, $K_{s,O\text{-}B}$ and $K_{s,O\text{-}CF}$ to the oxygen half-saturation constant in the butane and CF kinetic expression, X (mg$_{protein}$/L) to the cell concentration, Y (mg$_{protein}$/mg$_{butane}$) to the cell growth yield, b (1/day) to the endogenous decay rate, and $T_{c,CF}$ (mg$_{CF}$/mg$_{protein}$) to the chloroform transformation capacity, defined as the amount of chloroform transformed per unit amount of cells inactivated by the toxic transformation products or by the depletion of reducing energy (NADH). $k_{O,abio}$ represents the first-order constant of oxygen abiotic consumption, as a result of the oxidation of reduced species present in the soil or groundwater.

III – Flow-reactor tests. The third and main part of the study was conducted in a 2-m, 35-L column reactor filled with the same sandy soil utilized in the slurry microcosms and saturated with tap water (Fig. 2). The column is provided with 40 sampling ports that allow to monitor the concentration profiles of the different compounds. In order to operate a pulsed injection of primary substrate (butane) and oxygen, a reactor feeding system was set up, consisting of 3 gas-tight LDPE bags connected to the main feeding pipe by means of 2 electro-valves controlled by 2 timers. The bags were periodically filled with tap water; besides, one bag was nitrogen-stripped to remove the dissolved oxygen and additioned with butane; the second one was nitrogen-stripped but not supplied with butane; the third was additioned with pure oxygen so as to attain the desired oxygen concentration. After these additions, all the bags were supplied with CF. The column, operated at 30 °C, was fed by means of a peristaltic pump. Butane and CF concentrations were determined with an HP6890 gaschromatograph equipped with a FID and an ECD. The details of the gaschromatographic methods are described in reference [9]. Oxygen concentration was determined with an OXYSIS OxC polarographic electrode. To interpret the experimental data under flow conditions, the following mass-balance was utilized for butane, CF and oxygen, based on the hypothesis of equilibrium adsorption and of homogeneous conditions in each transversal section:

Fig. 2. Photo of the 2-m flow-reactor simulating a portion of saturated aquifer.

$$\delta_i \frac{\partial C_i}{\partial t} = -v_z \cdot \frac{\partial C_i}{\partial z} + D_{L,i} \cdot \frac{\partial^2 C_i}{\partial z^2} + R_i \qquad (5)$$

where δ_i indicates the retardation factor (expressing the compound's adsorption to the soil), v_z the interstitial velocity, equal to 0.49-0.50 m/d in all the tests, R_i the degradation term (given by eqs. 1, 2 and 4) and $D_{L,i}$ the longitudinal dispersion coefficient, given by $D_{mol,i} + \alpha_L v_z$ ($D_{mol,i}$ = molecular diffusion coefficient; α_L = longitudinal dispersivity). As for the biomass, we utilized a modified form of the mass balance, based on the following assumptions: i) there are two type of biomass: one, attached to the soil, is not subject to advection or dispersion (X_a, $mg_{protein}/kg_{dry\ soil}$); the other, suspended in the groundwater, is subject to both advection and dispersion (X_s, $mg_{protein}/L_{pore\ volume}$); ii) the concentrations of the two types of biomass are regulated by an equilibrium adsorption: $X_a = k_X X_s$; iii) both types of biomass are active on butane and CF and are exposed to the concentration of each compound present in the core of the liquid phase (*B*, *CF* and *O* in eqs. 1 and 2); therefore, the total biomass concentration (*X* in eqs. 1-3, $mg_{protein}/L_{pore\ volume}$) is given by $X_s + X_a\, \rho_b/\Phi$ (where ρ_b indicates the soil bulk density and Φ the porosity). The resulting biomass mass balance is therefore:

$$\left(1 + \frac{\rho_b k_X}{\Phi}\right) \frac{\partial X_s}{\partial t} = -v_z \cdot \frac{\partial X_s}{\partial z} + D_{L,X} \cdot \frac{\partial^2 X_s}{\partial z^2} + R_X \qquad (6)$$

A major difference between the pure-culture batch tests and the flow reactor tests is that the latter are characterized by the presence of a bacterial consortium. Preliminary pure-culture tests showed that, while all the culturable strains of the consortium proved able to grow on butane, only *Rhodococcus aetherovorans* BCP1 showed a significant CF degradation capacity. Therefore, the flow-reactor tests were modelled by assuming that, both in the suspended and in the attached biomass, only a fraction f_{CF} of the total biomass is active on CF degradation. If one assumes that the two types of biomass (CF-degrading and non CF-degrading) are equally inactivated by the toxic CF degradation products, and that they are characterized by the same butane-degrading capacity (same $q_{max,B}$, $K_{s,B}$ and Y), then the two biomasses follow the same kinetic equation of biomass growth and death/inactivation. Therefore, a single biomass kinetic equation (eq. 3) can be utilized, relative to the sum of the two types of biomass. Under these assumptions, the CF kinetic equations becomes:

$$R_{CF} = -\frac{q_{\max,CF} \cdot CF}{K_{s,CF} \cdot \left(1 + \frac{B}{K_{i,B}}\right) + CF} \cdot \frac{O}{K_{s,O-CF} + O} \cdot f_{CF} X \qquad (7)$$

Results and Discussion

I – Slurry microcosm tests. Butane utilization was observed in all the tests, with lag-times varying between 1.5 days (30 °C, 0.1 mg_{butane}/L) and 13 days (15 °C, 15 mg_{butane}/L). The butane lag-time showed an apparent tendency to increase with butane initial concentration, but it proved influenced mostly by temperature (at 1.5 mg/L, the butane lag increased from 3.0 days at 30 °C to 9.1 days at 15 °C) and CF concentration (at 1.5 mg/L and 30 °C, the butane lag increased from 3.0 days in the absence of CF to 9.4 days in the presence of 10 mg_{CF}/L). The CF lag-time proved significantly influenced by both temperature and concentration of butane-utilizing biomass at the time of CF addition. Indeed, in all the tests performed at 25 or 30 °C, when CF was added after 5-6 pulses of biomass growth on butane (group 1) its degradation began immediately. Conversely, when it was introduced before or simultaneously to the addition of butane (group 2) its degradation began after a lag-time (calculated from the time of butane addition) variable, at 25 or 30 °C, between 24 and 45 days. In the tests performed at 15 °C no CF degradation was observed within 2-3 months, independent of the biomass condition at the moment of CF addition. The immediate onset of CAH cometabolic degradation when the specialized biomass has previously grown on the primary substrate in the absence of CAHs is in agreement with analogous literature studies [9,19,20]. Conversely, when CF was introduced before supplying butane (group 2), the CF lag-times proved rather long (up to 45 days, at 30 °C). This, too, agrees with the results of another study of cometabolic CAH degradation [21]. These agreeing results suggest that the studied indigenous consortium contains a small fraction of bacteria capable of transforming the target CAHs via cometabolism. Consequently, CAH depletion becomes detectable only when this fraction has grown on the primary substrate up to a sufficient concentration. The minimum ratio of butane utilized to CF degraded (B/CF ratio) required to sustain the cometabolic process resulted equal to 1.6 mg_{butane}/mg_{CF}. The results of these tests are reported in detail in reference [10].

II – Pure-culture tests in liquid-gas microcosms. All the experimental data obtained in these tests resulted to be satisfactorily interpretable with the above-described kinetic model (in the absence of oxygen limitations). As an example, Fig. 3 shows the initial specific butane rates calculated from the assays of butane utilization as well as the initial specific CF rates calculated from the assays of chloroform degradation by butane-grown BCP1. The butane rates were interpreted by means of equation (1) with c_{CF} = 0, the CF rates by means of equation (2) with c_B = 0. The resulting best estimates of the butane and CF kinetic parameters are reported in Table 1. The CF degradation activity performed by BCP1 resulted in the complete conversion of the organic Cl into chloride ion. The results of these tests are reported in detail in reference [6].

Table 1 – Best estimates of the butane and CF kinetic parameters relative to the pure culture BCP1.

$q_{max,B}$	mg_B / (mg_{prot} day)	6.84
$K_{s,B}$	mg_B/L	1.16
$K_{i,B}$	mgB/L	0.435
Y	mg_{prot}/mg_B	0.21
b	1/d	0.007
$q_{max,CF}$	mg_{CF} / (mg_{prot} day)	2.63
$K_{s,CF}$	mg_{CF}/L	49.8
$K_{i,CF}$	mg_{CF}/L	49.8
$T_{c,CF}$	mg_{CF} / mg_{prot}	1.25

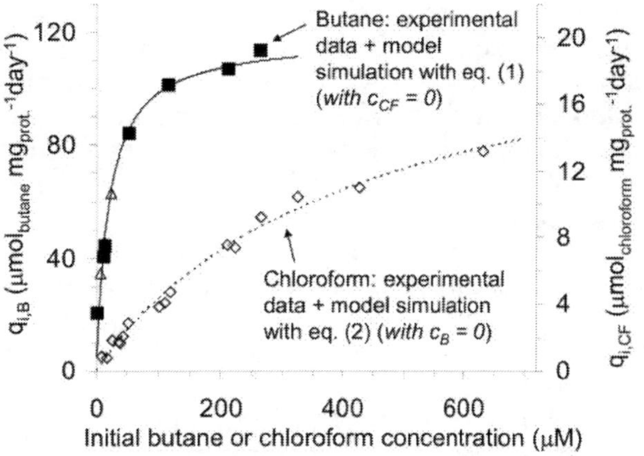

Fig. 3 - Initial specific butane utilization rate ($q_{i,B}$) and chloroform degradation rate ($q_{i,CF}$) versus initial butane and chloroform concentration for butane-grown BCP1 in the single-compound resting-cell tests.

III – Flow-reactor tests. A preliminary group of fluid-dynamic tests was aimed at determining the main fluid-dynamic parameters required for the modeling of the process: the reactor effective porosity Φ and longitudinal dispersivity α_L, and the retardation factors δ_i relative to butane and CF (assuming δ = 1 for oxygen). These tests consisted of pulses of only oxygen (with no butane or CF in the column), only butane or only CF. For each compound, different pulses at different interstitial velocities were operated. On the basis of preliminary simulations of the experimental data by means of a PDE solver (Comsol Multiphysics), the following best-estimates were obtained: Φ = 50%, α_L= 3.5 mm, δ_{butane} = 1.08, δ_{CF} = 1.00. In addition, a specific experiment yielded a best estimate of $k_{O,abio}$ = 0.14 1/day.

The column reactor was then run in continuous mode for about 6 months, subdivided into 3 periods of operation characterized by subsequent changes of the schedule of pulsed injection of oxygen and butane, with the goal to maximize CF degradation and to minimize butane consumption. The interstitial velocity was fixed at 0.49-0.50 m/day, while the CF inlet concentration varied between 0.2 and 0.4 mg/L.

The first period, which had an overall length of 105 days, began with a pulsed injection articulated on a 2-day cycle (CF + butane 11 mg/L for 15 h, only CF for 6 h, CF + oxygen 10.5 mg/L for 20 h, only CF for 7 h). This injection procedure led to an average CF removal equal to about 40%, calculated as $1 - CF_{IN}/\overline{CF}_{OUT}$, where $\overline{CF}_{OUT}$ indicates the average outlet concentration. A characteristic feature of the pulsed injection is that, in each instant, biomass growth occurs only in those parts of the reactor where an interface between an oxygen pulse and a butane pulse occurs. Thus, in order to increase biomass growth, in a modification of the initial injection schedule the duration of each pulse was halved, so as to double the number of oxygen-butane interfaces present in the 2-m reactor in each instant. A representative profile of the concentrations of butane, oxygen and CF over the column during this period is provided in Fig. 4, together with the corresponding model simulation performed with the kinetic parameters derived from the single-strain kinetic study and with the above-reported preliminary estimates of the fluid-dynamic parameters. In these preliminary simulations, the values of the other model parameters (biomass partition coefficient k_X, and oxygen half-saturation constants in the butane and CF kinetic expressions, $K_{s,O\text{-}B}$ and $K_{s,O\text{-}CF}$) were determined on the basis of a literature search. Therefore, the representative simulation reported in Fig. 4 does not represent a best-fit of the complete model on the experimental data. Indeed, the development of a suitable tool for performing the best-fit of the complete model is in progress. It can be observed in Fig. 4 that, at a given instant, CF cometabolic degradation occurs only in the reactor zones characterized - as a result of the pulsed injection - by the presence of oxygen. In these zones, CF degradation is further favored by the lack of butane inhibition. Due to convection and dispersion, these zones shift through the reactor. We therefore obtained a CF outlet concentration characterized by a fluctuating trend. Fig. 4 also shows that, with the modified injection schedule (1-day cycle), oxygen was completely depleted before the end of the 2-m column. Therefore, in a further modification of the pulsed injection the butane and oxygen concentrations were increased (butane 21-22 mg/L, oxygen 17 mg/L). As a result, each oxygen pulse crossed the entire column length before being completely depleted. With this modification, the CF degradation obtained with each passage through the column was raised to 50-55%. In the last portion of the 1st period of operation, the butane inlet concentration was reduced to about 10 mg/L. This change led to a further improvement of the average CF removal, which resulted equal to about 65%. It can thus be concluded that the last version of the pulsed injection allowed the development of a bioreactive zone over the entire length of the column (2 m) and prevented any measurable clogging of the porous medium. On the basis of the experimental data, the cometabolic process appeared to be sustainable under these conditions.

The second period of operation, which lasted 46 days, was characterized by a drastic reduction of the amount of butane supplied (about 33 mg/week, versus 200 mg/week in the last part of the first period), with a corresponding increase of the oxygen supplied. This change was obtained by re-defining the injection schedule on a 1-week cycle (CF + butane 15 mg/L for 12 h, only CF for 2.5 h, CF + oxygen 20 mg/L for 151 h, only CF for 2.5 h). Thanks to a significant increase of the extension of the aerobic portion of the column (longer oxygen pulse), a marked increase of the CF removal was obtained, the average CF removal during the first 20 days resulting equal to about 80%. However, after 20 days of operation under these conditions, the CF degradation rate decreased and the cometabolic process rapidly halted. Further experimental and modeling activities aimed at explaining this result are in progress. A possible explanation is that, as a result of the excessively low ratio of butane consumed to CF degraded, the cellular storage of NADH was progressively consumed.

In the third period of operation, which lasted 81 days, the duration of each pulse was equal to half the value characteristic of the 2nd period, so as to double the number of oxygen-butane interfaces present in the 2-m reactor in each instant. The resulting injection schedule was therefore based on a 3.5-day cycle. Under these conditions, the cometabolic process resumed rapidly and remained stable over time, with an average 80% CF removal. No sign of aquifer clogging was observed.

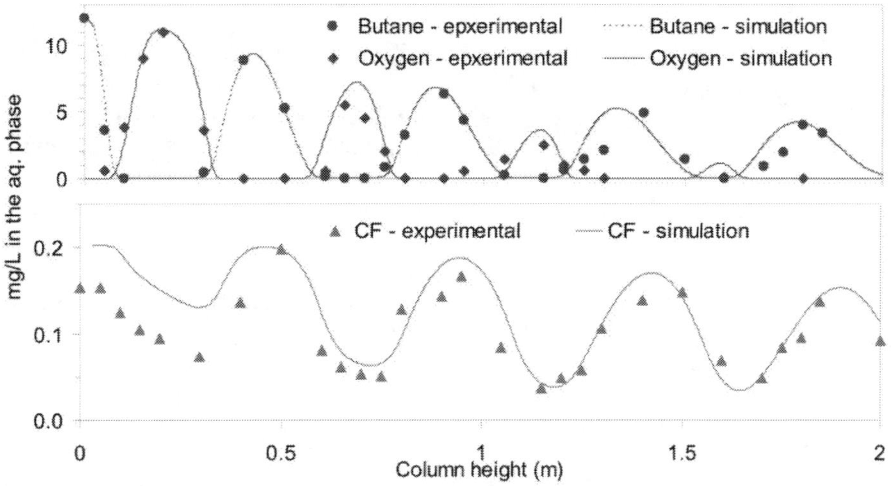

Fig. 4. Experimental and simulated profiles of butane, oxygen and CF in the flow reactor, during the first period of operation.

Conclusions

a) The preliminary batch tests showed that the indigenous biomass of the studied aquifer is capable to rapidly start to grow on butane, and that it contains at least one strain, named *Rhodococcus aetherovorans* BCP1, capable to degrade CF with a complete conversion of the organic Cl into chloride ion.
b) CF cometabolic degradation by BCP1 was satisfactorily interpreted with the model of aerobic cometabolism described by Chang and Alvarez-Cohen [15].
c) In the flow-reactor tests, the pulsed injection of growth substrate and oxygen proved an effective tool to prevent aquifer clogging as a result of an excessive biomass growth, and to attain a long bioreactive zone.
d) The kinetic parameters estimated in the batch, single-strain assays, combined with the fluid-dynamic parameters evaluated in the preliminary column tests, allowed a satisfactory interpretation of the experimental data obtained in the flow-reactor, mixed-culture tests.
e) During the 3 periods of operation of the flow-reactor, after several changes of the schedule of substrate and oxygen pulsed injection, a satisfactory injection scheme was attained,

which resulted in an 80% average CF removal and in an apparently sustainable cometabolic process.

Overall, this work provides encouraging indications on the successful application of aerobic cometabolism for the in-situ remediation of sites contaminated by a wide range of CAHs.

References

1. Chang HL, Alvarez-Cohen L. 1996. Biodegradation of individual and multiple chlorinated aliphatic hydrocarbons by methane-oxidizing cultures. Appl Environ Microbiol 62:3371-3377.
2. US EPA. 2000. Engineered approaches to in-situ bioremediation of chlorinated solvents: fundamentals and field applications.
3. Cupples AM, Spormann AM, McCarty PL. 2004. Comparative evaluation of chloroethene dechlorination to ethene by Dehalococcoides-like microorganisms. Environ Sci Technol 38:4768-4774.
4. Yu S, Dolan ME, Semprini L. 2005. Kinetics and inhibition of reductive dechlorination of chlorinated ethylenes by two different mixed cultures. Environ Sci Technol 39:195-205.
5. McCarty PL, Goltz MN, Hopkins GD, Allan JP, Dolan ME, Kawakami BT, Carrothers TJ. 1998. Full-scale evaluation of in-situ cometabolic degradation of trichloroethylene in groundwater through toluene injection. Environ Sci Technol 32:88-100.
6. Frascari D, Pinelli D, Nocentini M, Pii Y, Fedi S, Zannoni D. 2006. Chloroform degradation by butane-grown cells of Rhodococcus aetherovorans BCP1. Appl Microbiol Biotechnol 73:421-428.
7. Frascari D, Pinelli D, Nocentini M, Baleani E, Cappelletti M, Fedi S. 2008. A kinetic study of chlorinated solvent cometabolic biodegradation by propane-grown Rhodococcus sp. PB1. Biochem Eng J, 42:139-147.
8. Semprini L, Dolan ME, Mathias MA, Hopkins GD, McCarty PL. 2007. Laboratory, field and modeling studies of bioaugmentation of butane-utilizing microorganisms for the in-situ cometabolic treatment of 1,1-dichloroethene, 1,1-dichloroethane and 1,1,1-trichloroethane. Adv Wat Res 30:1528-1546.
9. Frascari D, Zannoni A, Fedi S, Pii Y, Zannoni D, Pinelli D, Nocentini M. 2005. Aerobic cometabolism of chloroform by butane-grown microorganisms: long-term monitoring of depletion rates and isolation of a high-performing strain. Biodegradation 16:147-158.
10. Frascari D, Zannoni A, Pinelli D, Nocentini M. 2007. Chloroform aerobic cometabolism by butane-utilizing bacteria in bioaugmented and non-bioaugmented soil/groundwater microcosms. Process Biochem 42:1218-1228.
11. Meza L, Cutright TJ, El-Zahab B, Wang P. 2003. Aerobic biodegradation of trichloroethylene using a consortium of five bacterial strains. Biotechnol Letters 25:1925-1932.
12. Gualandi G, Frascari D, Pinelli D, Nocentini M. 2007. Growth of chlorinated solvent-degrading consortia in fed-batch bioreactors and development of a double-substrate high-performing microbial inoculum. Engineering Life Sci 7:217-228.

13. Frascari D, Pinelli D, Nocentini M, Zannoni A, Fedi S, Baleani E, Zannoni D, Farneti A, Battistelli A. 2006. Long-term aerobic cometabolism of a chlorinated solvent mixture by vinyl chloride-, methane- and propane-utilizing biomasses. J Hazard Mater 138:29-39.
14. Alvarez-Cohen L, Speitel Jr. GE. 2001. Kinetics of aerobic cometabolism of chlorinated solvents. Biodegradation 12:105-126.
15. Chang HL, Alvarez-Cohen L. 1995. Model for the cometabolic biodegradation of chlorinated organics. Environ Sci Technol 29:2357-2367.
16. Arp DJ, Yeager CM, Hyman MR. 2001. Molecular and cellular fundamentals of aerobic cometabolism of trichloroethylene. Biodegradation 12:81-103.
17. Shim H, Ryoo D, Barbieri P, Wood TK. 2001. Aerobic degradation of mixtures of tetrachloroethylene, trichloroethylene, dichloroethylenes and vinyl chloride by toluene-o-xylene monooxygenase of Pseudomonas stutzeri OX1. Appl Microbiol Biotechnol 56:265-269.
18. Frascari D, Meniconi A, Nocentini M, Pinelli D. 2008. 1,1,2,2-tetrachloroethane aerobic cometabolic biodegradation by propane- and methane-utilizing consortia in slurry bioreactors. In: Proceedings of the 10th International UFZ-Deltares/TNO Conference on Soil-Water Systems – ConSoil 2008 (Milano, June 3-6, 2008), F&U Confirm Publisher, Leipzig, Germany, pp E617-E626 (on CD).
19. Frascari D, Kim Y, Dolan ME, Semprini L. 2003. A kinetic study of aerobic propane uptake and cometabolic degradation of chloroform, cis-dichloroethylene and trichloroethylene in microcosms with groundwater/aquifer solids. Water Air Soil Pollution: Focus 3:285-298.
20. Tovanabootr A, Semprini L. 1998. Comparison of TCE transformation abilities of methane- and propane-utilizing microorganisms. Bioremediation Journal 2:105-124.
21. Jitnuyanont P, Sayavedra LA, Semprini L. 2001. Bioaugmentation of butane-utilizing microorganisms to promote cometabolism of 1,1,1-trichloroethane in groundwater microcosms. Biodegardation 12:11-22.

SORPTION of Explosive Chemicals to Leaves and Leaf Litters of Tropical Plants FROM Hawaii

Matteo D'Alessio, Civil Environmental Engineering, University of Hawaii at Manoa, Honolulu, HI

Chittaranjan Ray, Civil Environmental Engineering, University of Hawaii at Manoa, Honolulu, HI

Abstract: Several areas in Hawaii contain unexploded ordnance (UXO) as a consequence of military training during the Second World War. This study was conducted to investigate the sorption of UXO chemicals to leaves and to examine their transport in the litter of different plants in Hawaii. The explosives studied, were selected based on their physico-chemical properties, frequency of occurrence in the environmental, and the environmental fate. Batch sorption experiments were conducted with 2,4,6-trinitrotoluene (TNT), 2,4-dinitrotoluene (2,4 DNT), hexahydro-1,3,5-trinitro-1,3,5-triazine (RDX), and 1,3,5,7-tetranitro-1,3,5,7-tetrazocane (HMX)using dried leaves of Casuarina, Araucaria heterophylla, and Eucalyptus and using litter, top soil, leaves and branches, of Casuarina, Araucaria and grass. Saturated column transport experiments were conducted with TNT, 2-4-DNT, RDX, HMX and bromide in the litters of Casuarina, Araucaria and grass. For all the UXO compounds, breakthrough curves were observed later than that for the conservative tracer indicating adsorption to the solid phase. Introduction and Background: Unexploded ordnance (UXO) are military munitions that have been prepared for action, fired, dropped or buried and remain undetonated, posing a hazard to operations, personnel, or material [1]. In the US, the Army alone has estimated that over 1.2 million tons of soil has been contaminated with explosives, and the impact of explosives contamination in other countries is of similar magnitude [2]. The primary release mechanism includes corrosion, leaking, or rupture of the containment system. Once the UXO containment system has been breached, dissolution of the explosive occurs. In the case of rupture and resulting leakage, the explosives may exist as a free product in the soil due to spillage or may be partially contained within the delivery system. After dissolution of the explosives has begun, fate and transport processes, such as advection, dispersion, adsorption-desorption, diffusion, biotic transformation, oxidation-reduction, covalent binding, polymerization, photolysis, infiltration and plant-root uptake, interact with the dissolved contaminant [1]. TNT and DNTs are classified as nitroaromatic explosives having ring structures, whereas RDX and HMX are nitramine explosives possessing N-nitro groups [3]. TNT and DNTs have higher octanol - water partition coefficients (Kow) values than RDX, suggesting that they are strongly bound to soil organic matter, whereas RDX is mobile due to poor sorption [4]. The environmental fate of these compounds is different. TNT tends to accumulate in plant roots (more then 75%), only a small portion, less than 25% is translocated into the aerial parts of plants. More than 80% of RDX taken up is translocated to aerial tissues. Several areas of Hawaii have significant UXO contamination as a consequence of military training. There is a continuing need for the military to be trained at live-fire ranges, and these military training lands are often located in tropical islands, for example, in Hawaii. The presence of unexploded ordnance and the release of chemical compounds are the main causes of environmental concern. At firing ranges, leaves and litter are particularly susceptible to contamination with energetic compounds such as nitroaromatic and nitroamine explosives. Contaminants present on the surface of leaves and on plant litter can be easily transported by natural means, for example birds, and animals and enter the food chain, soil, and water. Approach / Experimental: Chemical analysis of TNT, RDX, HMX and 2,4-DNT were performed using high performance liquid chromatography (HPLC) according to EPA Method 8330. The mobile phase was 43:57 methanol:water (vol:vol). Bromide breakthrough curves were measured by a Dionex DX-120 ion chromatography (IC) system. Kinetic experiments were conducted to determine the length of time necessary for the UXOs in solution to obtain sorptive equilibrium with samples of leaves or litters. The experiments were performed in batch reactors that were 40 mL glass vials stoppered with Teflon caps. Experiments were performed in triplicate for each kind of leaf and litters tested. For Casuarina leaves, 0.25 g of leaf was combined with a liquid phase of 20 mL of deionized water (1 mM $CaCl_2$) containing the UXO stock solution (concentration of 1 mg/L). After preparation, the reactors were placed on a rotating shaker at 10 rpm in room temperature. The 40 mL glass vials were withdrawn and tested at intervals of: 1, 2, 4, 8, 16, 24 and 48 h. Before analysis, each vial was centrifuged for 30 min at 5000 rpm (G force is 560 units) in order to separate the solid and liquid fractions for testing. Sorption and desorption isotherms for each explosive compound were determined for Casuarina, Araucaria and grass. The experiments were performed as described for the kinetic studies, except that different concentrations (1 mg/L, 0.75 mg/L, 0.5 mg/L, 0.25 mg/L and 0.1 mg/L) of UXOs solution were tested and the aqueous concentration of each was measured after the equilibrium time (16 hr) determined from the kinetic study. For the all the litters, 2 g of litter were mixed with 20 mL of deionized water (0.1 mM CaCl2). The same contact time of 16 hours was used for the three litters. Linear distribution coefficients (Kd) for the sorption of explosive compounds to the uncontaminated leaves and litters were determined for triplicate samples. Three flux-controlled flow-through columns were used in this study. They consisted of 4.70 cm internal diameter by 7.60 cm height stainless steel columns (Alltech HPLC Columns) and caps on the top and the bottom. An aqueous tracer (bromide) and UXO solution was introduced into the columns via a valve in the top. A Series I HPLC Digital Pump Lab Alliance (State College, PA), with Flow Accuracy 3% and Flow Precision 0.5 % RSD, was used to deliver the solution containing UXO on top of each column. Outflow samples were collected continuously into 20-mL vials using an automatic fraction collector. The fraction collector, model Retriever II from Teledyne Isco (Lincoln, NE), and the bottles containing background solution (1 mM CaCl2) and the contaminant solution were placed in a cooler. Low temperature was maintained to slow the degradation of chemicals in the influent and the effluent. Results and Discussion: For sorption of all the UXOs compounds to Casuarina equilibrium was achieved within 12 hours; however, in order to be certain that sorptive equilibrium was achieved during the experiments a contact time of 16 hours was used in all sorption tests performed. The Kd values for all the UXOs compounds were less than 0.5 mL/g for Casuarina, indicating that there was little interaction between the explosives and the leaves. A Kd of 3.2 mL/g was found for Eucalyptus for TNT, which could be partly due to degradation of TNT. No significant value of the linear adsorption coefficient was found in Araucaria. The Kd for all the litters was higher than 1 mL/g for HMX, RDX and 2-4DNT. The Kd values Grass and Araucaria heterophylla were similar, while Casuarina showed higher Kd value. RDX, HMX and 2,4-DNT had the same linear trend for all the litters. TNT degraded in all the litters. Breakthrough of the conservative tracer (bromide) occurred before one pore volume, indicating the absence of preferential pathways for flow. No air entrainment within the column was observed. For RDX and HMX breakthrough curves were observed later than for the conservative tracer indicating adsorption to the solid phase. No breakthrough curves where observed for TNT and DNT. One of the possible reasons is that the flow was stopped early. Summary and Conclusions: The study shows differing sorption behavior of the UXO compounds in litter and in leaves. Adsorption is stronger to litters than leaves. The Kd for HMX, RDX and 2,4-DNT follows the order: Grass > Araucaria heterophylla > Casuarina. The breaktrough curves of the three litters showed similar behavior. Acknowledgements: We acknowledge help of Mr. Joseph Lichwa of the Water Resources Research Center, University of Hawaii for helping in the analysis of the UXO chemicals. References: [[1] Conceptual Model and Process Descriptor Formulations for Fate and Transport of UXO, US Army Corps of Engineers, Technical Report IRRP-99-1 February 1999. [2] Lewis, T.A.; Newcombe, D.A.; Crawford, R.L. Bioremediation of soils contaminated with explosives. J. Environ. Mgmt. 2004, 70 (4), 291-307. [3] Hannink NK et al. (2002) Phytoremediation of explosives. Critical Reviews in Plant Sciences 21 511-538. [4] Jong Moon Yoon, David J. Oliver and Jacqueline V. Shanks Iowa State University Ames, IA

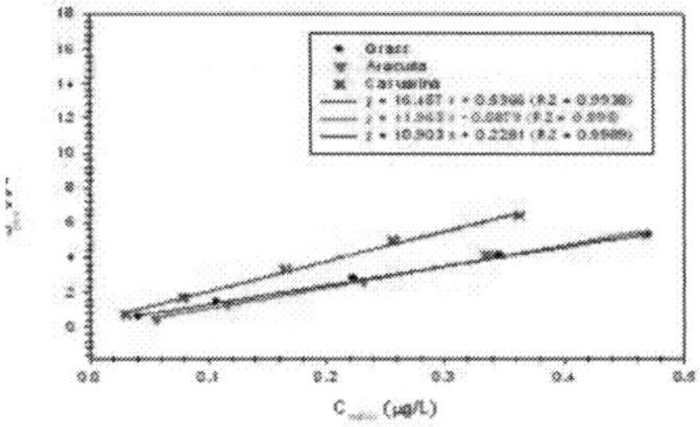

See more of this Session: Contaminant Transport and Site Remediation
See more of this Group/Topical: Environmental Division

Adsorption Studies of Endosulfan, Chloropyrifos and Monocrotophos Pesticides From Aqueous Solution Using Activated and Unactivated Cocoa Pod Husk

Jude C. Igwe, Industrial Chemistry, Abia State University, Uturu, Nigeria

Nnenna E. Okoronkwo, Industrial Chemistry, Abia State University, Uturu, Nigeria

Stephen Ihekweme, Industrial Chemistry, Abia State University, Uturu, Nigeria

Increased agricultural practice and pesticide application has resulted in the contamination of natural environments with different kinds of pesticides. The removal of toxic pesticides such as Endosulfan, Chloropyrifos and Monocrotophos by inexpensive biomaterial termed biosorption requires that the substrate displays high pesticide uptake and suitable mechanical properties for the applied remediation technique. The kinetics, intrparticulate diffusivities and adsorption isotherms for the adsorption of Endosulfan, Chloropyrifos and Monocrotophos onto activated and unactivated cocoa pod husk was investigated. The amount of the pesticides adsorbed increased as time and initial concentration was increased. Endosulfan and Chloropyrifos was maximally adsorbed at about 60 minutes with 37.0 and 37.25 mg/g respectively onto activated cocoa pod husk. The trend of the sorption capacity was found to be Chloropyrifos > Endosulfan> Monocrotophos for both activated and unactivated cocoa pod husk. Activation of the husk improved the sorption capacity. The mechanism of sorption of the pesticides followed particle diffusion model. The pseudo-first order and pseudo-second order kinetic equations were used to model the adsorption kinetics with the pseudo-second order giving a better fit to the sorption experimental data, as seen from the correlation coefficients (R2) which ranged from 0.2457 to 0.7913 for pseudo-first order and 0.9417 to 0.9980 for the pseudo-second order model. Among the three adsorption isotherms used, Dubinin-Radushkevich isotherm gave a better fit with R2 values ranging 0.6438 to 0.9639; followed by Freundlich isotherm (0.1101 to 0.9583) and then the Langmuir isotherm (0.0065 to 0.9797). The sorption process was found to be a chemisorption process as seen from the apparent energy of adsorption. Therefore, this study shows that activated and unactivated cocoa pod husk could be used to remove pesticides from aqueous solution and also established the mechanism of sorption.

Key words: Bioremediation, Adsorption, Pesticides, Kinetics, Isotherms.

See more of this Session: Contaminant Transport and Site Remediation
See more of this Group/Topical: Environmental Division

The Electrochemical Remediation of Pb-Contaminated Soils

Shih-Kai Chen, Chemical Engineering Department, National Taiwan University, Taipei, Taiwan

Yong-Fu Wu, Department of Chemical Engineering, Ming Chi University of Technology, Taipei, TN, Taiwan

Shi-Chern Yen, Chemical Engineering Department, National Taiwan University, Taipei, Taiwan

The electro-remediation for soils contaminated by heavy metals has been investigated in this study. The deasibility of applying an electric field with a constant or voltage across the soil specimen for removing Pb has been examined. As the electrolytic treatment is proceeded, H^+ and OH^- are generated at the anode and cathode, respectively, and a non-uniform pH distribution would be formed. Experimental results showed that the efficiencies of electro-treatments without buffering at the cathodic compartment were not high enough due to the pH variation across the soil specimen and the precipitation near the cathode. In order to control the pH distribution and improve the heavy-metal removal efficiency, a buffer solution was employed in the cathodic compartment. The Pb removal efficiency could attain about 85% after 32-day treatment.

See more of this Session: Contaminant Transport and Site Remediation
See more of this Group/Topical: Environmental Division

Characterization of the Effectiveness of CIAgent©, a Hydrocarbon Liquid Solidifier

Jola J. Solomon, Chemical Engineering, Auburn University, Auburn, AL

Thomas R. Hanley, Chemical Engineering, Auburn University, Auburn, AL

Solidifiers are composed of dry, high molecular weight polymers that have a porous matrix and a large oleophilic surface area. Solidifiers bond with the hydrocarbon liquids by molecular interaction (van der Waals forces, London forces, etc.) and are employed as a method of preventing spill propagation and dispersion in order to minimize the impact of hydrocarbon liquid spills.

CIAgent© is a non-toxic, proprietary polymer blend listed in EPS National Contingency Plan Product Schedule as "Solidifier" for use on oil spills in the navigable waterways of United States. CIAgent© can be used on oil, hydraulic fluid, transmission oil, or any petroleum-based fluids, solidifying the hydrocarbon liquid into a cohesive rubber-like, inert mass upon contact. Once solidified, the rubbery mass is easy to remove; keeping the spill contained and minimizing costs. The recovered solidified material can be reused as fuel or filler for other products or can be land filled.

Solidifier effectiveness is defined as the amount of agent required to solidify the hydrocarbon liquid under standard conditions. Hence, the solidifier needs to be characterized with respect to the effectiveness for the extensive hydrocarbon liquids that could be encountered.

This objective was achieved by measuring the heat of solidification using a solution calorimeter (Parr Instruments, Model No. 6755). Experiments were conducted at room temperature and atmospheric pressure. A Temperature-Time plot was obtained, and the temperature difference upon solidification was determined. The temperature change was correlated with the solubility parameters (dd, dp, and dh) of the hydrocarbon liquids and a three-dimensional map was generated to represent CIAgent© solidification effectiveness. The ease with which the CIAgent© solidifies the hydrocarbon liquid was determined from the calculated heat of solidification value. The release or adsorption of heat during the solidification process thus has implications for how CIAgent© applications should be conducted.

Keywords: solidification, three dimensional solubility parameter, solution calorimeter, temperature-time plot, heat of solidification

See more of this Session: Contaminant Transport and Site Remediation
See more of this Group/Topical: Environmental Division

Decontamination of Surfaces By Vapor Phase Hydrogen Peroxide

Marek Kuzma, Laboratory of Molecular Structure Characterization, Institute of Microbiology, Prague, Czech Republic, Jiří Švrček, Department of Organic Technology, ICT - Prague, Prague, Czech Republic, Jaroslav Červený, Laboratory of Molecular Structure Characterization, Institute of Microbiology, Prague, Czech Republic, Kamila Syslová, Laboratory of Molecular Structure Characterization, Institute of Microbiology, Prague, Czech Republic, Petr Kačer, Department of Organic Technology, ICT - Prague, Prague, Czech Republic, Libor Červený, Department of Organic Technology, ICT - Prague, Prague, Czech Republic, Libor Pánek, BLOCK a.s., Prague, Czech Republic

Introduction

In recent history, chemical and biological counter-terrorism have come to the forefront of defense due to the immense threat to public health and national security. In the last decade a number of chemical and/or biological attacks on civilian populations have occurred, including the release of sarin nerve gas in a Tokyo subway system (1995) and the release of anthrax spores through the mail system in the U.S.A. (2001). Advances in biotechnology have provided the capability for terrorist groups to produce bio-weapons in makeshift laboratories, secretly and inexpensively.

It is evident that after the release of any chemical and/or biological agent some level of decontamination is required before normal activities on the site affected can resume. There are several decontamination methods – wet, vapor and physical decontamination methods.

Particular decontamination methods were tested and approved for some applications and have their advantages and disadvantages. An application of wet decontamination method is limited to small places and/or small instruments. Physical methods based on ionizing radiation are dependant on such a beam source. On contrary, use of vapor decontamination methods is quite easy and can be applied for a wide space. This method is based on application of an appropriate decontamination agent in the vapor (gas) phase to clean spaces or equipment. Several decontamination agents exist, like chlorine dioxide, ethylene oxide, methyl bromide, hydrogen peroxide, etc. The effectiveness of decontamination procedure depends on many parameters, e.g. type of agent, its concentration, exposure time, reaction rate, temperature, humidity, etc. Some of the agents have considerable risks which make them unsuitable for use in indoor environments, for example ethylene oxide and propylene oxide are flammable at working concentrations and need to be used in carefully controlled chambers.

Vapor phase hydrogen peroxide is very progressive technology. This agent seems to be very active for decontamination and its elimination from decontaminated space is environmentally friendly, because it decomposes to water and oxygen.

There is rising interest in application of VPHP for decontamination of chemical warfare and biologically active compounds. But there is only limited information about influence of VPHP on chemical substances.

Using mass spectrometry (GC/MS, LC/MS) and NMR were studied decomposition products of chemical substances. Furthermore, using HPLC with ESI-MS/MS or GC/MS were monitoring kinetics of decontamination.

Experimental

Experimental equipment

Appropriate unique, sophisticated equipment, called "peroxybox", was developed for laboratory tests of VPHP for chemical substances decontamination (Fig. 1). It allows to do controlled decontamination of chemical substances by VPHP under various parameters (temperature, hydrogen peroxide concentration, relative humidity) and monitor the kinetics of decontamination of organic substances. Furthermore, it is possible to study the synergistic effect of VPHP with UV (wavelenght 254 nm) radiation too.

Decontamination procedure

The typical VPHP decontamination cycle consists of four phases in this peroxybox, namely: dehumidification, conditioning, decontamination and aeration. The tested compound is deposited on a microscopic slide and placed in the holder. The holder is inserted into the peroxybox, where is left for a defined period in the atmosphere of VPHP. The holder can be pushed out segment-by-segment without disrupting the atmosphere in the peroxybox to follow the kinetics of decomposition. The microscopic slide is afterwards washed with appropriate solvent and studied by GC/MS, and LC/MS or is evaporated to dryness and reconstituted in deuterated solvent for NMR analysis.

LC/MS

HPLC system consisted of dual pump Varian ProStar 210, degasser and autosampler Varian 410 (Varian Inc, USA) connected to a triple quadrupole mass spectrometer Varian 1200L (Varian Inc., USA).

GC/MS

It was used Varian Saturn 2000 (Varian Inc, USA) with ion trap allowing tandem mass spectrometry.

NMR

NMR spectra were measured on a Varian ^{UNITY}*Inova*-400 spectrometer (399.89 MHz for ^{1}H and 100.55 MHz for ^{13}C, Varian Inc., Palo Alto, CA, U.S.A.), Varian Mercury 300 (300.07 MHz for ^{1}H and 75.47 MHz, Varian Inc., Palo Alto, CA, U.S.A.), and Bruker AVANCE III 400 MHz (400.13 MHz for ^{1}H and 100.62 MHz for ^{13}C, Bruker BioSpin GmbH, Rheinstetten, Germany). Residual signal of solvent was used as an internal standard. ^{1}H NMR, ^{13}C NMR, COSY, TOCSY, ROESY, gHSQC, and gHMBC spectra were measured using standard manufacturers' software (VnmrJ 2.1, Varian Inc., Palo Alto, U.S.A., Topspin 2.1, Bruker BioSpin GmbH, Rheinstetten, Germany). 1D NMR spectra were zero filled to fourfold data points and multiplied by window function (for ^{1}H NMR with two-parameter double-exponential Lorentz-Gauss function, for ^{13}C NMR was applied line broadening 1 Hz) before Fourier transformation to improve resolution.

Results and discussion

We focus our research on mechanism and chemical structure effect on course of decontamination. Our goal is to explain relationship between structure of organic compounds and degree of decomposition. It was tested various aromatic compounds to get deeper information about synergic effect of various substituents. It was also applied VPHP for decontamination of various biological active compounds, like alkaloids.

The substituted benzaldehyde derivatives (Figure 1) were model substances for study of mechanism of VPHP decontamination.

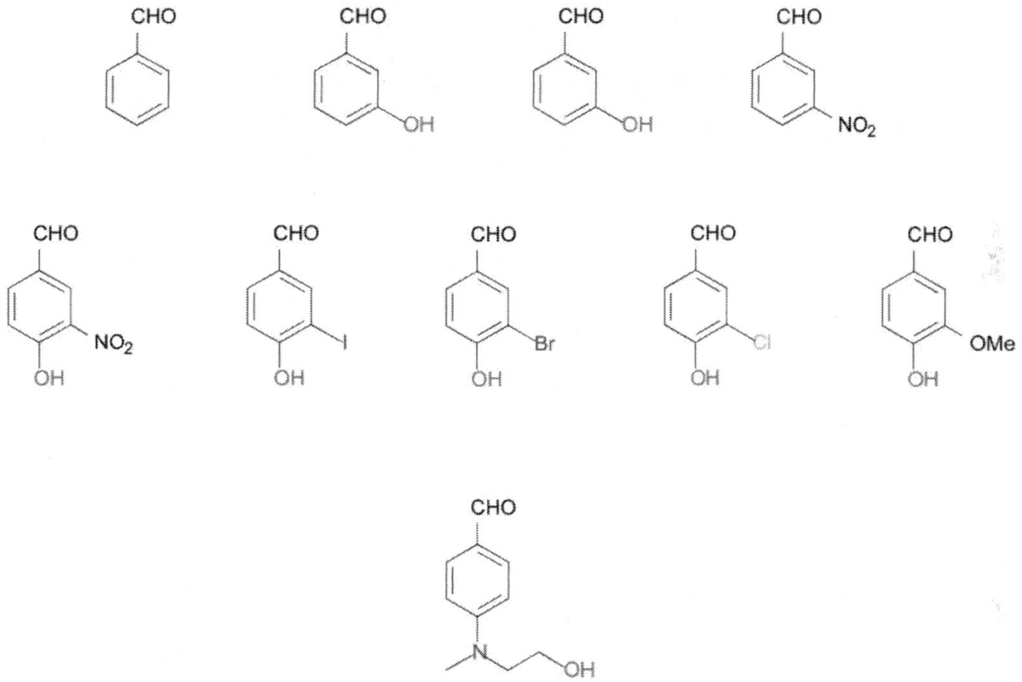

Figure 1: Studied benzaldehyde derivatives

The results indicate that several groups can have promotional effect on the decontamination. The best results were obtained when aliphatic nitrogen was introduced into the structure. But also nitro or methoxyl group have promoting effect. On contrary, halogens have negative effect on the course of degradation.

VPHP degradation was tested on a various ergot alkaloids - lisuride, pergolide, nicergoline, ergotamine and bromocryptine (Figure 2).

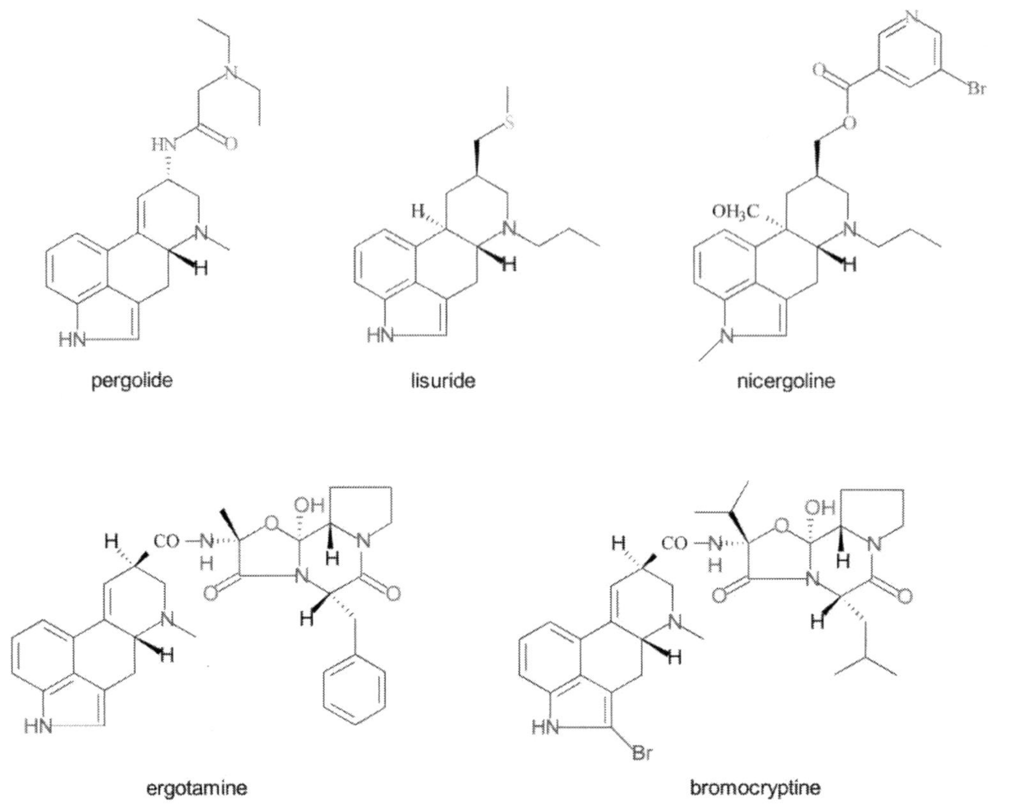

Figure 2: Studied ergot alkaloids derivatives

Ergot alkaloids are produced by the plant parasitic fungi *Claviceps purpurea*. The commercially available drugs are of natural or semisynthetic origin. They are important pharmaceuticals. The name ergot means 'spur' in French, since grains colonized with *Claviceps* spp. often resemble the spurs on the legs of a rooster. There are three main groups of ergot alkaloids, the clavine type, the water-soluble lysergic acid type, and the water-insoluble lysergic acid type or peptide ergot alkaloids. Ergot alkaloids or their derivatives were sensitive to VPHP decomposition. Their degradation is different between ergines and ergopeptines. It was observed complete degradation of non-peptidic part of ergopeptines. The final decontamination structure was a cyclol moiety (peptide part of ergopeptines). It can be explained by an N-oxide formation at nitrogen N-6 of ergoline or ergolene moiety. Nitrogens in the cyclol are amides which do not produce N-oxide and therefore this part remains untouched during decontamination.

Conclusion

This study was oriented on the decontamination of chemical and pharmaceutical substances by VPHP. On series of benzaldehyde derivatives was studied influence of various functional groups on the course of degradation and degradation products. The decontamination process was followed by GC/MS, LC/MS, and NMR to get information not only about the kinetics but to elucidate the structure of decontamination products. The results indicate substantial influence of the structure on the course of the decontamination and its products. Aliphatic nitrogen plays very positive role in the decontamination. It can be explained by N-oxide formation which can be the initial unstable intermediate. It well agrees with results of ergot-alkaloid decontamination.

Acknowledgements

The authors thank the Ministry of Industry and Trade of the Czech Republic (Grant No FT-TA 4/071), IMIC institutional research concept (AVOZ50200510), and Ministry of Education of Czech Republic (Grant CEZ: MSM 223 100001) for financial support.

eco·laboratory
Urban Cultivation

Imagine a vineyard. Each row is oriented, spaced, and maintained to maximize the amount of sunlight hitting the vines, in effect creating a sun catcher to grow grapes. Though "farming" and "harvesting" are seldom mentioned in the traditional architectural vocabulary, they hold crucial lessons in design. In the same way a vintner crafts elements of the sun, soil, and the ephemeral terrior to create wine, a designer can compose a building within a landscape to create place.

Eco-Laboratory, conceptually designed to meet the Living Building Challenge[1], won the 2008 Natural Talent Design Competition, awarded at GreenBuild (the U.S Green Building Council's annual national conference). The design team, sponsored by Weber Thompson, asserted that "there is no silver bullet for sustainability, no one solution for net-zero consumptive environments." Recognizing a living building's dependence on multiple systems, Eco-Laboratory is a synergy of economics, culture and environment working together in harmony. It is more than a building; its parts are inspired by a self-sustaining, diverse ecological system.

With this approach, our interdisciplinary team transcended preconceived notions to unite living site and living building. We began with unrealized potential: a parking lot, a shelter, and a neighborhood agricultural plot sharing a city block. We strove to combine and improve existing amenities, while maintaining a balance of dwelling, working, and mitigating ecological impacts.

Programmatically the building includes dwelling units, a work and training program, urban agriculture, and public education facilities. It mitigates the impact of the activities and demands on site. The designers thought of "harvesting" in various interpretations, looking for all opportunities to create symbiotic relationships. The project is designed to grow its own food, generate electricity, clean its air and water, and provide a place and purpose for the less fortunate in society. These systems bring home, work, shopping, gathering, food production, energy production, and waste disposal under one roof.

Integrated overlapping systems are at the heart of Eco-Laboratory: These systems facilitate the harvesting of available renewable resources.

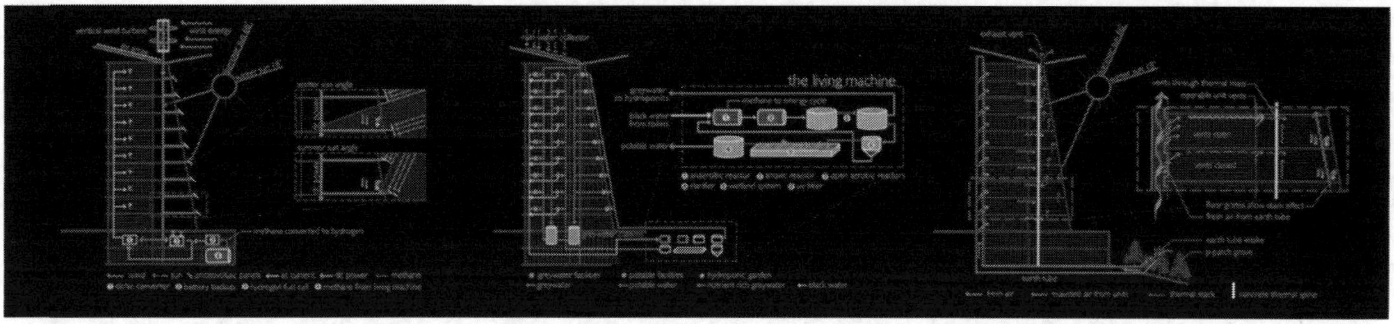

ENERGY To generate energy for Eco-Laboratory, a three-pronged approach was employed to capture the energy of the sun. Photovoltaics adorn the south and west façade, utilizing the direct energy from the sun's radiation. Wind, generated by the sun heating the Earth's atmosphere creating pressure differences, is captured by wind turbines. Methane, from the digestion of organic material grown on site, is "scrubbed" of its carbon, leaving H_2, the fuel source for the hydrogen fuel cell. The resulting electricity and heat generated by the fuel cell is also used to preheat the building's ventilation and water systems.

WATER Methane utilized in the energy cycle is a byproduct of the biological waste water treatment system. Black water and grey water from the building is cleaned in this simulated ecosystem. In a large tanks adjacent to the living machine, Tilapia (a fish common to aquaculture systems) are raised to provide food for residents. Excrement from the fish and nutrient-rich grey water is then pumped into the vertical greenhouses. There, hydroponic and aeroponic systems grow a wide variety of food for consumption by the residents and for sale in the farmers market. Additionally, rain water is captured on site and stored in large cisterns to supplement water loss through evaporation in the closed loop system.

INDOOR QUALITY The greenhouses act as a double-skinned façade, separating the residential units from direct sun. This space is integral for generating the thermal stack to draw air through the building, via underground earth tubes running from the community garden to the residential units. The plants grown in this space provide multiple benefits to the building and its occupants. Vegetation sequesters carbon, absorbs incoming solar radiation (reducing cooling load), humidifies the building, and removes harmful airborne contaminants. It also utilizes nutrients from the aquaculture system, cleans water, generates biomass for anaerobic digestion, grows food for the Tilapia and humans alike, necessitates jobs for maintenance, and even promotes occupant health in a biophilic sense.

Technically, this project is **"off the grid;"** contextually it is **completely connected.**

SOCIAL CONSCIOUSNESS

Eco-Laboratory is rooted to its rich site. An existing charity organization, dedicated to reintegrating people into society, was inspiration for creating a new training facility within the project. It is tasked with providing housing, daily amenities and food, and educating its residents to maintain the systems and cultivate the hydroponically grown crops. Longer-term affordable housing managed by the charity organization comingles with market-rate housing to sustain the complex financially. The existing community garden has also been expanded, an orchard has been added for produce, and to provide a location for community interaction. A public farmers' market is supplied by Eco-Laboratory's harvest connecting the residents to the larger community and providing a desirable neighborhood amenity.

CONCLUSION

A vigorous integrative design process significantly allowed ecology, economy, and equity to drive the building's design from initial concept diagram to the details and materials. Responding to the site's resources (in the same way a farmer might), the designers of Eco-Laboratory were inspired to grow an idea connected to site utilizing adjacencies and balancing ecological and social systems for a bountiful harvest. In this way the building, site, and ecological systems are intertwined and inseparable. The result is much more than the sum of its parts; it's a fundamental verification of a new type of thinking.

WEBER THOMPSON

Project Credits & Statistics

SPONSORED BY Weber Thompson
Architecture, Landscape, Interiors, Community
Seattle, WA
www.weberthompson.com
(206)-344-5700

AUTHORS Dan Albert – Associate ASLA, LEED AP
dalbert@weberthompson.com
Myer Harrell – Associate AIA, LEED AP
mharrell@weberthompson.com

TEAM MEMBERS Dan Albert – Associate ASLA, LEED AP
Myer Harrell – Associate AIA, LEED AP
Brian Geller – Associate AIA, LEED AP
Chris Dukehart – Associate AIA

LOCATION At the corner of Elliott Avenue and Vine Street
Seattle, WA

SCALE 21,600 s.f. – Lot Size
110,000 s.f. – Building
70+ Units
30,000 s.f. – Growing Space
5,000 s.f. – Classrooms

REFERENCES 1. Living Building Challenge
Developed by Cascadia Chapter of the United States Green Building Council
web site: http://www.cascadiagbc.org/lbc

Supplementary Photos

The building's South and West façades contain the greenhouse spaces and photovoltaic panels. The public plaza and living machine are oriented for maximum solar exposure. Earth tube intakes are visible between the trees in the community orchard.

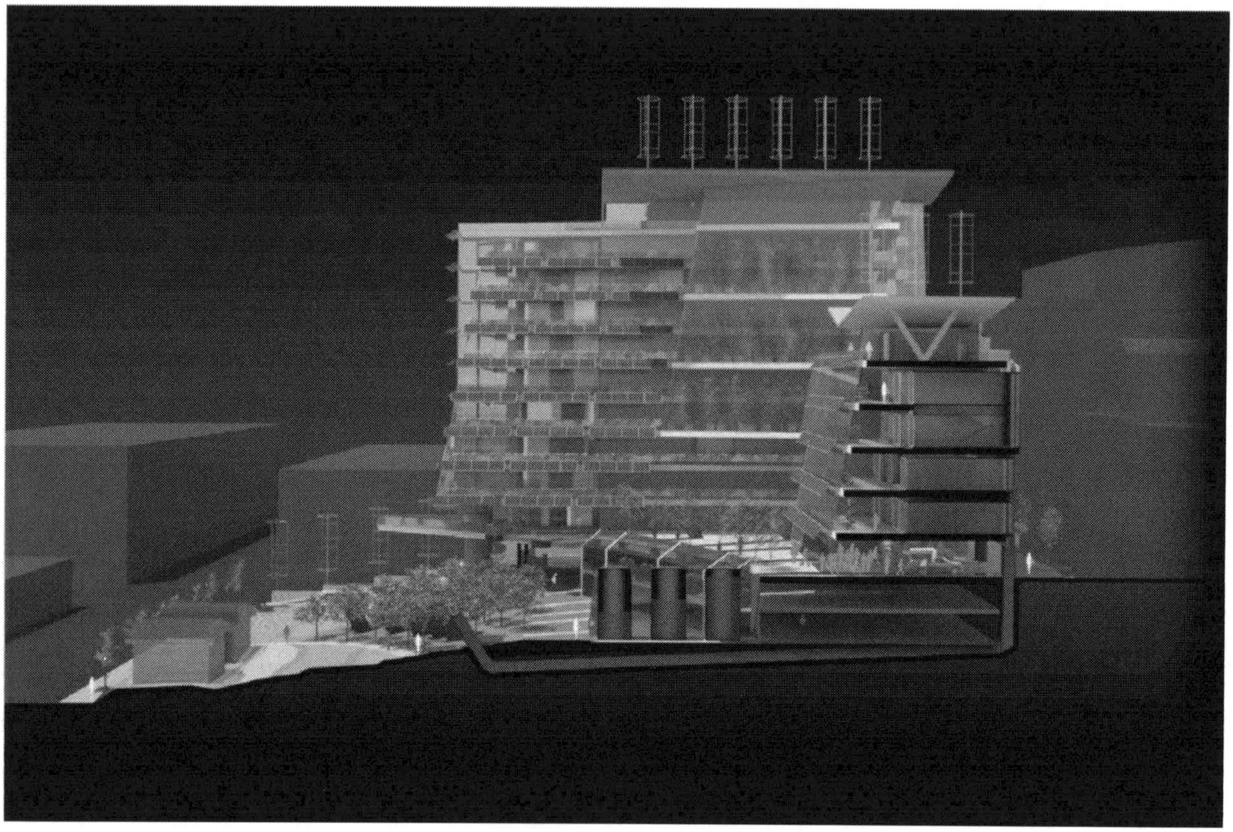

Greenhouse spaces with hydroponic and aeroponic systems are utilized to mitigate heat gain and grow produce to be sold in the market. Also visible are the cisterns, living machine, earth tubes, public plaza, community garden, and orchard.

North and East façades show the earth tubes (ventilation ducts using thermal mass from the ground), wind spires, photovoltaics, and rain water collection system. Re-purposed shipping containers make up units and add texture to the building.

The main entrance into the public plaza, and the heart of Eco-Laboratory, gracefully accents the street while creating a grand, sweeping vista of Puget Sound.

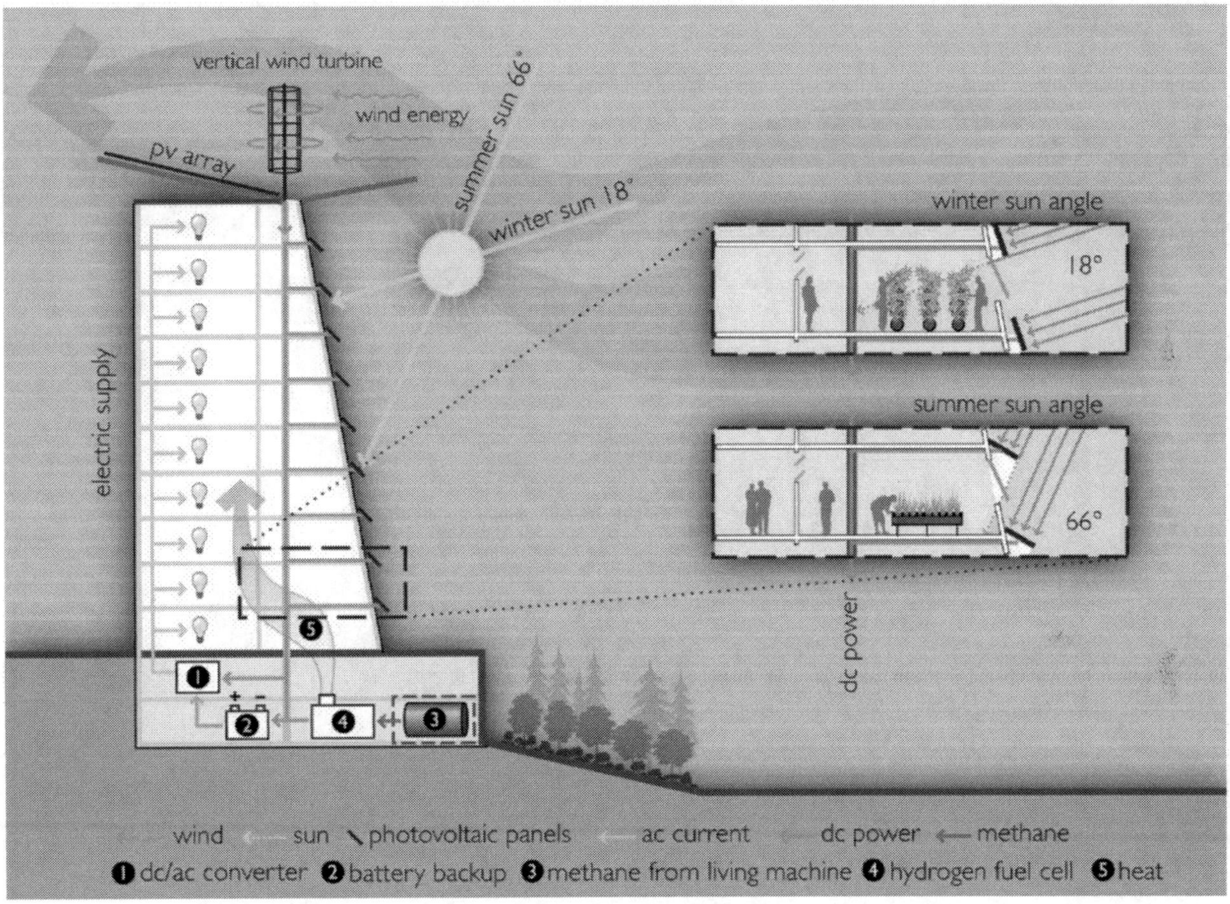

This project uses existing green technologies applied to proven environmental strategies. The building generates all its electricity by harvesting the sun, wind and methane.

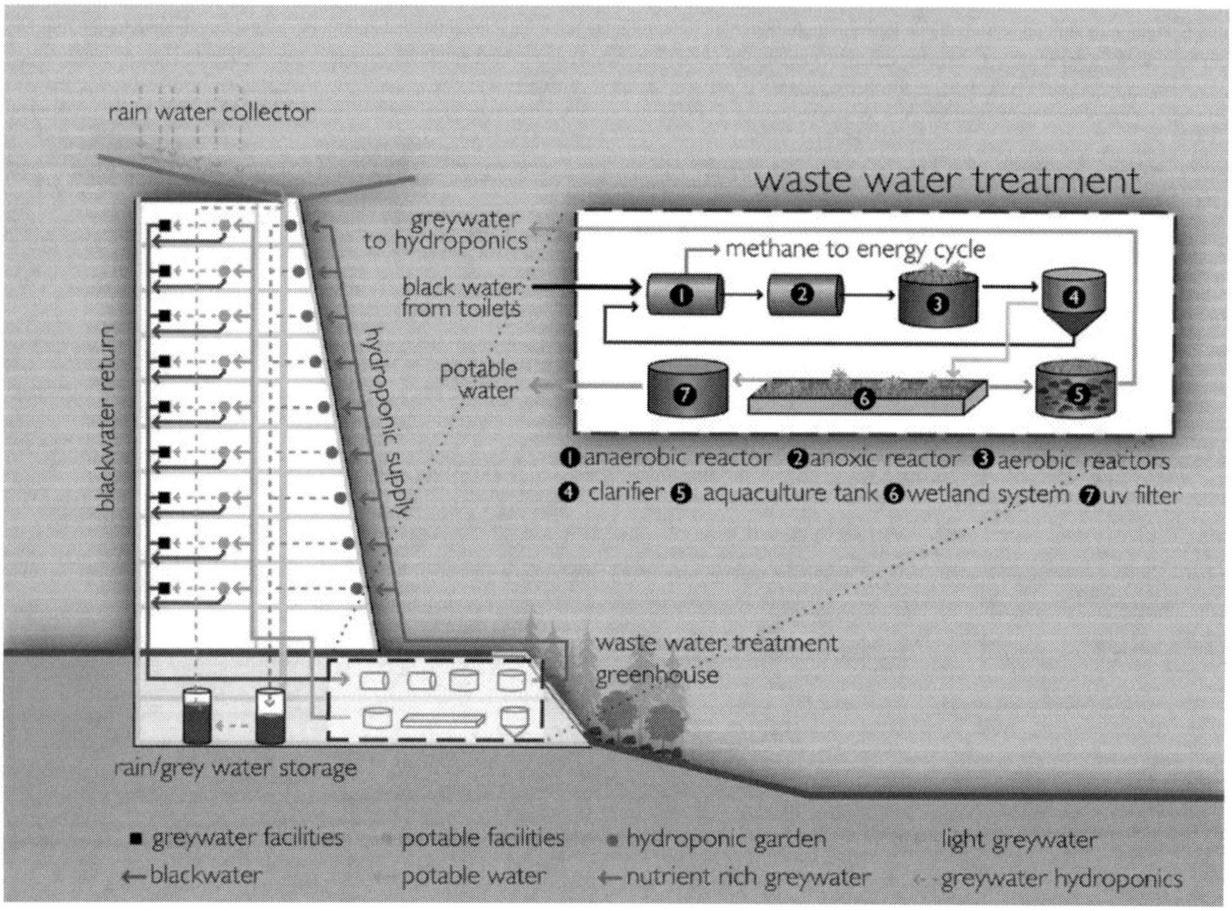

The water system has three components: grey water, drinking water, and nutrient rich water. The system is fed by rain, which is stored in cisterns and used to flush toilets. The black water is treated on site to drinking water quality. Additionally, nutrient rich water feeds the plants.

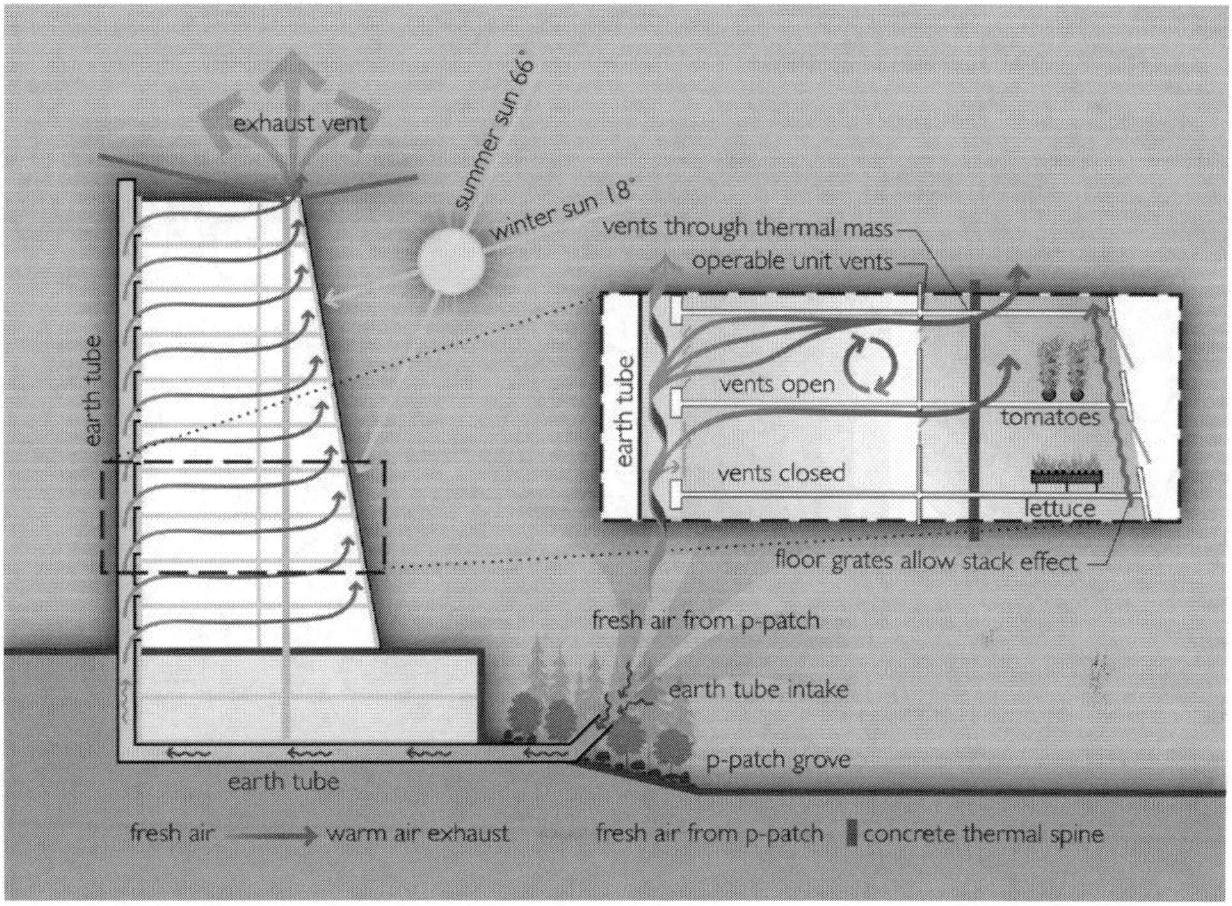

The building is passively ventilated. Warm air rises and the stack effect draws air through large underground ducts (called earth tubes) through the living units and exhausts in the warm greenhouse spaces.

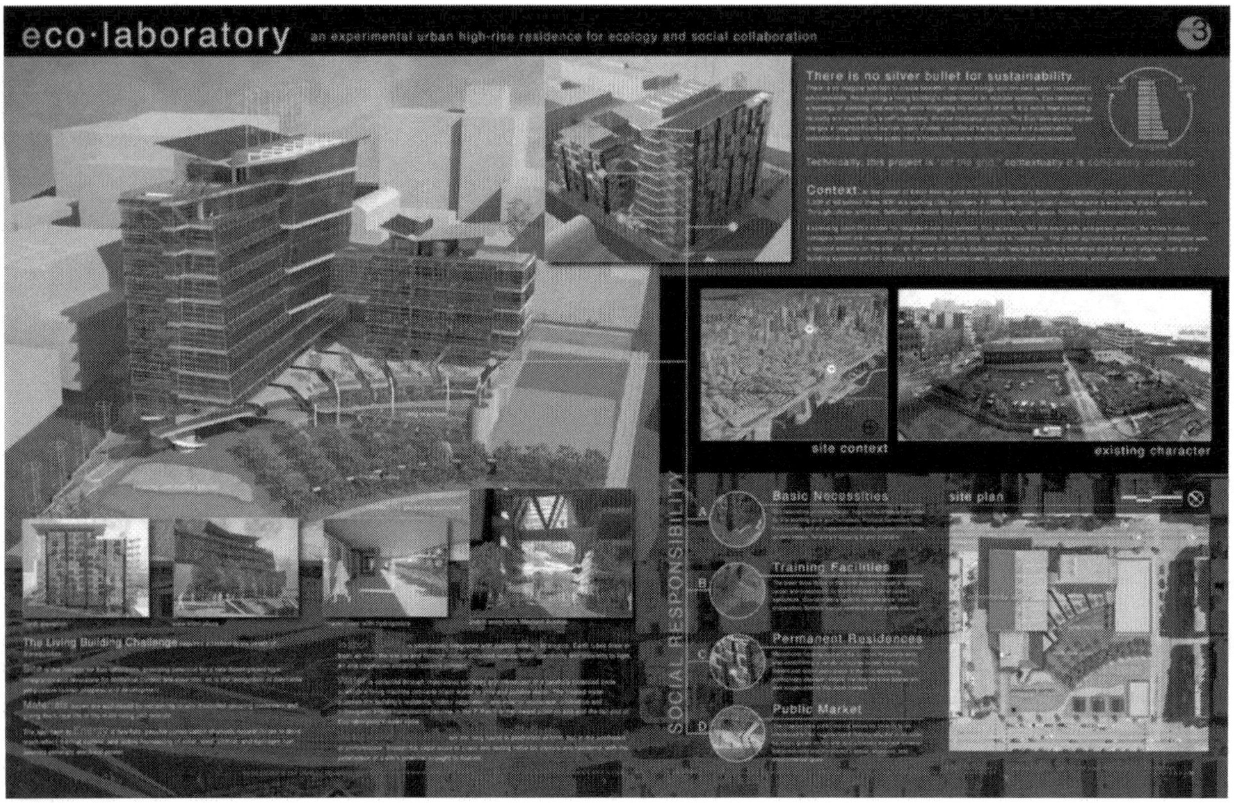

The first board of the National Design Submittal discusses context, concepts, the Living Building Challenge and the social components of the project using photos and architectural renderings to showcase the mood and feel of the project.

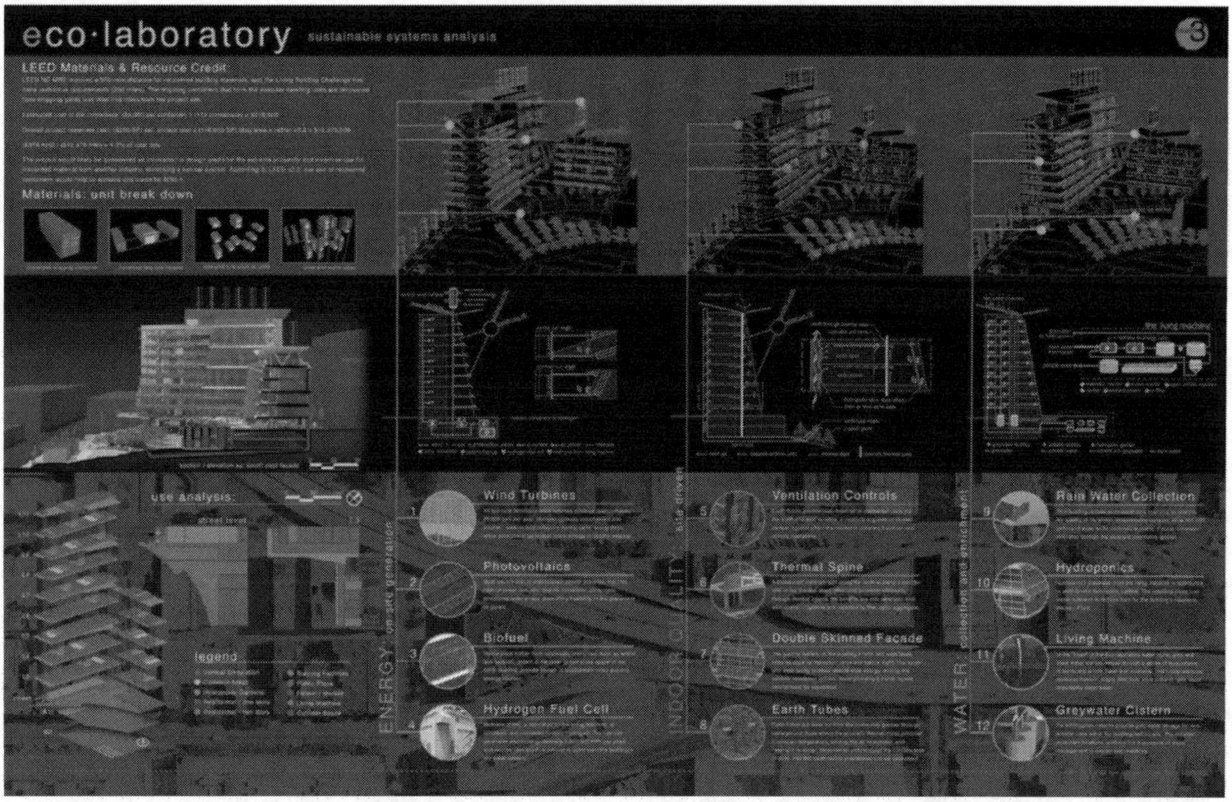

The second board looks deeper into the building's systems locating, diagramming and explaining systems within the project. Programmatic uses and materials are also diagramed and explained.

Evaluation of Designer Biochars to Ameliorate Select Chemical and Physical Characteristics of Degraded Soils

J.M. Novak, and W.J. Busscher*
USDA-ARS-Coastal Plains Research Center
2611 West Lucas Street, Florence, South Carolina, USA
**Corresponding author at: jeff.novak@ars.usda.gov.*

ABSTRACT

Sandy Coastal Plain soils of the southeastern US have low water holding capacities that can reduce crop yields. Water holding capacities and thus potential yields can be improved by amendments of soils with biochar if it has the right characteristics. Biochars were produced by slow and fast pyrolysis out of peanut hulls, poultry litter, pecan shells, hardwood and switchgrass at temperatures ranging from 250 to 700°C. Biochar mass recoveries were determined along with their physical and chemical characteristics. These were mixed at 2% (w/w) with Norfolk Ap loamy sand and were laboratory incubated for up to 120 days. Higher pyrolysis temperatures resulted in lower biochar mass recoveries and smaller O/C and H/C ratios, but greater percentages of ash, and higher pH values. Leaching soils with deionized water revealed that biochar type significantly influenced soil moisture contents. Biochar made from switchgrass at 500°C caused the Norfolk loamy sand to have the highest moisture content and to remain wetter for a longer time period. Biochars can be designed through alteration of feedstock and pyrolysis conditions to significantly improve the Norfolk's water holding capacity.

INTRODUCTION

In the sandy Coastal Plain of South Carolina, there is a long history (≈ 150 y) of row crop production particularly corn and cotton (Gray, 1933). It is common for crops grown in this region to experience severe water deficits because the sandy soil textures have low water capacities, and shallow high-strength root restricting subsurface layers (Busscher et al. 1986). These shallow sandy soils only hold about 0.08 m of water per meter of soil depth (Campbell et al., 1974). This amount of water is insufficient to sustain crop growth especially through frequent 5 to 20 d periods of drought (Sadler and Camp, 1986). The degree to which crop growth is limited depends upon the magnitude of these soil factors as well as rainfall amounts and distributions.

Physical strategies to reduce crop water stress in the Coastal Plain region include sub-soiling using shanks to loosen the subsurface hard layer allowing plant roots to penetrate to subsurface horizons that hold more water (Akinci et al., 2004). Subsoiling requires powerful tractors and specialized equipment (Karlen et al., 1991). The effects of deep tilling are not long lasting (Carter et al., 1996); these layers can re-cement so that deep tillage is usually preformed on an annual basis (Busscher et al., 2001). It would be desirable if a longer lasting solution were available.

One promising approach to establish significant, long-term chemical and physical improvements in degraded soils is to apply biochar as a soil supplements (Lehmann et al., 2006, Laird, 2008). Historically, biochar is a byproduct from the practice of 'slash and char' and from charred remnants of organic wastes produced by the South American Amerindian populations (Lehmann et al., 2006). The Amerindian inhabitants improved the infertile soils of the Amazon basin by applying biochar for several thousand years that resulted in the creation of anthropogenic soils called 'Terra Preta' (Glaser et al., 2002). Applying biochars to the Terra Preta soils improved their fertility and nutrient retention resulting in increased crop yields (Glaser et al., 2002).

More recently, interest in creating bioenergy products from organic feedstocks has spurred the manufacture of biochar. Biochar is a by-product of the pyrolysis process that converts organic feedstocks into bio-oil (Antal and Grønli, 2003). Feedstock selection and pyrolysis conditions can influence the biochars physical and chemical properties (Antal and Grønli, 2003). Biochar is composed of an assortment of C-containing structures, as well as other elements (i.e., N, P, Ca, K, etc.) that can significantly affect soil pH, supply plant nutrients (Novak et al., 2009) and increase soil moisture retention (Lehmann et al., 2006). The ability of biochar to increase a soils water holding capacity has not been fully investigated. The objectives of this study were to physically and chemically characterize biochars produced from four feedstocks pyrolyzed under two different temperature regimes, and to examine the effects of these different biochars on improving the water holding capacity of a sandy, southeastern US Coastal Plain soil.

MATERIALS AND METHODS

Biochars were produced by pyrolyzing five feedstocks -- a mixture of hardwoods (DCQ); peanut hulls (*Archis hypogaea*); pecan shells (*Carya illinoensis*); poultry litter (*Gallus domesticus*); switchgrass (*Panicum virgatum* L.). The hardwoods were pyrolyzed under a flash exposure of 1 s duration at ≈ 500°C. The other biochars were produced under slow pyrolysis at both a low (250 to 400°C) and high (500 to 700°C) temperature regimes. These two temperature regimes were chosen to obtain biochars with chemical compositional differences because -- cellulose and hemi-cellulose decomposition occurs at about 400°C (Antal and Grønli, 2003). The biochar mass recovery was calculated as a percentage of dry weight feedstock input (DW) and as biochar mass output (biochar mass DW/feedstock mass DW). For each raw feedstock (except DCQ) and for each biochar sample, a single estimate of the ash content on an air-dried, weight basis, and elemental (C, H, and O) content was determined on an oven-dried, weight basis by Hazen Research, Inc. (Golden, CO) following ASTM D 3172 and 3176 standard methods (ASTM, 2006). These results were used to calculate their atomic H/C and O/C ratios to examine their relative degree of hydrophobicity. The pH of each biochar was measured in triplicate in deionized water from a 1% (w/v) mixture after shaking at 200 rpm for 24 h.

The Norfolk soil used in this study is classified as a Fine-loamy, kaolinitic, thermic Typic Kandiudult. It was obtained from the Ap horizon (0 to 15 cm deep) from a field that has been under row crop production for over 20 years at the Clemson University Pee Dee Research and Educational Center, Florence, South Carolina. The Norfolk Ap was a loamy sand with a particle size distribution of 740, 250, and 10 g/kg, respectively, of sand, silt and clay. The Norfolk's clay fraction was composed mostly of kaolinite and the Ap horizon typically has an acidic pH value (5.5 to 6.0).

Biochar was mixed into the Norfolk Ap soil at 2% (w/w, equating to 40 to 44 mt/ha) and with sufficient deionized H_2O to bring the soil moisture content to 10% (w/w). The mixtures were transferred to open-top flower pots (10.3 cm id x 8.5 cm tall) and gently tapped on a bench top until soil settled down to a bulk density of 1.2 g/ cm^3. Each treatment was set up in four replicates and they were laboratory incubated for up to 120 days with the moisture content adjusted 3x per week to 10%. On day 28, all pots were leached with 1.2 pore volumes of deionized water; the leachate was collected over a 30 h period and later weighted. The mass of water leachate collected was measured and was used to calculate the percent soil moisture content (w/w) of each pot. Changes in soil moisture contents due to drying were then determined over an 11 day period past leaching by periodically weighing each pot. The mean values for the percent soil moisture contents on day 28 of incubation and on the 11th day past leaching were compared between treatments using a Holm-Sidek pair-wise comparison procedure at a $P = 0.05$. The soil moisture content for each treatment were correlated with the biochars H/C and O/C atomic ratios to determine if there was a relationships between these characteristics.

RESULTS AND DISCUSSION

The yields and chemical characteristics of the raw feedstocks and the high and low temperature pyrolyzed biochars are shown in Table 1.

Table 1. Mass yields and chemical properties of biochars.

Feedstock	Pyrolysis (°C)	Mass yield (%)	% ash (w/w)†	pH
Peanut hull	0	0	3	na‡
	400	40	8	7.9
	500	35	9	8.6
Pecan shell	0	0	2	na
	350	50	2	5.9
	700	30	5	7.2
Poultry litter	0	0	24	na
	350	57	26	8.7
	700	36	52	10.3
Switchgrass	0	0	2	na
	250	78	3	5.4
	500	29	8	8
DCQ	≈ 500	na	9	5.7

† based on air-dry weight
‡ where na = not available

More biochar was recovered at the lower pyrolysis temperatures. Below 350°C, the mass yield recovery was at least 50% but the yield declined to about 30% as the pyrolysis temperature increased up to 500 or 700°C because of the dehydration of hydroxyls, and thermal degradation of ligno-cellulose structures (Antal and Grønli, 2003). The ash contents for biochars made from peanut hulls, pecan shells, switchgrass, and DCQ were < 10%, while biochar made from poultry litter ranged from 24 to 52%. For all materials (except DCQ), ash contents rose with pyrolysis temperatures. Substantial increases in pH occurred at the higher temperatures because of the concentration of non-pyrolyzed inorganic elements in the feedstocks. Among the feedstocks, biochars produced from poultry litter had the highest pH values ranging from 8.7 to 10.3. These values were similar to Gaskin et al. (2008) and are probably due to un-assimilated nutrients in the poultry manure (Sistani and Novak, 2006) or to treatment of the poultry bedding material with chemicals to minimize ammonia emissions.

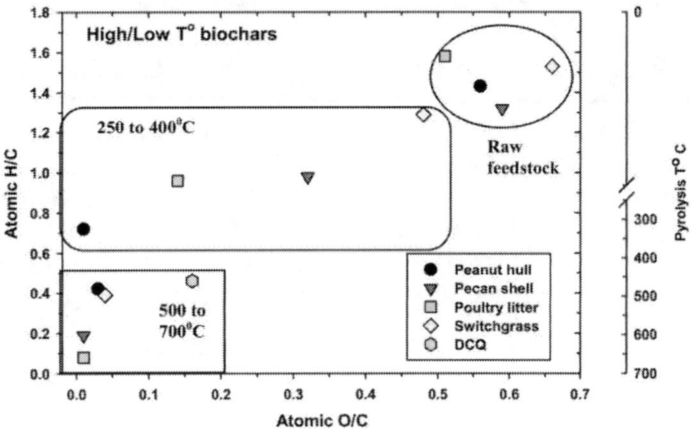

Fig. 1. Atomic O/C and H/C ratios of raw feedstocks, low (250 to 400°C) and high (500 to 700°C) temperature biochars.

The atomic O/C and H/C ratios of each feedstock and biochar are presented in Fig. 1. The biochars atomic ratios because of dissimilar O and H losses, varied considerably between feedstock and pyrolysis temperature, so that the biochars theoretically were most polar (high O/C) at the lower pyrolysis temperatures. Switchgrass biochar at 250°C was the most polar among the 8 biochars. Among the 4 feedstocks, the low temperature switchgrass biochar had the most polarity suggesting that it can be the most interactive with water.

After 28 days of incubation, during and after the leaching period, the moisture contents of the Norfolk Ap soil with and without biochar incorporation varied considerably (Fig. 2). A day after leaching, the control Norfolk Ap had a soil moisture content of 22.5%, and after 11 days, it declined to < 3%. At the same time, soil moisture contents of the Norfolk Ap with incorporation of 2% biochar increased significantly, but the effects varied among feedstocks and between pyrolysis temperatures. Among the low temperature biochars (Fig. 2A.), switchgrass, and peanut hull biochar significantly increased soil moisture contents up to 26%. This represents almost a 4% soil moisture content increase over the control. The largest significant increase in soil moisture content (33%) was measured after incorporating the high temperature switchgrass biochar – a relative soil moisture content increase of almost 11% over the control. Significant improvements in soil moisture contents were measured after incorporating the low temperature switchgrass, peanut hull, pecan shell and DCQ biochars, but the increases were only between 2 to 4% relative to the control. These results suggest that feedstock selection and pyrolysis conditions had an influence on the ability of each biochar to improve soil moisture contents.

Monitoring the soil moisture content of each treatment for 11 days past leaching revealed that there was still a significant improvement in the Norfolk's soil moisture content for all the low temperature biochars (Fig. 2A). The mean soil moisture content for the control dropped to 2.1% while the moisture contents for all other treatments ranged between 3.0 to almost 5%. Similarly, after 11 days of drying, the %soil moisture contents for treatments with high temperature biochars including DCQ were significantly higher than the control (2.1 vs. 3.3 to 11%, respectively). This implies that soil treated with high temperature switchgrass and DCQ biochar remained wetter for a longer time period than the control.

Mixing in high temperature pyrolyzed switchgrass biochar caused the most significant improvement in the Norfolk Ap soil moisture content. Correlation analyses revealed a poor relationship between the Norfolk's improved soil moisture content with the polar nature (O/C and H/C ratios) of the switchgrass and DCQ biochar ($P = 0.9$, $r^2 = -0.07$). It is speculated that another physical property of these biochars or a change is soil physical characteristic not considered caused these improvements. Water retention increases with soil amendments in a sandy soil have also been reported (Tyron, 1948; Gaskin et al., 2007), however, these increases required applications of between 4 to 45% charcoal, much higher than the 2% employed in this study.

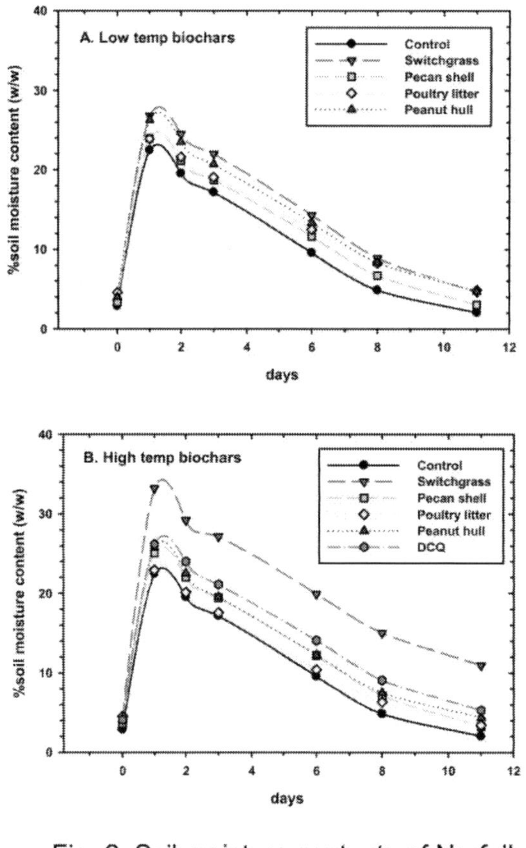

Fig. 2. Soil moisture contents of Norfolk control (+ 0% biochar) and Norfolk soil treated with 2% low (A) and high (B) temperature biochars.

CONCLUSIONS

Biochars made from different feedstocks and under contrasting pyrolysis temperatures were shown to have dissimilar char yields and physical properties. Biochar quality was a function of feedstock selection and pyrolysis conditions. Biochars incorporated into the Norfolk Ap loamy sand were found to significantly improve soil moisture contents; however, the effect was dissimilar among the 8 biochars. High temperature switchgrass biochar produced the biggest relative change in soil moisture contents and the effect was still significant after 11 days of soil drying. These results suggest that biochars can be produced by altering feedstock selection and pyrolysis conditions that can significantly improve a loamy sand's water holding capacity.

ACKNOWLEDGEMENTS

This publication is based on work supported by the US Department of Agriculture-Agriculture Research Service under the ARS-GRACEnet project. The authors thank all parties for assistance with biochar production, characterization and laboratory work.

REFERENCES

Akinci, I., E. Cakir, M. Topakci, M. Canakci, and O. Inan. 2004. The effects of subsoiling on soil resistance and cotton yield. Soil Till. Res. 77:203-210.

ASTM (American Society for Testing and Materials). Petroleum products. Lubricants, and Fossil Fuels; Coal, and Coke. 2006. West Conshohocken, PA. ASTM International.

Busscher, W. J., R.E. Sojka, and C.W. Doty. 1986. Residual effects of tillage on Coastal Plain soil strength. Soil Sci. 141:144-148.

Busscher, W.J., J.R. Frederick, and P.J. Bauer. 2001. Effect of penetration resistance and timing of rain on grain yield of narrow-row corn in a coastal plain loamy sand. Soil Till. Res. 63:15-24.

Campbell, R.B., D.C. Reicosky, and C.W. Doty. 1974. Physical properties and tillage of Paleudults in the southeastern Coastal plains. J. Soil Water Conserv. 29:220-224.

Carter, M.R., D.A. Holmstrom, L.M. Cochrane, P.C. Breton, D.R. Langille, W.G. Thomas, and J.A. van Roestel. Persistence of deep loosening of naturally compacted subsoils in Nova Scotia. Can. J. Soil Sci. 76:541-547.

Gaskin, J.W., A. Speir, L.M. Morris, K. Ogden, K. Harris, D. Lee, and K.C. Das. 2007. Potential for pyrolysis char to affect soil moisture and nutrient retention status of a loamy sand soil. In: Proceedings of the Georgia Water Resources conference. 27-29 March 2007. Inst. Ecology, Univ. GA, Athens, GA.

Gaskin, J.W., C. Steiner, K. Harris, K.C. Das, and B. Bibens. 2008. Effect of low-temperature pyrolysis conditions on biochar for agricultural use. Trans. ASABE 51:2061-2069.

Glaser, B., J. Lehmann, and W. Zech. 2002. Ameliorating physical and chemical properties of highly weathered soils in the tropics with charcoal-a review. Biol. Fertil. Soils 35:219-230.

Gray, L.C. 1933. History of Agriculture in the Southern United States to 1860. Carnegie Inst., Washington DC.

Karlen, D.L., W.J. Busscher, S.A. Hale, R.B. Dodd, E.E. Strickland, and T.H. Gardner. 1991. Drought condition energy requirements and subsoil effectiveness for selected deep tillage implements. Trans. ASAE 34:1967-1972.

Laird, D.A. 2008. The Charcoal vision: A win-win-win scenario for simultaneously producing bioenergy, permanently sequestering carbon, while improving soil and water quality. Agron. J. 100:178-181.

Lehmann, J., J. Gaunt, and M. Rondon. 2006. Bio-char sequestration in terrestrial ecosystems-a review. Mitig. Adapt. Strategy Global Change 11:403-427.

Liang, B., J. Lehmann, D. Solomon, J. Kinyangi, J. Grossman, B. O'Neill, J.O. Skjemstad, J. Thies, F.J. Luizão, J. Peterson, and E.G. Neves. 2006. Black carbon increases cation exchange capacity in soils. Soil Sci. Soc. Am. J. 70:1719-1730.

Novak, J.M., W.J. Busscher, D.L. Laird, M. Ahmedna, D.W. Watts, and M.A. Niandou. 2009. Impact of biochar on fertility of a southeastern Coastal Plain soil. Soil Sci. 174:105-112.

Sadler, E.J., and C.R. Camp. 1986. Crop water use data available from the southeastern USA. Trans. ASAE 29:1070-1079.

Sistani, K.R., and J.M. Novak. 2006. Trace metal accumulation, movement and remediation in soils receiving animal manure. In: M.N. Prasad, K.S. Sajwan, and R. Naidu, (eds).; Trace Elements in the Environment, Biogeochemistry, Biotechnology, and Bioremediation. CRC Press, Boca Raton, FL. pp. 689-706.

Tyron, E.H. 1948. Effect of charcoal on certain physical, chemical, and biological properties of forest soils. Ecol. Monogr. 18:81-115.

Algae Bio-Remediation, Environmental Restoration, and Biogas

Kelly M. Ogilvie, Blue Marble Energy Corporation, Seattle, WA

Introduction

Marine algae (or macroalgae) offer a vast renewable energy, bio-chemical, and fertilizer resource for countries throughout the world that have available coastline.

Large algae blooms are popping up at an alarming rate worldwide, fueled by unchecked human run-off. The loading of excess nutrient pollution from agricultural practices and inadequate wastewater treatment is creating massive algal blooms that choke oxygen from the water and reduce aquatic environments to underwater deserts. These blooms create dead zones that devastate marine ecosystems. Over 400 dead zones now exist, the highest number in recorded history [1].

Algae blooms naturally occur, but since the 1970's, phytoplankton blooms have grown abnormally large. According to a UN Environmental Program report, the phytoplankton blooms are due to skyrocketing marine levels of nutrients such as phosphorus and nitrogen from fertilizers, sewage, animal wastes, and pollution. These blooms are likely to persist worldwide because farmers are unlikely to transition away from high-nitrogen fertilizers as "organic" agriculture's production costs are higher and less efficient at improving crop yield.

A solution that removes these high levels of environmental toxins from our waterways is needed. Removal of excess algal growth, and cultivation of algae biomass to reduce nutrient impacts is explored in this paper. Algae has the potential to become an effective tool for remediation, environmental restoration, and feedstock for a myriad of renewable products that displace petroleum.

Cultivation

Blue Marble Energy has a methodology and patented process of growing algae within open environments in contained apparatuses. Cultivating algae biomass near sources

of nutrient loading can prove an effective bioremediation strategy. This methodology is used in aquaculture, and can be adapted for nutrient reduction of human impacts.

Intensive fish farming discharges large amount of nutrients, the majority of which are composed of dissolved nitrogen in ammonium form, which promotes eutrophication in

coastal waters. Macroalgae have been proven to effectively reduce the nutrients of fish farm effluents and at the same time increase the economic output of the aquaculture

system.

Bioremediation using macroalgae offers a unique opportunity to sequester high levels of toxins, as well as has great potential for biomass production, CO2 bioremediation, and biogas potential.

Macroalgae have high productivity, as great or greater than the most productive land plants, and do not compete with terrestrial crops for farm land. Biomass from macroalgae promises to provide environmentally and economically feasible alternatives to fossil fuels [14].

Marine Algae (Macroalgae)

Marine macroalgae are a critical part of the global food web. All terrestrial plants stem from these early plants, and they have existed on this planet for over 2 billion years. There are three types of marine algae: green, red, and brown, each of which has unique properties that make them suitable as environmental remediation and restoration tools. For the purposes of this paper, we will focus on Ulva lactuca macroalgae as an environmental remediation and restoration tool.

Nutrient Fixation (Nitrogen and Phosphorus)

Ammonia (nitrogen) naturally occurs in many freshwater and marine sediments. Like many other compounds, it is necessary as a nutrient for plant growth but can be toxic at

elevated concentrations [2, Ankley GT, Katko A, and Arthur JW. 1990]. Ammonia concentrations, normally naturally mediated in sediments by bacteria, are being

exacerbated by human impacts. These excess carbon inputs increase the depth of the anoxic zone in sediments, which in turn increases the population of ammonia-producing

bacteria. This, in conjunction with human agricultural run-off, sewage, and other industrial pollution are depositing high levels of nitrogen, phosphorus, and other

pollutants.

Ulva lactuca (green marine algae, sea lettuce) is an attached green seaweed capable of sequestering high environmental levels of nitrogen and phosphorus. Nitrogen

(ammonia) can be acutely toxic to marine organisms and is often found in dredged sediments from highly industrial areas or from areas with high carbon inputs. In target experiments, Ulva lactuca lowered concentrations of total ammonia from 75 to 17 mg/L in less than 8 hours in water-only exposures. This treatment also reduced 48

hour amphipod mortality from 75 to 20%. In whole sediments, U. lactuca lowered interstitial water concentrations of ammonia from 60 to 20 mg/L, while keeping overlying

water concentrations below detectable limits (<1 mg/L) [3, Kay T. Ho, Robert M. Burgess, and Andrea Helmstetter].

Phosphorus (P) is a primary driver for algal and other marine organism growth. Phosphorus is a necessary nutrient for all life forms and is generally considered the

limiting factor for algae blooms (both micro and macro) and excess marine plant growth. Phosphorus is produced globally from inorganic phosphate rock and is a constituent

ingredient of all fertilizers. The transfer of phosphorus from agricultural fields in the form of fertilizer into the open environments as a result of run-off is a primary driver of abnormal algae bloom growth. Removal and recovery of phosphorus and nitrogen is possible with algae biomass.

Heavy Metals

Ulva lactuca also is effective at removing heavy metals. Ulva generally uptakes 10 - 15% of metals present. In contrast, U. lactuca removed 77% of lindane and 95% of

fluoranthene. Results of testing conducted on sediments from an industrial marine harbor indicated U. lactuca was useful at removing environmental toxicants. Ulva

lactuca accumulation of metals-chromium (Cr), cadmium (Cd), copper (Cu), nickel (Ni), zinc (Zn), and lead (Pb)-and of two organic compounds-fluoranthene and lindane has been documented [3]. Because of U. lactuca s ability to uptake metals, it can be used as a bio-monitor for coastal contamination [4, Wen-Xiong Wang1 and R. C. H. Dei1].

Nonionic Organic Uptake

Ulva lactuca also has a high capacity to remove nonionic organic compounds from water. Studies indicate that U. lactuca can remediate high levels of compounds such as:

fluoranthene and lindane. Fluoranthene is a polycyclic aromatic hydrocarbon (PAH) consisting of a naphthalene

and a benzene. It is a member of the class of PAHs known as non-alternant PAHs because it has rings other than those with six carbon atoms. Fluoranthene is found in many combustion products, along with other PAHs. Its presence is an indicator of less efficient or lower-temperature combustion. Fluoranthene is one of the U.S. Environmental Protection Agency's 16 priority pollutant PAHs and is a carcinogen.

The basis for the uptake of ammonia by U. lactuca is supported in the literature [9,10,11]. Cohen and Neori reported values for ammonia uptake by U. lactuca to be

approximately 6.63 mg g-1 h-1. Based on these literature values [9], using 5 g U.lactuca with a 6-h exposure, the maximum uptake in our experiment could be approximately 200 mg. This data points to a significant potential for reducing environmental toxins using macroalgae.

Biogas (Anaerobic Digestion)

Marine algae produce the highest yield of natural gas from plants when converted through anaerobic digestion, are among the fastest growing plants on earth, and do not compete with terrestrial food crops. In the US alone, the area available for growth of marine plants is estimated at 2.3 million km2 (907,000 square miles or 58 x 107 acres; Snow et al., 1979). Assuming a growth yield of 5 dry kg/m2 (20 dry tons per acre) per year and a heating value of 10 MJ/kg (9 x 10 to the 6th btu/dry ton), that area could produce a biomass energy equivalent of approximately 100 EJ (exajoules or quadrillion Btu's). Nearly all of US energy needs could be produced on one million square miles of ocean.

Yield Potential Estimates

7 EJ (10 to the 18 joules or about 10 to the 15 Btu) for waste

30 EJ for terrestrial biomass

100 EJ for marine biomass

Potential Products from Macroalgae

Methane

Extraction of Chemicals (or chemical precursors)

Fertilizer

Animal Feed

Feedstock for thermal or biological processes

Alcohol fuels

The advantages of anaerobic digestion (bio-gasification) over thermal or other biological processes are:

1.) High moisture feeds can be processed (reduced costs)

2.) No pretreatment is required for degradation

3.) Pure culture not required

4.) Gas is easily separated from the medium

5.) Gas is primarily methane and carbon dioxide

6.) Product gas is useful with varied stages of purification and scales

7.) Process economics are good at small and large scales

Conclusion

Marine algae are an abundant and highly biodegradable renewable resource that are an excellent feedstock for methane, chemical products and precursors, and fertilizers. Utilizing the BME AGATE system, it is possible to attain higher than 80% of theoretical maximum methane yields. We hope to raise this threshold and continue optimization through our research and testing on additional hydrogen availability.

Since the Industrial Revolution, 525 billion tons of carbon dioxide (or half of all fossil emissions) have been absorbed by our aquatic ecosystems. The need for environmental remediation of high levels of excess nutrients and carbon inputs from human activity in our water ecosystems could never be greater. Ulva lactuca is capable of reducing environmental concentrations of ammonia, heavy metals, non ionic compounds, and improve the water quality for marine organisms. Removing algae

biomass from impacted coastlines, and cultivating algae at the source of nutrient pollution are solutions that require low capital cost, immediate environmental benefit, and opportunities for biofuel and renewable product generation.

See more of this Session: Sustainable Design: Materials/Water Conservation/Energy Efficiency
See more of this Group/Topical: Environmental Division

Comparing Evaporative Cooling to the Alternatives: Methodology, Data and Preliminary Results

Elaine T. Hale, Commercial Buildings Section, National Renewable Energy Laboratory, Golden, CO

Kristin Field, Commercial Buildings Section, National Renewable Energy Laboratory, Golden, CO

Paul Torcellini, Commercial Buildings Section, National Renewable Energy Laboratory, Golden, CO

In the United States, residential and commercial buildings account for 39% of total energy use and 73% of electricity use. About 13% of the primary energy used by buildings is consumed by space cooling, nearly all of which is used to produce electricity (DOE 2009). Evaporative cooling uses significantly less electricity than typical cooling systems; however, this class of technologies has been criticized for excessive water use. It remains an open question as to whether this criticism is valid.

An often overlooked aspect of electricity use is that a significant amount of water is evaporated by many power plants. Torcellini et al. (2003) found that, on average, U.S. thermoelectric plants use 0.47 gal (1.8 L), and hydroelectric plants use 18 gal (68 L) of fresh water per kWh of electricity consumed at the point of end use. That paper also establishes state and regional averages for consumptive water use by power plants--numbers that vary quite widely from the national metrics.

The minimum amount of water used by evaporative cooling is 1.37 gal/ton•h (1.47 L/kWh) of cooling. Inefficiencies, and the electricity to power fans and pumps adds to the total (Modera 2008). Thus, evaporative cooling may be competitive with typical cooling systems (direct expansion coils and chillers) with regards to water as well as energy use in some buildings and climates. By coupling the estimates of water consumption per kWh of Torcellini et al. (2003) and results from annual building energy simulations it is possible to directly compare evaporative technologies with the alternatives across both dimensions.

This talk will give an overview of Torcellini et al.'s 2003 work, describe how those data can be used in conjunction with the building energy simulation program EnergyPlus (DOE 2008) to calculate the site and source energy and water use of a given cooling technology, and present preliminary results concerning particular technologies and locations using the commercial building benchmark models (Deru et al. 2008).

Deru, M.; Griffith, B.; Long, N.; Halverson, M.; Winiarski, D.; Huang, J.; Crawley, D. (2008). "DOE Commercial Building Research Benchmarks for Commercial Buildings." In preparation. U.S. DOE.

DOE (2008). "EnergyPlus Energy Simulation Software." Washington, D.C., U.S. Department of Energy. http://www.eere.energy.gov/buildings/energyplus/.

DOE (2009). "2008 Buildings Energy Data Book." Prepared by D&R International, Ltd. for the U.S. DOE EERE Building Technologies Program under contract to the National Energy Technology Laboratory.

Modera, M. (2008). "Minimizing Water Use in Non-Compressor Cooling Applications." 2008 ASHRAE Annual Conference, Salt Lake City, Utah.

Torcellini, P.; Long, N.; Judkoff, R. (2003). "Consumptive Water Use for U.S. Power Production." NREL/TP-550-33905. Golden, Colorado: National Renewable Energy Laboratory.

See more of this Session: Sustainable Design: Materials/Water Conservation/Energy Efficiency
See more of this Group/Topical: Environmental Division

Thermochemical Characterization of Bran for Conversion in Energy

Yolanda Pliego -Bravo, A. Morales-González, G. Alvarado-Kinnell and E. Bolaños-Reynoso, Divisiôn de Estudios de Posgrado e Investigación, Instituto Tecnológico de Orizaba, Orizaba, Veracruz, Mexico

The biomass as a non-conventional energy source is being studied. The biomass material used was bran by- product from an industry brewer, the objective of this investigation was the thermochemical characterization of bran to evaluate its heating potential; it presents a high combustion energy (20482.1 KJ/Kg), a residual content of 5.28% and a content of humidity smaller than 10% in dry base.The bran can be considered as alternating source of energy to cover important part of the energy needs.

Introduction

The biomass is considered the source of renewable energy with a high potential to assist the energy demand of the modern society. The biomass is the fourth energy source in the world, it gives around 14% of primary energy (Demirbas, 2006), it presents a closed cycle of CO2, null sulfur emissions, good opportunities for the use of ashy and potential to save the resources of fossil energy.

The industries should modify their processes so that their residuals can be recycled and to use them as matter prevails for other processes to reduce the contamination, the industry brewer generates big quantities of by-products and residuals being the trash of malt (bran) the most abundant since corresponds around 85% of the total of generated by-products, this is around 31% of the original weight of malt and it represents 20 kg approximately for each 100 liters of produced beer (Mussatto et al., 2006).

The thermochemical characterization of bran to evaluate its heating potential as alternating source of energy to cover important part of the energy is the objetive.

The gasification of biomass has attracted the biggest interest among the technologies of thermochemical conversion; it converts the biomass through a partial oxidation toward a gas mixture (synthesis gas) (Wang et al. 2008), this gas possesses a heating value that can replace the fossil fuels with a high efficiency to generate heat (Yassin et at the 2008).

Experimental

Chemical, Thermal and Lignocellulosic Characterization.

The chemical characterization was carried out by means of a bromathological analysis, with application of the procedures ASTM to determine humidity content (D 2867-99), ash (E1755-01), volatile matter (D2866-83) and fixed coal (by difference), likewise it was carried out a thermal analysis to know the variation of different properties of the matter as function of the temperature carrying out a thermogravimetry study using a thermobalance Pyris TG/DTA Perkin Elmer, the method marks to weigh the sample in alumina panels, heating from 25 to 800ºC, using three heating conditions at 10, 15 and 20 ºC/min. The kinetics of decomposition was obtained using a thermobalance Mettler Toledo 851, with the same conditions applying the ASTM E698.

The elementary analysis was carried out under the ASTM D5373/D1989 specific conditions. This is a destructive technique, a quantity is weighed between 2 and 4 mg is subjected to a thermal oxidation among 1.600-1.800 ºC, in atmosphere of oxygen, with the total and quantitative conversion of the components was gotten in CO2 (Carbon), H2O (Hydrogen) and N2 (Nitrogen).

The lignocellulosic analysis was made whit procedures of TAPPI to determine the cellulose content (T 203 os-74), lignin (T 222 om-88) and hemicellulose (for difference) carrying out a previous extraction of the waxes (T-12 -75), it was determined in sample virgin and 300 ºC, 400 ºC and 500 ºC. These temperatures were extracted from DSC results.

Results

The bromathological analysis was carried out triplicated for all the samples, it is presented in Fig. 1, the average of the obtained results.

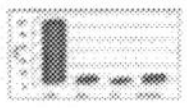

Fig. 1. Bromathological analysis

A content of volatile material was observed bigger than 80% and a smaller percentage than 10% of ashy, these characteristics present the appropriate bran to use it in a gasifier.

Differential scanning calorimery (DSC)

The obtained results by means of a DSC, they represent the differential heat contributed to a sample (regarding the one contributed to the reference), expressed as flow of heat (dQ/dt) in function generally of the temperature. The results obtained of the analysis they are in Table 1.

Table 1. DSC results

Temperature (°C)	Δ H (J/g)
113.96	56.55
254.49	63.02
371.61	-85.64
460.04	1148.08

The temperatures were determined for the analysis of the lignocellulosic content of the bran, it was demonstrated that at temperatures lower than 200 °C comes off the humidity of the sample, of 220 at 315 °C the decomposition of the hemicellulose is observed, of 315 at 400 °C the decomposition of the cellulose is presented and to temperatures higher than 400 °C the decomposition of the lignin is observed. Data compared with the contained lignocelluosic of the palm oil (Yang et al., 2005)

Elementary Analysis

It was determinated the content of Carbon, Hydrogen, Nitrogen, Sulfur and Oxygen; since it influences in the proportion agent-oxidizer / residual-optima, besides defining the production of pollutants of the type of nitrogen oxides and / or sulfur, or chloride of hydrogen. It is shown in the Fig. 2

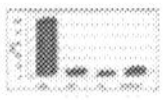

Fig. 2 . Elementary analysis

Combustion energy

It's gross energy of combustion that possesses the bran which is transferred in form of heat, it is denominated to be able of heating, it is determined in a theoretical way using the obtained values of the carried out elementary analysis using the following formula:

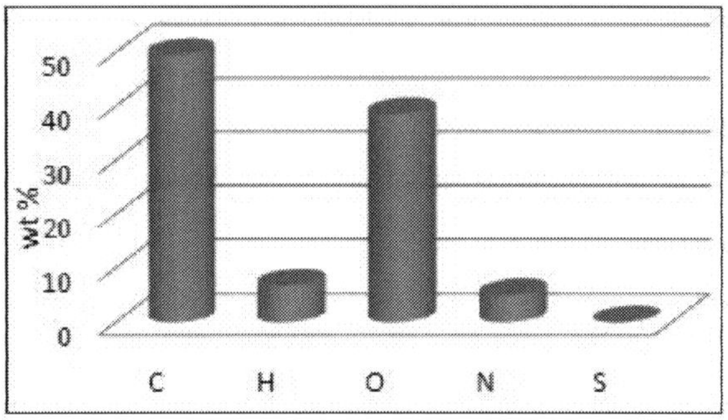

It presents 2.1 % of the value measured to the calculated value (Reed and Das, 1988).

Thermogravimetric Analysis

This measure is the variation of the mass of a sample when it undergoes a change of temperature in a controlled atmosphere. The record of these changes gives information on the sample breaks down or reaction with other components generating volatile components, that originate physical and chemical changes. For most of these chemical reactions it has been found that the dependent factor of the temperature is adjusted to the equation of Arrhenius (2)

where K0 is the factor of frequency and Ea is the energy of activation of the reaction. This expression is adjusted to the experimental results in a wide range of temperatures and it is considered like a first appropriate approach for the study of the effect of the temperature on the kinetic equation. Of the previous expression it can be seen that to constant temperature as much as bigger is the Ea, smaller it will be the constant of speed and therefore slower it will be the reaction speed. The obtained value of the kinetics of the decomposition of first order, it is shown in Table 2, which corresponds to the obtained determination of the application of the Star software version 6.2 of the thermobalance used.

Table 2. Kinetics results

Material	**ln (Ko)**	**Ea (KJ/mol)**	**n**
Trash of malt (Bran)	46.33	236.27	1.00

It was observed that to more heating speed (20 °C/min) it has a quick lost of weight approximately of 49 % at 320 °C at time of 17 min, in a range of 400-700 °C it had lost 88. 63% in 37 min. The experiment concluded leaving in 38.5 min as residual 5.25% at 800 °C.

Lignocellulosic Analysis

The realization of this analysis used information collected in the thermal analysis, with the purpose of using the appropriate temperatures. This can be seen in Fig. 3

Fig. 3. Lignocellulosic Analysis

It was observed that the cellulose content disappeared in the calcined sample at 500 °C, however it was remarkable the increase of the lignine content among the samples in virgin state and the calcined samples at 300 °C, following its increment in smaller proportion at 400 °C and 500 °C. The hemicellulose content, it presents a remarkable decrease at 300 °C having an increment at 400 and 500 °C.

Conclusions

The bran can cover an important part of our energy needs, when considering the advantages of its use, since it presents high combustion energy (20482.1 KJ/Kg), a residual content of 5.28% and a content of humidity smaller than 10% in dry base, it is considered that to achieve a future development of the use of these alternative energy, it is important to diffuse the existent possibilities and to develop technologies for its use.

References

1.Yang H., Yan R.,Chen H., Zheng. 2005. In-Depth Investigation of Biomass Pyrolysis Based on Three Major Components: Hemicellulose, Cellulose and Lignin. *National Laboratory of Coal Combustion, Huazhong University of Science and Technology, Wuhan, 430074, P. R. China.*

2.González, J.F., Gañan, J., Ramiro, A., González-García, C.M., Encinar, J.M., Sabio, E., Román, S. 2006. Almond residues gasification plant for generation of electric power. Preliminary study. *Fuel Proccesing Technolog,* 87, 149-155.

3.Mussatto, S.I., Dragone G., Roberto, L.C. 2006. Brewers' spent grain: generation, characteristics and potential applications. *Journal of Cereal Science,* 43, 1-14.

4.Wang, L. C., Weller, L., Jones, D.D., Hanna, M.A. 2008. Contemporary issues in thermal gasification of biomass and its application to electricity and fuel production. *Biomass and Bioenergy,* 32, 573-581.

5.Yassin, L., Lettieri, P., Stefaan, J.R., Germana, A. 2009. Techno-economic performance of energy from-waste fluidized bed combustion and gasification processes in the UK context. *Chemical Engineering Journal,* 146, 3, 315-327

GROUNDS OF COFFEE FOR THE GENERATION OF ENERGY

Y. Pliego-Bravo, M.De Felipe- Ramírez, R. Cerecero-Enríquez, G.Alvarado-Kinnell
Instituto Tecnológico de Orizaba, Orizaba, Veracruz, 94320, México.
ypliego2002@yahoo.com; marcedefelipe@hotmail.com; cereceros@yahoo.com

Introduction

The biomass is considered one of the most important energy sources due to its renewable character and non pollutant, this can be used as energy.The energy use of the biomass does not count in the increase of the gases of effect green house, because the balance of emissions of CO_2 to the atmosphere is neutral. Indeed, the CO_2 generated in the combustion of the biomass is reabsorbed by means of the photosynthesis in the growth of plants, therefore, there are not emissions that increase the quantity of CO_2 present in the atmosphere. (Spitzer, 1996)

Wang, L. et al.(2007) affirm that the combustion, pyrolysis and gasification are the main methods of conversion of thermochemistry, by means of which the biomass is burnt to provide heat to the processes of industry.The heating power of the materials, solids, liquids or gases can be determined experimentally in a calorimetric bomb or to be calculated starting from their composition or some properties using a mathematical model. Several models have intended to predict the heating value of many types of materials of diverse composition. (Sheng, J.et al., 2007).

The main aim of this work was to characterize the coffee grounds, considered a waste product from the coffe agroindustry and it can be used to contribute with a very important calorific content.

Experimental

The chemical characterization was carried out applying the ASTM methods for the bromathological analysis, humidity content (D2867-99), ash (E1755-01) and volatile matter (D2866-83). Likewise the elementary analysis (D5373/D1989) and the x-ray fluorescence.This continued with lignocellulosic analysis using the TAPPI procedures for the cellulose determination (T203 os-74), lignine (T222 om-88) and hemicellulose, these were performed in raw coffee grounds and calcinated samples at 300, 400 and 500 °C.

Results

Bromathological analysis. It was carried out triplicated for all the samples, the average of the obtained results is humidity 5.9022 %, ash 1.2014 %, volatile matter 97.1315 % and fixed carbon1.6683 %, these results present the suitable matter to be used in a thermochemical process. Low content of ash and high content of volatile matter.

Elementary analysis.It was determined the average content C 58.00 %, H_2 6.95 %, N_2 2.34 % and S 0%. In order to come to know the composition of the ashes, the analysis by x-ray fluorescence was performed.The results are shown in Table 1.

Table 1. Analysis by x-ray fluorescence

Coffee grounds	% Si	% P	% Cl	% K	% Ca	% Fe	% Na	% Al	% Mn
1	1.66	1.71	1.00	13.58	13.58	32.41	0	0	0

Lignocellulosic analysis. The cellulose and hemicellulose were disappearing, and the lignine was like reorganizeing the liberated carbons. It is in Table 2.

Table 2 . Analysis lignocellulosic

Type of sample	% Gums	% Cellulose	% Lignine	% Hemicelulose
Raw material	14.00	17.58	38.91	29.51
300 °C	3.27	2.22	87.94	6.58
400 °C	1.40	2.63	84.73	11.24
500 °C	0.35	0.0	87.13	14.47

The thermal analysis was performed in a Thermobalance Mettler Toledo, it can be appreciated a quick loss of weigth and the sample does not go further in temperature

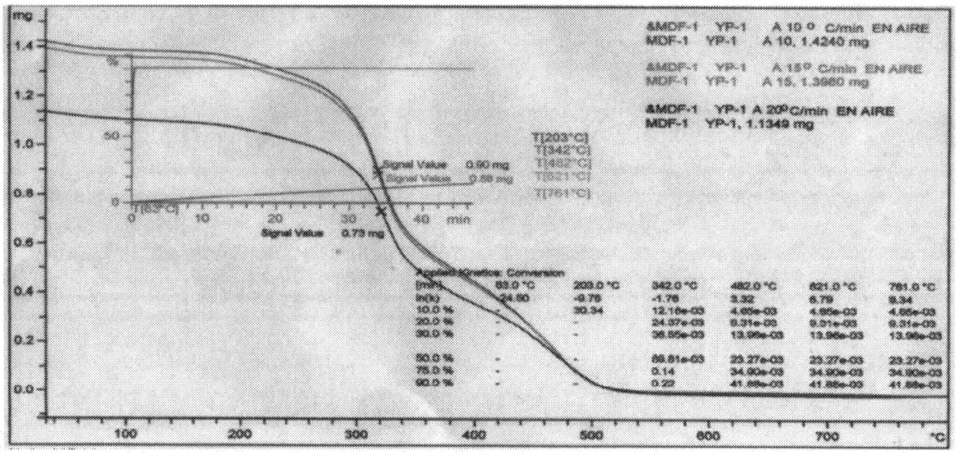

Figure 1. Thermogram of decomposition of coffee grounds

Conclusions

The characterization of the matter is an important factor because the obtained parameters will be a factor that affects the design of the thermochemical process directly, which will be adapted to exclusive conditions of the coffee grounds.The design of the thermochemical process will be according to the conditions of the matter making of the prototype an unique item for the use of the waste product of the coffee industry.

References

Ondrej, M. Miki, K. and Sou, H. (2008) A study on pyrolytic gasification of coffee grounds and implications to allothermal gasification. Biomass and Bioenergy. 32, pp 78-89

Ragnar W. (2000). Gasification of biomass: comparison of fixed bed and fluidized bed gasifier. Biomass and Bioenergy. 18, 6, pp 441-527

Feik, C. Carpenter, D. y Gaston, K. (2008). Pilot-Scale evaluation of fluidizable reforming catalyst for biomass syngas cleanup, National Bioenergy Center, National Renewable Energy Laboratory, Golden, CO.Power-Gen Europe.

Babu, B. y Sheth, P. (2006). Modeling and simulation of reduction zone of downdraft biomass gasifier: Effect of char reactivity factor. Energy Conversion and Management. 47, 15-16, pp 2602-2611

Baratieri, M. Baggio, P. y Fiori. (2008). Biomass as an energy source: Thermodynamic constraints on the performance of the conversion process. Bioresource Technology. 99, 15, pp 7063-7073

Changdong, S., Azevedo, J.L. (2007) Estimating the higher heating value of biomass fuels from basic analysis data. Biomass and Bioenergy, 28, pp 499-507.

Stability of Pyrolysis Oils Produced From Auger and Entrained Flow Reactors

Javeed Mohammad, Chemical Engineering, Mississippi State University, Misissippi State, MS

R. Mark Bricka, Chemical Engineering, Mississippi State University, Mississippi State, MS

Pyrolysis oil or bio-oil currently has tremendous potential to be used as a heavy fuel oil (ASTM No. 4) substitute meeting the 21st century energy demand at least partially. Pyrolysis oil can be produced from divergent feed stocks including waste biomass such as forestry residues, saw dust, bark, chicken litter, demolition wood, used tires, and others. The increased fuel energy density compared to raw biomass accompanied by negligible waste production from pyrolysis provides attractive options to the pyrolysis oil producers. Environmental benefits of pyrolysis oil include significantly lower green house gas (GHG) emissions as compared to the fossil fuels. Furthermore, the utility of pyrolysis oil as a light fuel oil (ASTM No. 2) can be likely achieved by upgrading techniques such as catalytic hydrodeoxygenation, hot vapor filtration, and solvent addition. Being a novel liquid fuel in the today's market both chemical and physical properties (density, viscosity, pH, water content, flash point etc) of pyrolysis oil need to be studied extensively before its engine application. During the authors' investigation, solvent addition technique was explored to increase the storage stability of pyrolysis oils, and possibly further the understanding of their chemical mechanisms. As such, pyrolysis oils are widely known to undergo polymeric chemical reactions which are detrimental to their long-term (ASTM definition ≥ 1 yr) storage stability. Hence, a detailed study was undertaken to investigate the storage stability of pyrolysis oils by varying feedstock, reactor, and storage conditions. A wide range of chemical additives with optimized concentrations were evaluated to stabilize the pyrolysis oils. The entire investigation was conducted in four phases and the results obtained during these phases will be discussed in detail at the forthcoming annual AICHE meeting.

See more of this Session: Environmental Aspects of Biofuels: III-Soil/Biomass Pyrolysis
See more of this Group/Topical: Environmental Division

Modification and Testing of a Natural Gas Combustion Unit to Accept Biomass as a Fuel Source

R. Mark Bricka, Chemical Engineering, Mississippi State University, Mississippi State, MS

Heather S. Thomas, Chemical Engineering, Mississippi State University, Mississippi State, MS

A green house operation in north-east Mississippi utilized a hot water system to provide heating for their greenhouses. This hot water system consisted of a series of natural gas fired boilers and a pressurized forced water distribution system to circulate heated water to the houses. Recently the spike in natural gas prices more than tripled the heating bill for the green house operation. As a result of this cost increase technologies were sought to lower fuel expenses.

A study was initiated to evaluate the effectiveness of different technologies and a variety of fuel sources to assist with cost saving effort. Originally package small wood burring boilers were installed and evaluated but proved to have high operational and maintenance costs. It was decided that an existing gas boiler could be retrofitted to accommodate solid fuel feeds. The system was constructed and studies were initiated to evaluate the use of coal, wood, waste paper, waste wood and a mixture of wood and glycerol as fuel sources.

Technical support was provided to improve the design of the solid feed combustion unit and stack sampling was undertaken. Mass and energy balances were conducted to evaluate the efficiency of the modified boiler and various feed stocks.

This paper will discussed the fabrication and modifications made to the boiler, the stack gas sampling efforts, the effects of the various feed stocks. The solution for final design implementation will also be discussed.

See more of this Session: Environmental Aspects of Biofuels: III-Soil/Biomass Pyrolysis
See more of this Group/Topical: Environmental Division

Laboratory-Scale Pyrolysis of CCA-Treated Wood Waste for Metals Recycle

Amy M. Parker, Dave C. Swalm School of Chemical Engineering, Mississippi State University, Mississippi State, MS

R. Mark Bricka, Chemical Engineering, Mississippi State University, Mississippi State, MS

For over 30 years chromated copper arsenate (CCA) has been the most widely used wood preservative. However, concerns over possible exposure to the toxic metals has lead to the restriction of CCA-treated wood to industrial applications since 2003. The restricted use policy has lead to an increased volume of CCA-treated wood slated for disposal, with some estimates as high as 24 million tons of CCA available for disposal by 2020. Although landfilling the out of service material has been the acceptable means of disposal, concern over soil and groundwater contamination, linked to the leaching of CCA-metals from landfills, has created the need for an alternative disposal method for CCA-treated wood waste.

This research focuses on employing fast pyrolysis as a means of recycling CCA-treated wood waste. A primary goal is to concentrate the metal components in the bio-oil, for re-use in wood preservatives, while recovering the energy value of the wood. Pyrolysis experiments were performed in a laboratory-scale pyrolysis system at temperatures ranging from 400°C to 550°C, at varying carrier gas flow rates, under atmospheric conditions, and each pyrolysis product was analyzed for CCA-metal content. Subsequent analysis of the bio-oil product included the determination of viscosity, water content, and pH. The char fraction additionally underwent electrokinetic (EK) treatment to extract the remaining CCA-metals. The results of this study and details of on-going research will be presented in this paper.

See more of this Session: Environmental Aspects of Biofuels: III-Soil/Biomass Pyrolysis
See more of this Group/Topical: Environmental Division

Design and Operation of a Laboratory-Scale Pyrolysis System

Melissa Cook, Dave C. Swalm School of Chemical Engineering, Mississippi State University, Mississippi State, MS

Amy M. Parker, Dave C. Swalm School of Chemical Engineering, Mississippi State University, Mississippi State, MS

R. Mark Bricka, Chemical Engineering, Mississippi State University, Mississippi State, MS

Since the early 1970's, the most widely used wood preservative has been chromated copper arsenate (CCA), resulting in nearly 80% of all treated wood products in North America being treated with CCA. By the end of 2003 CCA-treated wood was restricted to industrial applications, resulting in a considerable increase in the volume of CCA-treated wood slated for disposal. Landfilling was considered an acceptable means of discarding CCA-treated wood products until recently, as there have been instances of the toxic metals leaching from the landfills and contaminating the surrounding soil and groundwater. It is clear that traditional disposal methods are not adequate and that a safe and efficient disposal method for CCA-treated wood must be developed.

Fast pyrolysis, the heating of biomass at temperatures between 400°C and 650°C in the absence of oxygen, is a promising technology that can be applied to the disposal of CCA-impregnated wood waste. Pyrolysis of lignocellulosic material produces char, liquid condensate (bio-oil), and non-condensing gases. The focus of this research is on removing the CCA-metals from the treated wood waste while recovering the energy value of the wood. This is accomplished by concentrating the CCA-metals in the bio-oil, for possible re-use in wood preservatives, during pyrolysis. A laboratory scale pyrolysis system, capable of operating in the desired temperature range under atmospheric and vacuum conditions, has been designed. The system is also designed to enable the collection of each pyrolysis product so that complete mass balances on the metals can be performed, tracking the fate of the CCA components. This paper discusses the process of designing and operating the laboratory scale pyrolysis system, as well as preliminary experimental results.

See more of this Session: Environmental Aspects of Biofuels: III-Soil/Biomass Pyrolysis
See more of this Group/Topical: Environmental Division

A Improved Kinetics Model of Biosorption and Its Application in Bio-Electro Tower Reactor

Wang Tao[1], Liu Bo-tan[2]*, Zhou Yun[3]*

Abstract: A 2-D numerical kinetic model considering flow velocity and adsorption is developed to simulate the bio-electro tower reactor (BETR). This new model considers the adsorbed amount when equilibrium q_e as transient variable, which is superior to the old pseudo-first-order and the pseudo-second-order model which regards q_e as a constant. We researched the intensifying effect of electric field upon heavy metal ions adsorption process. The calculation result fits well with the experimental data. BETR is a coupling technique; the mechanic of BETR is that outer electric field can enhance the mass transfer rate when the solute is metal ions. Two kinds of carriers, pottery ball (PB) and 3-dimensional electrode (3DE), were used to support the biofilm layer. And organic wastewater containing Zn^{2+} is selected as a sample to validate the model. The 3DE carriers can be polarized by outer electric field, but PB cannot. It is found that Zn^{2+} transfer faster in 3DE carriers than that in PB and a intensifying coefficient$\eta$is introduced to stand for this effect in BETR.

Keywords: numerical model, biosorption, 3-dimensional electrode, mass transfer, BETR

(**First author**: The Key Lab of Hollow Fiber Membrane Materials and Membrane Process of Ministry of Education , Tianjin Polytechnic University , Tianjin 300160 , China.

***Correspondence author2**: State Key Laboratory of Chemical Engineering, Tianjin University, Tianjin 300072, China; Email: liubotan@126.com;

***Correspondence author3**: Tianjin Lonwin Network Science & Technology Development Corp. No. 15, Longtan road, Hedong district, Tianjin 300181, Email: zhouy@lonwin.com.cn)

Introduction

Nowadays, industrial utilization of heavy metals such as metal plating, tanning, preparation of catalysts, generates large quantity of aqueous effluent containing high levels of heavy metal (Volesky, 1990; Hatifiend et.al, 1996; Hatifiend et.al, 1998) . It is necessary to develop harmless novel process for efficiently treating this sort of wastewater (Li et.al, 2001; Ahluwalia et.al, 2007; Zaki et.al, 2007 ;). Generally, biosorption has been thought a better way than the conventional methods. Further more, bio-electro reactor, coupling electrochemistry technique with microbial way, has been proved a more effective novel process (Li et.al, 2006). The BETR (bio-electro tower reactor) that we invented (Li et.al, 2006) has an anode in the center and cathode on the tower wall. It fills 3-dimensional electrode materials (granule or crumb-shaped) and made them electrified by outer electric field. These 3DE materials form the third electrodes. On the surface of the 3DE some electrochemistry reactions take place, and biofilm can be obtained after culturing and acclimating microorganism under low direct current.

Adsorption/biosorption equilibria information is most important for analysis or design of the reactor. There has been large research effort on the adsorption equilibria in heavy metals biosorption (Dursun, 2006; Deng et.al, 2007; Hanif et.al, 2007; Preetha et.al, 2007). The equilibrium concentrations are a function of temperature. Therefore, the adsorption equilibrium relationship at a given temperature is referred as adsorption isotherm. Several adsorption isotherms were adopted to correlate the biosorbent-assisted heavy metals removal, among them Freundlich and Langmuir equations are the most widely used. However，Kinetic studies in biosorption of heavy metals are far from perfect. Several adsorption kinetic models have been established to understand the adsorption kinetics. The pseudo-first-order and the pseudo-second-order kinetic model are the most frequently used to quantify the extent of uptake in biosorption (Djeribi 2008; Kargi et.al, 2007; Pamukoglu et.al, 2007; Miretzky et.al, 2008; Guo et.al, 2008).

The rate of adsorption is expressed as

$$\frac{dq}{dt} = k_1(q_e - q) \quad (1)$$

The pseudo-first-order kinetic expression is deduced from integration of Eq. (1)

$$q = q_e(1 - \exp(-k_1 t)) \quad (2)$$

with the boundary conditions as follow: t=0, q = 0, and at $t = t$, $q = q$.

Similarly, pseudo-second-order model is derived from the rate of adsorption

$$\frac{dq}{dt} = k_2(q_e - q)^2 \quad (3)$$

Integration of Eq. (3) with the boundary conditions t=0, q = 0, and at $t = t$, $q = q$, gives

$$\frac{t}{q} = \frac{t}{q_e} + \frac{1}{k_2 q_e^2} \quad (4)$$

Here, both the models hypothetically regard q_e, the adsorbed amount when equilibrium, as a constant during the whole process. However, it is obviously logical contradiction because q_e actually complies with the adsorption isotherm equation. As q_e is a time-dependent variable, the pseudo-first-order and pseudo-second-order model are not correct.

In this paper, Zn^{2+} was selected as simple ion to conduct the experiment and a numeric kinetic model considering mass transfer inside biofilm is established originally which firstly includes the isotherm equation. Langmuir equation is introduced for adsorption isotherm according our previous work (Zhang 2005) because it is consistent with our experiment data. The model is numerically solved by calculating q_e at every time step, avoiding regarding q_e as a constant. Meanwhile, it is also original that the model is applied to study the biosorption under electric field. The calculated results are compared with experimental data. The experiment in this work includes two parts: 1. organic wastewater purifying in the fixed bed made of pottery ball and 2. of 3DE granule. Theoretically, the mass transfer is enhanced by the electric field when the fixed bed is made of 3DE granule. For this reason, different mass transfer parameters $K\alpha$ are regressed from experimental data respectively and used as important parameter in the simulation.

Method

1. Numeric model:

We set up the numeric model of electric-biofilm system inside the reactor tower under pole coordinate. It is assumed that:

1) the supporting particles on which the microbial biofilm is located can be treated as porous media;

2) the adsorption of metal ions on the biofilm is of Langmuir style;

3) the wall effect of the tower is neglected;

4) the effect of acceleration on the ions movement due to the DC electric field only affects the mass transfer coefficient K.

So the governing equations of the model include:

continuity equation:

$$\nabla u = 0 \qquad (5)$$

and momentum equation:

$$\frac{\partial u}{\partial t} + u \cdot \nabla u = -\frac{1}{\rho}\nabla p + \nu\nabla^2 u - \Delta p \qquad (6)$$

Here the last item on the right is the drag due to the particles. In this work, d = 0.1m, u = 7.42×10^{-5}m·s^{-1}. The Re number is roughly estimated to be 7.42×10^{-3} by substitution the density and viscosity of the metal ions solution with those of water, thus the flow pattern is laminar inside the tower.

Therefore, the Blake-Kozeny equation (Ergun, 1952) which is appropriate for laminar flow is applied to acquire the pressure drop.

$$\frac{|\Delta p|}{L} = \frac{150\mu}{D_p^2}\frac{(1-\varepsilon)^2}{\varepsilon^3}u \qquad (7)$$

The governing equation of the ions concentration is as follows:

$$\frac{\partial c_L}{\partial t} + u \cdot \nabla c_L = D\nabla^2 c_L - K\alpha(c_L - c^*) \qquad (8)$$

K is the mass transfer coefficient, and α, the surface area of unit volume. c^* is the equilibrium ion concentration corresponding to the concentration inside the biofilm, c_s.

The concentration inside the biofilm c_s is the total ions accumulated during the time form 0 to t.

$$c_s = \int_0^t K\alpha(c^* - c_L)dt \qquad (9)$$

If the equilibrium equation is $c_e = f(c_L)$, then the c^* will be

$$c^* = f^{-1}(c_s) \qquad (10)$$

The boundary conditions are: at the inlet of reactor: $u_{in} = u_0$, $c_L=c_0$; at the outlet of reactor: $\partial u / \partial z = 0$, $\partial c_L / \partial z = 0$; and on the wall: u=0, $\partial c_L / \partial r = 0$. Obviously c^* is similar to q_e in Eq(2, 4), but in another form.

The equilibrium equation $f(c_L)$ is the adsorption isotherm. In this work, according to the previous study of our group (Wang, 2006), the adsorption model is of Langmuir:

$$c_{se} = \frac{Q^0 bc_{Le}}{1 + bc_{Le}} \qquad (11)$$

Here the parameter Q^0=251.1mg·L^{-1}, b=0.46511.

2. Experiment

In the electro-biofilm reactor pictured in Fig. 1, functional bacteria were enriched and inoculated on polypropylene packing to formed biofilm. After cultured and acclimated, the functional bacteria can endure $Zn^{2+}$of 250 mg.L^{-1}. The biofilm would come into being in two weeks.

In this work, solution is at condition of pH=7.0, temperature at 20°C and $ZnSO_4$ (analytical grade) of 150 mg·L^{-1}. Direct current regulated power supply of WYJ—0 type was supplied by Quanli Electrical Appliance Co., Shanghai. Zn^{2+} was measured by the Polarized Zeeman Atomic Adsorption Spectrophotometer (HITACHI 180-80, Japan).

The experiment equipment is given in Fig. 1.

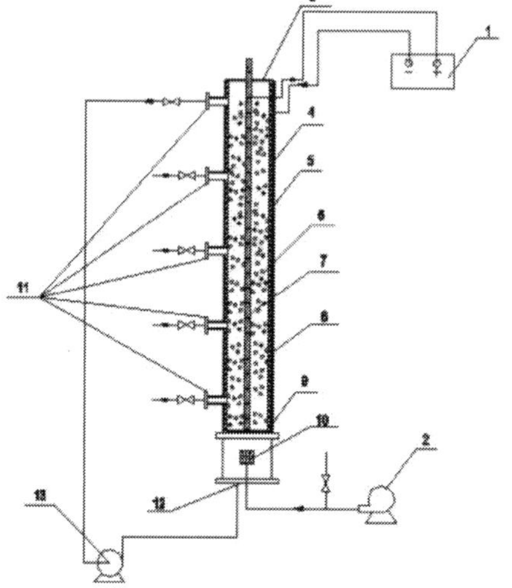

1 DC power 2 air pump 3 porous plane 4 insulation layer 5 Plexiglas tower 6 filler
7 stainless steel anode 8 active carbon fiber cathode 9 porous plane 10 aeration nozzle
11sampling nozzle 12 liquid inlet 13 peristaltic pump

Fig.1 Electro-biofilm tower reactor

As showed in Fig.1, electro-biofilm tower reactor has inner diameter of 100mm, it is divided into two parts by a porous plane. The upper section is 1000mm high and the down section is 120mm. In the up section, a stainless steel anode is located in tower center, and the activated carbon fiber cathode clings to the tower wall, outside the tower wall covers a layer of insulated material. The reactor is filled with pottery balls or 3DE material as biofilm carriers, and the electric field is formed between anode and cathode by an adjustable direct current regulated power supply(WYJ—0 type, Quanli Electrical Appliance Co., Shanghai, measure range: 0-15V, 0-2A). In the center of lower section there is an air distributor. The air supplied by air pump enters reactor from the bottom and generates a lot of little bubbles symmetrically through air distributor. Little bubbles blend adequately with organic solution containing Zn^{2+} that come from wastewater container and come into upper section of reactor through the porous board. After bio-electro reaction, solution is depurated and flow back to wastewater container again. This flowchart is called interim circle process.

The biofilm is formed by the activated sludge which can endure Zn^{2+} solution at least $150mg{\cdot}L^{-1}$ after cultured and acclimated.

The experiment is conducted under following conditions: direct current 2.8 voltage, air supply $0.75m^3.h^{-1}$, electrode immerge 1000mm in depth, stack volume of pottery balls (3DE material) 3.5L, temperature 25℃ and initial concentration of Zn^{2+} is 150 $mg{\cdot}L^{-1}$. The adsorption process apply interim circle method and 4L Zn^{2+} solution was infused into reactor, velocity of circle flow is of $35ml{\cdot}min^{-1}$. We measure concentration of Zn^{2+} at fixed time interval.

Results

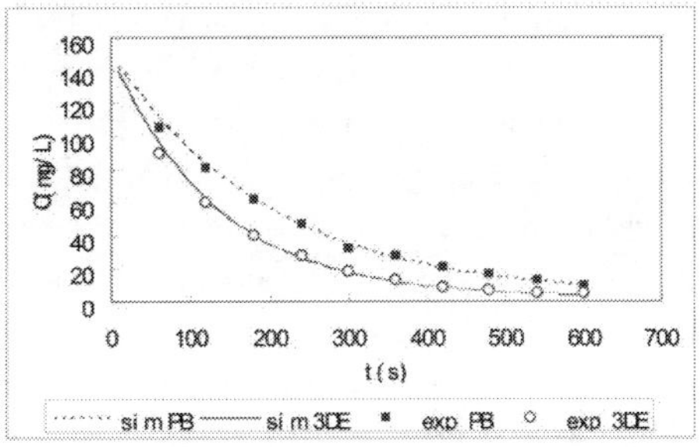

Fig.2 The experimental and simulated data of solution ions concentration at different times

The model is calculated using ADI algorithm. In the experiment the concentration is measured once per minute until the concentration be unchanged. Then the model is calculated and through regression the $K\alpha$, the very important parameter of mass transfer is obtained. Fig. 2 gives the experimental and simulated concentration of Zn^{2+} at different time. It can be seen that the simulated curves fit the experimental data very well. The electric field can enhance the mass transfer rate when the solute is metal ions. The 3DE material can be polarized by outer electric field, but PB cannot. So the ions theoretically move faster in 3DE case. Fig.2 also indicates that the Zn2+ concentration of 3DE reached the equilibrium faster. For the pottery ball, $K\alpha$ is regressed to be 7.51 $mg \cdot L^{-1}s^{-1}$, and for the 3DE particle, $K\alpha$ is 5.20 $mg \cdot L^{-1}s^{-1}$. Here we define: $\eta=(K\alpha)_{3de}/(K\alpha)_{pb}$ as the intensifying factor, therefore, $\eta=1.44$. This suggests that the ions transfer 1.44 times faster in the 3DE carrier than that in the pottery ball carrier.

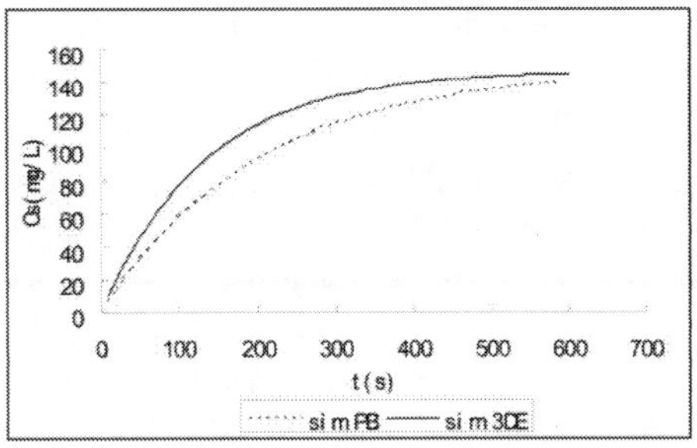

Fig.3 The simulated data of ions concentration inside biofilm at different times

Fig.3 gives the transient concentration of Zn^{2+} ions inside the biofilm. This is calculated through Eq(9) above. The concentration curves indicate typical Langmiur adsorption type, but rarely seen in the previous literatures because the transient concentration data within the sorbent can hardly be measured.

Discussion

In this study, a numerical kinetic model for bio-sorption of metal ions inside tower reactor is developed originally. This model is superior to the pseudo-first-order and the pseudo-second-order kinetic model. Firstly, both the pseudo-first-order and the pseudo-second-order kinetic model assume that q_e, the adsorbed amount when equilibrium, be a constant, thus the Eq(2,4) can be acquired simply by integration. These two approaches are easy to understand but hardly consistent with the actual process for the impetus of adsorption is always changing. While the model in this study includes the isotherm equation so as to avoid regarding q_e as a constant, and can be numerically solved by calculating q_e at every time step. Secondly, the model developed here is a "field" model, i.e. this approach can give the local concentration and velocity information at any position. This means the approach can be used to assist reactor designing.

To validate the model in this work, the model is applied to simulate the adsorption in the biofilm tower reactor. Zn^{2+} is selected as solute in the liquid. The reactor is filled with pottery ball and 3-d electrode material as carrier respectively. Because of the extremely small inlet liquid velocity, the flow field is not important in this experiment, but the transient concentration field becomes very typical. By regression from experimental data the mass transfer parameter $K\alpha$ is acquired and the simulated result gives typical concentration variation of adsorption process. Both the simulated results of pottery ball carrier and the 3-d electrode carrier fit well with the experimental data under the model hypothesis. The result that intensifying factor $\eta=(K\alpha)_{3de}/(K\alpha)_{pb}=1.44$ showed 3-d electric field can accelerate mass transfer of heavy metal ions and improve bio-adsorption of Zn^{2+}.

The the pseudo-first-order and the pseudo-second-order kinetic model assume q_e, the adsorbed amount at equilibrium, as constant, and the parameter q_e, k_1 or k_2 are usually regressed from the experimental data. Now we consider two cases in which the adsorption equilibriums are same but the mass transfer rates are different. What if we use, for example, the pseudo-first-order kinetic model to approach these two experimental data? It might be the case that two different q_e were regressed from the data respectively. That means the equilibriums of these two cases are different, which is contrary to our assumption. So we can see clearly that the former models are inappropriate. In this work, we use PB and 3DE as carrier respectively, the mass transfer rate is different in these two cases. So, as discussed above, the the pseudo-first-order and the pseudo-second-order kinetic model are not proper to approach the experiment in this work. In other words, the experiment in this paper has exhibited the limitation of the former models. In the present model, we think q_e as a changing parameter instead of a constant. By introducing Langmuir equation in calculation, this new approach can treat the experiment data correctly and precisely. This indicates that the numerical model of this study is successful. Furthermore, we give the intensifying factor $\eta=1.44$. The metal ions transfer faster in 3DE material carriers than that in pottery ball carrier. The intensifying factor η is defined as $\eta=(K\alpha)_{3de}/(K\alpha)_{pb}$ to give a precise description.

Conclusion

The pseudo-first-order and the pseudo-second-order model are two kinetic models to assess the extent of uptake in biosorption. These two models have limitation as they regard q_e

hypothetically as a constant. The parameter q_e, k_1 or k_2 are usually regressed from the experimental data, but in this paper, we use a novel type BETR to study treatment of heavy metal ions under electric field, with pottery ball and the 3DE as carrier respectively. In this case, neither pseudo-first-order model nor the pseudo-second-order model can approach the experiment correctly because that may lead to some confusion as we have discussed in detail.

So we develop a novel kinetic model which takes q_e as a time-dependent variable in this paper. Using Zn^{2+} containing solution as sample, bio-adsorption and mass transfer process in BETR were simulated. The result of model calculation is consistent with the experimental data. It is demonstrated that the metal ions transfer faster in 3DE carriers than that in pottery balls. And the intensifying factor η=1.44 which gives a precise description. The numeric model in this study is proved successful and its application in simulation of BETR is a valuable example.

Acknowledgments The research was supported and by Nature Science Fund of Tianjin, China (07JCZDJC01400) and Nature Science Fund of China (50808130)

Nomenclature

b - parameter of Langmiur equation

c_e - equilibrium ion concentration in liquid, mg·L^{-1}

c^* - equilibrium ion concentration corresponding to c_s, mg·L^{-1}

c_L - ion concentration in liquid, mg·L^{-1}

c_{Lin} - ion concentration in liquid at inlet, mg·L^{-1}

c_s - ion concentration inside biofilm, mg·L^{-1}

d - diameter, m

D - turbulent mass diffusion coefficient, m^2·s^{-1}

D_p—diameter of carrier particles, m

Ka - multiple of mass transfer coefficient and the surface area per unit volume, s^{-1}

Q^0 - parameter of Langmiur equation

q_e—adsorption amount when equilibrium

q—adsorbed amount

Re - renold number

u - liquid velocity, $m{\cdot}s^{-1}$

u_{in} - liquid velocity at inlet, $m{\cdot}s^{-1}$

ρ - density, $kg{\cdot}m^{-3}$

μ - viscosity, $kg{\cdot}m^{2}{\cdot}s^{-1}$

ε - porosity

α - surface area per unit volume, $m^{2}{\cdot}g^{-1}$

η - intensifying factor of mass transfer due to electric field

References

Ahluwalia S S, Goyal D(2007), Microbial and plants derived biomass for removal of heavy metals from wastewater. Bioresour Tech. 98: 2243-2257

Deng L, Zhu X, Wang X., et. al(2007), Biosorption of copper(II) from aqueous solutions by green alga *Cladophora fascicularis*, Biodegrad. 18: 393–402.

Djeribi R, Hamdaoui O(2008), Sorption of copper(II) from aqueous solutions by cedar sawdust and crushed brick, Desalin . 225: 95–112.

Dursun A Y(2006), A comparative study on determination of the equilibrium, kinetic and thermodynamic parameters of biosorption of copper(II) and lead(II) ions onto pretreated Aspergillus niger, Biochemical Engineering Journal 28: 187–195.

Ergun S (1952). Fluid flow through packed columns, Chem. Eng. Prog, 48: 89-94

Guo X Y, Zhang A Z, Shan X Q(2008), Adsorption of metal ions on lignin, J. Haz. Mater. 151: 134–142.

Hanif M A, Nadeem R, Bhatti H N et. al(2007), Ni(II) biosorption by *Cassia fistula* (Golden Shower) biomass, J Haz. Mater. B. 139: 345–355.

Hatifiend T L, Kleven T L, Pierce D T(1996), Electrochemical remediation of metal-bearing wastewaters part□ : Copper removal from simulated mine drainage waters, Journal of Appl. Electrochem., 26:567-574

Hatifiend T L, Pierce D T(1998), Electrochemical remediation of metal-bearing wastewaters part : Corrosion-based Inhibition of Copper Removal by Iron (□), Journal of Appl. Electrochem., 28:397-403

Kargi F, Cikla S(2007), Kinetics of zinc(II) ion biosorption onto powdered waste sludge (PWS) at different operating conditions, Env. Eng. Sci . 24: 687–695.

Li X G, Cao H B, Wu J C, et. al (2001). Inhibition of the metabolism of nitrifying bacteria by direct electric current. Biotechnol Lett, 23:705-709

Li X G, Wang T, Jiang B, et al(2006). A bio-electro tower reactor dealing with wastewater include heavy metal ions and difficult biodegradation organic substances, CN1749177

Li X G, Wang T, Sun J S , et. al(2006), Biodegradation of high concentration phenol containing heavy metal ions by functional biofilm in bio-electro reactor. J Envir Sci, 18: 639-643

Miretzky P, Munoz C, A. Carrillo-Chavez(2008), Experimental binding of lead to a low cost on biosorbent: Nopal

(*Opuntia streptacantha*), Bioresour. Technol. 99: 1211–1217

Pamukoglu M Y, Kargi F(2007), Effects of operating parameters on kinetics of copper(II) ion biosorption onto pre-treated powdered waste sludge (PWS), Enzy. & Microb Technol. 42: 76–82

Preetha B, Viruthagiri T, Batch and continuous biosorption of chromium(VI) by *Rhizopus arrhizus*, Sepa. & Puri. Techno. 57: 126–133

Volesky B(1990), Biosorption of Heavy Metals, Department of Chemical Engineering, Montreal Quebec, Canada: McGill University, pp.14-19

Wang T(2006), Treatment of wastewater wontaining heavy metal ions and difficult biodegradation organic matter by electro-microbe coupling technics, PhD Thesis. China: Tainjin University, pp.74-76.

Zaki M M, Nirdosh I, Sedahmed G H (2007), Mass transfer characteristics of reciprocating screen stack electrochemical reactor in relation to heavy metals removal from dilute solutions, Chem Eng J. 126: 67-77

Zhang J(2005),Treatment of Plating Wastewater with Heavy Metal ions by micro-eletrolysis and Biofilm Process. Master Thesis. China: Tainjin University, pp.57-59

Density Functional Theory Studies On the Relative Reactivity of Chloroethenes On Fe(110)

Dong-Hee Lim, Civil and Environmental Engineering, University of Michigan, Ann Arbor, MI

Christian M. Lastoskie, Civil & Environmental Engineering, University of Michigan, Ann Arbor, MI

The adsorption and dissociation of perchloroethene (PCE), trichloroethene (TCE), and cis-dichloroethene (cis-DCE) on Fe(110) were investigated using density functional theory (DFT) and the generalized gradient approximation (GGA) to evaluate hypotheses concerning the relative reactivity of these compounds on zerovalent iron. Of adsorption sites investigated, an atop site, where the chloroethene C=C bond straddles a surface iron atom, was the most energetically favorable site for the adsorption of all three chloroethenes. Electronic structure and property analyses provide an indication of the extent of sp2-sp3 hybridization. The strong hybridization of the π-bonding orbital between the chloroethene C=C bond and the iron surface suggests that adsorbed chloroethenes are strongly activated on Fe(110) and are likely precursors for subsequent chloroethene dissociation on the Fe surface. The climbing image nudged elastic band (CI-NEB) method was employed to calculate activation energies of the chloroethene compounds according to the principal dechlorination mechanism of β-elimination. Chloroethenes with a higher number of chlorine atoms were found to have lower activation energies than those with fewer chlorine atoms. The activation energies of PCE, TCE, and cis-DCE at their rate-limiting steps in the gas phase are 9.9, 16.6, and 23.8 kJ/mol, respectively. Energy profiles along the reaction coordinate for the dechlorination paths are presented. The relative reactivity order among the chlorothenes was found to be PCE > TCE > cis-DCE. Furthermore, defects of the iron surface will be employed in the same simulation study to better understand how the degree of chlorination affects the reactivity of chloroethenes on iron in the remediation environment.

See more of this Session: Fundamentals of Environmental Kinetics and Reaction Engineering
See more of this Group/Topical: Environmental Division

Membrane Bioreactor Process for Purification of Groundwater Contaminated with Petroleum Hydrocarbons: Process Design and Upscaling

Varadarajan Ravindran, University of Southern California, Los Angeles, CA

Massoud Pirbazari, University of Southern California, Los Angeles, CA

The present investigation relates to the application the membrane bioreactor (MBR) technology for the purification of groundwater contaminated with petroleum organics. It represents a continuation of our previous study, where a predictive model was developed and used for performance forecasting and design of the MBR process under a variety of operating and process conditions. The petroleum organics in question included aromatics such as benzene, toluene, ethylbenzene and xylenes (BTEX), and aliphatic oxygenated additives exemplified by methyl-tert-butyl ether (MTBE). The MBR system discussed here employed a powder activated carbon (PAC) slurry with microorganisms acclimated to petroleum hydrocarbons. The process was operated as a closed system to prevent the volatilization of gasoline contaminants from the aqueous phase into the atmosphere.

This is an on-going study using a modeling approach wherein the total concentration of petroleum hydrocarbons was represented by total organic carbon as a lumped parameter for MBR process design and upcaling. The model considered various phenomena pertaining to pollutant transport, sorption and biochemical degradation. With reference to substrate transport, the model considered film transfer from bulk liquid phase to the biofilm-liquid interface, biofilm diffusion, and diffusion into the adsorbent particle. The biochemical reaction occurred in the biofilm immobilized on the PAC adsorbent particle as well as in the aqueous suspended phase. These reactions described biofilm and biomass growth and decay within the MBR system, and were governed by Monod kinetics. The adsorption equilibrium and kinetic relationships controlled the sorption uptake from the aqueous phase, and release of contaminants from the adsorbent to the biofilm immobilized on the PAC as well as to the microorganisms in the suspended phase. Thus, the contaminant removal mechanisms in the process included a combination of biofilm degradation, suspended phase biodegradation, and adsorption. The MBR model parameters were determined from independent laboratory-scale experiments and correlation techniques. The adsorption equilibrium and rate parameters for the model were determined by independent batch reactor studies, while the biokinetic parameters were estimated from chemostat studies. The MBR studies involved laboratory-scale experiments with and without PAC slurry, and organic contaminant removals as well as permeate fluxes were measured in each case. These experiments established the effectiveness of the MBR technology and provided the means for model verification under different process conditions.

The present study focuses on the application of the above model in upscaling the process from laboratory-scale to pilot-scale and eventually to full-scale. Scale-up of biological sorption processes and membrane filtration processes pose considerable challenge in the realm of process design. An integrated hybrid MBR process would therefore pose a bigger challenge owing to the combination of bio-sorption and membrane systems. The modeling and upscaling technique proposed in this presentation is an efficient time-saving and cost-effective strategy for design of real processes from bench-scale or laboratory-scale systems. The scale-up process shall involve some dimensional analysis and similitude techniques, wherein the scale invariance of dimensional groups will be investigated. The process upscaling protocol proposed here takes into consideration several aspects: biodegradation of the target contaminants in the biofilm phase and/or the bulk liquid phase, sorption of the contaminants on the adsorbent particles, and membrane filtration and separation. The dimensionless groups are developed specifically for the adsorption and biodegradation phenomena associated with the MBR process. The upscaling will involve dimensionless groups that compare the relative rates of biofilm kinetics and sorption kinetics. These will involve a combination of Monod kinetics and adsorbent diffusion transfer, and liquid film transport of the target contaminant(s). Various dimensionless numbers will be employed to examine the scale invariance or variance of different phenomenon. For instance, Damkohler numbers would relate the ratio of biochemical reaction rates with the rates of film transfer, adsorbent particle diffusion and biofilm diffusion of the contaminant(s). The Sherwood number would establish the relationship between liquid film transport rate outside the adsorbent, and diffusion rate into the adsorbent (pore and/or surface diffusion). The Schmidt number would denote the ratio of mass diffusive transport to momentum transport relating the diffusion rate of the contaminant(s) to the dynamic viscosity of the fluid. Several other dimensionless groups will be used to describe the relative rates of biochemical reactions and mass-transfer. The fluid dynamic regime will be characterized by the Reynolds number, and this shall be compared to the relative influence of other non-dimensional groups of importance. Another important aspect of upscaling the MBR process shall relate to membrane module design, essentially based on permeate flux, reactor residence time, process throughput, trans-membrane pressure, and contaminant levels.

Model simulations of the MBR process will be employed to predict the dynamics of the upscaled system under different conditions based on flow characteristics, biomass requirements, adsorbent requirements, contaminant removals, and membrane permeate fluxes. The demonstration of this upscaling protocol will be based on real MBR data for laboratory-scale operations, comparison with model simulations, and subsequent upscaling to a hypothetical pilot-scale or full-scale plant.

See more of this Session: Fundamentals of Environmental Kinetics and Reaction Engineering
See more of this Group/Topical: Environmental Division

Computational Study of the Interactions Between Ionic Liquids and Volatile Organic Compounds

Tingting Gao, Civil, Environmental and Sustainable Engineering, Arizona State University, Tempe, AZ

Jean M. Andino, Chemical Engineering, Arizona State University, Tempe, AZ

Geometry optimization and single point energy calculations were applied to investigate the interactions of ionic liquids with different kinds of volatile organic compounds. Methanol, formaldehyde, ethane and ethylene were used to represent alcohols, aldehydes, alkanes and alkenes, respectively. The results of the computational studies show that C_4minCl likely interacts with alcohols as compared to other classes of volatile organic compounds that were studied in this work. The interactions of methanol with different kinds of ionic liquids were also studied. In comparing C_4minCl, C_4minBr,and C_4minBF_4, the computational results suggest that C_4minCl is more likely to interact with methanol. Therefore, C_4minCl was proposed as the absorbent to develop a gas-phase alcohol sensor. The mechanisms of interaction between C_4minCl and ionic liquids are also explained from the electronic perspective. In addition, the computational result also suggests that there is a dissociative reaction between C_4minBF_4 and methanol, with the products being H_2O and CH_3•. Therefore, C_4minBF_4 may potentially be applied as a solvent to remove alcohols from a surrounding matrix. The findings in this work provide information for future experimental design.

See more of this Session: Fundamentals of Environmental Kinetics and Reaction Engineering
See more of this Group/Topical: Environmental Division

Natural Gas Based Hydrogen Production with Ethylene and Formic Acid Co-Production

Jorge Pena Lopez, Chemical Engineering Department, University of California at Los Angeles, Los Angeles, CA

Vasilios I. Manousiouthakis, Chemical Engineering Department, UCLA, Los Angeles, CA

The air quality in major metropolitan areas follows a deterioration trend, chiefly due to the large population of small and medium transportation vehicles. Given this conditions, scientists and engineers around the world are committed to use all tools available to reduce and finally stop this path to environmental destruction. It has previously proposed the use of hydrogen as an effective energy carrier for small and medium vehicles. This option has strengthened with the advancement in the development of fuel cells manufacturing and hydrogen storage. This growth is also reflected in the development of new hydrogen production designs that effectively surpass existing technologies. Several options exist for hydrogen production; one of the most explored has been the reforming of natural gas given that methane contains potentially two hydrogen gas molecules. The main problem with conventional natural gas reformers is the emission of carbon dioxide, a greenhouse whose emissions have led to an increase in its concentration in the atmosphere. Transforming carbon dioxide into a useful product prevents its emission into the atmosphere, however an inadequate process design might largely increase the cost of producing hydrogen, making it uncompetitive with gasoline.

In this work we present a hydrogen production process in which hydrogen and ethylene are co-generated, while no carbon dioxide is emitted. Natural gas undergoes oxidative coupling to produce ethylene and water. This mixture is separated and the water produced undergoes reforming along with natural gas to produce a hydrogen containing mixture. Hydrogen is purified through a pressure swing adsoption, a fraction of which goes through a series of compressors and becomes the final product. The rest of it is sent to a formic acid production system in which reacts with carbon dioxide producing formic acid. The overall process goes through heat and power integration techniques that allow the maximum utilization of energy in favor of hydrogen production, maintaining a self-powered plant and reducing the carbon dioxide emissions to zero. We show that there exists an optimal themodynamical and operational point for the operation of the plant.

See more of this Session: Fundamentals of Hydrogen Production
See more of this Group/Topical: Environmental Division

Identification of Low Temperature Carbon-Free Hydrogen Production Methods through Thermochemical Cycle Synthesis

Ryan J. Andress, Chemical and Biological Engineering, Rensselaer Polytechnic Institute, Troy, NY

Lealon L. Martin, Chemical and Biological Engineering, Rensselaer Polytechnic Institute, Troy, NY

The hydrogen economy presents itself as a means for reducing our energy requirements and overall environmental impact. Although energy conversions via hydrogen fuel cells are highly efficient, environmentally friendly, cost effective hydrogen production has been largely unrealized. Here we present a systems-engineering approach to identifying, screening, and evaluating thermochemical cycles, in which low temperature, carbon-free hydrogen production is achieved through emulating water-splitting. It is shown that with the presented methodology we can quickly and efficiently synthesize and evaluate complete sets of hydrogen producing reaction clusters based on thermodynamic feasibility and overall cycle efficiency.

See more of this Session: Fundamentals of Hydrogen Production
See more of this Group/Topical: Environmental Division

BIOLOGICAL HYDROGEN PRODUCTION FROM CORN-SYRUP WASTE USING A NOVEL SYSTEM

Hisham Hafez[1], George Nakhla[1] and Hesham El Naggar[1]
[1]Civil and Environmental Engineering Department, University of Western Ontario, London, Ontario, Canada
hhafez@uwo.ca, gnakhla@eng.uwo.ca, helnaggar@eng.uwo.ca

ABSTRACT

The patent-pending system comprises a novel biohydrogen reactor with a gravity settler for decoupling of solids retention time (SRT) from hydraulic retention time (HRT). The biohydrogenator was operated for 100 days at 37° C, hydraulic retention time 8 h and solids retention time 2.5 days. The feed was a corn-syrup waste from an industrial facility for bioethanol production located at South Western Ontario, Canada. The system was initially started up with a synthetic feed containing glucose at concentration of 8 g/L and other essential in-organics. Anaerobicaly-digested sludge from the St. Marys wastewater treatment plant (St.Mary, Ontario, Canada) was used as the seed. After 10 days, when the hydrogen production was steady, the corn-syrup waste was introduced to the system. Glucose was the main constituent in the corn-syrup; its concentration was varied over a period of 90 days. The first 30 days after start-up, glucose concentration was 8 g/L then it was increased to 16 g/L for 30 days and 25 g/L for the last 30 days. The change in glucose concentration was used to study the impact of variable organic loading on the stability of hydrogen production in the biohydrogenator. The hydrogen production rate increased from 10 LH_2/L /d to 32 LH_2/L /d with the increase of glucose concentration, while a maximum hydrogen yield of 3.2 mol H_2 /mol glucose was achieved in the system with an overall average of 2.9 mol H_2 /mol glucose and a standard deviation less than 10 %.

1. Introduction

Hydrogen does not contribute to the greenhouse effect and has a high energy yield of 142 kJ/g, 2.75 times more than that of any hydrocarbon [1]. Biological hydrogen production is potentially regarded as one of the most promising alternatives for sustainable green energy production, despite the feasibility of hydrogen production through water electrophoresis and chemical cracking of hydrocarbons. Among different biological processes for hydrogen production, dark fermentation is the most attractive one because of its potential of direct use of wastewater streams and organic wastes and its higher rate of hydrogen production in comparison with photo-fermentative processes.

The conversion of organic waste into hydrogen is attractive both from pollution control and energy recovery points of view. However, only a few studies have been conducted for hydrogen production from real wastewater due to challenges associated with inhibition and microbial shifts. Table 1 shows the process conditions and system performances of selected literature studies using wastewaters from rice winery [2], noodle [3], sugar [4,5], sugar beet [6] and molasses manufacturing [7], food processing [8], and filtered leachate of municipal solid wastes [9]. The highest hydrogen yield of 321 mL H_2/gCOD was demonstrated by Ueno *et al.* [4] for the treatment of sugar factory wastewater in a continuous stirred tank reactor (CSTR) with 63% glucose conversion efficiency. With the exception of the two batch studies with soil microorganisms as seed [7,8], most of the studies achieved around 200 mL H_2/gCOD. Furthermore, Table 1 does not show any advantages of packed bed reactors (PBR) over continuously stirred tank reactors (CSTR) in terms of hydrogen production.

The maximum specific growth rate (μ_{max}) for mixed cultures of hydrogen producing bacteria of 0.333 h^{-1} [10] corresponds to a minimum solids retention time (SRT_{min}) of 3.0 h and thus CSTRs operated for hydrogen production are characterized by hydraulic retention time (HRT) of 3–8 h. However, high dilution rates result in a marked decrease in biomass content in the reactor due to severe cell washout and system failure ensues [11]. Although, fill and draw (fed-batch) reactors have been used for hydrogen production, they invariably suffered from inconsistent hydrogen production [12] and methane production [13]. In order to overcome biomass washout in hydrogen reactors, decoupling of SRT from HRT in hydrogen bioreactors has been achieved primarily by using biofilms on several media including synthetic plastic media and treated anaerobic granular sludge [14], activated carbon, expanded clay and loofah sponge [15], glass beads [16] and membranes [17]. Problems with the development of methanogenic biofilms on the carrier media adversely impact process stability, which is critical for sustained hydrogen production. Moreover, membranes have not shown many advantages in terms of volumetric hydrogen yield and are also prone to fouling in such a reductive environment.

An extensive literature search using Scifinder Scholar has revealed that the concept of using a clarifier for decoupling SRT from HRT has not been explored. Thus, in this innovative research, the use of a clarifier after a hydrogen reactor (Bioydrogenator) [18] for decoupling SRT from HRT through sludge recirculation has been investigated for the first time using corn syrup wastewater. Moreover, the paper will focus on the performance of the biohydrogenator under variable organic loadings, highlighting the various mechanisms contributing to hydrogen production.

Table 1. Process and performance parameters for actual wastewaters.

Feedstock	Reactor type	Seed sludge	pH	Temperature (°C)	HRT (h)	Hydrogen Content (%)	Yield (mL H_2/ gCOD)	Maximum volumetric rate (L H_2/L·d)	Ref.
Sugar factory wastewater	CSTR	Compost	6.8	60	12	64	321	4.8	[4]
Wastewater containing sugar and ethyl alcohol	PBR	ADS	6.0–6.5	37	8	60	-	1.8	[5]
Molasses	Batch	Soil	6.0	26	-	-	102	-	[7]
Noodle manufacturing wastewater	CSTR	ADS	5.2	35	18	-	187	-	[3]
Rice winery wastewater	PBR	AS	5.5	55	2	61	272	3.8	[2]

Filtered leachate of waste biosolids	Batch	Waste biosolids	6.7–6.9	35	-	-	184	-	[9]
Sugar beet wastewater	CSTR	ADS	5.2	32	15	57	216	3.0	[6]
Food processing wastewater	Batch	Soil	4.0–6.4	23	-	60	100	3	[8]

Notes: CSTR, continuous stirred tank reactor; PBR, packed-bed reactor; ADS, anaerobic digested sludge; AS, acclimated sludge.

2. Experimental Section

2.1. Systems setup and operation

The biohydrogenator was operated for 100 days at 37 °C (Figure 1). A summary of the operational conditions is shown in Table 2. The system comprised a CSTR for biological hydrogen production with a 5 L working volume, followed by an 8 L gravity settler. The system effluent was monitored every two days for total chemical oxygen demand (TCOD), soluble COD, volatile fatty acids (VFA), ethanol, lactate, glucose, volatile suspended solids (VSS), total suspended solids (TSS) and daily for biogas composition including hydrogen, methane and nitrogen. Samples were filtered through a 0.45 micron filter paper (Whatman, 7141-104, Japan) prior to measurement of VFAs, ethanol, lactate, and glucose.

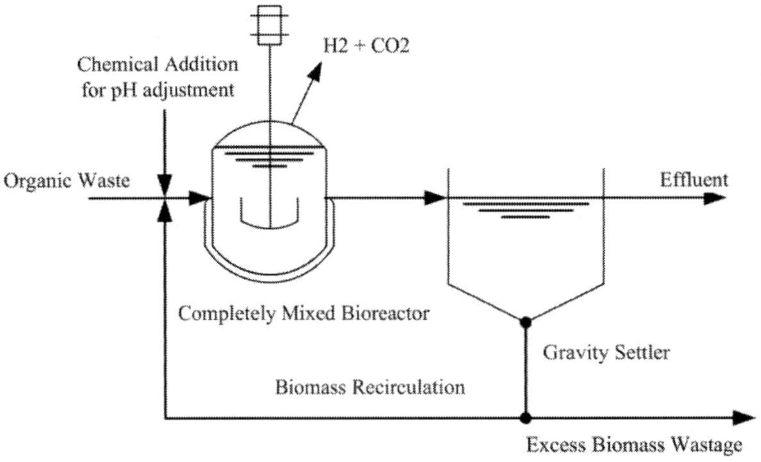

Figure 1. Experimental setup for the biohydrogenator.

Table 2. Operational conditions of the biohydrogenator.

Phase	HRT (h)	SRT (d)	Sludge wastage (L/d)	Waste sludge biomass VSS (g/L)	OLR (gCOD/L·d)	pH
1	8	2.3	0.6	28	26	5.5–6.5
2	8	2.5	0.4	25	52	5.5–6.5
3	8	2.2	No wastage	-	81	5.5–6.5

2.2. Inocula and media composition

Anaerobicaly-digested sludge from the St. Mary's wastewater treatment plant (St. Mary, Ontario, Canada) was used as the seed. In order to enrich hydrogen producing bacteria, the seed sludge was heat treated at 70 °C for 30 min. The system was seeded with 5 L of sludge and started up in a continuous mode with the feed containing 8 g/L glucose. Solids retention time was controlled by sludge wastage from the clarifier. The feed contained sufficient inorganics (mg/L): $NaHCO_3$, 4000; $CaCl_2$, 140; $MgCl_2{\cdot}6H_2O$, 160; NH_4HCO_3, 600; $MgSO_4{\cdot}7H_2O$, 160; urea, 500; Na_2CO_3, 124; $KHCO_3$, 156; K_2HPO_4, 15; trace mineral solution, 500; H_3PO_4, 250. After 10 days, when the system reached steady state, the feed was switched to corn-syrup waste generated as a byproduct from an industrial facility for bioethanol production located in southwestern Ontario, Canada. The corn-syrup was characterized by TSS of 400–500 g/L; SCOD of 100–350 g/L; glucose of 80–300 g/L and a pH of 3.5–4.5. The experimental period consists of three consecutive phases of 30 days each. The waste was diluted to approximately 1:10 according to the desired OLR for each phase of the experiment (see Table 2). The organic loading rate was increased in a stepwise fashion from 26 kg $COD/m^3{\cdot}d$ in phase 1 to 52 and 81 kg $COD/m^3{\cdot}d$ in phases 2 and 3, respectively. Based on the influent flow rate of 15 L/d, the bioreactor HRT was maintained constant at 8 h in all three phases of operation while the SRT varied narrowly from 2.2 to 2.5 days due to variations in system effluent VSS. The SRT was estimated according to the amount of VSS (g) in the hydrogen bioreactor (excluding biomass in the clarifier) divided by the summation of the amount of VSS (g/d) leaving the system in both the clarifier liquid effluent and waste sludge.

2.3. Analytical methods

Biogas composition including hydrogen, methane and nitrogen was determined by a gas chromatograph (Model 310, SRI Instruments, Torrance, CA, USA) equipped with a thermal conductivity detector (TCD) and a molecular sieve column (Molesieve 5A, mesh 80/100, 6 ft × 1/8 in.). Argon was used as carrier gas at a flow rate of 30 mL/min. The temperatures of the column and the TCD detector were 90 and 105 °C, respectively. The concentrations of volatile fatty acids (VFAs) were analyzed using a gas chromatograph (Varian 8500, Varian Inc., Toronto, Canada) with a flame ionization detector (FID) equipped with a fused silica column (30 m × 0.32 mm). Helium was used as carrier gas at a flow rate of 5 mL/min. The temperatures of the column and detector were 110 and 250 °C, respectively. Lactic acid concentrations were measured using a high-performance liquid chromatography system (1200 series, Agilent Technologies) equipped with Aminex HPX-87H ion exclusion column (300 mm × 7.8 mm i.d.; BIO-RAD), and a UV-detector at 210 nm. The column temperature was adjusted to 30 °C. The same instrument with a refractive index detector (RID) was used to measure the concentrations of glucose. The temperature of the RID detector was set to 35 °C. The amount of volatile suspended solids (VSS) and chemical oxygen demand (COD) were measured according to standard methods [19].

3. Results and Discussion

Figure 2 shows the diurnal variation of volumetric hydrogen production rate for the biohydrogenator, with the steady-state data summarized in Table 3. After 10 days of operation on the synthetic glucose solution, when the hydrogen production rate stabilized at 9.6±0.7 L H_2/L·d, the feed was switched to corn syrup wastewater as phase 1 of operation. An organic loading rate of 26 gCOD/L·d was maintained for 30 days. As apparent from Figure 1, the change in the feed from synthetic to real waste at the same OLR did not adversely impact hydrogen production, corroborating both the success of the acclimatization phase and the lack of inhibitors in the corn syrup. The system steadily produced hydrogen at a rate of 9.8±0.6 L H_2/L·d. The average hydrogen content in the biogas was 68±4%, with carbon dioxide as the balance. At the end of phase 1 the organic loading was doubled to 52 gCOD/L·d. The hydrogen production rate increased gradually to 20 L H_2/L·d over a period of 6 days, with the average hydrogen production rate and hydrogen content of the biogas phase 2 of 19.3±1.1 L H_2/L·d and 60± 4%, respectively. For the last 30 days of operation (phase 3), the organic loading rate was increased to 81 gCOD/L·d. After 14 days of operation the system reached a maximum hydrogen rate of 34 L H_2/L·d. In phase 3, the average hydrogen production rate and hydrogen content of the biogas produced were 32±2.3 L H_2/L·d and 62±3%, respectively. As depicted in Figure 3, hydrogen production rate increased from 10 to 34 L H_2/L·d with the increase of OLR from 26 to 81 gCOD/L·d.

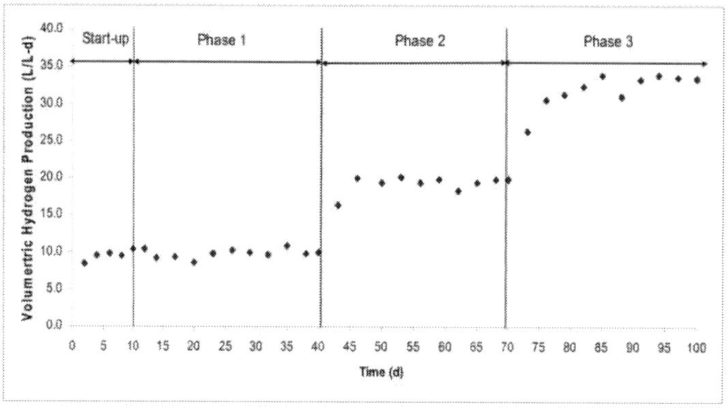

Figure 2. Temporal variation of volumetric hydrogen production rate.

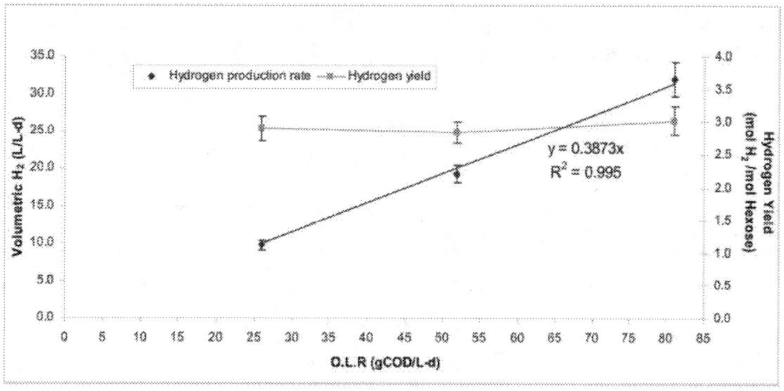

Figure 3. Relationship between hydrogen production rate and biomass yield versus OLR.

The linear relationship between hydrogen production rate and organic loading rate evident from Figure 3 emphasizes the lack of substrate inhibition as well as other inhibitors in the corn syrup at OLR as high as 81 gCOD/L·d. The highest hydrogen yield achieved throughout the experimental period was 3.2 mol H_2/mol hexose. As shown in Figure 4, the maximum hydrogen yield calculated as (mL H_2/gCOD converted) was 430 mL/gCOD higher than 321 mL/gCOD which was reported by Ueno *et al.* [4]. It must be asserted that as evident from Table 3, the influent glucose was essentially completely removed in all three phases of the study. Assuming a maximum conversion of 544 mL H_2/g hexose at 25 °C [20], the average conversion efficiency of glucose to hydrogen was 73% in the three phases.

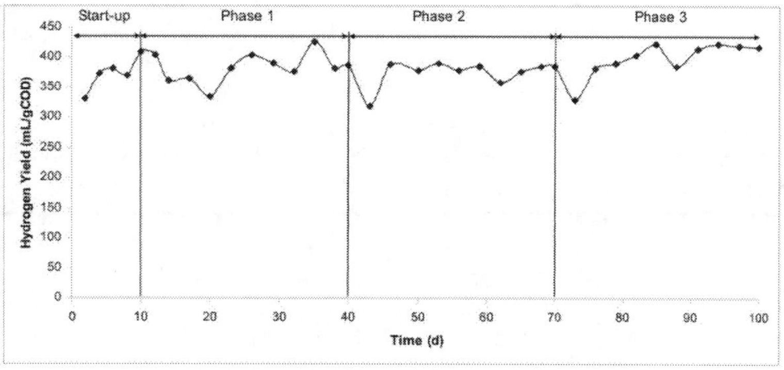

Figure 4. Volumetric hydrogen yield profile.

COD reduction efficiencies in phases 1, 2 and 3 were 28%, 31% and 32%, respectively compared to the theoretical value of 33% and 12% calculated from Equations 1 and 2, respectively [21]. The COD mass balance for the three phases, computed considering the influent and effluent CODs, and the equivalent CODs for both gas and biomass is shown in Table 3. The closure of COD mass balances at 109–113% validates the reliability of the data.

$$C_6H_{12}O_6 + 2H_2O \rightarrow 2CH_3COOH + 2CO_2 + 4H_2 \quad (1)$$

$$C_6H_{12}O_6 \rightarrow CH_3CH_2CH_2COOH + 2CO_2 + 2H_2 \quad (2)$$

Table 3. Metabolites and COD mass balance.

	Phase 1	Phase 2	Phase 3
VSS (mg/L)	17226±2496	23540±1997	25287±1879
VSS out (mg/L)	1343±208	2446±228	3747±705
SCOD out (mg/L)	6125±399	11753±787	18189±1335
Acetate (mg/L)	2647±274	5139±385	9060±1352
Propionate (mg/L)	36±16	159±24	87±62
Isobutyrate (mg/L)	0	0	0
Butyrate (mg/L)	1730±178	2747±315	3793±671
Isovalerate (mg/L)	3±3	50±18	63±25
Valerte (mg/L)	0	0	0
Ethanol (mg/L)	23±9	78±22	77±31
Lactate (mg/L)	0	0	0
VFA (mgCOD/L)	6087±579	11001±529	17015±2781
Glucose Out (mg/L)	0	0	0
Hydrogen Gas (L/d)	49±3	96±6	160±12
Hydrogen Gas (gCOD/d) *	31±2	61±4	101±7.5
COD balance (%) **	112±6	109±5	113±7

* Based on 8 g COD/g H_2. ** Sample of calculation phase 1: COD balance (%) = (1343 × 1.42 × 15 / 1000 + 31 + 6125 × 15 / 1000) / (135 × 100)

The distribution of metabolites formed during biological hydrogen production is often a crucial signal in assessing the efficiency of hydrogen-producing cultures [22]. The composition of the VFAs produced during the three phases is shown in Table 3. It must be asserted that the concentrations of soluble metabolites i.e. acetate, butyrate, propionate, isovalerate, and ethanol in the corn syrup were less than 1% than those measured in the reactor and shown in Table 3, and thus were neglected in the calculation of production rates. The VFA analysis revealed high concentrations of acetate and butyrate, while propionate, isovalerate, ethanol were present in minor amounts, and no detection of lactate, implying that the acid-forming pathway dominated the metabolic electron flow. Figure 5 shows the number of moles for acetate and butyrate throughout the experimental period. The steady-state average molar ratios of acetate/butyrate were 2.2, 2.8 and 3.5 for phases 1, 2 and 3, respectively. The theoretical hydrogen yield from hexose with acetate formation is 4 mol H_2 / mol hexose, which is twice as high as that of butyrate formation, 2 mol H_2/mol hexose. Previous studies indicate that the hydrogen yield increases with the molar ratio of acetate/butyrate [22,23]. According to the measured concentrations of acetate and butyrate and using Equations 2 and 3, the contribution of the two pathways was estimated. In phases 1 and 2, 68% and 32% of the hydrogen produced were through the acetate and butyrate pathways, respectively. While, in phase 3 approximately 78% and 22% of the hydrogen yield through the acetate and butyrate pathways, respectively. The increased contribution of the high hydrogen producing acetate pathway in phase 3 relative to phase 1 is consistent with the aforementioned increase in molar acetate/butyrate ratio.

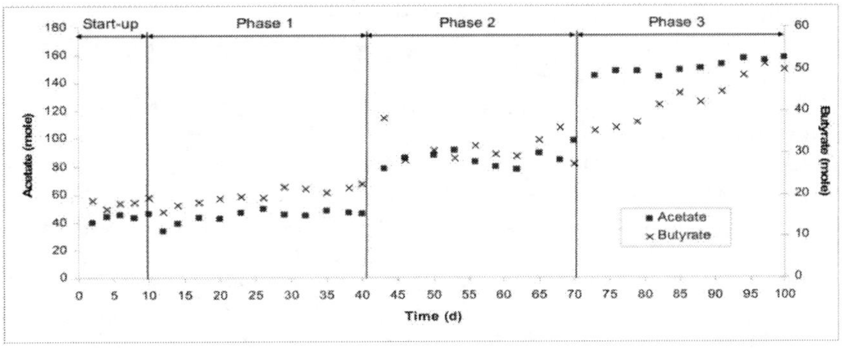

Figure 5. Temporal variation of molar acetate and butyrate production.

4. Summary and Conclusions

The overall results of this experiment show that biological hydrogen production from corn-syrup using heat pre-treated anaerobicaly digested sludge can be achieved in the biohydrogenator. The hydrogen production rate was a function of the organic loading rate; it increased from 10 to 34 L H_2/L·d with the increase of OLR from 26 to 81 gCOD/L·d. No inhibition of hydrogen production was observed at loadings as high as 81 gCOD/L·d. The highest hydrogen yield achieved throughout the experimental period was 3.2 mol H_2/mol hexose corresponds to 430 mL H_2/gCOD and an overall average glucose to hydrogen conversion efficiency of 73%. The relatively high hydrogen yield was verified by the high molar ratio of acetate to butyrate and the COD mass balance. The contribution of the high hydrogen producing acetate pathway increased from 68% of the overall hydrogen at 26 gCOD/L·d to 78% at 81 gCOD/L·d. The decoupling of SRT from HRT in biohydrogen production systems facilitated by the superior sludge settling characteristics of hydrogen producers, evaluated in this work, validated the promise of using a gravity settler after a CSTR to maintain high biomass retention in the system and decrease biomass washout, thus improving hydrogen yield and sustainability of hydrogen production, rendering the system as a competitor for biological hydrogen production from waste.

References

1. Cai, M.L.; Liu. J.X.; Wei. Y.S. Enhanced biohydrogen production from sewage sludge with alkaline pretreatment. *Environ. Sci. Technol.* **2004**, *38*, 3195–3202.
2. Yu, H.Q.; Zhu, Z.H.; Hu, W.R.; Zhang, H.S. Hydrogen production from rice winery wastewater in an upflow anaerobic reactor by using mixed anaerobic cultures. *Int. J. Hydrogen Energy* **2002**, *27*, 1359–1365.
3. Noike, T. Biological hydrogen production of organic wastes - Development of the two-phase hydrogen production process. In *Proceedings of International Symposium on Hydrogen and Methane Fermentation of Organic Waste*, Tokyo, Japan, March 2002; pp. 31–39.
4. Ueno, Y.; Otsuka, S.; Morimoto, M. Hydrogen production from industrial wastewater by anaerobic microflora in chemostat culture, *J. Ferment. Bioeng.* **1996**, *82*, 194–197.

5. Kim, M.S. An integrated system for the biological hydrogen production from organic wastes and waste-waters. In *International Symposium on Hydrogen and Methane Fermentation of Organic Waste*, Tokyo, Japan, 2002; pp. 11–18.
6. Hussy, I.; Hawkes, F.R.; Dinsdale, R.; Hawkes, D.L. Continuous fermentative hydrogen production from sucrose and sugarbeet. *Int. J. Hydrogen Energy* **2005**, *30*, 471–483.
7. Logan, B.; Oh, S.E.; Kim, I.K.; van Ginkel, S.W. Biological hydrogen production measured in batch anaerobic respirometers. *Environ. Sci. Technol.* **2002**, *36*, 2530–2535.
8. Van Ginkel S.W.; Oh, S.E.; Logan, B.E. Biohydrogen gas production from food processing and domestic wastewaters. *Int. J. Hydrogen Energy* **2005**, *30*, 1535–1542.
9. Wang, C.C.; Chang, C.W.; Chu, C.P.; Lee, D.J.; Chang, B.V.; Liao, C.S.; Tay, J.H. Using filtrate of waste biosolids to effectively produce bio-hydrogen by anaerobic fermentation. *Water Res.* **2003**, *37*, 2789–2793.
10. Horiuchi, J.I.; Shimizu, T.; Tada, K.; Kanno, T.; Kobayashi, M. Selective production of organic acids in anaerobic acid reactor by pH control. *Bioresour. Technol.* **2002**, *82*, 209–213.
11. Wu, S.Y.; Hung, C.H.; Lin, C.Y.; Lin, P.J.; Lee, K.S.; Lin, C.N., Chang, F.Y.; Chang, J.S. HRT-dependent hydrogen production and bacterial community structure of mixed anaerobic microflora in suspended, granular and immobilized sludge systems using glucose as the carbon substrate. *Int. J. Hydrogen Energy* **2008**, *33*, 1542–1549.
12. Lin, C.Y.; Jo, C.H. Hydrogen production from sucrose using an anaerobic sequencing batch reactor process. *J. Chem. Technol. Biotechnol.* **2003**, *78*, 678–684.
13. Youn, J.H.; Shin, H.S. Comparative performance between temperature-phased and conventional mesophilic two-phased processes in terms of anaerobically produced bioenergy from food waste. *Waste Manage Res.* **2005**, *23*, 32–38.
14. Das, D.; Khanna, N.; Veziroglu, T.N. Recent developments in biological hydrogen production processes. *CI&CEQ* **2008**, *14*, 57–67.
15. Chang, J.S.; Lee, K.S.; Lin, P.J. Biohydrogen production with fixed-bed bioreactors. *Int. J. Hydrogen Energy.* **2002**, *27*, 1167–1174.
16. Zhang, H.S.; Bruns, M.A.; Logan, B.E. Biological hydrogen production by clostridium acetobutylicum in an unsaturated flow reactor. *Water Res.* **2006**, *40*, 728–734.
17. Vallero, M.V.G.; Lettinga, G.; Lens, P.N.L. High rate sulfate reduction in a submerged anaerobic membrane bioreactor (SAMBaR) at high salinity. *J. Membr. Sci.* **2005**, 253, 217–232.
18. Hafez, H.; Nakhla, G.; El Naggar, H. Integrated system for hydrogen and methane production during landfill leachate treatment. Pending patent, US Patent and Trademark Office, 2009, US 61/202, 137.
19. *Standard methods for examination of water and wastewater*, 19th ed.; American Public Health Association: New York, NY, USA, 1995.
20. Li, C.; Fang, H.H.P. Fermentative hydrogen production from wastewater and solid wastes by mixed cultures. *Environ. Sci. Technol.* **2007**, 37, 1–39.
21. Hawkes, F.R.; Dinsdale, R.; Hawkes, D.L.; Hussy, I. Sustainable fermentative hydrogen production: challenges for process optimization. *Int. J. Hydrogen Energy* **2002**, *27*, 1339–1347.
22. O-Thong, S.; Prasertsan, P.; Birkeland, N. Evaluation of methods for preparing hydrogen-producing seed inocula under thermophilic condition by process performance and microbial community analysis. *Bioresour. Technol.* **2008**, *100*, 909–918.
23. Chen, C.C.; Lin, C.Y.; Lin, M.C. Acid–base enrichment enhances anaerobic hydrogen production process. *Appl. Microbiol. Biotechnol.* **2002**, *58*, 224–228.

Hydrogen production from oleic acid and starch over supported metal catalysts in supercritical water

Emhemmed. A. Youssef, George Nakhla, and Paul Charpentier*
Department of Chemical and Biochemical Engineering
The University of Western Ontario
London, Ontario, Canada, N6A 5B9
Tel. 519-661-2111 ext. 81134
Fax. 519-661-3498
Email: pcharpentier@eng.uwo.ca

ABSTRACT

Supercritical water gasification (SCWG) is a process that is currently being developed for the gasification of biomass in supercritical water. Hydrogen production from sewage sludge may be a solution for cleaner fuel and the sewage sludge disposal problem. The catalytic supercritical water partial oxidation process principle is to utilize the drastic changes of the physical and chemical properties of water above its critical conditions. These properties allow for a nearly complete conversion of sludges into energy-rich fuel gases, namely hydrogen, methane, and carbon dioxide. Because sewage sludge usually contains plenty of lipids, and carbohydrate and to understand the sewage sludge model compounds behavior at supercritical water conditions, Oleic acid and Starch were used as the model compounds of lipids, and carbohydrate respectively. The catalytic supercritical water gasification (SCWG) of oleic acid and starch in the presence of several commercial catalysts, namely; Pd (Palladium)/AC, Pt (Platinum)/AC, Ni/silica-alumina, and hydrogenation catalyst (KATALCO) has been investigated in a heated batch reactor. To avoid the reactions in subcritical conditions, de-ionized water and the catalyst were first heated and pressurized to desired temperature. The reactants were then injected into the reactor against it pressure at the reaction supercritical conditions by employing syringe pump. This paper presents the experimental results obtained at reaction conditions of 500℃, 28 MPa, and 30 minute residence time. The main gas components were carbon dioxide (CO_2), carbon monoxide (CO), hydrogen (H_2), and methane (CH_4). The influence of using different catalyst for the gasification and hydrogen gas yield were studied and discussed. The liquid after reaction product characteristics were studied by comparing both the Chemical Oxygen Demand (COD) and COD reduction efficiency with the quality standard of domestic wastewater. In addition, the effect of catalysts on the Chemical Oxygen Demand (COD) destruction efficiency were also investigated and reported.

1. INTRODUCTION:

With increasing public awareness about the environmental impact of fossil fuels as well as the depletion, hydrogen production from biomass is considered as one of the effective routes and an excellent alternative. However, the water content of biomass is generally high, in a range of 90 % or even higher, and the thermochemical conversion needs a prior drying process which, in turn, consumes a large amount of energy. Supercritical water gasification (SCW) is an emerging technology that has been developed to treat hazardous wastewater streams as well as producing green gases such as hydrogen. Supercritical water can dissolve most organic substances and gases and has low viscosity and strong transport ability. In the oxidation above critical condition of water (374°C, 22.13 MPa) all organic compounds are present in a single, dense fluid phase, minimizing mass-transfer resistance and provides rapid reaction rates. Gaseous product composition of waste biomass model compounds from supercritical water gasification significantly depends on the reactant concentration (Yu and Antal, 1993) and temperature (Holgate, 1995; Xu, 1996; Lee et al., 2002).In last decades, there have been a lot of studies carried out for gasification of wet biomass (Yu and Antal, 1993; Xu, 1996; Antal et al., 2000; Lu et al., 2006; Furusawa et al., 2007; Lee et al., 2009) and aqueous organic wastes (Blasi et al., 2007; Yan et al., 2007; García Jarana et al., 2008) in supercritical water. Enhancement of biomass conversion through oxidation in supercritical water (Holgate and Tester, 1995; Calvo, 2002) or partial oxidation (Adschiri et al., 1998; Watanabe et al., 2003; Yoshida, 2004) have been also studied.

Several studies have been reported concerning waste biomass or biomass model compounds gasification for H_2 production. These studies employed both heterogeneous and homogeneous catalysts (such as NaOH, KOH, activated carbon, metallic catalysts). (Osada et al., 2003) reported that supported ruthenium, rhodium, platinum, palladium, and nickel catalysts are active in decomposition of aromatic compounds.

Catalyst plays an important role in hydrogen production from waste biomass gasification in supercritical water. (Watanabe et al., 2003) studied the effect of different catalysts on the gasification of biomass model compounds in batch reactor at a temperature range of 400-440°C. However, (Yu and Antal. 1993) reported that the achievement of 95 % or above in supercritical water requires a reaction temperature above 600°C. (Courson et al. 2000) and

(Wang et al., 2000) reported that Nickel catalyst is expected to crack tar and promotes water-gas shift reaction, methanation, and hydrogenation reaction.

Sewage sludge typically consists of 41% protein, 25% lipid, 14% carbohydrate, and rest is ash and biodegradable and recalcitrant organic compounds. A better understanding of sewage sludges behavior in supercritical water requires an extensive experimental and theoretical investigation. This has motivated several research groups to study model compounds for both biomass and sewage sludges. In this work, we report the supercritical water gasification experimental results of the Oleic acid and Starch as the model compounds of lipids, and carbohydrate respectively. Oleic acid is a long chain fatty acid can serve as a model compound for lipids, and it has the formula $C_{18}H_{34}O_2$. Starch with a chemical formula $(C_6H_{10}O_5)_n$ is a polysaccharide carbohydrate consisting of a large number of glucose units joined together by glycosidic bonds, and thus starch severs as carbohydrate model compound. The influence of catalysts on the gas yield and composition of products as well as the chemical oxygen demand destruction efficiency were the main aims of the study.

2. MATERIALS AND METHODS:

2.1 Materials

Oleic acid and Starch were obtained from Sigma-Aldrich chemical company and used without further purification. De-ionized water obtained from a compact ultrapure water system (EASY pure LF, Mandel Scientific co, model BDI-D7381). KATALCO high temperature hydrogenation catalyst was purchased from Alfa Aesar (Ward Hill, MA, USA). 5% Palladium supported on activated carbon, 5% Platinum supported on activated carbon, and Reduced commercial nickel on silica alumina (63 wt % nickel) catalysts were obtained from Sigma-Aldrich Canada ltd (Oakville, Ontario, Canada).

2.2 Experimental setup

Experiments were performed in the main reactor body which was obtained from the Autoclave engineers, (Erie, Penna, U.S.A. The reactor was constructed of Hastelloy C-276 with a capacity of 600 ml, with a maximum pressure rating of 41.4 MPa and maximum temperature rating of 343℃. From the pressure-temperature rating of the Hastelloy alloy published on the manufacturer's web site, the reactor was modified to be able to sustain higher temperature of maximum 500℃ at a lower pressure level of 36 MPa. This facilitated operation of the reactor above the critical conditions of water (647°K, 22.1 3 MPa). The reactor was heated with a 1.5 kW electrical furnace that is surrounded its main body supplied by the same manufacturer. Figure 1 portrays a schematic diagram of the experimental system setup.

The experimental procedure consisted of several steps started by opening and washing thoroughly with distilled water to remove any residual carbon from previous experiments. The catalyst and 70 ml of de-ionized water was added to the reactor after which it is closed and purged with helium gas at a constant pressure rate of 0.2 MPa for 20 minutes to drive away all the air and oxygen that may be presented in the system. After purging with helium, the outlet valve (VO1) is closed and the pressure in the reactor increased to 0.7 MPa to prevent water evaporation through heating phase.

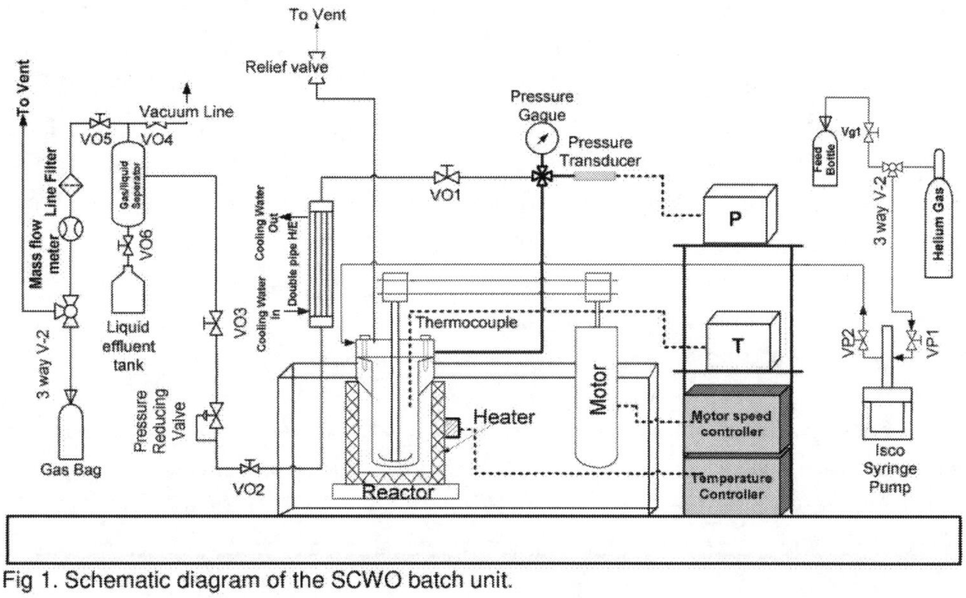

Fig 1. Schematic diagram of the SCWO batch unit.

The reactor was heated to desired temperature, and the pressure was increased accordingly to about 22.8 MPa. After reaching the desired temperature, the reactor left for 5 minutes to stabilize the temperature. The feed then injected to the reactor against its pressure by employing an ISCO syringe pump (Model 100 DX, Lincoln NE, USA). As soon as the feed fully injected, the reaction time (t) started counting. After 30 minutes, the valve (VO1) is opened to allow for effluent gases to pass through condenser (Double pipe H/E) where it cooled depressurized using a Swagelok piston-sensing high pressure reducing regulator (KHP series Solon, OH, USA). The cooled depressurized effluent passes to a gas liquid separator where gases and liquid products are separated.

2.3 Gas & liquid analysis

The gases product left the separator to pass through in line filter to remove any moisture before it passes through an OMEGA mass flow meter (FMA 1700/1800 series 0-2 L/min, Laval (Quebec), Canada). The mass flow meter is equipped with totalizer that utilizes a K-factor to relate the mass flow rate of an actual gas to Nitrogen as a calibrated reference gas. The actual gas flow rate is calculated by determining the average K factor for the produced gas by mean of the mole fraction of each gas in the stream. After passing through the mass flow meter, the product gases were collected in 3L volume Tedlar gas sampling bag. As soon as the gas bag filled, the flow of product gases was redirected to the fume hood where it is vented.The gas products were analyzed by gas chromatograph (Shimadzu, GC-2014) equipped with a thermal conductivity detector and 120/80 D Hayesep stainless steel Nickel packed column (Grace Davidson) having dimensions of (6.2 m x 3.18 mm). Helium was used as the carrier gas. The gas chromatograph was calibrated using a standard gas mixture of known composition. The analysis was performed manually using 1ml SGE gas tight syringe (Model number 008100, Reno, NV USA) by collecting the sample from the gas bag. The organic content of the process liquid effluent was examined by measuring the total and soluble chemical oxygen demand (TCOD, SCOD) using HACH methods and test kits (HACH Odyssey DR/2500.

For Oleic acid, the calculation of product gas yield and carbon balance was performed by the same procedure used by (Yu and Antal, 1993). The authors measured the gas yields (yield = (mol of gas species produced)/ (mol of glucose in feed)). For starch, the yield was measured and reported as ml/gCODremoved (that is, yield= volume of gas species produced/grams of COD removed). The gCODremoved= gCOD fed to the reactor – gCOD measured for the liquid effluent.

3. RESULTS AND DISSCUSION:

3.1 Oleic acid gasification

Figure 2 portrays the gas product distribution from the catalytic gasification of oleic acid on the each catalyst employed. The total gas yield increased with the presence of catalysts and the product gas mixture was composed of hydrogen, Carbon dioxide and methane as major components and Carbon monoxide, ethane, ethene, and propane were as minors. In the case of control experiment i.e. no catalyst used, methane was the main product followed by carbon dioxide, hydrogen and carbon monoxide. However, employing catalysts led to a diminution of the carbon monoxide in the gas phase and a corresponding increase in hydrogen and carbon dioxide yield. This can be attributed to the improvement of the water-gas shift reaction give by $CO + H_2O \rightarrow CO_2 + H_2$ (Antal et al. 2000 and Kruse et al.2000). From an energy efficiency point of view, the maximum rich fuel yield (H_2 & CH4) was obtained when the Ni/silica-alumina catalyst was employed. However, we can not speculate that the aforementioned catalyst is the optimum among the other catalysts employed. The increase of H_2 in the product gas was coinciding with an increase in the CH_4 which suggest that the part of the H_2 produced was consumed by the methanation reaction through the equation ($CO + 3H_2 \rightarrow CH_4 + H_2O$).

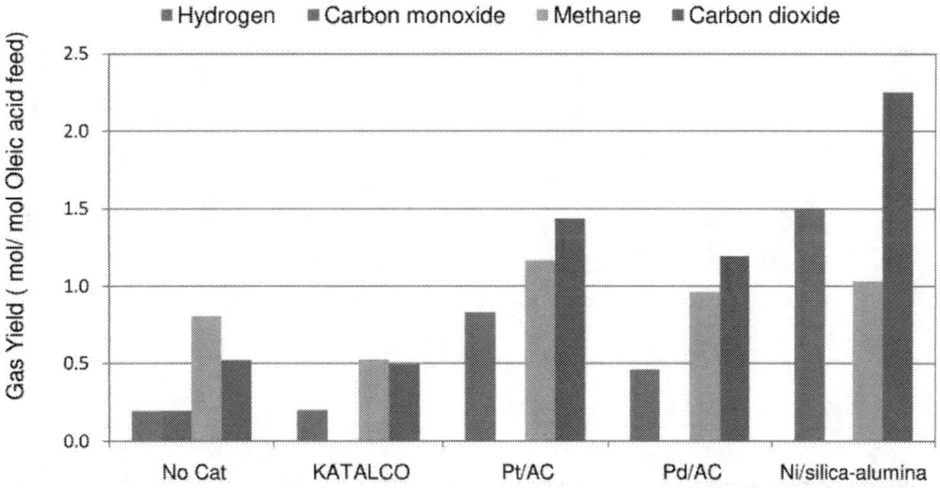

Fig 2. Gas distribution from Oleic acid experiments

3.2 Starch gasification

Figure3 shows the product gas product distribution from the starch catalytic gasification with the employed catalysts. The same trend observed with regard to carbon monoxide in oleic acid gasification was repeated in the case of starch experiments. Again, this is evidence that the water-gas shift reaction proceeded, consumed most of the CO and eventually produced more

H_2. However, a significant amount of CO of 7ml/gCODremoved was measured in the case where Ni/silica-alumina was employed as a catalyst. Nickel as a hydrogenation catalyst is expected to increase the relative methane and hydrogen yield (Kruse et al.2000). However, a drastic deviation on gas-phase composition from this rule was observed with the use of Ni/silica-alumina catalyst. It was also noticed that starch was easier to gasify that oleic acid. This was confirmed by the lower values of the effluent (COD) in the starch experiments than oleic acid ones.

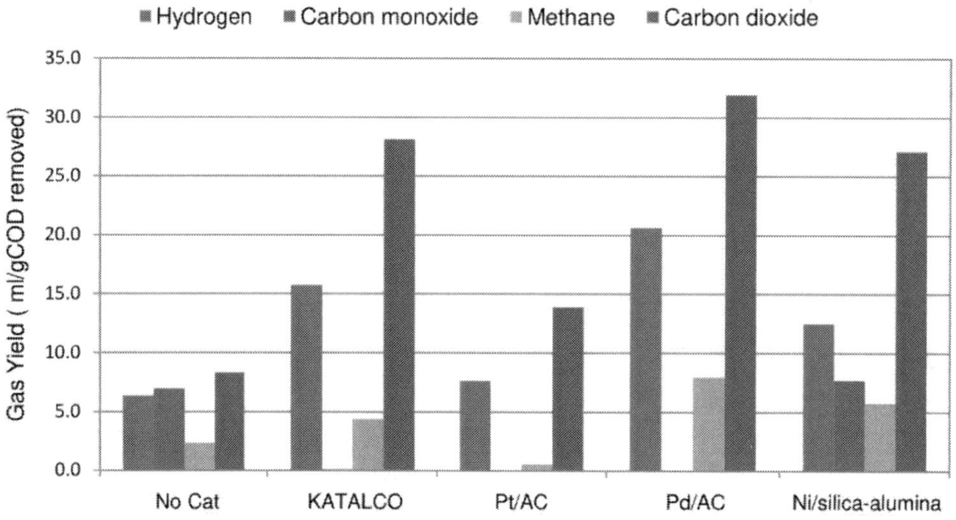

Fig 3. Gas distribution from Starch experiments

3.3 Liquid effluent quality

Chemical oxygen demand (COD) is the common wastewater pollution measurement parameter from environmental point of view, the COD destruction efficiency defined as $[((COD_{in} - COD_{out})/(COD_{out})) \times 100]$ was selected as an indication parameter of the liquid effluent quality. Table 1 reports both the TCOD and % COD reduction efficiency in both oleic acid and starch experiments. The employed catalysts led to decrease in the effluent Liquid COD in all oleic acid experiments whereas the trend was different in the starch experiments. Although starch was easier to gasify, there was also significant conversion of the starch intermediate products into char rather than into the more desired product of hydrogen gas (Kruse et al 2000). A visual comparison between the liquid effluent in both oleic acid and starch revealed that a significant amount of tar and char was produced in the starch experiments.

Table 1 Chemical oxygen demand (COD) values and % reduction efficiency

Catalyst type	TCOD effluent		% COD reduction	
	Oleic acid	Starch	Oleic acid	Starch
No catalyst	16401.91	3374	66	93
KATALCO	11876	1813	75	96
Pt/AC	8651	1974	82	96
Pd/AC	10364	3686	78	92
Ni/silica-alumina	8297	5982	83	88

CONCLUSION:

Catalytic supercritical water gasification of oleic acid and starch was performed in the presence of different metallic catalysts at 500°C and 28 MPa in a batch reactor. Results showed that the gas yield increased significantly with the use of catalysts. Starch gasification was easier than oleic acid although starch was more prone to produce tar and char. The COD reduction efficiency reached as maximum as 96% for starch and 83% for oleic acid.

AKNOWLEDGMENT:

The authors gratefully acknowledge support from Ontario Centers of Excellence (OCE). The authors grateful for the financial support in the form of Discovery Grant from the Natural Science and Engineering Research Council of Canada (NSERC).

REFERENCES:

Adschiri, T. S., R.; Sato, R.; Watanabe, M.; Arai, K. (1998). "Catalytic Hydrodesulphurization of Dibenzothiophen through Partial Oxidation and Water-Gas Shift reaction in Supercritical Water." Ind. Eng. Chem. Res. **37**: 2634.

Antal, M. J. J. A., S. G.; Schulman, D.; Xu, X. (2000). "Biomass Gasification in Supercritical water." Ind. Eng. Chem. Res. **39**: 4040.

Au, C. T. W., H.Y.; Wan, H.L. (1996). "Mechanistic Studies of CH_4/O_2 Conversion over SiO_2-Supported Nickel and Copper Catalysts." J. of Catalysis **158**: 343.

Blasi, C. D. B., C.; Galgano, A.; Meier, D.; Brodzinski, I.; Malmros, O. (2007). "Supercritical Gasification of Wastewater from Updraft Wood Gasifiers." Biomass Bioenergy **31**: 802.

Calvo, L. V., D. (2002). "Formation of organic acids during the hydrolysis and oxidation of several wastes in sub- and supercritical water." Ind. Eng. Chem. Res. **41**: 6503.

Osada, M.; Sato, O.; Hiyoshi, N.; M.; Arai, K.; Shirai, M. Catalytic effects of NaOH and ZrO_2 for partial oxidative gasification of n-hexadecane and lignin in supercritical water. *Fuel.* 82 (2003) 545–552.

García Jarana, M. B. S.-O., J.; Portela, J. R.; Nebot Sanz, E.; Martínez de la Ossa, E. J. (2008). "Supercritical Water Gasification of Industrial Organic wastes." J. Supercrit. Fluids **46**: 329.

Holgate, H. R. M., J.C.; Tester, W.J. (1995). "Glucose Hydrolysis and Oxidation in Supercritical water." AlChE J. **41**: 637.

Lee, I. G. I., S.K. (2009). "Catalytic Gasification of Glucose over Ni/Activated Charcoal in Supercritical Water." Ind. Eng. Chem. Res. **48**: 1435.

Lee, I. G. K., M.S.; Ihm, S.K. (2002). "Gasification of Glucose in Supercritical water." Ind. Eng. Chem. Res. **41**: 1182.

Lu, Y. J. G., L. J.; Ji, C. M.; Zhang, X. M.; Hao, X. H.; Yan, Q. H. (2006). "Hydrogen production by biomass Gasification in Supercritical water: A parametric study." Int. J. Hydrogen Energy **31**: 1597.

Watanabe, M. I., H.; Osada, M.; Sato, T.; Adschiri, T.; Arai, K.; (2003). "Catalytic Effects of NaOH and ZrO_2 for partial oxidative gasification of n-hexadecane and lignin in Supercritical Water in Supercritical Water." Fuel **82**: 545.

Xu, X. M., Y.; Stenberg, J.; Antal, M. J. Jr. (1996). "Carbon catalyzed Gasification of Organic Feedstocks in Supercritical Water." Ind. Eng. Chem. Res. **35**: 2522.

Yan, B. W., C.H.; Hu, C. S.; Xie, C.; Wu, J. Z. (2007). "Hydrogen Generation from Polyvinyl Alcohol-Contaminated Wastewater by a Process of Supercritical Water Gasification." J. Environ. Sci. **19**: 1424.

Yoshida, T. O., Y. (2004). "Partial Oxidative and Catalytic Biomass Gasification in Supercritical Water: A Promising Flow Reactor System." Ind. Eng. Chem. Res. **43**: 4097.

Yu, D. A., M.; Antal, M. J. Jr. (1993). "Hydrogen Production by Steam Reforming Glucose in Supercritical Water." Energy Fuels **7**: 574.

Yanik, J. Ebale, S. Kruse, A. Saglam, M. Yüksel, and M. "Biomass gasification in supercritical water: Part 1. Effect of the nature of biomass" Fuel 86 (2007) 2410–2415.

Kruse, A. Meier, D. Rimbrecht, P. Schacht, M. "Gasification of Pyrocatechol in Supercritical Water in the Presence of Potassium Hydroxide". *Ind. Eng. Chem. Res*. 2000, *39*, 4842-4848.

$Ba_{0.5}Sr_{0.5}Co_{0.8}Fe_{0.2}O_x$ Mixed Conducting Ceramic Membranes for Syngas Production

Sedigheh Faraji, Karen J.Nordheden, and Susan M. Stagg-Williams
Department of Chemical & Petroleum Engineering, University of Kansas, Lawrence, KS 66045

There is an increasing interest in the development of a process for producing syngas ($CO+H_2$) from methane using non-porous oxygen-conducting ceramic membranes with mixed ionic and electronic conductivity (MIECs) as an alternative to the conventional expensive process based on air separation plant [1, 2]. Our previous studies, which focused on the use of $SrFeCo_{0.5}O_x$ (SFC) and $Ba_{0.5}Sr_{0.5}Co_{0.8}Fe_{0.2}O_x$ (BSCF) ceramic membranes for the carbon dioxide reforming of methane at 800 °C, have shown that the methane conversion on high oxygen flux asymmetric BSCF membrane is higher than that on low oxygen flux dense SFC membrane [3]. However, the roles of ceramic membrane material in the enhancement of this reaction have not yet been investigated. In this work, the role of a $Ba_{0.5}Sr_{0.5}Co_{0.8}Fe_{0.2}O_x$ (BSCF) membrane material in the conversion of hydrocarbons has been studied.

BSCF powder was prepared by Citrate-EDTA method [4] and was pressed and sintered in flowing air for 5 hours. The SFC powder (from Praxair Specialty Ceramics) was passed through a 60-mesh sieve and coated with 1 wt% ethylcellulose binder prior to pressing. SFC powder was then pressed and sintered in flowing air for 10 hours. Finally, both membranes were crushed and passed through a sieve. The catalyst $Pt/CeZrO_2$ was prepared by depositing Pt on commercially available $CeZrO_2$ support material (MEI CHEM) using the incipient wetness impregnation technique (0.5 wt% Pt supported on $CeZrO_2$). The catalyst was characterized by conducting ICP and TEM studies. Ni/Al_2O_3 catalyst (15 wt% Ni supported on Al_2O_3) was prepared by incipient wetness impregnation method.

Raman spectroscopy and temperature programmed desorption (TPD) studies have been conducted in order to better understand the surface reactions happening during carbon dioxide reforming of hydrocarbons. Also, TGA technique has been used to investigate the oxygen release and uptake of SFC and BSCF membranes. For CO, CO_2, and H_2 adsorption studies, a plug flow reactor was used in which the crushed membranes were exposed to 50 μl pulses of CO, CO_2, and H_2.

Pulse injections of CO_2 over crushed membranes at 800 °C have shown more CO_2 dissociation on BSCF membrane than SFC membrane, resulting in higher CO formation on BSCF membrane. This could be due to higher number of oxygen vacancies in BSCF. Similar to the CO_2 pulses, when CO was injected on the samples at 800 °C, CO_2 production was seen and CO_2 production was higher on BSCF than SFC. The crushed membranes were heated to 800°C in argon and then exposed to pulses of hydrogen while monitoring the effluent using a mass spectrometer. Hydrogen was able to interact with the oxygen in the membrane to form water. It was found out that hydrogen consumption on BSCF particles is 24 times higher than that on SFC particles. Furthermore, Raman spectroscopy and temperature programmed desorption studies of CO and CO_2 showed a higher CO and CO_2 adsorption at different temperatures ranging from room temperature to 600°C on BSCF compared to the SFC membrane.

During thermogravimetric analysis, the onset temperature of oxygen release for BSCF was observed to be lower than that for SFC while the amount of oxygen release was significantly greater. CO_2 reforming reactions on BSCF and SFC dense membranes, which were performed in a two-chambered concentric quartz tube membrane reactor, showed higher methane conversion and H_2/CO ratio on BSCF than SFC in the presence of the $Pt/CeZrO_2$ catalyst. This high conversion and H_2/CO ratio could be ascribed to higher CO, CO_2, and H_2 adsorption on BSCF than SFC. Commercial Ni/Al_2O_3 catalyst has shown a very good and stable performance (for 14 hr of reaction) on BSCF membrane for CO_2 reforming of methane. Carbon is continuously removed from catalyst surface by oxygen from BSCF; thus, no carbon deposition was detected on this catalyst. Figure 1 shows the superior performance of Ni/Al_2O_3 catalyst compared to $Pt/CeZrO_2$ catalyst. These studies suggest that BSCF might be suitable for hydrocarbon conversion reactions at much lower temperatures than conventional dense ceramic membranes.

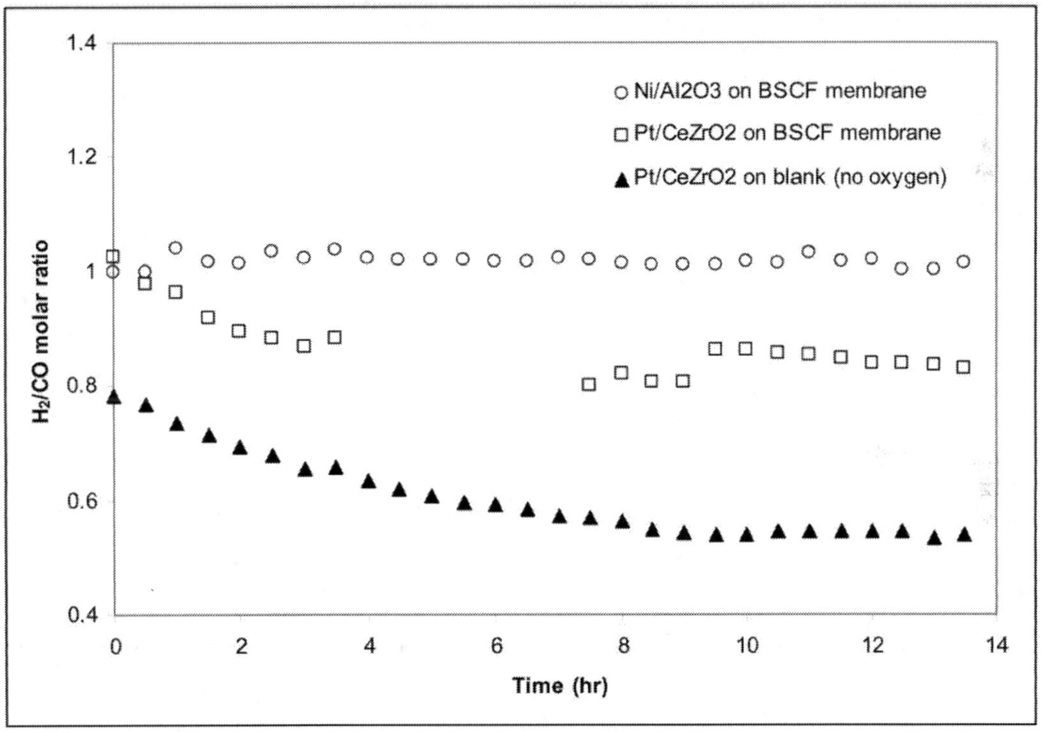

Figure 1: The H_2/CO molar ratio resulted from CO_2 reforming of methane at 800 °C in membrane reactor.

Acknowledgement:

Financial support for this project was provided by the Office of Naval Research (N00014-03-1-0601) and grant DTOS59-06-G-0047 by the US Department of Transportation Research and Innovative Technology Administration.

References:

[1] H. J. Bouwmeester, "Dense ceramic membranes for methane conversion," *Catalysis Today*, vol. 82, pp. 141-150, 2003.

[2] D. A. Slade, A. M. Duncan, K. J. Nordheden, and S. M. Stagg-Williams, "Mixed-conducting oxygen permeable ceramic membranes for the carbon dioxide reforming of methane," *Green Chemistry*, vol. 9, pp. 577-581, 2007.

[3] Q.Jiang, S.Faraji, K.J. Nordheden, and S.M. Stagg-Williams, "Application of Ba0.5Sr0.5Co0.8Fe0.2Ox Asymmetric Oxygen-permeable Membranes for Hydrocarbon Conversion Reactions", 2009 North American Meeting (NAM), June 10, 2009, San Francisco, CA.

[4] H. H. Wang, Y. Cong, and W. S. Yang, "Investigation on the partial oxidation of methane to syngas in a tubular Ba0.5Sr0.5Co0.8Fe0.2O3-delta membrane reactor," *Catalysis Today*, vol. 82, pp. 157-166, 2003.

Hybrid Ion Exchange in Environmental Separation: From Decontamination to Desalination

Arup K. Sengupta, Department of Civil & Environmental Engineering, Lehigh University, Bethlehem, PA

See more of this Session: Lawrence K Cecil Award Lecture
See more of this Group/Topical: Environmental Division

ASSESSMENT OF FRESHWATER REQUIREMENTS AND CONSERVATION OPTIONS FOR THE PRODUCTION OF LIGNOCELLULOSIC BIOFUELS AND THERMOELECTRICITY

Bala P. Lingaraju[1], Joo-Youp Lee[1,], Tim C. Keener[2], Y. Jeffery Yang[3]*

[1] *Department of Chemical and Materials Engineering, University of Cincinnati, Cincinnati, OH*
[2] *Department of Civil and Environmental Engineering, University of Cincinnati, Cincinnati, OH*
[3] *National Risk Management Research Laboratories, U.S. Environmental Protection Agency, Cincinnati, OH*

Introduction

Water is an essential element for energy production [1]. Freshwater withdrawals for thermoelectric power generation account for 39 percent of the total fresh water in the USA [2]. Even in non-drought conditions in the west, demand for water exceeds the supply [3]. Water shortages are expected in 36 states in the next 10 years [4]. Population growth makes the problem difficult to tackle. The Energy Information Administration (EIA) projects that the electricity demand will grow by 50% due to increase in the population of U.S by about 70 million in the next 25 years [5]. Study of availability of water in the future cannot be done with certainty due to potential effects of climate change [1]. Transportation fuels use relatively a small amount of water compared to thermoelectricity generation. But, as imported transportation fuels like petroleum and natural gas are replaced by biofuels like ethanol, and biodiesel from algae demand for water to produce transportation fuels can grow significantly. The assessment of water requirements for unit operations can pave the way for an effective strategy in the energy sector. In this study, a survey of 191 coal-fired power plants was conducted. Power plants withdrawing exceptionally high percentage of water from the river were identified. The cooling water system used by the coal-fired power plants is analysed and water saving options were recommended. An estimation of freshwater requirements for cellulosic ethanol production has been made for upcoming planning and management of water resources. The critical areas that need attention of the scientific fraternity in the upcoming algae industry for biofuels production were also identified.

Experimental Procedure

Water Requirements for Thermo-electric Power Plants

Data Source

Coal Power Plant Database version 2007 released by the Office of Fossil Energy's National Energy Technology Laboratory (NETL) has been used for this study. It contains various information including power production capacity, water withdrawal, consumption for all coal-fired power plants.

Data Analysis

The distribution of coal-fired power plants and annual average cooling water withdrawal rates in the U.S. is carried out in order to study the freshwater consumption and availability. Once-through, wet recirculation, closed-loop, dry cooling, and cooling ponds are the widely used cooling options in coal-fired power plants [6]. Once-through cooling systems consume more fresh water. The percentage of power plants having the once-through cooling system which consumes more water is calculated. The percentage of water withdrawn out of the total river flow at a point source by the coal-fired power plant was estimated by obtaining the river flow data from U.S geological survey and the cooling water annual average withdrawal rate from the coal power plant database. The formula used for calculating the percentage of water with drawn is

$$\text{Percentage of water withdrawn from the river} = \frac{\text{Cooling water annual average withdrawal rate}}{\text{River flow annual average}} \times 100$$

Cellulosic Ethanol Water Requirement

Potential for energy efficient cellulosic material to displace corn and reduce the environmental footprint, lead to increased research and investment in cellulosic feedstock [7]. The amounts of water required for cultivation is ~70% and the water required for processing the lignocellulosic feedstock to generate ethanol (industrial water) is ~30% of the total water requirements [8]. The demand of industrial water on local water resources is also significant. For example, an ethanol plant which requested use of 800,000 gallons of fresh water each day faced resistance from the people of the city of Tampa [9]. Hence, the shortage of water could be the vulnerable issue of the ethanol industry [10]. The current study focuses on the analysis of industrial water, as no water is spent for the sole purpose of growing cellulosic feedstock.

Case Selection

Hardwood, corn stover, and switchgrass are the most widely used feedstocks for cellulosic ethanol production. Four cases were selected for this water requirement analysis. The variables among the four cases are feedstock, pretreatment and fermentation process, and water network design. A detailed description of the four cases selected for water analysis is shown in Table 1.

National Renewable Energy Laboratory (NREL) established the process for the conversion of cellulosic feedstock into ethanol, the details of which were reported by Wooley et al. (1999) and Aden et al. (2002). The schematic representation of the various unit operations in the plant is shown in Figure 1. In the current study, the freshwater requirement of each unit operation is addressed. Unit operations which consume the majority of the fresh water were identified.

Table 1. Estimation of Water Requirements for Four Cases from the Literature

	Feedstock	Pretreatment	Water Requirement Estimation	Type of Water	Authors of the Estimate	Reference
Case 1	Hardwood	Dilute Acid	NA	NA	NA	Wooley et al. [11]
Case 2	Corn Stover	Dilute Acid	6 gal water/gal	Make-up water	Aden et al. [12]	Aden et al. [13]
			2.6-46 gal water/mile	Water re-circulated	King et al. [14]	
			5.6-63 gal water/mile	Once through water	King et al. [14]	
			4.5-9 gal water/mile	Make-up water	NRC report [15]	
Case 3	Switchgrass	Dilute Acid	5.79-16.68 lit water/lit gasoline eq	Make-up water	Laser et al. [16]	Laser et al. [16]
Case 4	Switchgrass	AFEX	NA	NA	NA	Laser et al. [16]

Notes:
gal water/gal: gallons of water per gallon of ethanol
gal water/mile: gallons of water spent per each mile traveled by a vehicle with ethanol
lit water/lit eq: liters of water per liter of gasoline equivalent of ethanol
AFEX: Ammonia Fiber Expansion

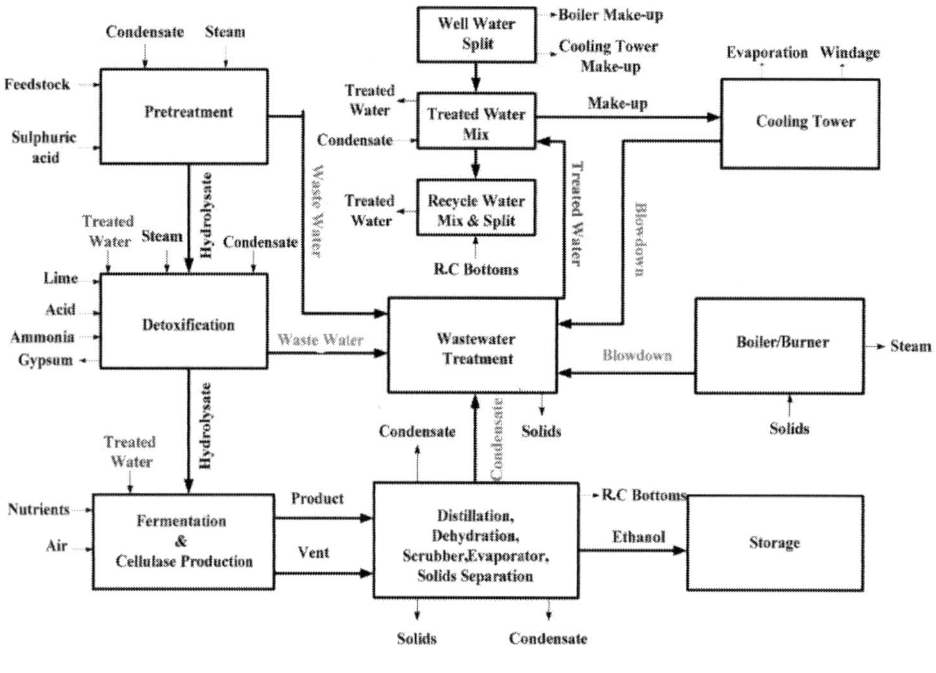

Figure 1. Schematic overview of a cellulosic ethanol plant.

Water Quantity and Quality

Fresh water is withdrawn by a cellulosic ethanol plant as make-up water for the water lost in consumption, vent, evaporation and windage. The water quantity is estimated in gallons of fresh water withdrawn per gallon of ethanol produced. This unit is chosen for the estimation of the quantity of water. If the production capacity of an ethanol plant is known, the amount of water required for the ethanol facility can be easily estimated if gallons of fresh water withdrawn per gallon of ethanol are known. The freshwater flow rate and production capacity of the plant are taken from the sources mentioned in Table 1. Annual freshwater requirements were estimated based on the production capacities of the ethanol plants listed at the Renewable Fuel Association (RFA) website [17].

In a cellulosic ethanol plant, all the waste streams generated by individual unit operations of the cellulosic ethanol plant are directed towards the wastewater treatment plant. The model cellulosic ethanol plants in the references listed in Table 1 do not discharge any waste water. Chemical Oxygen Demand (COD), insoluble solids (IS) and soluble solids (SS) are the three parameters used for indicating the quality of water. The COD of the streams in kg COD/ hr is obtained by multiplying the mass flow rate of each stream by a COD factor as illustrated by Merrick et al., 1998. The components in each stream are classified as soluble solids and insoluble solids by Merrick et al., 1998. According to this classification, the estimation of percentage soluble solids and insoluble solids is made and the COD loading estimation is made.

Results and Discussion

Thermoelectric Power Plants

Data on coal-fired power plants in the U.S., cooling water usage, and cooling tower technologies were analyzed from the Coal-Fired Power Plant Database. The distribution of coal-fired power plants which withdraw 15-40% of the river water at the point source are indicated in a map in Figure 2. It can be observed from the map that some of the power plants with high withdrawal rate overlap with drought-affected areas. This might severe the local water shortage.

The primary cooling water systems used in the U.S. coal-fired power plants were analyzed as shown in Figure 3. It shows that the majority (43%) of the power plants use once-through system using fresh water, which consumes more water compared to the other cooling options like dry-cooling and closed loop cooling.

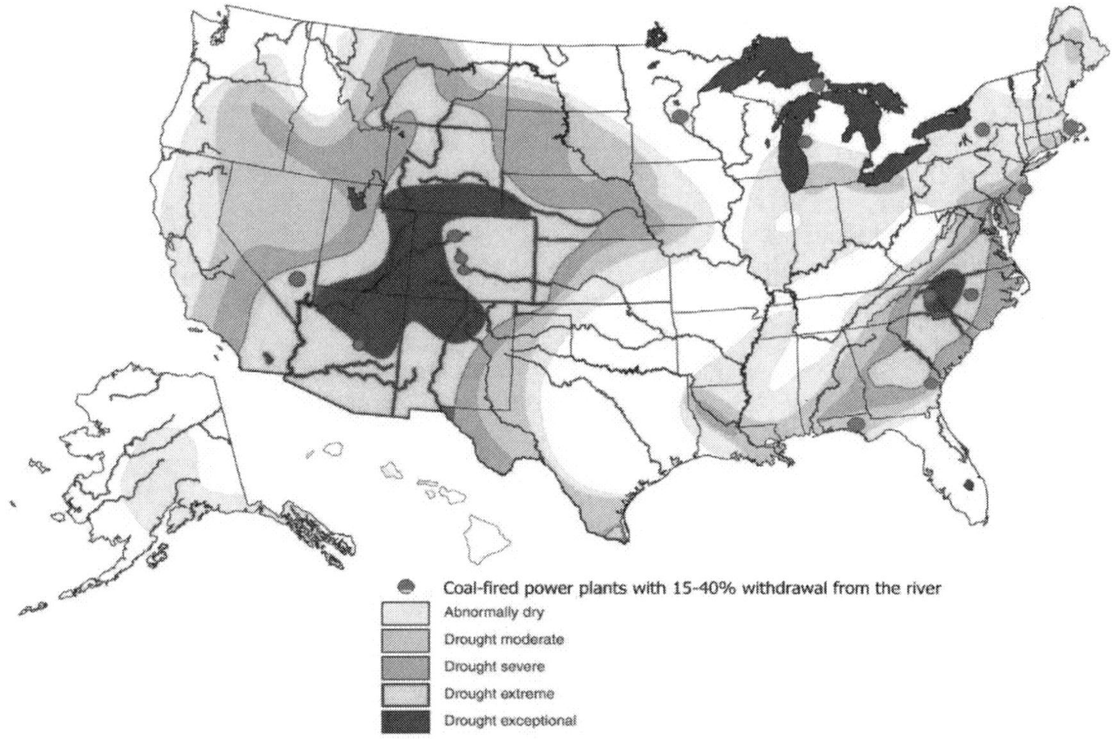

Source of the drought map: National Drought Mitigation center, University of Nebraska-Lincoln; U.S Department of Agriculture; and National Oceanic and Atmospheric Administration

Figure 2. Coal-fired power plants with high percentage of withdrawal from the river.

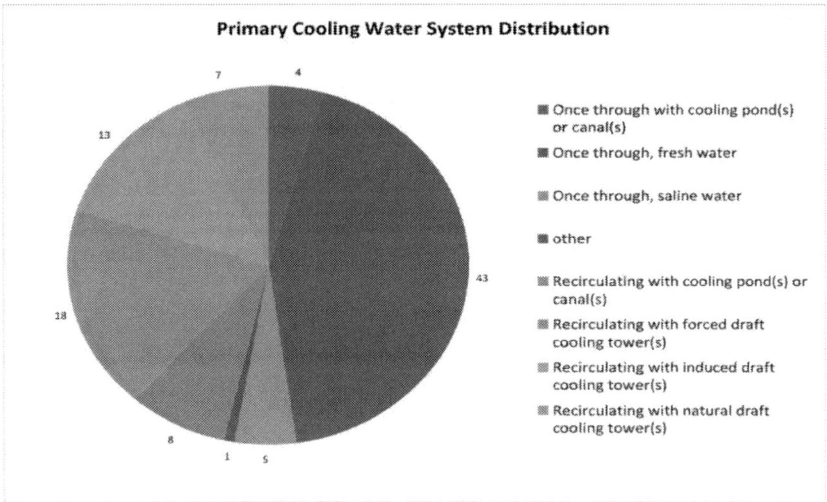

Figure 3. Distribution of primary cooling water systems used by the U.S. power plants.

Cellulosic Ethanol

Water Quantity

The quantity of fresh water has been estimated for the four Cases listed in Table 1, and the results are summarized in Table 2. The water requirement ranges from 9.4 to 10.7 gallons of fresh water per gallon of ethanol produced, which is close to the values estimated by the NRC report [15]. The freshwater requirement per each unit operation is also presented. It can be observed from the results in Table 2 that the cooling tower consumes a maximum amount of fresh water as make-up water. 75% of the make-up water is evaporated or lost in windage from the cooling tower, and 25% of the make-up water goes into blowdown. The freshwater requirement can be reduced by implementing innovative water-saving cooling tower technologies.

The freshwater requirement for each of the four cases is different for the pretreatment and detoxification section. The difference is not because of the difference in feedstock or pretreatment process but because of the water network design used in each of the four cases. For example, in Case 2, only recycled water is used for pretreatment and detoxification and no fresh water is used. Hence the freshwater requirement for that case is theoretically zero for the pretreatment and detoxification section. However, the distinctly high freshwater requirement of the pretreatment and detoxification section of Case 4 can be attributed to the Ammonia Fiber Expansion (AFEX) pretreatment process, which requires chilled water for the recovery of ammonia used in the pretreatment [16]. Hence water network design is a very important factor, and effective water recycling can result in a change of 10% in the usage of water as it is observed from the values of the total freshwater requirement of each of the four cases.

Regional water requirements for future cellulosic ethanol plants are calculated, and a bar graph showing the water required by each state is shown in Figure 4. According to a previous report [15], the water requirement for corn ethanol plants is at least 3 gallons of water per gallon of ethanol. In this study, at least 10.7 gallons of water per gallon of ethanol is estimated for cellulosic ethanol production. It can be observed that the state of Iowa has the maximum freshwater requirement. Because of the high availability of corn crops in Iowa, there are many cellulosic ethanol plants in Iowa. The fresh water withdrawn in Iowa is 1,226,400 million gallons per year. The maximum freshwater requirement is 27,081.7 million gallons per year for the upcoming cellulosic ethanol plants.

Table 2. Estimation of Freshwater Requirements for Four Cases

Unit	Case 1 Hardwood + Dilute Acid	Case 2 Corn Stover + Dilute Acid	Case 3 Switchgrass + Dilute Acid	Case 4 Switchgrass + AFEX
Pretreatment and Detoxification	2.5	0	4	7.9
Fermentation	0.3	0	0.4	0
Product Recovery	0.7	3.8	0.5	1
Residue Processing and Steam	3.8	1.5	0	0
Cooling Tower	3.4	4.3	4.5	1.4
Total	**10.7**	**9.7**	**9.4**	**10.3**

Notes:
AFEX: Ammonia Fiber Expansion.

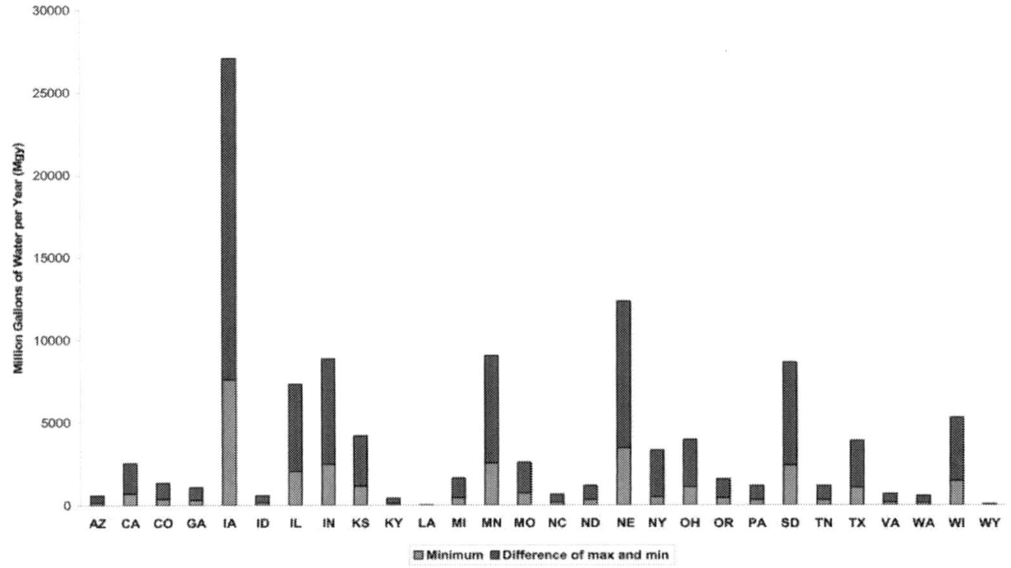

Figure 4. Water Requirements of Future Ethanol Industries in the U.S.

Water Quality

Anaerobic digestion followed by aerobic treatment is the best wastewater treatment (WWT) strategy for converting lignocellulosic biomass to fuel ethanol [18]. All the four Cases selected in this study use the same strategy. The quality of the waste streams discharged from the individual unit operations and channeled to the wastewater treatment units are shown in Table 3. The wastewater loading is about 44,147 Mg COD/L to 1,323,100 Mg COD/L with less than 0.1% of soluble and insoluble solids. After the wastewater treatment, the COD is reduced by 94–99 %. The temperature of the wastewater streams sent to the wastewater treatment unit ranges between 28 and 309 °C. After wastewater treatment, the temperature of the streams comes down to 21 °C. The COD of the water streams recycled without wastewater treatment is about 18,000 Mg COD/L with a temperature of 40 to 121 °C. All the four Cases selected to operate with a zero discharge of waste water. Pure well water is used for the vent scrubber and as boiler make-up water according to the layout shown in Figure 1, which represents the schematic of the water network used by the cellulosic ethanol plants in all the four Cases. The boiler make-up water used for steam generation has to be pure water in order to avoid corrosion of the boilers. The cooling tower make-up water contains 66% recycled water and 34% [18]. The process water used for the pretreatment and detoxification process and fermentation process is either treated water from the wastewater treatment plant or water recycled without treatment according to the data obtained from Merrick et al., 1998. Hence, high quality water is required for utility purpose and recycled water can be used for other unit operations.

Table 3. Water Quality for Four Cases

	Case 1 Hardwood + Dilute Acid	**Case 2 Corn Stover + Dilute Acid**	**Case 3 Switchgrass + Dilute Acid**	**Case 4 Switchgrass + AFEX**
Capacity of the plant (Dry ton per day)	2,000.00	2,000.00	4,535.00	4,535.00
Quality of input water to WWT				
Total input to WWT (kg/hr)	179,346.00	97,265.00	187,603.00	1,323,100.00
COD (Mg COD/L)	44,148.00	37,811.00	68,254.00	1,056,793.00
% SS	< 0.1	< 0.1	<0.1	<0.1
% IS	< 0.1	< 0.1	< 1.5	<1.5
Temperature (°C)	28 C - 309	28 C - 309	28 C - 309	28 C - 309
Quality of output water from WWT				
Total output from WWT (kg/hr)	173,154.00	95,805.00	181,822.00	1,225,845.00
COD (Mg COD/L)	2,237.00	1,315.00	1,747.00	13,052.03
% SS	< 0.1	< 0.1	< 0.1	< 0.1
% IS	< 0.1	< 0.1	< 0.1	< 1.5
Temperature (°C)	21	21	25	21
Quality of recycled without WWT				
Average COD (kg/hr)	198,259.00	47,637.00	NA	NA
Average COD (Mg COD/L)	18036.4	18310.5	NA	NA
% SS	< 3.6	< 0.3	< 0.3	< 0.3
% IS	< 0.9	< 0.3	< 0.3	< 0.3
Temperature (°C)	40-121	40-121	40-121	40-121

Notes:
COD: Chemical Oxygen Demand
IS: Insoluble Solids
SS: Soluble Solids

WWT: Waste Water Treatment

Options for Saving Water

From the water requirement analysis of the cellulosic ethanol plant and the literature survey of water requirement for thermo-electric power plants, a major share of make-up water withdrawn is used as cooling water. Hence, exploring options to reduce the cooling water consumption and alternative water sources can reduce the total freshwater requirement. Water saving options and other alternatives which are available in the literature to reduce overall water use in the energy industry are briefed in Table 4.

Table 4. Review of water-saving options

Water Reduction Technology	Water Reduction Claimed	Reference
• Hardness and silica removal process to reduce cooling tower blowdown.	20%	Owens, S.R.[19]
• Patented dry cooling technology to save cooling water by using cool air instead of water.	20%	Wurtz, W.[20]
• Closed loop evaporative wet surface air coolers by maximizing heat transfer efficiency.	50%	Demakos, P.E.[21]
Water Re-use and Alternative Sources		
• Re-use of non-traditional waters (ash basin water, cooling water, flue gas desulphurization water, produced water)	NA	Rodgers Jr., J.H.[22]
• Water extraction from flue gas	NA	Feeley III., T.J.[23]
• Desalination of brackish waters to fresh-water	NA	[1]

Conclusion

Assessment of the water requirement of cellulosic ethanol plants revealed that the quantity and quality of utility water is much higher than the process water requirement. Cooling tower and steam generation section consume 13-50% of the total fresh water withdrawn. From the assessment of water requirement of future ethanol plants, it is found that Iowa has the highest number of ethanol plants. As the ethanol industry grows, other states which have a smaller number of ethanol plants might encounter the same situation and the numbers will grow in the future. It can be concluded that for effective water-resource planning, the optimization of water use is required in the cellulosic ethanol industry.

The distribution of coal-fired power plants which withdraw 15-40% of the river water at the point source overlap with drought-affected areas. This might severe the local water shortage.The once through cooling water system which consumes more water when compared to the other cooling options e.g., dry-cooling ,closed loop cooling should be replaced with the new water saving cooling technology.

References

1. Energy Demands on Water Resources: Report to Congress on the Interdependence of Energy and Water; U.S. Department of Energy; Washington DC, 2006; available at http://www.docstoc.com/docs/629446/Energy-Demands-on-Water-Resources-Report-to-Congress-on-the-Interdependency-of-Energy-and-Water-Dec (accessed 2009).

2. Hutson, S.S., et al. Use of Water in the United States in 2000; USGS Circular 1268; available at http://water.usgs.gov/pubs/circ/2004/circ1268/index.html (accessed 2009).

3. Statement before the Senate Energy and Natural Resources Committee Oversite Hearing on Water in the West. March 9, 2004; available at www.doi.gov/water2025/raley3-9-05.doc. (Accessed 2009).

4. Freshwater Supply: States' Views of How Federal Agencies Could Help Them Meet the Challenges of Expected Shortages; United States General Accounting Office; GAO Highlights,2003; available at www.gao.gov/new.items/d03514.pdf (accessed 2009).

5. EIA, Energy Information Administration. Annual Energy Outlook 2009: With Projections to 2030, Washington, DC, 2009; available at http://www.eia.doe.gov/oiaf/aeo (accessed 2009).

6. Stiegel, G.J.; McNemar, A.; Nemeth, M.; Schimmoller, B.; Murphy, J.; Manfredo, L.; Estimating Fresh Water Needs to Meet Future Thermoelectric Generation Requirements. Research and Development Solution, LLC. Report prepared for the Department of Energy, DOE, 2008; available at http://www.netl.doe.gov/technologies/coalpower/ewr/pubs/2008 Water Needs Analysis-Final 10-2-2008.pdf (accessed 2009).

7. Dias De Oliveira, M.E.; Vaughn, B.E.; Rykiel, Jr.E.J; Ethanol as Fuel: Energy, Carbon Dioxide Balances, and Ecological Footprint. *Biosciences*. **2005**, 55, 593-601.

8. US Department of Agriculture. The Economic Feasibility of Ethanol Production from Sugar in the United States. United States Department of Agriculture. Washington, DC, July 2006.
9. Zink, J. Water-use estimates for ethanol plant go up 60 percent. Tampa Bay, FL, 2007; available at http://www.sptimes.com/2007/03/09/Hillsborough/Water_use_estimates_f.shtml (accessed 2009).

10. Keeney, D.; Water Use by Ethanol Plants: Potential Challenges. Institute of Agriculture and Trade Policy Report. Minneapolis, MN. October, 2006.

11. Wooley, R.; Ruth, M.; Sheehan, J.; Ibsen, K. Lignocellulosic Biomass to Ethanol Process Design and Economics Utilizing Co-Current Dilute Acid Prehydrolysis and Enzymatic Hydrolysis Current and Futuristic Scenarios. NREL report TP-580-26157, July 1999; available

at http://www.ott.doe.gov/biofuels/process_engineering.html (accessed 2009).

12. Aden, A. Water Usage for Current and Future Ethanol Production. *Southwest Hydrology*. **2007**, 6, 22-23.

13. Aden, A.; Ruth, M.; Ibsen, K.; Jechura, J.; Neeves, K.; Sheehan,J.; Wallace, B.; Montague, L.; Slayton, A.; Lukas, J. Lignocellulosic Biomass to Ethanol Process Design and Economics Utilizing Co-Current Dilute Acid Prehydrolysis and Enzymatic Hydrolysis for Corn Stover ; NREL report NREL/TP-510-32438; 2002.

14. King, C. W.; Webber, M. E. Water Intensity of Transportation; *Environ. Sci. Technol.*, **2008**, 42 (21), 7866-7872.

15. Water Implications of Biofuels Production in the United States; Committee on Water Implications of Biofuels Production in the United States: Water Science and Technology Board, Division on Earth and Life Studies, National Research Council; The National Academies: Washington DC, 2008.

16. Laser, M.; Jin, H.; Jayawardhana, K.; Lynd, L.R. Coproduction of Ethanol and Power from Switchgrass. *Biofuels, Bioprod. Bioref.* **2009,** 3, 195–218.

17. Renewable Fuels Association Website: U.S. Fuel Ethanol Industry Biorefineries and Production Capacity; available at http://www.ethanolrfa.org/industry/locations.html. (accessed 2009).

18. Wastewater Treatment Options for the Biomass-to-Ethanol Process*; Report No* AXE-8-18020-01; Final Report of Merrick & Company to National Renewable Energy Laboratory Merrick & Company: Golden, CO, 1998.

19. Owens, S. Reduce Cooling Tower Water Consumption by 20%; Ethanol Producer Magazine; June 2007; available at www.ethanolproducer.com (accessed 2009).

20. Wurtz, W. Air Cooled Condensers Eliminate Plant Water Use. Power Magazine; September 2008; available at http://www.powermag.com/issues/features/1361.html (accessed 2009).

21. Demakos, P.G. Wet Surface Air Coolers Minimize Water Use by Maximizing Heat Transfer Efficiency; Power Magazine; September 2008; available at http://www.powermag.com/water/Wet-surface-air-coolers-minimize-water-use-by-maximizing-heat-transfer-efficiency_1363_p2.html. (accessed 2009).

22. Rodgers, J.H.; Castle, J.W. An Innovative System for the Efficient and Effective Treatment of Non-traditional Waters for Reuse in Thermoelectric Power Generation; Final Technical report to U.S. Department of Energy; DE-FG26-05NT42535; November, 2008.

23. Feeley, T.J.; Carney B. Innovative Approaches and Technologies For Improved Power Plant Water Management; U.S. Department of Energy, Office of Fossil Energy, National Energy Technology Laboratory.

Challenges to Water Quality Sustainability from Chronic PAH Pollution in an Urban Estuary

Leslie M. Shor[a,*], Sandra N. Valle[b], Gabriela Munoz[b], Marta A. Panero[b,†] Lisa A. Rodenburg[c]

[a]Department of Chemical, Materials and Biomolecular Engineering and Center for Environmental Science and Engineering (CESE), University of Connecticut, Storrs CT 06269

[b]New York Academy of Sciences, 250 Greenwich Street, New York, NY 10007

[c]Department of Environmental Sciences, Rutgers University, New Brunswick, NJ

*Corresponding author, Email: leslie.shor@uconn.edu, Phone: (860) 486-3136

†Current affiliation: NY-NJ Harbor Estuary Program/New England Interstate Water Pollution Control Commission, New York, NY

Abstract

High-density development offers several sustainability advantages, including reduced per-capita energy consumption, but the impacts of urbanization on chronic water pollution must also be considered. The objective of this research is to provide a systematic analysis of the chronic sources of polycyclic aromatic hydrocarbon (PAH) contamination of the New York/New Jersey Harbor estuary as a function of demographics, land use, and infrastructure. First, primary emissions of PAHs were determined for both point and non-point sources to air, land, and water for 15 Priority PAHs for each of the 42 counties in the New York Harbor watershed. The results show that the dominant primary emission sources are all atmospheric, including residential fireplaces, motor vehicles, and volatilization from creosote-treated wood. For example, over 95% of primary benzo[a]pyrene emissions in the region originate from domestic fireplaces, with the great majority of these emissions coming from predominantly rural counties a great distance from the harbor. Appreciating that rates of contaminant transmission will vary greatly by compound, distance, and media, chronic loadings of PAHs to the harbor were modeled by combining estimates of primary emission with individualized measures of regional transmission potential. Results show harbor loadings originate as a combination of diffuse primary sources to the atmosphere (50%), land (40%), and water (10%). The leading sources for harbor loadings include transportation-related sources; volatilization, abrasion, and wear of creosote- and coal tar-containing products; and residential wood combustion. A comparison of predicted and measured PAH concentrations in regional air and water and provide an external validation on the analysis. Substantial differences in primary emission patterns and transmission rates highlight the need for systems-based modeling approach. This work offers a tool to identify pollution prevention and transmission control strategies that will most effectively promote the sustainable management of urban water resources.

Introduction

The growing world population and accompanying changes in per capita consumption patterns will place increasing demands on global water, energy, and food resources. Demands on these essential resources are interdependent. Some of the most promising "green" energy sources including ethanol and biodiesel require significant quantities of water per BTU, while water treatment and desalination technologies can be very energy intensive. Today, more than 2B people are currently experiencing severe water stress, where >40% of river discharges is already allocated to agriculture, industry, and domestic purposes (1). Feedbacks associated with global climate change are predicted to rapidly intensify regional water shortages in the future (2). High-density populations can lead to efficiencies in per capita consumption patterns; yet cities can also strain water resource quality locally and regionally. Pharmaceuticals and other organic contaminants are ubiquitous in drinking water sources (3). Safeguarding water resource quality is essential to ensuring the long-term sustainability of urban centers.

The objective of this study is to identify the activities, emission sources, and transmission routes that most contribute to diffuse polycyclic aromatic hydrocarbons (PAHs) loadings to the New York/New Jersey (NY/NJ) Harbor (Figure 1). PAHs are ubiquitous pollutants that result from nearly any combustion process. PAHs are also found in petroleum-based materials such as coal tars and creosote. Several PAHs are suspected carcinogens, and one in particular, benzo[a]pyrene, is on the EPA priority list of persistent and bioaccumulative toxins. PAHs are hydrophobic and bind to particles in environmental media and therefore tend to accumulate in the sediments underlying lakes, rivers, and estuaries.

In this study, primary emissions to air, land, and water were estimated for each of the 16 EPA Priority PAHs by source and county throughout the 42,128 km^2 NY/NJ Harbor watershed. Next, a regional fate and transport model was used to estimate transmission potential for primary emission to the harbor boundary. Finally, predicted loadings were compared with loadings determined from environmental monitoring data (4). The end result is a validated, spatially resolved, activity-based model of both point source and non-point source pollution in a large urban estuary.

Approach

A PAH emissions inventory for the NY/NJ Harbor watershed was developed using Substance Flow Accounting (SFA). This form of industrial ecology uses established inventory data to develop a material balances for substances through economic processes. Our SFA research quantified PAHs released into various environmental compartments within the Watershed as a result of importation, generation, or consumption of PAH-containing or PAH-evolving materials in the regional economy. Inherent in the approach is a systematic, comprehensive, consistent treatment of all primary emissions, of all compounds, from all sources, to all media, from all locations. This approach allowed an objective comparison of the magnitude of emission from different activities throughout a large geographical area.

The PAHs considered in this study were the 16 EPA priority PAHs (naphthalene, acenaphthylene, acenaphthene, fluorene, phenanthrene, anthracene, fluoranthene, pyrene, benz[a]anthracene, chrysene, benzo[b]fluoranthene, benzo[k]fluoranthene, benzo[a]pyrene, dibenzo[a,h]anthracene, benzo[g,h,i]perylene, and indeno[1,2,3-cd]pyrene).

The environment of concern in this study was the NY/NJ Harbor. With the exception of direct releases to the Harbor, including atmospheric deposition to the Harbor surface area, transport of PAHs to the Harbor is dictated by fluvial systems and is therefore constrained by the regional watershed area. The Harbor watershed region covers approximately 42,128 km^2 of land area, including 12 counties of northern New Jersey and 30 counties in eastern New York in the northeastern United States. The U.S. Census reported in 2004 that approximately 14.6 million people lived in the watershed regio. Atmospheric emissions of PAHs from outside the watershed boundary have the potential to reach the Watershed as well. However, given the high population density within the region, and the generally more rural areas beyond the watershed boundary, net atmospheric imports of PAHs into the region are likely to be exceeded by net exports of PAHs outside the watershed borders.

Releases to air, water, and land were calculated for over forty separate point and nonpoint (or diffuse, disperse) sources for each PAH in each county in the region. Nonpoint sources include both stationary (e.g., residential heating) and mobile (e.g., vehicle exhaust) sources. Sources were identified through a literature search by consultation with experts from industry, government, and academia. Emissions from nonpoint sources were calculated by multiplying an emission factor, which is a measure of emissions of a particular compound from a specific activity, by the activity level occurring in a localized area. For example, regional primary emissions of the benzo[a]pyrene (BAP) from combustion of wood in home woodstoves was calculated by multiplying the emission factor for burning wood in a particular type of wood stove (mg BAP/kg wood) by the total kilograms of wood consumed in that type of stove in a given county each year. PAHs releases from point sources were calculated by applying compound-specific emission factors to facility-specific annual activity level data (e.g., lbs of fuel consumed/yr or tons of material produced/yr). If emission factors were not available for a point source, reported emissions (e.g., to the U.S. EPA Toxics Release Inventory database). See also Valle et al. (5).

Our modeling objective was to estimate the potential for individual PAH compounds released to air, land, or water throughout the region to be transmitted to the NY/NJ Harbor. Harbor input routes included direct atmospheric deposition, direct gas absrobtion, or tributary, stormwater, or combined sewer overflow flows. The regional transport model considered the following environmental processes: photodegradation, absorption of gaseous PAHs to the Harbor's water surface, wet or dry particle deposition to land surfaces, runoff from impervious surfaces, transmission via stormwater discharges to surface waters, particle settling, biodegradation, and tributary transport to the Harbor boundary. Within compartments, the transmission model accounts for multiphase partitioning between air, water, and particle phases, and makes use of localized information on land use, stormwater system design and operation, and regional hydrogeographical parameters. PAH fate and transport within the Harbor boundary is not considered. Modeling parameters for a single parameter value could be determined with high confidence, such as physiochemical properties of PAHs K_{ow}, octanol-water partition coefficient, or the combined sewer overflow versus separately-sewered land coverage for a given county, were treated deterministically. By contrast, distributed parameters were used for quantities likely to take on a range of values, such as suspended solids or dissolved organic carbon concentration in surface waters as per previously published studies. Central tendencies and confidence intervals for PAH loadings were computed using at least 10,000 iterations of a Monte Carlo simulation performed using Microsoft Excel.

Results

Major Sources of PAH Loadings. Atmospheric sources account for 80% of all PAH emissions in the region and 50% of total PAH loadings. While only about 1% of primary atmospheric emissions reach Harbor directly, deposition and re-suspension of atmospheric particles is an important transmission route for higher molecular weight PAHs such as BAP. Land- and water-based sources, many of which have higher transmission rates, contribute 40% and 10% of loadings, respectively. Key atmospheric sources contributing to Harbor loadings of total PAHs are residential wood combustion (2,182 kg/yr), emissions from on-road (1,750 kg/yr) and off-road (924 kg/yr) vehicles, and volatilization from creosote-treated utility poles (1,613 kg/yr) and railroad ties (1,239 kg/yr) (Figure 2). Other important sources to Harbor loadings are land-based sources involving emission of PAHs to impervious surfaces, including motor oil leaks (2,139 kg/y), tire wear (1,171 kg/yr), and use of coal tar-based driveway sealants (726 kg/yr). Leaching of PAHs from marine pilings contributes 1,385 kg/yr to total loadings.

Comparing the relative contribution of classes of sources to Harbor loadings, transportation-related activity including on-road and off-road vehicle exhaust, tire wear, and leakage and improper disposal of motor oil account for about 40% of all PAH loadings to the Harbor. Other important sources are creosote treated wood (40%), wood combustion (14%), coal tar sealants (5%) and oil spills (1%).

Major Trends by Compound. PAH physical properties span several orders of magnitude, causing PAHs to be transported in the environment at different rates. The leading source for naphthalene loadings are motor vehicle exhaust (45%, combined on-road and off-road), volatilization from creosote-treated wood (23%), motor oil leaks (13%), wood combustion (10%), and leaching from creosote treated pilings (7.5%). Major loadings sources for BAP are entirely different. Coal tar sealants are the largest source of loadings, with 73 kg/yr loadings resulting from 276 kg/yr primary emissions, which is 30% of total BAP loadings. Together, the transportation sources (tailpipe emissions, leaking oil, and tire wear) contribute 31% to Harbor BAP loadings, while emissions from creosote-treated wood and residential wood combustion each contribute 18%.

Comparison with Mass Balance. Harbor loadings from this study were compared with harbor loadings from a separate analysis based on measured PAH concentration in the atmosphere, surface waters, runoff, and overflows (4). The consistency between emission-based loadings and measurement-based loadings given the size of the region and the number of parameters involved provides an external validation of the emissions and transmission computations (Figure 3). Low estimates (such as BAP in the NJ tributaries) may indicate missing emission factors for certain compounds, and/or the emission factor, activity factor, and/or transmission potential was too low. Specific sources known to be missing in the emissions inventory include migration of PAHs from historically-contaminated sites, industrial or residential PAH inputs to wastewater, and several other sources for which emission factors or activity factors were not. For some important sources of PAHs, including tire wear, there are no emission factors available for several PAH compounds. The central tendency of the predicted loading for fluoranthene and pyrene was above the high end of the measured range for the Hudson River and the New Jersey tributaries, respectively. This may indicate emissions estimates for these compounds were too high. The most important source for fluoranthene and pyrene identified in this study is volatilization or leaching from creosote-treated products

Conclusions

We find the transportation sector is an important source of PAH loadings to the Harbor. Public policies aimed at increasing the use of public transportation or pedestrian access and encouraging awareness about the environmental impact of oil leaks and improper disposal of used motor oil, especially in the core harbor counties, could have the greatest impact at reducing loadings to the Harbor. Preventing BAP inputs to the Harbor could be achieved by restricting the use of coal tar-based parking lot and drive way sealants. Product bans for these products are already in effect or under consideration in some municipalities, including Austin, TX and Madison, WI (5). Since coal tar sealants are used primarily for aesthetic reasons, decision-makers elsewhere may wish to consider similar action.

On equal footing with pollution prevention is the need for community planning and engineering controls that limit the transmission of contaminated particles in urban or sub-urban areas. By county, the factors most responsible for high transmission rates are the existence of CSO versus MS4 sewers and the percent impervious land area in a given county. Policies aimed at increasing the use of pervious surfaces and reducing stormwater runoff may have the greatest impact on reducing chronic PAH loadings to the Harbor.

This work represents a systems-based, population- and economic activity-grounded regional flow analysis of a complex and important group of environmental contaminants. Using a similar approach, the key sources of other compounds in other locations can be assessed, and the future impacts of population changes, technological innovations, and policy changes on nonpoint source pollution may be evaluated in urban watersheds worldwide.

References

1. Oki T & Kanae S (2006) Global Hydrological Cycles and World Water Resources. *Science* 313(5790):1068 - 1072.
2. Vörösmarty CJ, Green P, Salisbury J, & Lammers RB (2000) Global Water Resources: Vulnerability from Climate Change and Population Growth. *Science* 284:284-288.
3. Focazio MJ, *et al.* (2002) A national reconnaissance for pharmaceuticals and other organic wastewater contaminants in the United States — II) Untreated drinking water sources. *Science of The Total Environment* 402(2-3):201-216
4. Rodenburg LA, Valle S, Panero M, & Shor L (2009) Mass Balances on Selected Polycyclic Aromatic Hydrocarbons (PAHs) in the NY/NJ Harbor. *In Review.*
5. Valle S, Panero MA, & Shor L (2007) Pollution prevention and management strategies for polycyclic aromatic hydrocarbons in the New York/New Jersey Harbor. (New York Academy of Sciences report, New York, NY).

Figures

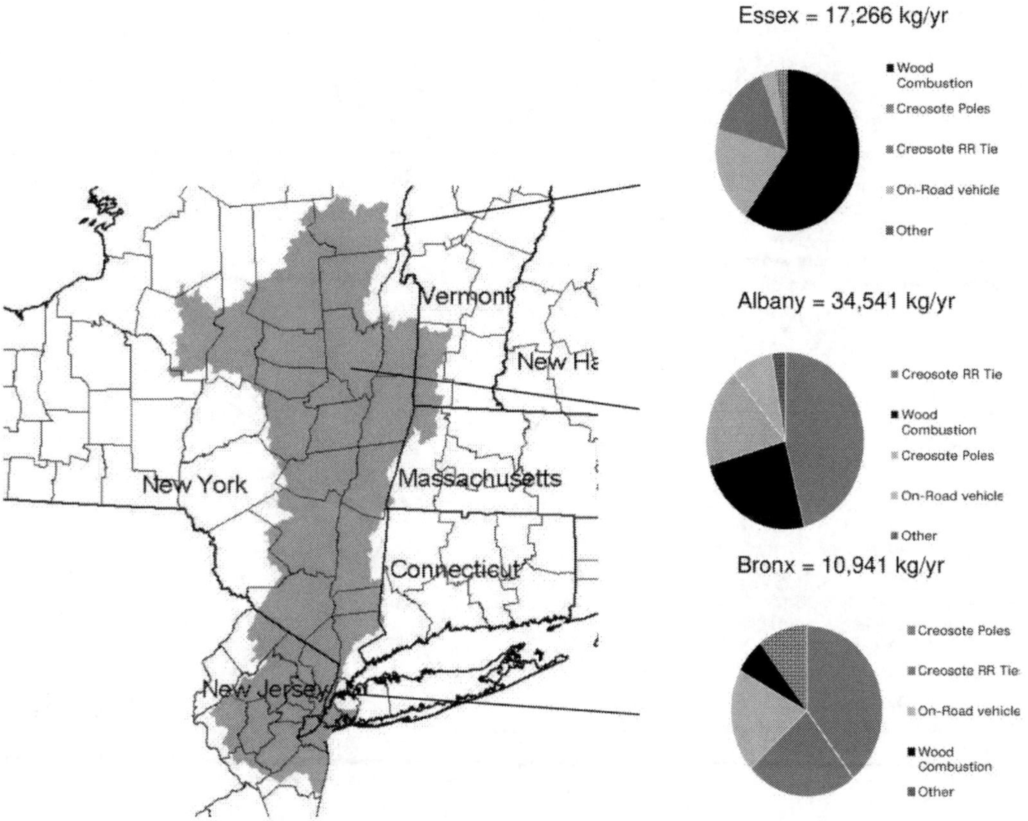

Figure 1. Emissions of PAHs from the 11 major sources by NY/NJ Harbor Watershed Counties (watershed area shown on map) and relative emissions by source for three of the counties (kg/yr).

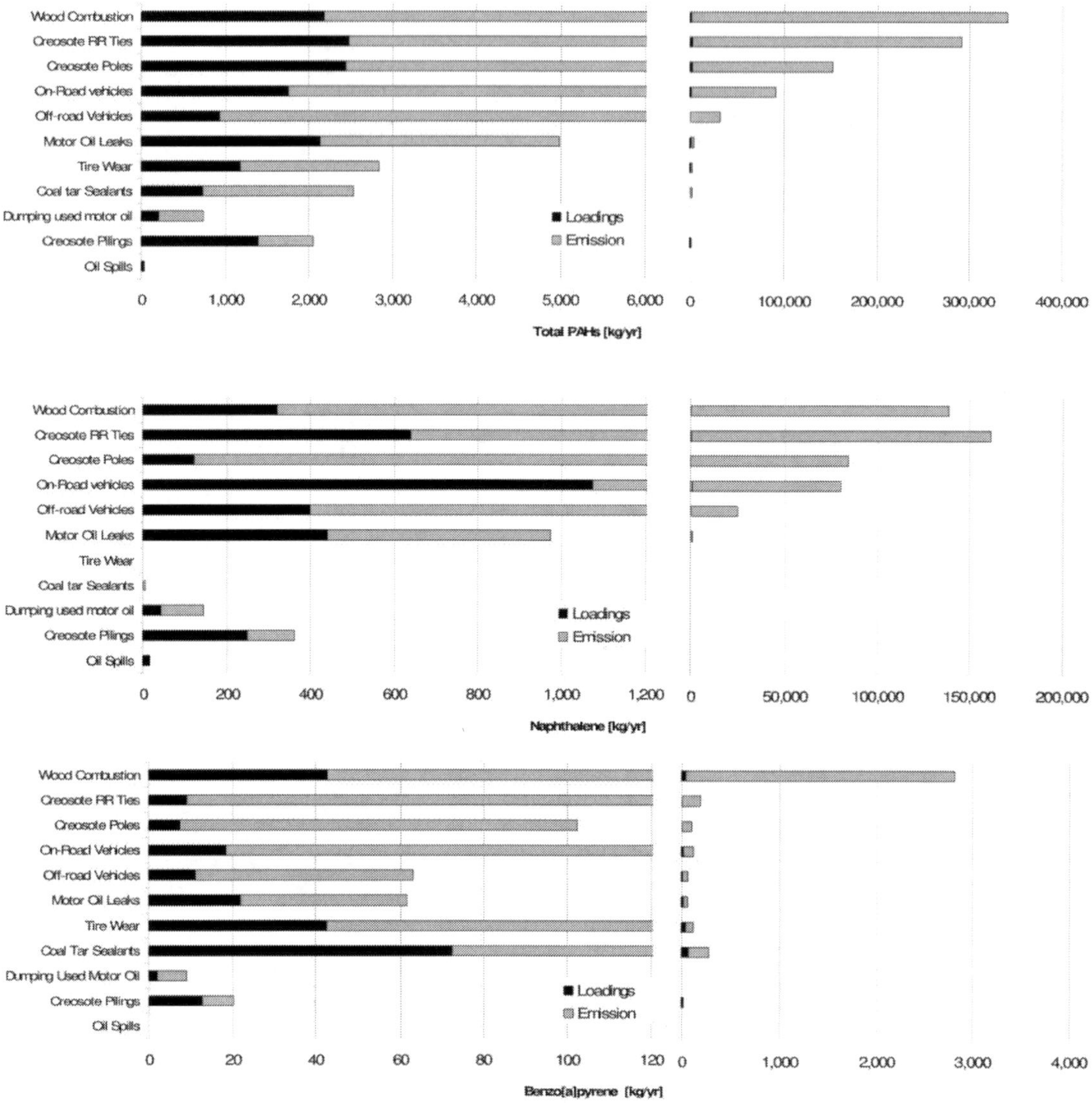

Figure 2. Comparison between total emissions and harbor loadings (kg/yr) for the 11 major sources for (a) total PAHs, (b) naphthalene, and (c) benzo[a]pyrene. To show both the detail and the overall magnitude, the data is plotted on 2 separate scales for each major source.

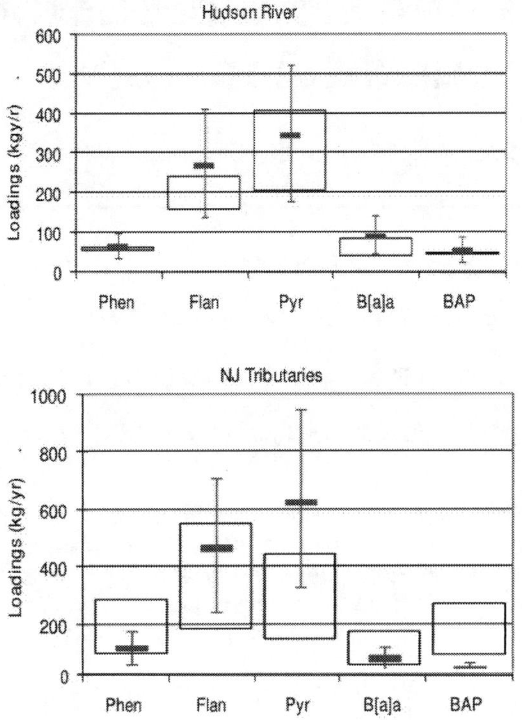

Figure 3. Example comparison between loadings calculated from monitoring data (open boxes showing range of estimate) and Monte Carlo model of primary emissions and regional transmission (red lines, with 95% inner confidence interval) for loadings from the Hudson River and the NJ tributaries. Abbreviations are Nap, naphthalene; Flan, fluoranthene; Pyr, pyrene; B[a]a, benz[a]anthracene; and BAP, benzo[a]pyrene.

Evaluating the Fate of Several Micro-Pollutants in Wastewater Using Membrane Bioreactor and Conventional Activated Sludge Reactor

Qiang Zhang, Civil and Environmental Engineering, University of Cincinnati, Cincinnati, OH

Rendahandi G. Silva, Shaw Environmental, Inc., Cincinnati, OH

Ruth Marfil-Vega, Civil and Environmental Engineering, University of Cincinnati, Cincinnati, OH

Ali Medella, Civil and Environmental Engineering, University of Cincinnati, Cincinnati, OH

Makram Suidan, Department of Environmental Engineering, University of Cincinnati, Cincinnati, OH

Jeff Yang, Mail Stop 690, USEPA National Risk Management Research Laboratory WSWRD / WQMB, Cincinnati, OH

Safe water reuse depends on the understanding and control of micro-pollutants in wastewater treatment and effluent reuse practices. Micro-pollutant impacts on human health and other living organisms have promulgated extensive research on wastewater treatment technologies with varying levels of success. While each type of wastewater treatment systems has its own merit, they vary dramatically in their energy requirements, cost-benefits, sensitivity and reliability in removing a wide range of pollutants. In the reported study, membrane bioreactor (MBR) with relatively greater pore size and conventional activated sludge reactor (CAS) were evaluated for their ability to remove a few classes of micro-pollutants together. Two solid retention times (SRTs) of 9 and 18 days were used to investigate the removal efficiency of micro-pollutants. Micro-pollutants belong to pesticide (aldicarb), herbicide (atrazine and glyposate), anti-inflammatory drug (Ibuprofen) and psychoactive stimulant drug (caffeine) were introduced to influent at a 200-ìg L-1 concentration. Results showed that both aldicarb and caffeine biodegraded effectively (90-100%) under both SRTs with no obvious difference between reactors. For anti-inflammatory drug, ibuprofen, biodegradation rate fluctuated during lower and higher SRTs. The slightly greater removal efficiency at lower SRT (35%) disappeared at higher SRT (20%) in CAS probably be attributed to limited biomass settling characteristics at higher SRT in that reactor. Ibuprofen biodegradation efficiency was similar and about 40% at both SRTs in MBR. Minimal biodegradation of atrazine (<2%) was observed in CAS at 9 day SRT and did not improve even at higher SRT. Atrazine biodegradation was approximately 20% and 30% in MBR for lower and higher SRTs, respectively. Glyposate degradation fluctuated significantly between sampling times with no specific trend in both reactors. Among the less degradable pollutants atrazine and ibuprofen had high affinity to water while glyposate was biphasic. Despite the irregularity in herbicide biodegradation, the MBR system was comparatively more effective in removing atrazine than the CAS and generally offers smaller footprint and more stable performances for the tested micro-pollutants. These preliminary results demonstrate the challenge in removing recalcitrant micro-pollutants in biological treatment systems, and also point to the directions for future research and development of the treatment technologies to assure safe water reuse.

See more of this Session: Sustainability of Water Use
See more of this Group/Topical: Sustainable Engineering Forum

Bio-Remediation of Waste Water Streams Using Soil Bio-Technology (SBT)

Chandrashekar Shankar, Chemical Engineering, University of Pittsburgh, Pittsburgh, PA

H. S Shankar, Indian Institute of Technology Bombay, Mumbai, Maharashtra, India

Urban, industrial and agricultural liquid waste streams are chief pollutants of planetary aqueous ecosystems as evidenced by large scale fish and marine die-offs reported the worldwide. Pollutants such as sewage and agricultural fertilizer runoffs have resulted in accelerated algal blooms depriving aquatic ecosystems essential dissolved oxygen. Acidic industrial waste streams have also resulted in leaching of heavy metals into such aquatic ecosystems. We show bio-remediation of such waste streams by a novel eco-friendly, energy efficient method called soil biotechnology (SBT). SBT is based on a bio-conversion process where fundamental reactions of nature, namely respiration, photosynthesis and mineral weathering, take place in a media housing micro- and macro-organisms which bring about the desired purification. SBT is an oxygen supplying biological engine, so the process can treat all types of waste water: domestic municipal and industrial with a salinity $<$ 2500 mg/l. Due to its near neutral buffered pH of about 7.5 SBT also proves efficient in precipitation of toxic heavy metals. We show results from nearly 10 years of commercial operations of SBT plants delivering treatment solutions for a variety of waste streams such as sewage, restaurant waste and tobacco processing spent wash. Other applications include on-board sewage treatment and recovery for cruise ships, transportable solutions for remote locations and venues, oxygenation and waste removal for commercial fish farms. SBT technology is scalable and can be integrated with into the present urban landscapes via aesthetically pleasing gardenlike environments.

See more of this Session: Sustainability of Water Use
See more of this Group/Topical: Sustainable Engineering Forum

Formation of Halogenated Organic Compounds in Seawater

Ahmed Abdel-Wahab, Chemical Engineering, Texas A&M University at Qatar, Doha, Qatar

Ahmed Khodary, Chemical Engineering, Texas A&M University at Qatar, Doha, Qatar

Patrick Linke, Chemical Engineering, Texas A&M University at Qatar, Doha, Qatar

Bill Batchelor, Civil Engineering, Texas A&M University, College Station, TX

Mahmoud El-Halwagi, Department of Chemical Engineering, Texas A&M University, College Station, TX

The use of seawater in industrial cooling is a common practice in many parts of the world that have limited fresh-water resources. One of the primary operational problems of using seawater in cooling is biofouling. Biofouling can result from growth of microorganisms on surfaces where they form biofilms or from the growth of macro-organism such as clams. Biofilms tend to stick to heat-exchange surfaces, thereby significantly reducing heat-transfer coefficients. In some cases, excessive bio-fouling can lead to plugging of heat exchangers. There are several techniques for preventing biofouling of both types, but application of chlorine-based biocide is most common.

Chlorine is added to the seawater to control biofouling of the cooling system. The added chlorine reacts with bromide and other compounds in the water to produce a wide range of chemical oxidants. These include conversion of bromide ion to hypobromous acid and other reactive forms of bromine. These brominated products are the active forms of biocide in seawater systems and their relative concentrations change on time scales from fractions of seconds to days. Understanding their behavior is critical to insuring effective control of biofouling within the plant and minimizing environmental impacts outside the plant. Additionally, the brominated residual biocide also can react with natural organic matter in the seawater to form a number of halogenated organic compounds that are toxic to aquatic life as well as humans. Brominated forms of biocide are much more effective in producing halogenated organics than chlorine, but have not been studied as extensively.

Understanding the reaction scheme of chlorine, brominated products and natural organic matter is critical to optimizing performance of biofouling control systems while minimizing impacts to the aquatic environment and to human health. Studies by Shams et al. (1991) on Umm Al Nar seawater desalination plant in Abu Dhabi showed that bromoform represented 95% of the trihalomethanes (THM) that were formed. Ali and Riley (1986) reported that THM concentrations as high as 90 μg/L were observed in the vicinity of discharges from combined power/desalination plants in Kuwait. Many other reports are available on the production of halogenated organics in freshwater systems such as water treatment plants and water distribution systems. However, limited data are available on biocides chemistry and their reaction by-products in seawater.

This purpose of this study is to develop the understanding of biocide chemistry in seawater and apply that understanding to develop methods to quantitatively predict changes in concentrations of biocides and reaction products in cooling water during use and after release to the environment. This will provide the basis for developing a full kinetic model to describe the biocide chemistry in seawater and for developing a simulation tool to determine the fate and transport of biocides and their reaction products in the environment.

Kinetic experiments were conducted to evaluate the kinetics of formation of halogenated compounds in seawater. All experiments were conducted using seawater sampled from the Arabian Gulf in Qatar. In order to study the effect of chlorine dose on the kinetics of trihalomethanes formation, various chlorine doses (1, 2, 3, and 4 mg/L) were reacted with the seawater. Samples were taken at times 0.0, 0.5, 1, 6, 18, 24, 48,

72, 96, 120 and 168 hours and were analyzed for trihalomethane compounds using Gas Chromatography (GC) equipped with electron capture detector (ECD). Results of these experiments showed that concentrations of trihalomethanes increased rapidly during the first half hour of the reaction. Bromoform was found to be the dominant halogenated compound formed in seawater.

References:

Ali, M. and Riley, P. (1986) The distribution of halomethanes in the coastal waters of Kuwait, Marine Pollution Bulletin 17, 409-414, 1986.

Batchelor, B., (1989) A Kinetic Model for Formation of Disinfection By-Products, Fellowship Report, American Academy for the Advancement of Science/Environmental Protection Agency Fellowship Program, Washington, DC.

Goodman P.D. (1987) Effect of chlorine on materials for sea water cooling systems: A review of chemical reactions. British Corrosion Journal 22: 56-62.

Johnson, J.D., Inman, G.W., Trofe, T.W., (1982) Cooling Water Chlorination: The Kinetics of Chlorine, Bromine, and Ammonia in Sea Water", NUREG/CR-1522 RE, Office of Nuclear Regulatory Research, U. S. Nuclear Regulatory Commission, Washington, DC.

Johnson, J.D., Overby, R. (1971) Bromine and Bromamine Disinfection Chemistry, Journal of the Sanitary Engineering Division, ASCE, 97(SA5): 617-628, 1971.

Shams El Din, A. M., Rasheed, A. A. and Hammoud A. A. (1991) A contribution to the problem of Trihalomethane formation from the Arabian Gulf Water, Desalination 85, 13-32.

See more of this Session: Sustainability of Water Use
See more of this Group/Topical: Sustainable Engineering Forum

Design of Water Networks Using Rigorous Models

Debora C. Faria, School of Chemical Engineering and Material Science, University of Oklahoma, Norman, OK

Miguel J. Bagajewicz, Chemical, Biological and Materials Engineering, University of Oklahoma, Norman, OK

One of the weaknesses in currently methods to solve the water allocation problem (WAP) in process plants is a lack of accurate modeling of water regeneration and water using processes. In general, two kinds of model assumptions are made: regeneration processes have a fixed outlet concentrations (Koppol et al., 2003); and, regeneration processes have a fixed rate of removal (Takama et al, 1980; Guanaratnam, 2005; Karuppiah and Grossmann, 2006; Alva-Argáez, 2007). In reality, outlet concentration and/or rate of removal of regeneration process may vary with inlet concentrations and flowrates for each regeneration process design and also with different flowrates and inlet conditions. Similar assumptions are made in water using processes: fixed loads are used, when these loads are in many cases a function of inlet water conditions and flowrates.

Empirical equations or equations obtained from mathematical models and simulations can be used to better describe regeneration process. We use these equations to determine true outlet concentrations in water using units and to design the regeneration processes together with determining the network structure. Thus, the operating conditions like flowrate and inlet concentrations are taken into account when the design parameters of the regeneration processes are chosen. We propose a mathematical programming framework where the regeneration processes equations are transformed in discrete parameters and binary variables are used to select the optimum design.

Alva-Argaéz, A., Kokossis, A.C., Smith, R. (2007). A conceptual decomposition of MINLP models for the design of water-using systems. Int. J. Environment and Pollution. 29, 177.

Gunaratnam, M.S. (2003). Total water system design. Ph.D. Thesis, University of Manchester Institute of Science and Technology - UK.

Karuppiah, R., Grossmann, I.E. (2006). Global optimization for the synthesis of integrated water systems in chemical processes. Computers and Chemical Engineering, 30, 650-673.

Koppol, A.P.R., Bagajewicz, M.J., Dericks, B.J. and Savelski, M.J. (2003). On zero water discharge solutions in the process industry. Advances in Environmental Research, 8, 151-171.

Takama, N., Kuriyama, T., Shiroko, K., Umeda, T. (1980). Optimal water allocation in a petroleum refinery. Computers & Chemical Engineering, 4, 251-258.

See more of this Session: Sustainability of Water Use
See more of this Group/Topical: Sustainable Engineering Forum

EVALUATING WATER QUALITY CHANGES IN WASTEWATER REUSE USING PILOT-SCALE EXPERIMENTS

Paper No. 171g

Donald A. Schupp, Shaw Environmental & Infrastructure, Inc., Cincinnati, OH
E. Radha Krishnan, Shaw Environmental & Infrastructure, Inc., Cincinnati, OH
Y. Jeffrey Yang, U.S. EPA, Cincinnati, OH
Young-Shin Jun, Washington University in St. Louis, St. Louis, MO

Abstract

The pilot-scale experimental study described in this paper was designed to evaluate water quality changes in wastewater reuse applications and, thus, to determine the water quality treatment requirements for environmentally safe applications. For this evaluation, two types of pilot-scale systems (an activated sludge treatment unit and two soil columns) were set up at the U.S. EPA Test & Evaluation (T&E) Facility in Cincinnati, Ohio. The activated sludge treatment unit was used to simulate contaminant behaviors in conventional municipal wastewater treatment systems, while the soil columns were constructed to investigate contaminant fate and transport in land applications for wastewater reuse. Aldicarb, a pesticide that is sometimes found in low concentrations in groundwater and surface water was used as the target contaminant to evaluate its fate and transport in both the activated sludge treatment unit and the soil columns. Atomic force microscopy (AFM) was used to observe the surface morphologies of extracellular polysaccharides (biofilms) on quartz coupons inserted within the soil columns. Fourier Transform Infrared (FTIR) spectroscopy experiments were also conducted to provide information regarding the mechanism of binding/transformation that may be occurring inside the soil columns.

Disclaimer

The U.S. Environmental Protection Agency, through its Office of Research and Development, funded and managed the research described herein. Shaw Environmental, Inc. conducted the work under Contract No. EP-C-04-034. It has been subjected to the Agency's peer and administrative review and has been approved for external publication. Any opinions expressed in this paper are those of the author(s) and do not necessarily reflect the views of the Agency, therefore, no official endorsement should be inferred. Any mention of trade names or commercial products does not constitute endorsement or recommendation for use.

Background/Objectives

In FY2007, U.S. EPA began to implement a new science and engineering research program on the Nation's aging water and wastewater infrastructure. The program is aimed to meet the future water needs for the Nation and, in doing so, protect human health and the environment. One program component is to evaluate water (wastewater, stormwater, and graywater) reuse and further develop innovative water reuse applications, establish water quality criteria, and evaluate infrastructure requirements for expanded water reuse in the nation. Currently, rehabilitation and capital improvement of the aged water infrastructure is urgent in order to provide adequate water supply and wastewater management in the face of climate change and socioeconomic development in the 21st century. This also provides a good opportunity to incorporate water reuse practice into the overall sustainable water resources development.

A pilot-scale experimental study was designed and conducted in 2008 to evaluate water quality changes in wastewater reuse applications and, thus, to determine the water quality treatment requirements for environmentally safe applications. For this purpose, two types of pilot-scale systems (an activated sludge treatment unit and two soil columns) were set up at the U.S. EPA Test & Evaluation (T&E) Facility in Cincinnati, Ohio. The activated sludge (AS) treatment unit was used to simulate contaminant behaviors in conventional municipal wastewater treatment systems, while the soil columns were constructed to investigate contaminant fate and transport in land applications for wastewater reuse. Aldicarb, a pesticide that is sometimes found in low concentrations in groundwater and surface water (EPA, 2008) was used as the target contaminant to evaluate its fate and transport in the pilot-scale systems. The pilot-scale water reuse investigations were designed to:

1) Determine the extent of aldicarb and macronutrients removal in the AS unit. Chemical oxygen demand (COD) was selected as the macronutrient parameter for the investigation.
2) Evaluate aldicarb transport in the soil columns under simulated land applications of treated wastewater effluent.
3) Assess the mechanism/pathway for aldicarb degradation in the soil columns.
4) Determine the role of microorganisms attached to soil particles of the soil column in aldicarb degradations.

Materials and Methods

Activated Sludge Treatment Unit

<u>Bench-Scale Tests</u>

Bench-scale studies were performed using a synthetic wastewater (Zhang, 2006) with varying concentrations of aldicarb: 100 μg/l, 500 μg/l, and 1,000 μg/l. The tests were conducted for 24 hours in 2-L glass beakers to determine the reaction kinetics of aldicarb degradation by AS. The synthetic wastewater included 6 organic compounds and 10 nutrient chemicals, and the composition was formulated to simulate constituents in raw wastewater at wastewater treatment plants (WWTPs). The synthetic wastewater composition is shown in Table 1. Nine different conditions were evaluated in the bench-scale tests (Table 2).

The beakers representing the different run conditions were continuously agitated using a magnetic stirrer. Air was introduced into the bottom of each beaker through a coarse bubble diffuser at a flow rate of 2 L/min. Samples were collected hourly for the first 6 hours and every 2 hours for the remainder of the 24-hour study period. Aldicarb and aldicarb sulfone analyses were performed using extraction with methylene chloride and analysis by high-pressure liquid chromatography (HPLC) according to EPA Method 8318. The COD samples were digested and analyzed by Hach Method 10067.

<u>Pilot-Scale Tests</u>

A pilot-scale AS treatment unit, consisting of a 25-L primary clarifier, a 213-L aeration basin, and a 52-L secondary clarifier, was fabricated from Plexiglas® and set up at the T&E

Facility (Figure 1). The system was seeded with AS from the Greater Cincinnati Metropolitan Sewer District (MSD) WWTP located adjacent to the T&E Facility.

Table 1. Composition of Synthetic Wastewater Used for Activated Sludge Tests

Material	**Concentration (mg/l)**
Organic Compounds	
Casein	47.0
Tryptone	47.0
Starch	84.4
Sodium Acetate	31.9
Glycerol	12.0
Caproic Acid	11.6
Nutrients	
Ammonium Sulfate	116.0
Magnesium Sulfate	69.6
Calcium Chloride	22.5
Potassium Phosphate	27.6
Iron Chloride	11.0
Cupric Sulfate	0.09
Sodium Molybdate	0.15
Manganese Sulfate	0.13
Zinc Chloride	0.23
Cobalt Chloride	0.42

Table 2. Conditions Employed in the Activated Sludge Bench-Scale Tests

Run Number	**Test Name**	**Aldicarb Concentration (μg/l)**	**Activated Sludge (ml)**	**Synthetic Wastewater (ml)**
1	Aeration Control	100	0	1500
2	Aeration Control	500	0	1500
3	Aeration Control	1000	0	1500
4	Aldicarb	100	150	1350
5	Aldicarb	500	150	1350
6	Aldicarb	1000	150	1350
7	Aldicarb	100	450	1050
8	Aldicarb	500	450	1050
9	Aldicarb	1000	450	1050

The synthetic wastewater stream was prepared, and the target contaminant (aldicarb) was introduced into the feed of the primary clarifier for the AS treatment unit. The COD of the synthetic wastewater was approximately 300 mg/l. Synthetic wastewater was fed to the AS treatment unit at 0.5 L/min, resulting in hydraulic residence times (HRTs) of approximately 0.84 hrs, 7.09 hrs, and 1.75 hrs in the primary clarifier, aeration basin, and secondary clarifier, respectively. Pilot-scale tests were performed by spiking aldicarb into the pilot-scale system at a target concentration of 100 μg/l for 56 hours. The analytical procedures employed for the pilot-scale tests were identical to those for the bench-scale tests.

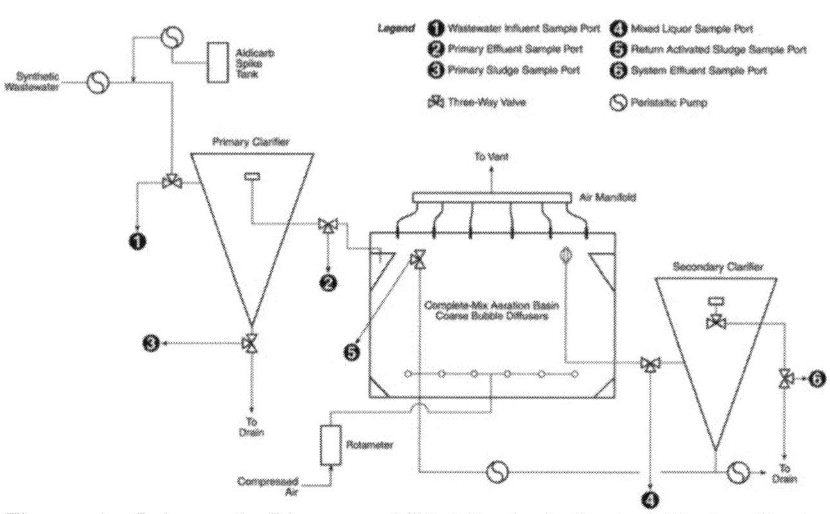

Figure 1. Schematic Diagram of Pilot-Scale Activated Sludge System (Alvarez, 1999)

Soil Columns

Two vertical soil columns, each 3 meters tall and 40.6 centimeters in diameter were fabricated from Sch. 80 PVC pipe and assembled at the T&E Facility for this water reuse application. Sample ports were located at 45.7-centimeter intervals along the height of each column. The columns were filled with "sandy" soil obtained from Preble County, Ohio. A gradation sieve analysis showed that ninety percent of the soil particles were in the range of 0.075 mm to 4.75 mm. One column was operated under saturated conditions under a pressure of 10 psig, at which the feed water flow rate was at approximately 350 ml/min. The second column was operated under unsaturated conditions with the feed water flow rate at approximately 35 ml/min and atmospheric pressure.

In order to evaluate solute transport in the soil columns, tracer studies were conducted by introducing 5 L and 10 L, respectively, of a 10 wt.-percent sodium chloride solution into the top of the saturated and unsaturated columns. The salt solution eluted from the two columns at 350 ml/min and 35 ml/min, respectively. The conductivity of the column effluents from the saturated and unsaturated columns was monitored over 18 hours and 7 days, respectively, using a YSI 556 multisonde probe.

Following the tracer studies, pilot-scale studies were conducted on both the soil columns by introducing non-chlorinated secondary effluent obtained from the MSD WWTP spiked with 100 μg/l aldicarb into the saturated and unsaturated columns at approximately 350 ml/min and 35 ml/min, respectively. The study was conducted over a 1-month period, and samples were collected approximately every 5 days. Aldicarb and aldicarb sulfone were analyzed in the same manner as in the AS studies.

During the pilot-scale soil column study, eight quartz coupons were inserted into the saturated soil column to investigate the mechanism of microorganism attachment to the soil particles. These coupons were located at two depths in the column (top and middle), at the center and wall of the column, and were oriented perpendicular and parallel to the water flow through the column. Atomic force microscopy (AFM) was used to observe the surface morphologies of the quartz coupons and the presence of extracellular polysaccharides (biofilms) before and after the 1-month study period. The changes in the quantity of biofilm formation and microbial growth were also studied to estimate the potential risk of blocking the soil pores in the wastewater reuse systems.

Fourier Transform Infrared (FTIR) Spectroscopy experiments were also conducted to provide information regarding the mechanism of binding/transformation inside the soil column. FTIR spectroscopy provided us fundamental information about 1) whether the chemicals bind with soil particles, 2) how strongly they bind to the soil surface, and 3) if the chemicals undergo transformations.

Results and Discussion

Activated Sludge Treatment Unit

<u>Bench-Scale Tests</u>

Figure 2 shows the COD and aldicarb concentration versus time for each of the bench-scale AS test runs. COD reductions between 57 percent and 81 percent were observed, depending on the amount of AS used in the bench-scale tests. Higher amounts of AS resulted in faster and more COD reduction; lower amounts of AS resulted in slower and less COD reduction. When no AS was used (Aeration Control – Runs 1 to 3 in Table 2), some COD reduction was still observed, but the reduction was slow at the beginning of the test. The majority of the COD reduction occurred later in the 24-hour study period.

Little to no degradation of aldicarb was observed during the 24-hour test period in the bench-scale tests. The initial aldicarb concentration did not appear to have an effect on the aldicarb degradation. Furthermore, the aeration control experiments (no AS addition) did not show any aldicarb degradation. No aldicarb sulfone was detected during the bench-scale AS tests.

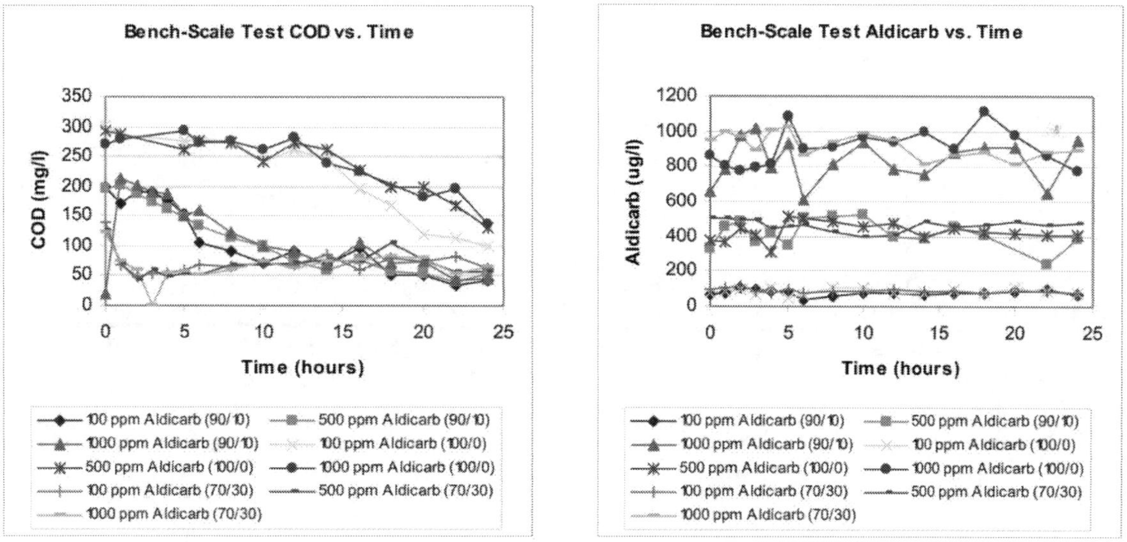

Figure 2. COD and Aldicarb Concentration vs. Time for All Bench-Scale AS Test Runs (numbers in parentheses represent the percentage of wastewater and sludge, respectively)

Pilot-Scale Tests

Figure 3 presents the COD and aldicarb concentration versus time for the pilot-scale AS test. COD concentrations decreased from an average of 210 mg/l to 54 mg/l, representing a 74 percent reduction in COD in the pilot-scale AS treatment unit. The largest decrease was between Ports 2 and 4, where the activated sludge in the aeration basin decreased organic compounds. COD reductions were consistent throughout the 56-hour study period, confirming that the AS treatment unit was at steady-state when the aldicarb was introduced into the system, and that the aldicarb did not interfere with the normal operation of the AS treatment unit. The COD reductions were comparable to those obtained towards the end of the 24-hour study period in the bench-scale tests.

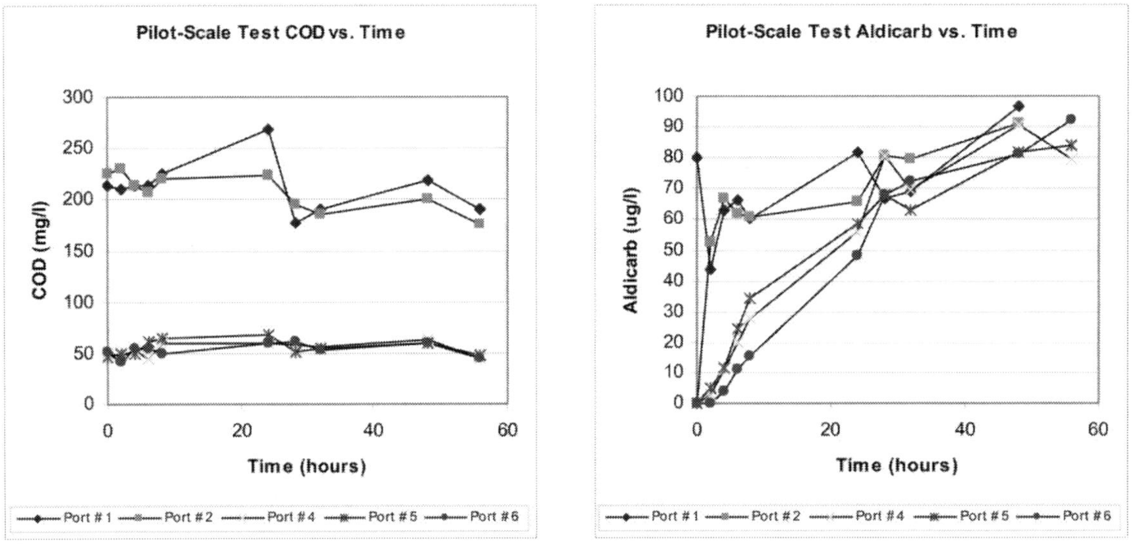

Figure 3. COD and Aldicarb Concentration vs. Time for Pilot-Scale AS Test (Port 1 = System Influent; Port 2 = Primary Clarifier Effluent / Aeration Basin Influent; Port 4 = Aeration Basin Effluent / Secondary Clarifier Influent; Port 5 = Secondary Clarifier Recycle / Waste; Port 6 = System Effluent)

The aldicarb concentrations increased in each of the effluent samples within the AS treatment unit throughout the study period. By the end of the study period, the aldicarb concentration in the effluent samples approached the 100 μg/l aldicarb target concentration that was spiked into the synthetic wastewater. A steady-state aldicarb concentration was not reached during the study period. Confirming the bench-scale tests, little to no degradation of aldicarb was observed during the pilot-scale tests. As in the results from the bench-scale tests, no aldicarb sulfone was detected in any of the samples.

Soil Columns

Figure 4 shows the conductivity versus time plot for the tracer studies in both the saturated and the unsaturated soil columns. The conductivity in the saturated column began to increase after 4 hours, and the maximum conductivity was measured approximately 6 hours after introducing the salt solution. The majority of the salt solution passed through the saturated column within 12 hours.

The conductivity in the unsaturated column began to increase after approximately 25 hours, and the maximum conductivity was measured approximately 49 hours after introducing the salt solution. The majority of the salt solution passed through the unsaturated column within 89 hours. In both columns, the effluent conductivity formed a classic tracer elution curve from which the hydraulic properties of the soil column were determined.

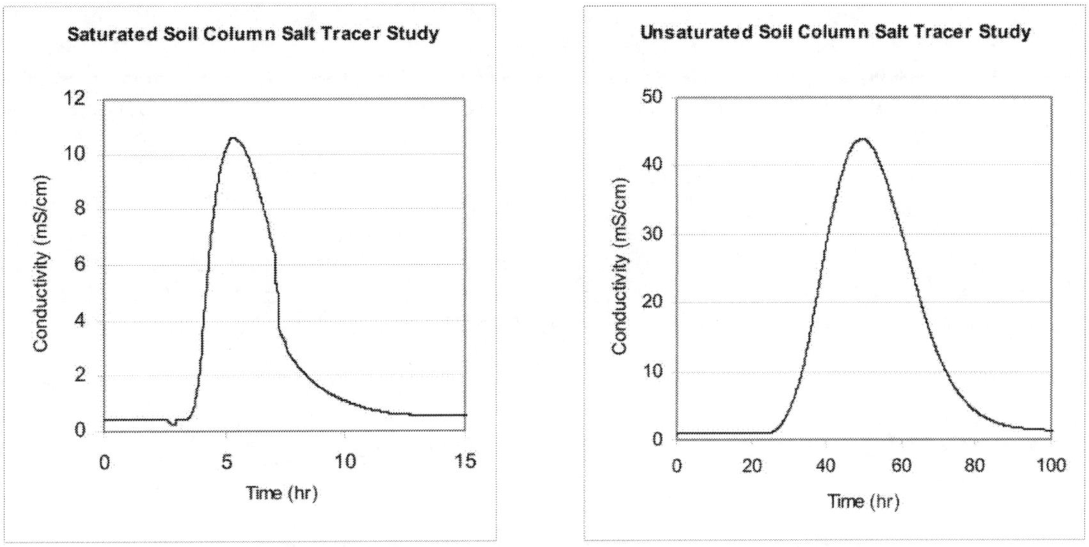

Figure 4. Soil Column Tracer Study

Table 3 shows the aldicarb and aldicarb sulfone concentrations measured at various locations in both the saturated and the unsaturated soil columns. The data show that aldicarb was present in the influent and the upper sample port for the saturated column. However, by the time the water reached the middle sample port, most of the aldicarb was not present in the aqueous phase. Little to no aldicarb was detected in the column effluent. No aldicarb sulfone was detected in the saturated column.

Table 3. Aldicarb and Aldicarb Sulfone Concentrations in Soil Column Pilot-Scale Experiments

	Day 0	Day 9	Day 15	Day 20	Day 26	Day 30
Saturated Column Aldicarb Concentration (μg/l)						
Feed	64.6	94.9	111.0	127.9	93.3	107.3
Upper Port	BDL	93.0	116.8	120.8	162.8	121.9
Middle Port	BDL	7.0	BDL	3.6	2.6	3.2
Effluent	BDL	8.6	BDL	BDL	3.0	BDL
Unsaturated Column Aldicarb Concentration (μg/l)						
Feed	68.2	--	133.5	149.0	110.1	139.0
Effluent	BDL	14.2	11.6	55.4	91.8	107.0
Unsaturated Column Aldicarb Sulfone Concentration (μg/l)						
Effluent	BDL	3.5	1.9	16.8	20.6	18.2

BDL = Below Detection Limit (approximately 1.5 μg/l)

In the unsaturated column, the aldicarb concentration in the effluent increased throughout the 1-month study period. Also, the aldicarb sulfone concentration in the effluent increased with time. The appearance of aldicarb sulfone is a strong indication of biodegration in the oxygen-saturated unsaturated column.

The dissolved oxygen concentrations in the saturated column and unsaturated column effluents were less than 0.1 mg/l approximately 4 mg/l, respectively.

From the saturated soil column, quartz coupons were taken one month after the aldicarb spike and were analyzed using AFM tapping mode scan. By mapping the phase of the cantilever oscillation during the tapping mode scan, phase imaging goes beyond simple topographical mapping to detect variations in composition, adhesion, friction, viscoelasticity, and other properties. This technique can identify contaminants, map different components in composite materials, and differentiate regions of high and low surface adhesion or hardness.

Figure 5 shows the AFM images of quartz samples which have vertical and horizontal orientations at the upper port in the soil column. Morphologies of microorganisms are significantly different between two different orientations. The occurrence density of microorganisms and the biofilm density in horizontal orientation are much higher than those in vertical orientation. The biofilm density was calculated based on the phase images. At the middle port in soil column, microorganism occurrence was decreased significantly. This aspect can be related to decreases available concentrations of aldicarb in Table 3. The AFM results indicated that the microorganisms and density of the biofilms were closely related to the location (depth in the column) as well as the orientation of the coupons with respect to the flow of the wastewater.

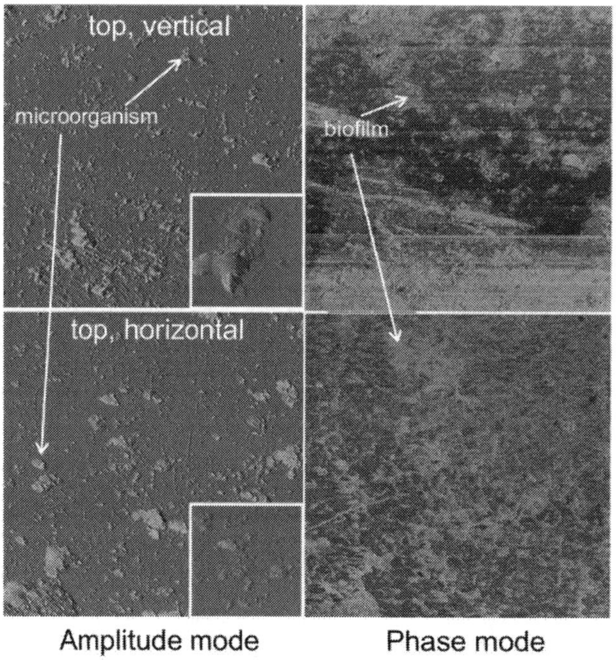

Figure 5. Atomic force microscopy images of quartz coupons from soil column after 30 days exposure to aldicarb-containing wastewater. Scan size: 10 μm x 10 μm. Insets show enlarged images (3 μm x 3 μm) for emphasizing microorganism morphologies.

In addition, FTIR spectroscopy allowed us to investigate the mechanism of aldicarb binding to the soil. Quartz and calcite components of soil were mainly responsible for the aldicarb binding to the soil particles. The aldicarb sorption to soils reached equilibrium after 1 day, and the binding nature was not changed even four days later. This result suggests that there is no significant chemical transformation of aldicarb after the sorption to the soil.

Conclusions

The pilot-scale experiments show that aldicarb is not readily degraded under aerobic conditions, either in an activated sludge treatment system or in an unsaturated soil column, which simulates an unsaturated aquifer. This aldicarb behavior of fate and transport in treatment and water reuse applications has implications on the safe and sustainable wastewater reuse in landscape irrigation, aquifer storage and recovery, bank storage, and other beneficial reuse applications.

Future Work

Future studies will investigate aquifer storage, treatment, and recovery (AST/R) to improve the understanding of the effect of secondary effluent from a wastewater treatment plant on the oxidation and dissolution of arsenic-containing pyrite.

Future testing of water reuse applications is also expected to include additional contaminants (pesticides, pharmaceuticals, EDCs) as well as tertiary treatment processes (chlorination, advanced oxidation, membrane biological reactors) to determine treatment effectiveness and the economics associated with these advanced treatment processes.

References

Alvarez, Franklin R., Glenn M. Shaul, E. Radha Krishnan, Daniel L. Perrin, and Maqsud Rahman (1999), "Fate of Terpene Compounds in Activated Sludge Wastewater Treatment Systems," *J. AWMA,* June, pp. 734-739.

U.S. Environmental Protection Agency (EPA) (2008), "Summary of Nominations for the Third Contaminant Candidate List - Draft," Office of Water, 815-R-08-008.

Zhang, Qiang, Makram T. Suidan, Kai Zhang, Daniel B. Oerther, and Albert D. Venosa (2006), "Comparison of Biomass Selection Between a Novel Membrane Bioreactor and Activated Sludge Process," *WEFTEC 06*, pp. 5030-5037.

Catalytic Ozone Oxidation Processes for Waste Gases Treatment from Kraft Pulping Mills

E. Sahle-Demessie[1], C. B. Almquist[2]

[1]U.S. Environmental Protection Agency, Office of Research and Development, NRMRL, 26 W. Martin Luther King Drive (MS 443), Cincinnati, Ohio 45268.

[2]Paper Science and Chemical Engineering Department, Miami University, 246 Gaskill Hall, Oxford, OH 45056

Abstract

Pulp and paper industry consumes vast quantities water and energy resources and generates large amounts of pollutants including an average of 4.5×10^5 lb toxic substances per facility released to air, water, and land. The combined air and water "cluster rule," provided individual mills with incentives to adopt Advance Pollution Control technologies for reduction in annual emissions of hazardous air pollutants (HAPs), volatile organic compounds (VOCs) and total reduced sulfur compounds (TRSs). Emerging technologies have major impacts in moving the pulp and paper industry to a cleaner and sustainable future.

The typical pollutants from Kraft pulp mill are H_2S, CH_3SH, CH_3SCH_3, CH_3SSCH_3 and methanol. We have conducted bench and pilot scale studies to develop, demonstrate and evaluate the performance and the feasibility of utilizing catalytic oxidation of these are air pollutants that are found at high concentrations. Ozone is used as an oxidant and Cu, Mo and Mn oxides, and mixed oxide catalysts supported on γ-alumina at temperatures ranges 100-200°C. Maximum conversion of DMS was obtained using 10-wt% $MnO_2/\gamma\text{-}Al_2O_3$ as a catalyst, however, the selectivity towards partial oxidation products such as dimethyl sulfoxide and dimethyl sulfone are high (≈20%). The best catalytic activity is obtained over 10-wt% CuO-10- wt% MoO_3/γ-alumina catalyst with 100 % DMS conversion, high selectivity (≈96%) towards complete oxidation products such as CO_2 and SO_2. Although, $MoO_3/\gamma\text{-}Al_2O_3$ catalyst shows poor catalytic activity, MoO_3 acts as a good promoter when used with CuO.

The catalyst was not poisoned by the sulfur compounds and remained active even at high concentrations of water (~50%). Catalytic oxidation occurs at lower temperatures and has key environmental advantages over incineration, including the minimal formation of nitrogen oxides (NO_x) and lower energy requirements. The process converts methanol, a highly flammable toxin produced by the pulping of wood, into methyl formate and completely removes odor causing TRSs. We have achieved 98+ % removal of individual hazardous air pollutants, and total reduced sulfur (TRS) compounds. The new technology was tested using a pilot-scale unit at a commercial pulp and paper mill, Domtar Corporation pulp mill in Hawseville, Kentucky.

1. Introduction

Existing and anticipated air pollution laws have forced industries to reduce air emissions. The integrated multimedia regulation, Cluster rule, for the pulp and paper industry is intended to reduce toxic pollutant releases to the air and water with far-reaching impacts for the industry using Maximum Achievable Control Technologies. The technology standards in the rule cut toxic air pollutant emissions by almost 60 percent from 1998 levels. The rule also provides individual mills with incentives to adopt Advanced Pollution Control Technologies that will further reduce pollutant discharges beyond current discharge limits of volatile organic compounds (VOCs) by nearly 50 percent [1-4]. Currently, the paper industry collects non-condensable waste gas streams from the pulp mill through extensive ductwork from various sources and burn them in a thermal oxidizer. Although effective on many of the pollutants of concern, there are significant costs associated with this method of air pollution control, including capital and maintenance costs for the extensive ductwork, the costs of energy to move the air emissions from their sources to the incinerator, and the costs of fuel to maintain sufficiently high temperatures in the incinerator to oxidize the organic and TRS emissions.

The goal of this study is to demonstrate the use of ozone-enhanced catalytic oxidation for the treatment gaseous emissions from a Kraft mill using mobile pilot-scale system to eliminate non condensable pollutants at much higher concentrations. Previously, research on ozone-enhanced catalytic oxidation has been conducted on a bench-top scale, and pollutants (methanol and dimethyl sulfide) that are typically present in Kraft pulp mill non-condensable gas streams were successfully oxidized [5-9]. As a follow-on to the previous bench-scale work, a large scale test was conducted using a mobile pilot-scale test system of the ozone-enhanced catalytic oxidation in a field study to treat a slip stream of non-condensable waste gas from a Kraft mill. The reliability and performance of the technology and the residual waste by-products from the system under actual operating conditions will be determined.

2. Experimental Methods and Parameters

2.1 Project Site

The field project site was Domtar's Kraft mill in Hawesville, Kentucky. This mill is an integrated, with both pulping and paper machine operations located at the site. The Mill has an annual paper production capacity of 634,000 tons and an annual pulp production capacity of 455,000 tones bleached Kraft hardwood pulp (http://www.domtar.com/en/contacts/3522.asp). The waste gas line that was used for this study was a high volume low concentration (HVLC) stream located at the top of the blow tank (approximately 200 feet above ground), which separates the pulp from the black liquor and volatile gases. The temperature and pressure of the

HVLC stream was approximately 65 °C and 10 Bar gauge pressure. The waste gas line that was located at the blow tank, in which the pulp is separated from the cooking liquor directly following the digester. Pictures of process line and process pipes where digester and sample were collected are shown in Figures 1. Some of the organic compounds in the waste gas line

Figure 2. Continuous Digester and Blow tank at Domtar Mill, Howesvill, KY.

from the blow tank are by-products that were formed in the digester as the cooking liquor chemicals react with the lignin in the wood chips. The stream is saturated with water vapor and the composition of the gas is given in Table 1. The use of this gas a feed stream allows in testing the feasibility and limitations of ozone-enhanced catalytic oxidation for the treatment of HVLC waste gases at a Kraft mill were realized.

Table 1 Waste gas composition in the exit stream of Blow Tank

Pollutant	**Average Concentration, ppm**
Methanol	359 ± 80
Dimethyl Sulfide	4,384 ± 673
Methane Thiol	ND
Dimethyl Disulfide	152 ± 25
Hydrogen sulfide	350 ±
CO	517 ± 45
SO_2	92 ± 45

2.2 The Test System

A skid-mounted test system was constructed to treat up to 30 L per min of waste gas. A schematic of the test system is shown in Figure 2. A compressor was used to pull the waste gas from a process line at a Kraft mill and pressurize it to approximately 8.6 Bar. A condensate trap was placed between the mill's process line and pilot system's compressor. The compressed gas was then sent through the test system at a controlled mass flow rate via mass flow controllers and rotameters. A second compressor was used to compress ambient air to approximately 8.6 Bar. The compressed ambient air, then, was used to dilute the process gas during experimental trials and to purge the test system of the waste gas following each experimental trial. In the test system, resistance heaters were used to heat the gas up to 150 °C prior to the reactor. Thermocouples were placed prior to, within, and following the heated zone to monitor the temperature of the gas in the test system. Temperature controllers were used to control the temperature in the heated zone during the experiments.

Figure 2. Schematic of Ozone-Enhanced Catalytic Oxidation Test System

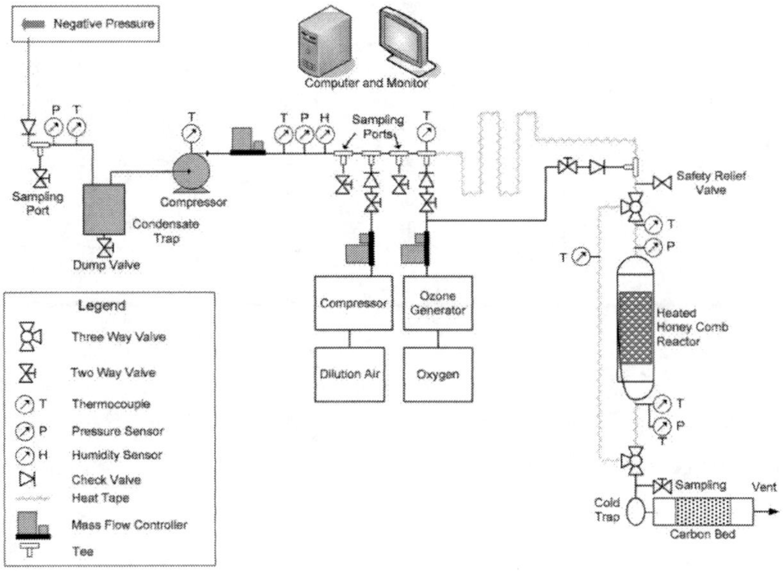

Experimental Design and sample collections

Multilevel factorial experimental design was used in testing the effectiveness of ozone-enhanced catalytic oxidation for treating waste gas effluents from the pulp and paper industry (Table 2). The design is to be run in two blocks, fully randomized to protect lurking variables and the factors include temperature, flow rate, and flow composition and the type of catalyst. The experimental approach used was to set the reaction parameters to their desired values and then characterize the gas stream prior to and following the ozone-enhanced catalytic oxidation reactor. The response were conversions of feed pollutants and product selectivity

Table 2. Experimental Design for Field Tests

Operating Condition	**Slip Stream V1, L/min**	**Diluent (Air) V2, L/min**	**Oxyge n V3, L/min**	**Total Flow**	**Space Velocity (hr^{-1})**	**Ozone, %**	**Furnace Set Temp., oC**	**Catalyst**
1	7	1	2	10	968	4.30%	250	Cu-Mo
2	14	2	4	20	1935	1.80%	250	Cu-Mo
3	21	3	6	30	2903	1.20%	250	Cu-Mo
4	7	1	2	10	968	5.50%	200	Cu-MO
5	14	2	4	20	1935	3.00%	200	Cu-Mo
6	21	3	6	30	2903	2.00%	200	Cu-Mo
7	7	1	2	10	968	4.90%	250	ViO_2/TiO_2
8	14	2	4	20	1935	2.90%	250	ViO_2/TiO_2
9	21	3	6	30	2903	2.00%	250	ViO_2/TiO_2
10	10	0	10	20	1935	1.30%	250	Cu-Mo
11	13	0	7	20	1935	1.60%	250	Cu-Mo
12	6	0	14	20	1935	0.90%	250	Cu-Mo

Analysis of VOCs and TRS compounds

The conversions of all volatile and semi volatile organic compounds including waste gas components the reaction by-product compounds were calculated based on their disappearance from the inlet flow. Gas tight syringes and for some cases Tedlar bags were used to collect three gas samples every hour from different points in the pilot system for GC-FID analysis Concentrations of all organic compounds were measured with Agilent 6890 Series GC, equipped with a split/splitless injection port, Supelco-Q^{TM}-plot (30m x 0.32 mm # 33577-04A) capillary column and FID detector. Samples were taken from the flowing influent and effluent streams and introduced into the GC injection port by using sampling valve. Quantification of the all target analytes was done by performing the calibration of all the reactants and reactions products. All volatile and semi volatile organic compounds were analyzed using gas GC/FID. The GC/FID was calibrated using standard solutions of known analyte concentrations.

Results and Discussion

The composition of the process gas being delivered to the pilot-scaled test system over two weeks is summarized in Figure 3. Noted is that the predominant emission in the HVLC waste gas stream are dimethyl sulfide (nearly 6000 ppm) and methanol. However, the methanol was contained predominantly in the condensate formed as the HVLC waste gas stream cooled in the drop line from the top of the blow tank to ground level where the pilot-scale system was located. The flow stream was cooled to about ambient temperature followed by a compression stage to 10 bars prior to feeding to the test system. The process gas stream was diluted with ambient air and/or oxygen according to the experimental design shown in Table 2, prior flowing to the reactor. The approximate water vapor concentration in the reactor feed stream based upon vapor pressure, temperature, and pressure in the compressor was 3000 ppm.

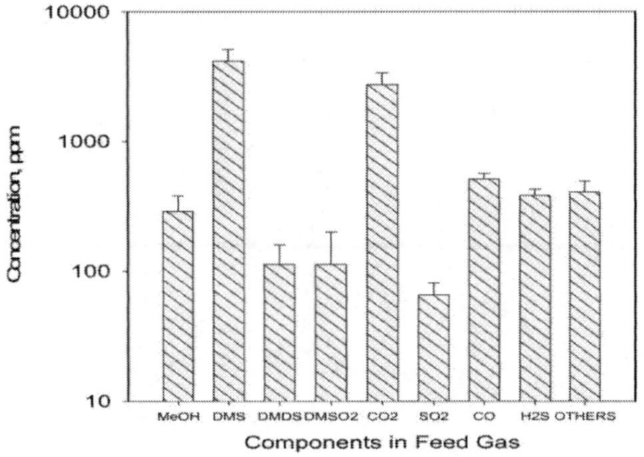

Figure 3. Process gas composition flowing into the pilot-scale test system, following condensation and compression but prior to dilution. The balance of the process gas stream predominantly consists of nitrogen, oxygen and water vapor.

Process Variables

Figure 4 shows the concentrations of each species analyzed in the outlet stream relative to its concentration in the inlet stream at a reaction temperature of 250 °C. The effect of space velocity and reaction temperature on DMS conversion over the CuO/MoO_3 catalyst in the pilot-scale test system is shown on Figure 5. The DMS conversion decreased with increasing space velocity and decreasing reaction temperature as was observed in the bench scale tests. Increased ozone/sulfur ratio results in greater DMS conversion and greater selectivity toward SO_2. The DMS conversion increased to exceed 90 % only when the ozone-to-sulfur mole ratio was about 2, which is a lot higher amount than needed in the bench scale tests. At higher ozone concentrations the DMS conversion was relatively independent of the ozone-to-sulfur ratio. The selectivity to SO_2, however, is rather independent of the ratio until the ozone-to-sulfur ratio of 2 or more, the selectivity to SO_2 increases significantly. The selectivity to CO also increases with ozone-to-sulfur ratio greater than 2. This result was rather unexpected, since it would be more intuitive if the CO oxidized to CO_2 at higher ozone/sulfur ratios.

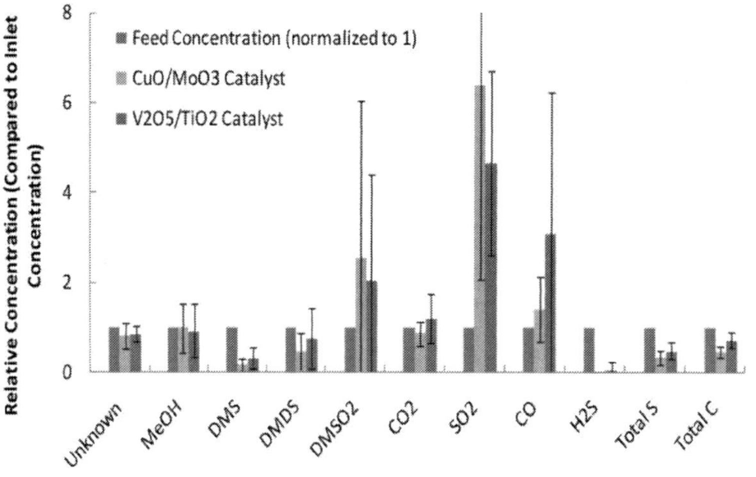

Figure 4. Concentrations of each species analyzed in the outlet stream relative to its concentration in the inlet stream at a reaction temperature of 250 C.

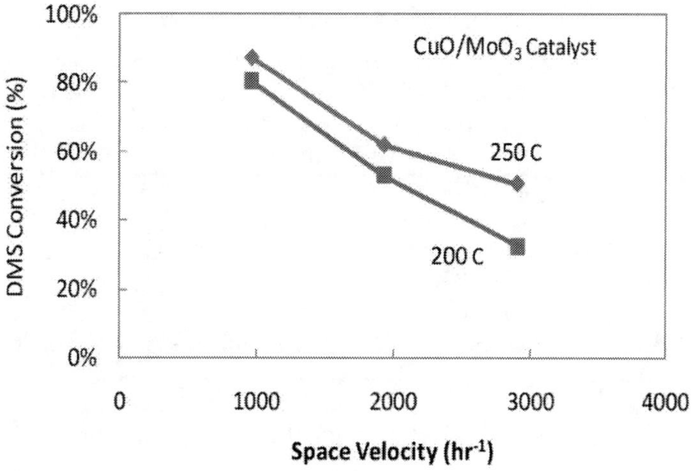

Figure 5. Effect of space velocity and reaction temperature on DMS conversion over CuO/MoO_3 catalyst at 200 °C and 250 °C.

Conclusion

The pilot-scale field study conducted at the Domtar Pulp Mill using a slip stream from a HVLC waste gas stream at an integrated Kraft pulp and paper mill allowed us to successfully assess the feasibility of ozone-enhanced catalytic oxidation for the destruction of organic and TRS constituents at their source. Although the technology appeared to be feasible at the lab scale [5-10], several obstacles, as noted above, must be overcome prior to this technology being implemented in large-scale. Insufficient destruction efficiencies for the organics and TRS compounds were observed in the field study for this technology to be considered as a replacement for incineration. The presence of value-added materials from partial oxidation products was not observed. Ozone / sulfur ratios > 2, low space velocities < 1000 hr^{-1}, and reaction temperatures > 200 °C are required to achieve >90% DMS destruction.

References

1. United States Environmental Protection Agency. (1997). Fact Sheet: EPA's Final Pulp, Paper, and Paperboard "Cluster Rule" – Overview. EPA-821-F-97-010.

2. World Bank Group, (1998). "Pollution Prevention and Abatement: Pulp and Paper Mills." Technical Background Document. Environment Department, Washington, D.C.

3. Springer, A.M. (2000). Industrial Environmental Control: Pulp and Paper Industry, 3rd Ed. TAPPI Press, Atlanta, GA, p. 497,498.

4. Sahle-Demessie, E., Devulapelli, V. G. Applied Catalysis A: *Volume 361, Issues 1-2*, 20 June **2009**, 72-80.

5. Devulapelli, V.G., E. Sahle-Demessie, *Applied Catalysis A: General* Volume 348, Issue 1, 30 September **2008**, 86-93.

6. Sahle-Demessie, E., Devulapelli, V. G. Applied Catalysis B: Environmental 84 (**2008**) 408–419.

7. Almquist, C. B., Sahle-Demessie, E., Sridara Chandra Sehker, and Sowash, J. Methanol Oxidation Using Ozone on Titania-Supported Vanadia *Environ. Science & Technology*, **2007;** *41*(13) pp 4754 – 4760.

8. Sahle-Demessie, E., Sridara Chandra Sehker, Catherine B. Almquist The Role of Catalyst Properties on Methanol Oxidation over V_2O_5-TiO_2 using Ozone, Water, Air & Soil Pollution, 8:297-309,**2007**.

9. Burgess, TL, Gibson, AG, Furstein, SJ, Wachs, I.E., Converting waste gases from pulp mills into value-added chemicals, *Env., Progress,* 21, 3, 137-141, **2002.**

10. Hunter, P. H., Oyama, S.T., 2000. Control of Volatile Compound Emissions, Conventional and Emerging Technologies, John Wiley & Sons, New York.

Durability Analysis of a Field-Aged Diesel Particulate Filter Used in a School Bus Retrofit Program

Teresa L. Barone, John M. E. Storey and Norberto Domingo, Fuels, Engines, and Emissions Research Center, Oak Ridge National Laboratory, Knoxville, TN

Abstract

A field-aged, passive diesel particulate filter (DPF) employed in a school bus retrofit program was evaluated for emission reductions of particulate matter (PM) mass, PM number and total hydrocarbon (HC) concentration. In addition, the effect of the DPF on nitrogen oxides (NO_x) emissions, including the fraction of NO_2, was measured. As part of the Environmental Protection Agency (EPA) Clean Diesel Program, the DPF accumulated several years of use in the Wissahickon School District in Pennsylvania. It was removed from the exhaust system and mounted on a test engine of similar make and model as that of the school bus for bench dynamometer tests. The performance of the DPF was evaluated as-received and after regeneration over the heavy-duty federal test procedure (FTP) transient and 13-mode steady-state cycles. For the PM mass measurements, filter samples were collected with a BG-3 partial flow sampling system (Sierra Instruments Inc.) and analyzed gravimetrically. For the PM number concentration measurements, the exhaust was sampled using a micro-dilution tunnel. Total number concentration and number-size distributions were measured by a TSI Inc. 3025 condensation particle counter (CPC) and 3936 scanning mobility particle sizer (SMPS), respectively.

The results of the evaluation show that, after multi-year use, the DPF was effective in PM removal. The PM mass emissions rate was reduced on average by about 95 ± 2% for the hot and cold FTP cycles. The PM number concentration was reduced approximately 10,000-fold over the 13-mode steady state cycle. PM mass emissions decreased after DPF regeneration. However the number concentration of DPF-emitted particles increased by about a factor of 20; this is comparing the average number concentration during hot FTPs before and after regeneration. The number concentration returned to pre-regeneration values after about 2 hrs of transient cycle operation. For the gaseous species, the effects were less substantial. The HC emissions reduction was about 45%. The concentration of NO_x emitted by the DPF remained the same as that of the engine baseline. The fraction of NO_2 was elevated from about 0-8% at the DPF inlet to 3-32% at the DPF outlet over the 13-mode cycle. Overall, the study provides new information on field-aged DPF PM emissions before, during and after regeneration and supports evidence of more than adequate durability of retrofit DPFs.

Introduction

Emissions control devices are often retrofit to older diesel engines and vehicles for ambient air pollution abatement. Recently, the state of California passed legislation requiring most diesel trucks and buses to be equipped with diesel particulate filters (DPFs) by 2014 (*1*). Thus, understanding the performance of these devices over time is critical. Several studies have evaluated passive DPF durability for diesel truck and bus retrofits *(2-11)*. A study by Chatterjee et al. *(5)* regarding DPF durability on New York City transit buses included PM mass and number-size distribution measurements. The passive DPFs were field-aged on urban buses for 9 – 12 months. The results of heavy-duty chassis dynamometer tests for the Central Business District (CBD) and New York Bus cycles showed 87% and 92% reductions in PM mass, respectively. With SMPS measurements of 10 nm and 100 nm particles, 100- to 1000-fold reduction in number concentration was found over the

CBD cycle. Few other examinations of DPF durability included PM number concentration measurements *(3, 4, 8)*.

PM emissions during passive regeneration have been characterized for an uncatalyzed DPF, which was called a particle trap *(12)*. The beginning of regeneration was associated with the release of nanoparticles that formed a prominent nuclei mode in the number-size distribution. By the end of regeneration, the nuclei mode was absent. (For a definition of typical modes of the number-size distribution, see *(13)*.) The study also compared the emissions of a clean and loaded particle trap. The most abundant particles emitted by the loaded trap were between 100 – 200 nm, while those of the clean trap were 20 – 60 nm. This implies the soot cake of a loaded trap helps to filter ultrafine particles (diameter < 100 nm).

For a catalyzed DPF, ultrafine particle emissions are affected by the temperature at which regeneration occurs, the speed of regeneration, and the catalyst formulation *(14, 15)*. Cauda et al. *(15)* investigated nanoparticle emissions during the regeneration of a DPF prepared with a $CoCr_2O_4$ catalyst. During a fast regeneration event at 540 °C, an abundance of particles with diameter less than 20 nm were emitted. Their findings suggest the emissions may have resulted from the oxidative fragmentation of soot aggregates, the formation of sulfuric acid droplets, and/or the condensation of hydrocarbons released or generated during the oxidation process. This study provides information on mechanisms which lead to the emission of particles during DPF regeneration.

The present study provides an analysis of field-aged DPF performance before, during, and after regeneration. The objective of the study is to evaluate the reduction of PM mass and number concentration by an ECS Purifilter DPF, which had been used on a school bus for almost four years. The DPF-emitted particle characteristics are compared to those of the engine baseline (without aftertreatment).

Experimental Methods

DPF, Engine, and Fuel Specifications

School buses in the Wissahickon School District in southeastern Pennsylvania were retrofit with DPFs as part of the Environmental Protection Agency's (EPA) National Clean Diesel Campaign. An ECS SC-13 Purifilter used in the retrofit program for almost 4 years was selected. The Purifilter DPF is a verified retrofit technology with the EPA *(16)* and California Air Resources Board *(17)*. It requires an engine exhaust temperature of at least 280 °C for at least 25% of the duty cycle for regeneration *(16)*.

The DPF was removed from the school bus and mounted as received on an engine for bench dynamometer tests at the Fuels, Engines, and Emissions Research Center of Oak Ridge National Laboratory. For the study, a 1999 Cummins B5.9 diesel engine was used. The engine emissions were originally certified at 0.1 g/hp-hr PM and 4.0 g/hp-hr NO_x. This engine is of similar model and emissions calibration as that of the school bus. The engine was connected to a 300 hp DC motoring dynamometer (General Electric) and fueled with ultra-low sulfur certification diesel from Chevron-Phillips Specialty Chemical Company with less than 10 ppm sulfur. The DPF was evaluated over the federal test procedure (FTP) heavy-duty transient cycle and the discreet mode cycle Supplemental Emissions Test (SET). The FTP is defined in the Code of Federal Regulations (CFR) Title 40, Part 86.1333-90 *(18)*. The SET is a 13-mode steady-state cycle equivalent to the European Stationary Cycle defined in CFR Title 40, Part 86.1360-2007 *(19)*. The engine speed and load for each mode of the SET are given in Table 1.

Sampling Systems and Analytical Methods

The partial flow sampling system (BG-3; Sierra Instruments Inc.) was employed for PM mass sample dilution and collection. The BG-3 controls the exhaust sample flow rate, such that, it is proportional to the total exhaust flow rate. For experiments during steady-state engine operation, a dilution ratio of 13 was used. For transient engine operation, a 0.075 sample fraction was used. The sample fraction is the ratio of the exhaust flow rate entering the BG-3 tunnel to the total exhaust flow rate of the engine. Following dilution, PM was collected on 47 mm TX40 Teflon-coated quartz-fiber filters (Pall Inc.). The filter samples were analyzed gravimetrically in a temperature and humidity controlled environmental chamber (Electro-tech Systems Inc.). The microbalance (Model UMX2, Mettler-Toledo Inc.) had a 0.1 µg resolution.

A micro-tunnel dilution system was used to prepare PM samples for number concentration and size distribution measurements. The system is based on an ejector pump dilution design by Abdul-Khalek et al. *(20)*. During steady-state engine operation, number-size distributions were measured by a SMPS (Model 3936, TSI Inc.) with long differential mobility analyzer (Model 3080, TSI Inc.) and CPC (Model 3025, TSI Inc.). During transient cycles, second by second measurements of total number concentration were made by the CPC.

Raw exhaust was sampled for HC and NO_x concentration measurements using a heated filter and heated Teflon line maintained at 190 °C. The exhaust passed through an insulated manifold before analysis of HC concentration by a heated flame ionization detector (Model 300, California Analytical Instruments Inc.) and analysis of NO_x and NO_2 concentration by dual heated chemiluminescence detectors (Model 400, California Analytical Instruments Inc.).

Table 1. Engine Speed and Load for the 13-Mode Steady-State Supplemental Emissions Test

Mode	Speed (RPM)	Load (ft lbs)
1	800	0
2	1526	307
3	1808	149
4	1808	222
5	1526	150
6	1526	222
7	1526	75
8	1808	303
9	1808	74
10	2090	315
11	2090	150
12	2090	222
13	2090	75

Results and Discussion

HC and NO_x Emissions

The total HC concentration present in the exhaust was measured using a HFID. The average HC concentration during the hot FTP for the engine baseline and DPF-treated exhaust is shown in Figure 1. The results are based on an average of at least six repetitions. The DPF-treated average includes measurements made before and after DPF regeneration. A significant difference in HC reduction before and after regeneration was not found. The overall HC reduction was approximately 45 ± 13%. This value would probably be greater for higher exhaust temperatures and hence more rigorous duty cycles *(2)*. Nevertheless, a previous study has shown that the HC reduction can decline over time *(5)*. For a new DPF retrofit to a city bus, the average HC reduction was 75%; this decreased to 58% after nine to twelve months of field use *(5)*. The HC emission reduction was possibly affected by a decrease in catalyst activity. In contrast, the PM mass emission reduction can improve with field use *(5, 21)*. This suggests that HC emission reduction is more dependent on catalyst aging.

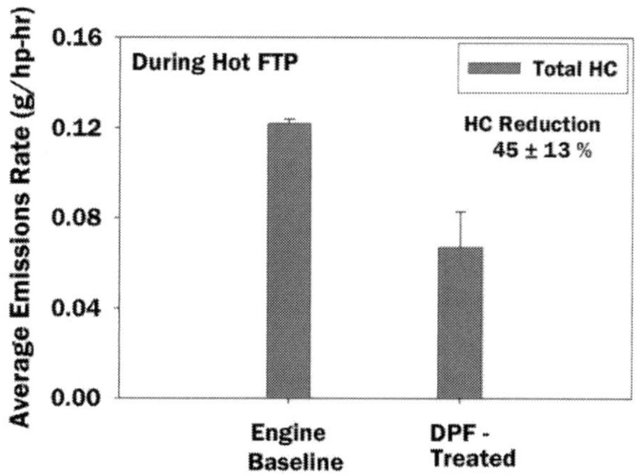

Figure 1. Total HC concentration present in the engine baseline and DPF-treated exhaust. The DPF-treated data are based on an average of measurements made before and after regeneration.

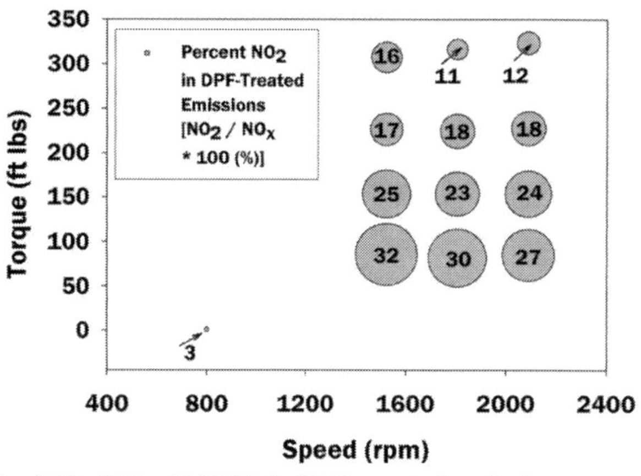

Figure 2. Percent NO_2 (NO_2/NO_x *100%) in DPF-treated emissions over the 13-mode SET. The NO_2 emissions were low at high speed and load modes. In addition, the concentration was low at the near-idle mode (800 rpm).

The concentration of NO_x and NO_2 present in the engine baseline and DPF-treated exhaust were measured by HCLDs. The average concentration of NO_x and NO_2 during the hot FTP remained approximately the same following DPF treatment. However, the fraction of NO_2 can be elevated, since it is generated by the catalyst to facilitate soot oxidation. For the 13-mode cycle, the fraction of

NO_2 in the exhaust at the DPF inlet was 0 – 8%. At the DPF outlet, the range was 3 – 32% (Figure 2). Thus, the fraction of NO_2 was elevated for part of the 13-mode cycle. The fraction was small at the lowest speed and load point probably because the exhaust temperature limited the NO to NO_2 conversion by the catalyst *(22)*. In addition, the fraction was small at the highest speed and load points. This may have resulted since the equilibrium of the following reaction results in greater NO concentration at high temperature: $NO + \frac{1}{2} O_2 \leftrightarrow NO_2$ *(22)*.

PM Mass Emissions

The engine baseline PM mass emissions were compared to the DPF-treated emissions. The DPF-treated PM mass emissions were 0.005 ± 0.002 g/hp-hr and 0.004 ± 0.002 g/hp-hr for the cold and hot FTPs, respectively. Thus, the DPF reduced the engine baseline emissions by 95 ± 2 % (cold FTP) and 96 ± 2 % (hot FTP). These are overall values based on measurements made before and after DPF regeneration, and include the results of 5 cold and 11 hot FTPs. Individual results with respect to regeneration will be shown later in this section. The overall PM mass reduction is consistent with previous studies of passive DPF durability. In previous studies, field-aging occurred between 6 months and 6 years and greater than 85% reduction in PM mass was reported *(2-7, 9-11)*. The DPF should provide at least 85% reduction to meet the California Air Resources Board requirement for Level 3 Diesel Emission Control Systems *(21)*. The EPA requirements are less stringent *(21)*.

The DPF can exert backpressure on the engine, which may alter the engine emission characteristics. The effect of increased backpressure on PM mass emissions was studied by simulating the presence of the DPF. A valve was installed in the exhaust pipe and constricted, such that the backpressure was 28 kPa at 1400 rpm (70% of rated speed). (This value is a little lower than the average backpressure experienced by the school bus engine with a loaded DPF (31 kPa).) Two cold and six hot FTPs were carried out with the flow restricting valve. The results for increased backpressure were compared to the engine baseline and DPF-treated emissions. As shown in Figure 3, engine PM mass emissions were elevated with increased backpressure and can be twice as much as the baseline. This suggests that the actual PM mass reduction was greater than that calculated from engine baseline measurements.

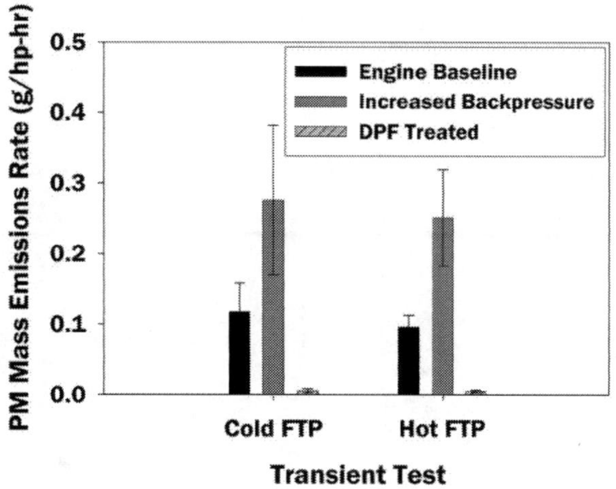

Figure 3. PM mass emissions for the engine baseline, increased backpressure and DPF-treated conditions. The backpressure was imposed by means of a flow restricting valve in the exhaust pipe.

The DPF-treated PM mass emissions shown above are an average of measurements made before and after DPF regeneration. The individual results with respect to regeneration are as follows. For the cold FTP, the DPF-treated PM mass emissions were 0.007 ± 0.002 g/hp-hr before regeneration and 0.003 ± 0.001 g/hp-hr after. For the hot FTP, the emissions were 0.006 ± 0.001 g/hp-hr before regeneration and 0.003 ± 0.002 g/hp-hr after. The PM mass emissions were somewhat greater before regeneration. Perhaps semi-volatile material had adsorbed and accumulated on the soot cake layer of the DPF during school bus operation. During high temperature events of the FTP, semi-volatiles may have evaporated and re-condensed on the sample filter. The stored semi-volatiles in the DPF may have been released or oxidized during regeneration and thus not present in samples afterward. For a new DPF, the results of Geichaskiel et al. *(23)* indicate that the PM mass emissions are slightly lower after regeneration. The improved PM mass reduction after regeneration suggests the DPF filter substrate was not cracked or compromised after several years of use. This is because the soot cake layer in the DPF can help filter PM *(12)* and the soot was partially removed during regeneration. So particle penetration resulting from defects in the filter substrate would more likely occur after regeneration.

PM Number Concentration Emissions

Although the PM mass concentration decreased after regeneration, the particle number concentration increased. This suggests that ultrafine particles, which have a minor contribution to the total PM mass, are emitted after regeneration. Probably the absence of the soot cake layer resulted in a lower trapping efficiency for ultrafine PM *(12)*.

The average particle number concentration emissions for the DPF-treated exhaust are summarized in Figure 4. One cold and two hot FTP sets are shown before and after regeneration. The average number concentration emissions were in the following order: Cold FTP > Hot FTP 1 > Hot FTP 2. The emissions during regeneration were almost two orders of magnitude greater than the emissions before regeneration. The average particle number concentration of the first hot FTP after regeneration was 20 times greater than that immediately before regeneration; the concentration was a factor 20 greater 30% of the time and a factor 10 greater 50% of the time during the FTP. As the DPF became loaded with PM, the number concentration emissions decreased. The value for the second FTP set after regeneration was similar to that of the first set before regeneration. These findings are consistent with the results of a DPF-equipped light-duty diesel vehicle study *(24)*.

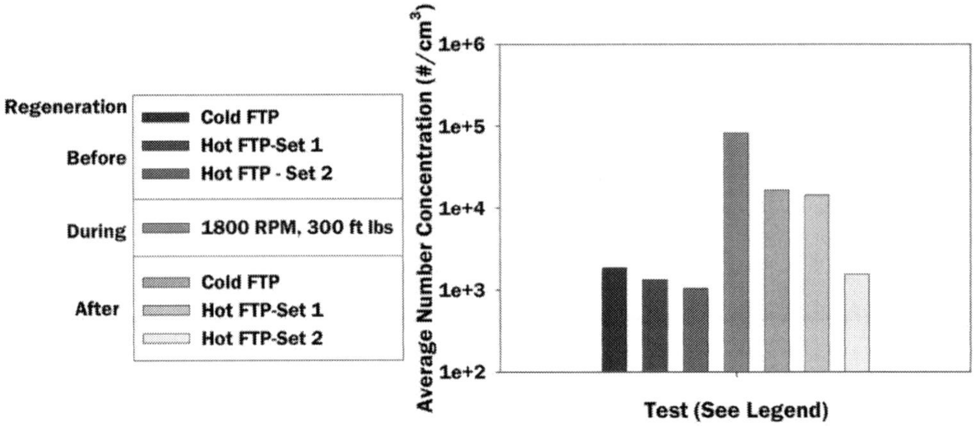

Figure 4. Summary of average number concentration emissions before, during and after DPF regeneration. The ordinate is on a log scale. The emissions increased during regeneration, and gradually decreased afterward.

The engine baseline and DPF-treated particle number concentration emissions were compared over the 13-mode SET. For the engine-emitted PM, the test was completed before DPF installation.

For the DPF-treated emissions, the 13-mode cycle was carried out immediately after regeneration. The number concentration emissions were higher than would have been if the test was done before regeneration. Thus, the emission reduction calculation was a conservative estimate. For most modes of the cycle, the particle number concentration was reduced 10^4 times. The values ranged from 4.5 x 10^3 to 8.6 x 10^4 times. This indicates good DPF performance even immediately after regeneration, when some of the soot cake was not present to help filter ultrafine PM.

References

1. Statewide Truck and Bus Regulation of 2008. Resolution 08-43, 2008; http://www.arb.ca.gov/regact/2008/truckbus08/res0843.pdf
2. Allansson, R., Cooper, B. J., Thoss, J. E.; Uusimäki, A., Walker, A. P., and Warren, J. P. European Experience of High Mileage Durability of Continuously Regenerating Diesel Particulate Filter Technology; SAE Paper no. 2000-01-0480; Society of Automotive Engineers: Warrendale, PA, 2000.
3. Lanni, T., Chatterjee, S., Conway, R., Windawi, H., Rosenblatt, D., Bush, C., Lowell, D., Evans, J., and Mclean, R. Performance and Durability Evaluation of Continuously Regenerating Particulate Filters on diesel Powered Urban Buses at NY City Transit; SAE Paper no. 2001-01-0511; Society of Automotive Engineers: Warrendale, PA, 2001.
4. Chatterjee, S., McDonald, C., Conway, R., Windawi, H., Vertin, K., LeTavec, C. A., Clark, N. N., and Gautam, M. Emission Reductions and Operational Experiences with Heavy Duty Diesel Fleet Vehicles Retrofitted with Continuously Regenerated Diesel Particulate Filters in Southern California; SAE Paper no. 2001-01-0512; Society of Automotive Engineers: Warrendale, PA, 2001.
5. Chatterjee, S., Conway, R., Lanni, T., Frank, B., Tang, S., Rosenblatt, D., Bush, C., Lowell, D., Evans, J., McLean, R., and Levy, S. Performance and Durability Evaluation of Continuously Regenerating Particulate Filters on Diesel Powered Urban Buses at NY City Transit – Part II; SAE Paper no. 2002-01-0430; Society of Automotive Engineers: Warrendale, PA, 2002.
6. Lev-On, M., Le Tavec, C., Uihlein, J., Alleman, T. L., Lawson, D. R., Vertin, K., Thompson, G. J., Gautam, M., Wayne, S., Okamoto, R., Rieger, P., Yee, G., Ospital, J., Zielinska, B., Sagebiel, J., Chatterjee, S., and Hallstrom, K. Chemical Speciation of Exhaust Emissions from Trucks and Buses Fueled on Ultra-low Sulfur Diesel and CNG; SAE Paper no. 2002-01-0432; Society of Automotive Engineers: Warrendale, PA, 2002.
7. LeTavec, C., Uihlein, J., Vertin, K., Chatterjee, S., Wayne, S., Clark, N., Gautam, M., Thompson, G., Lyons, D., Hallstrom, K., Chandler, K., and Coburn, T. Year-Long Evaluation of Trucks and Buses Equipped with Passive Diesel Particulate Filters; SAE Paper no. 2002-01-0433; Society of Automotive Engineers: Warrendale, PA, 2002.
8. Mayer, A., Czerwinski, J., Pétermann, J-L., Wyser, M., and Legerer, F. Reliability of DPF-Systems: Experience with 6000 Applications of the Swiss Retrofit Fleet; SAE Paper no. 2004-01-0076; Society of Automotive Engineers: Warrendale, PA, 2004.
9. Ball, B., Hully, D., Lausseur, P., and Lyons, N. Experience of Continuously Regenerating Particulate Traps on City Buses in Europe; SAE Paper no. 2004-01-0078; Society of Automotive Engineers: Warrendale, PA, 2004.
10. Kimura, K., Alleman, T. L., Chatterjee, S., and Hallstrom, K. Long-term Durability of Passive Diesel Particulate Filters on Heavy-Duty Vehicles; SAE Paper no. 2004-01-0079; Society of Automotive Engineers: Warrendale, PA, 2004.

11. Thornton, M., Tatur, M., Tyrer, H., Tomazic, D., Weber, P., and Webb, C. In 11th Diesel Engine Emissions Reduction (DEER) Conference; Chicago, IL, U.S.A., 2005.
12. Abdul-Khalek, I. S., Kittelson, D. B., and Brear, F., Diesel Trap Performance: Particle Size Measurements and Trends; SAE Paper no. 982599; Society of Automotive Engineers: Warrendale, PA, 1998.
13. Kittelson, D. B. (1998), "Engines and Nanoparticles: A Review," *J. Aerosol Sci.* Vol. 29, pp. 575–588.
14. Cauda, E., Hernandez, S., Fino, D., Saracco, G., and Specchia, V. (2006), "PM0.1 Emissions During Diesel Trap Regeneration," *Environ. Sci. Technol.* 40, pp. 5532–5537.
15. Cauda, E., Fino, D., Saracco, G., and Specchia, V. (2007), "Secondary Nanoparticle Emissions During Diesel Particulate Trap Regeneration," *Topics In Catalysis*. Vol. 42, pp. 253–257.
16. U.S. Environmental Protection Agency, Verified Retrofit Technologies from Engine Control Systems, www.epa.gov/OMS/retrofit/techlist-ecs.htm
17. California Air Resources Board, Currently Verified Level 3 Technologies with ≥ 85% PM Reduction, www.arb.ca.gov/diesel/verdev/level3/level3.htm
18. Code of Federal Regulations, Section 86.1360-2007, Title 40, p 255.
19. Code of Federal Regulations, Section 86.1333-90, Title 40, p 223.
20. Abdul-Khalek, I. S., Kittelson, D. B., Graskow, B. R., Wei, Q., and Brear, F. Diesel Exhaust Particle Size: Measurement Issues and Trends; SAE Paper no. 980525; Society of Automotive Engineers: Warrendale, PA, 1998.
21. Mooney, J. J. (2007), "Toxic Solid Nanoparticles, The Importance of Retrofitting Diesel Engine Particle Emission Control Systems to Older In-Use Diesel Engines, and Available Methods," *Österreichische Ingenieur- und Architekten-Zeitschrift*. Vol. 152, pp. 1–3.
22. Kandylas, I. P. and Koltsakis, G. C. (2002) "NO2-Assisted Regeneration of Diesel Particulate Filters: A Modeling Study," *Ind. Eng. Chem. Res.*, Vol. 41, pp. 2115-2123.
23. Giechaskiel, B., Munoz-Bueno, R., Rubino, L., Manfredi, U., Dilara, P., De Santi, G., and Andersson, J. Particle Measurement Programme (PMP): Particle Size and Number Emissions Before, During and After Regeneration Events of a Euro 4 DPF Equipped Light-Duty Diesel Vehicle; SAE Paper no. 2007-01-1944; Society of Automotive Engineers: Warrendale, PA, 2007.
24. Montajir, R. M., Kusaka, T., Kaori, I., Kihara, N., Asano, I., Adachi, M., and Wei, Q. Soot Emission Behavior from Diverse Vehicles and Catalytic Technologies Measured by a Solid Particle Counting System; SAE Paper no. 2007-01-0317; Society of Automotive Engineers: Warrendale, PA, 2007.

CFD Modeling of Pollutant NOx Dispersion & Absorption From Vehicular Exhaust Pipe

Anjan Kumar Tula, Chemical Engg Dept, Lamar University, Beaumont, TX

Kanwar Devesh Singh, Chemical Engineering, Lamar University, Beaumont, TX

Dr. Helen Lou, Chemical Engineering, Lamar University, Beaumont, TX

Dr. Daniel Chen, Chemical Engineering, Lamar University, Beaumont, TX

The major emissions from vehicular tailpipes are ozone (O3), Particulate Matter (PM), Nitrogen Oxides (NOx), Sulfur Oxides (SOx), Carbon Monoxide (CO), and Volatile Organic Compounds (VOCs). Many methods have been implemented to reduce the NOx emissions, including the budding green technology of Photo-Catalytic Oxidation (PCO) techniques for NOx control. PCO uses a photo catalyst under UV/visible light illumination to oxidize pollutants to harmless compounds. In this control technique, NOx is oxidized to nitric acid and VOCs are oxidized to water, CO2, chlorine and other compounds.

The objective of the project is to simulate the NOx dispersion from the exhaust pipe, interaction of this plume with road surfaces, and the effects of wind speed using the help of Computational Fluid Dynamics (CFD). The gases coming out of the exhaust pipe go down to the catalyst coated road surface and get adsorbed there due to photo catalytic oxidation . But, the amount nitrogen oxides adsorbed varies with velocity, geometry, direction of tail pipe and other factors. The present study aims at modeling the NOx dispersion from the exhaust pipe of a car using the RANS approach and the k-ε turbulence model. The turbulence-chemistry interaction between the NOx and the atmospheric air, as well as, the photo catalyst surface is into consideration.

The CFD simulation package selected for this purpose is FLUENT. The NOx emitted from single vehicle source is considered and the dispersion of NOx plume in the environment is studied. The variation of concentration profile versus distance from the exit after NOx is emitted to the atmosphere having a catalytic absorbing surface is studied. With known exhaust gas velocity, reaction kinetics & exhaust pipe geometry, the simulation was done. To study further, three different cases were taken with various tail pipe directions i.e. horizontal, vertical and inclined to ground at 45 degrees. For all the three cases the dispersion from a vehicle at low idle condition (when the vehicle's engine is running but the vehicle is still, for example toll gates or at traffic signals, when the signal is 'red') and high idle condition (when the vehicle starts to move after the signal turns green or when the toll is paid) are simulated . The CFD simulations were first conducted in 2-dimensional geometries then in 3-dimensional geometry. The simulation results will facilitate the development of novel NOx reduction technologies.

See more of this Session: Technologies for Reductions of Emissions
See more of this Group/Topical: Environmental Division

Treatment of a dynamic mixture of n-hexane and benzene in a trickle-bed air biofilter integrated with cyclic adsorption/desorption beds

*Ashraf Aly Hassan, Abderrahman Zehraoui, George Sorial**

** Civil and Environmental Engineering, University of Cincinnati, P.O. Box 21007, Cincinnati, OH 45221-0071, Office: 765G Baldwin Hall Tel: (513)556-2987, Fax: (513) 556-2599*

e-mail: George.Sorial@uc.edu

One of the main challenges facing biological treatment of contaminated gas streams is the inconsistency of contaminants' concentrations in the gas feed. The microorganisms responsible of the biodegradation process are very delicate to influent concentration changes. This adds an extra constraint for the other difficulties encountered in the biodegradation process.

Previous research in our laboratories focused on enhancing the degradation of hydrophobic contaminants. A mixture of two compounds, namely, n-hexane (hydrophobic compound) and benzene (sparingly soluble compound), are fed into two parallel trickling bed air biofilters (TBABs). TBAB "A" was preceded by an integrated system of a cyclic 2-bed adsorption/desorption unit. A schematic of the unit could be found in figure 1. The primary goal of the study was to maintain long-term, stable consistent high performance of volatile organic compounds (VOCs) degradation in the TBAB. A schematic of the TBAB could be found in figure 2.

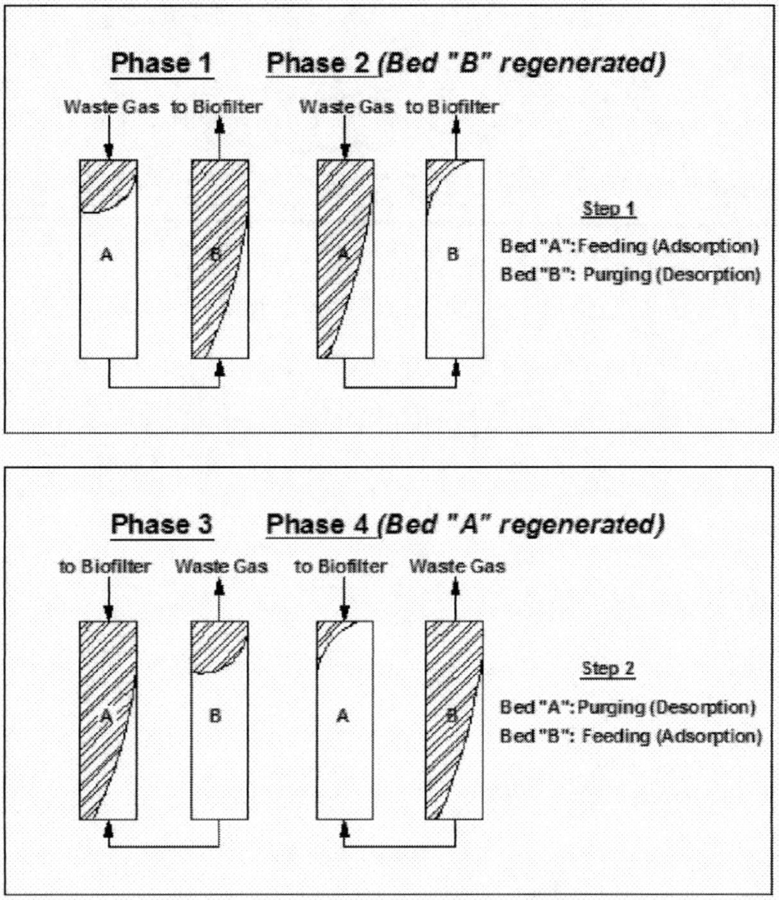

Figure 1: Schematic of the two-bed adsorption/desorption system

Both TBABs were fed with a square wave change in the influent concentration. The average concentration of VOCs in the feed applied to both TBABs was within the allowable limits as studied in an earlier study. TBAB "A" was receiving a constant feed of the average value of the square wave as its influent, while TBAB "B" (the control) was directly subjected to the high and low peaks.

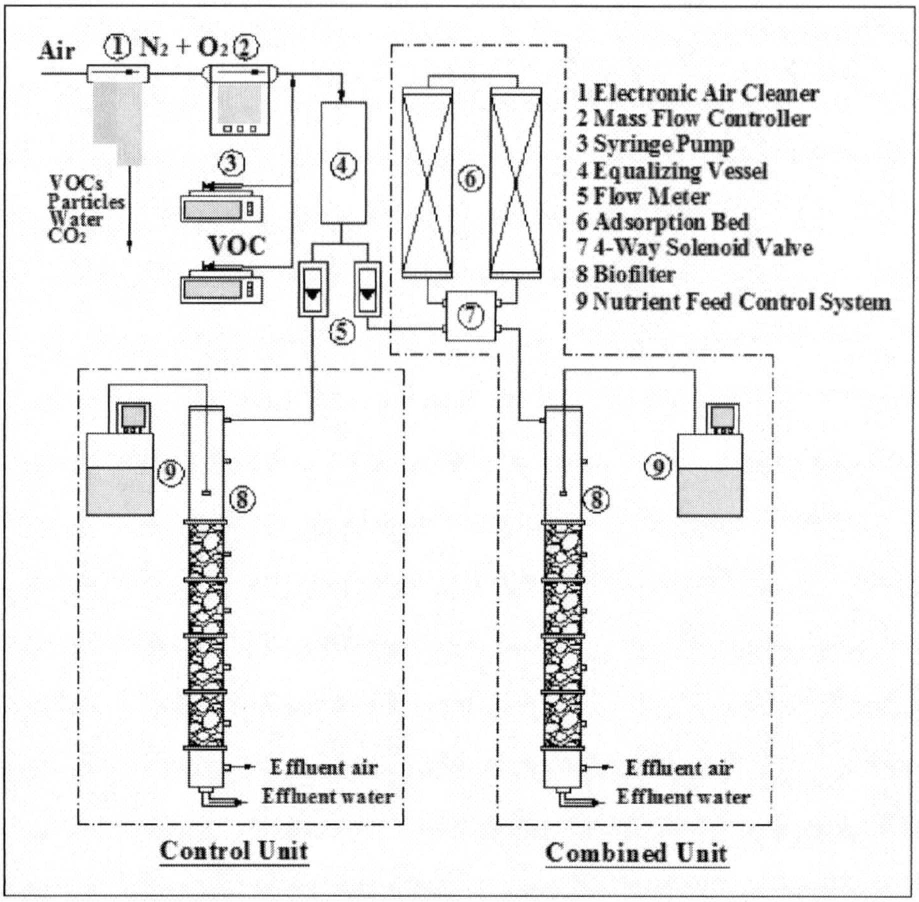

Figure 2: Schematic of the TBAB

Experimental results have shown that TBAB "A" was operating at a constant acceptable behavior, while TBAB "B" did not show satisfactory performance even at the time of low peak. The results further revealed that the 2-bed adsorption unit can serve as: 1) polishing unit during initial acclimation of the biofilter, 2) buffering unit under fluctuating volatile organic compounds (VOCs) loading, and 3) feeding unit during non-emission periods encountered during shut down in industries.

Performance Evaluation of Aqueous Ammonia Process with Regeneration at High Temperature and Pressure

Kwang B. Yi, Greenhous Gas Research Center, Korea Institute of Energy Research, Daejeon, South Korea

Chang Hyun Ko, Greenhouse gas research center, Korea Institute of Energy Research, Daejeon, South Korea

Jong-Ho Park, Korea Institute of Energy Research, Daejeon, South Korea

Jong-Nam Kim, Korea Institute of Energy Research, Daejeon, South Korea

Technologies targeting removal of carbon dioxide from power plants' flue gas are under intensive development. The most proximal example of the technology is chemical solvent absorption process capturing carbon dioxide from power plant flue gas. Even though amine absorption is still considered as the most advanced chemical absorption technology, electricity producing industries are reluctant to adopt it because of its high operational cost resulting in dramatic efficiency reduction in electricity production. Recently, NETL/PowerSpan, EPRI/Alstom in US and KIER in Korea have gone through intensive study about aqueous ammonia process as cost-effective carbon dioxide capture technology. Aqueous ammonia process has numerous advantages over the amine process and its real scale performance and economic viability are prepared to be under evaluation through construction and operation of large demonstration plants across the Europe and North America. In this study, we would like to introduce the recent research results on KIER's aqueous ammonia process and recent results from preliminary study on low temperature absorption and high pressure regeneration

See more of this Session: Technologies for Reductions of Emissions
See more of this Group/Topical: Environmental Division

Plasma soot removal system for diesel exhaust gas consist of grid electrodes

Mohsen Sadeghi[a,], Abbasali Khodadadi[a], Seyyed Shamsodin Mohajerzadeh[b], Saeed Shokrollahi[a]*

[a] *Catalyst and Reaction Engineering Research Lab, Department of Chemical Engineering, University of Tehran, Tehran, I.R.Iran*

[b] *Thin Film Lab, Department of Electrical and Computer Engineering, University of Tehran, Tehran, I.R.Iran*

Abstract

In this work, efficient method using self-consistent non-thermal plasma system was developed and tested to eliminating of fine particulate matter from diesel exhaust gas. For the purpose of this research, numerous removal efficiency measurements performed for various geometric and electric parameter settings under laboratory and real condition.

The results of experiments carried out to optimize electrode configuration using six kinds of electrodes, indicated that the grid electrode was more effective in respect of plane and punched electrodes. The data obtained also showed that the removal efficiency of grid electrodes, increased with hole size decreasing. Other investigations, demonstrated that the removal efficiency of selected grid electrode weakly affected by soot concentration. In addition, for the case of without precharger, the system showed a negligible and satisfactory variation in removal efficiency vs. time in 35 min. interval.

Under real condition, experimental results of fabricated system indicated that the removal efficiency about 83% was achieved with Minibus Benz Model O508D exhaust gas operating in steady-state condition and in the present of precharger.

The data suggest that, the non-thermal plasma system consist of grid electrodes can be considered to apply for efficient elimination of fine particles from diesel exhaust.

Keywords: Diesel exhaust; Soot; Non-thermal plasma; Plasma system; Grid electrode;

*Corresponding author.
E-mail address: m_sadeghi@cape.iust.ac.ir

Objectives

Diesel engines are being increasingly used worldwide due to their higher thermal efficiency, longer durability, increased reliability and lower operating costs compared with gasoline engines. In return for these benefits, they emit 2–20 times more nitrogen oxides and roughly 30–100 fold higher particulate matter than gasoline engines. The particles present in diesel exhaust are composed of a center core of elemental carbon and adsorbed organic compounds [1]. These particles have a large surface area, which makes them an excellent medium for adsorbing organics. In addition, their small size makes them highly respirable and able to reach the deep lung. Many of the organic compounds present on the particles are individually known to have mutagenic and/or carcinogenic properties [2, 3].

Several techniques have been developed for decreasing particulate emissions from diesel engines such as engine modification, fuel additives, alternative fuels, and after-treatment systems. Mentioned approaches are in fact complementary and must be followed simultaneously [1, 4].

One of the most effective and commonly after-treatment devices is non-thermal plasma systems, which are used extensively since give high removal efficiency, low-pressure drop and compact construction [5]. The present work is focused on presenting and developing efficient method using self-consistent non-thermal plasma system consist of grid electrodes, which can be applied for eliminating of fine particulate matter from diesel exhaust gas.

Experimental

For the purpose of this research, numerous removal efficiency measurements performed for various geometric and electric parameter settings under laboratory and real conditions.

The carbon black particles (in range of 0.4-1.2 μm) were considered in this study to simulate soot particle behaviors in the system [6]. Loadings depend on experiment ranged from 0.8 to 5.0 g/m^3. The average residence time in the system was about 0.32 second and was kept constant for all tests presented here.

The concentration of particles was measured at downstream of the system with a particle counter capable of measuring the number of particles in real time. The net removal efficiency was calculated then as $\varepsilon = (1 - C_{on}/ C_{off}) \times 100$ where C_{on} and C_{off} were the concentrations at the system outlet when the voltage was turn on and off, respectively. This measuring strategy allows relating the obtained fractional efficiencies exclusively to the electric and geometric setting.

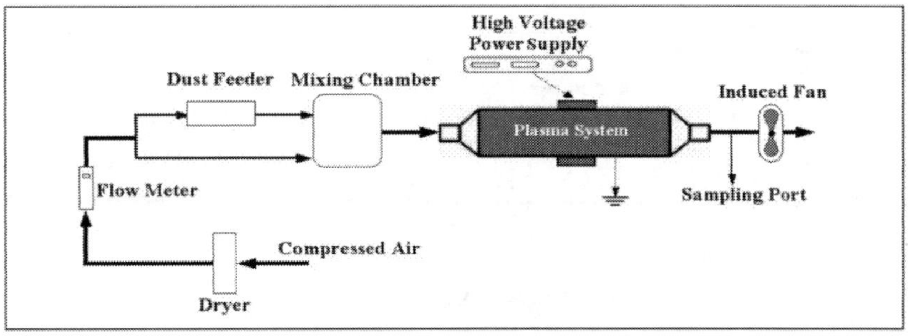

Fig. 1. Schematic diagram of experimental setup.

Experimental investigations were performed with a laboratory scale plasma system Equipped with 16 parallel electrodes with a width of 50 mm, a height of 40 mm and an interelectrode spacing of 8 mm. Parallel electrodes act as corona discharge and collecting electrodes simultaneously.

To perform investigation on the effect of electrode geometry on removal efficiency, six kinds of electrodes consist of plane electrode (A0), punched electrodes (A1, A2) and grid electrodes (B, C, D) used. For the improvement of particle removal efficiency under short residence times and also increase the performance of system for ultra-fine to submicron particle elimination, a corona type precharger employed and located before of the parallel electrodes. The mentioned precharger consists of 13 needle electrodes (in two rows) made of stainless steel and a flat electrode in dimension of 15×50 mm made of aluminum. Minus 3.2 kV DC was applied to the needle electrodes through 2 MΩ resistors, which results in a high concentration of negatively charged ions near the needles tip.

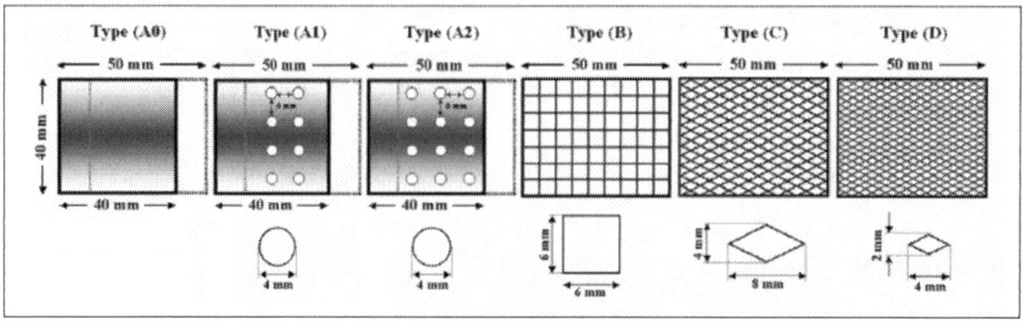

Fig. 2. Structure and shape of six kinds of electrodes used by plasma system:
(A0) Plane electrode, (A1, A2) Punched electrodes and (B, C, D) Grid electrodes type.

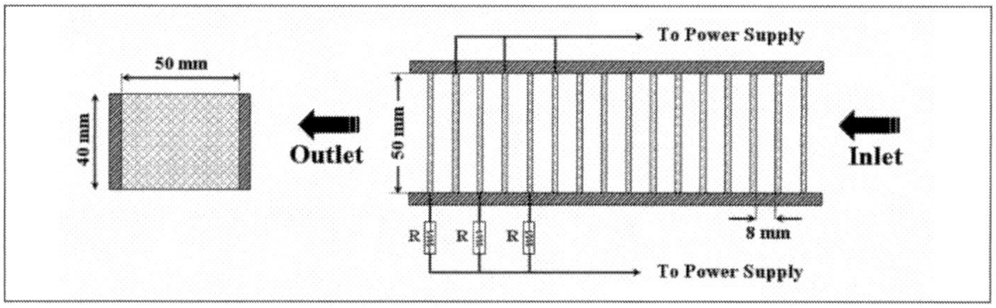

Fig. 3. Electrical configuration and Arrangement of electrodes used by plasma system.

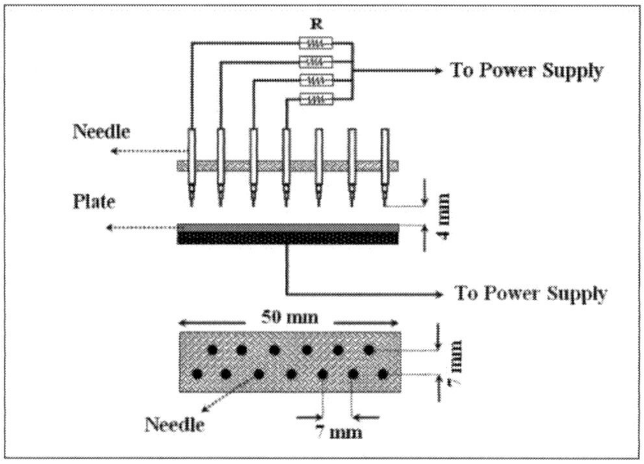

Fig. 4. Electrical configuration of corona type precharger
used by plasma system.

Results and Discussion

The results of research carried out to optimize electrode configuration using six kinds of electrodes, indicate that the grid electrode was more effective in particles elimination in respect of plane and punched electrodes probably as a result of the stronger electric field at the knife-edges or sharp points of any grid which greatly facilitates the charge and removing of particles.

In addition the results show that the removal efficiency of grid electrodes, increased with hole size decreasing. Nevertheless, the increase in removal efficiency, slows down with decrease hole size probably due to the particle re-entrainment on electrodes and the homogenization of the electric field which reduces the field strength at the electrodes surface. Base this phenomenon the removal efficiency of Type-C and D electrodes are almost identical despite of the fact that the hole size of Type-D is smaller considerably (Fig. 5).

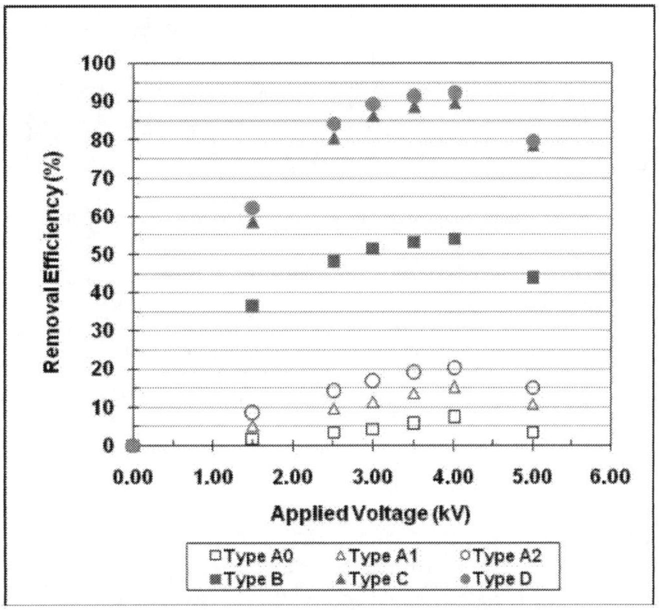

Fig. 5. Removal efficiency of used electrodes as a function of applied voltage for the case of without precharger and C_{soot}: 5 g/m^3.

Experimental results on the variation of removal efficiency of electrode C with applied voltage and soot concentration (from 0.8 to 5.0 g/m^3) demonstrate that the removal efficiency weakly affected by soot concentration probably due to increase contribution of fine particles in the lower concentration. The results also show that the drop in removal efficiency slowed down with increase in applied voltage and decrease in soot concentration.

As expected, Investigation of removal efficiency of electrode C under a constant soot concentration (1 g/m^3), for the case of with and without precharger, indicates that the use of precharger caused increase of removal efficiency.

Other experiments, which performed to determine the effect of operation Time, show that when precharger was not employed, the removal efficiency had no obvious variation with time up to 35 min and it was kept stable at around 81%. However, it diminished gradually after this time due to the contamination of the electrodes and the particles re-entrainment [7].

For the case of with precharger, removal efficiency was kept stable at around 93% up to 20 min of operation time and then dropped sensibly during the test period. In this case, it reduces gently to the value of without precharger case (Fig. 6).

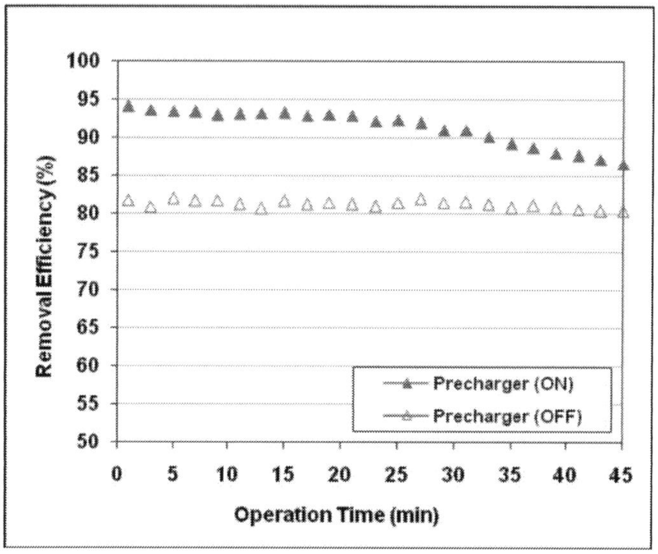

Fig. 6. Removal efficiency of electrode C as a function of operation time in present of precharger and without it.
C_{soot}: 1 g/m^3 and Applied voltage: 3.5 kV.

Under real condition, experimental results of fabricated system consist of Type-C electrode, indicate that the removal efficiency about 83% was achieved with Minibus Benz Model O508D exhaust gas operating in steady-state condition and in the present of the precharger. The dissimilarity occurred between the results obtained from laboratory and real conditions may be due to the difference condition in diesel exhaust such as temperature, humid and soot chemical and physical property.

Conclusion

Presented results suggest that, the non-thermal plasma system consist of grid electrodes can be considered to apply for efficient elimination of fine particulate matter from diesel exhaust gas. However, it seems that some modification must be done for improvement of removal efficiency under real condition.

It should be noted that experimental investigations for developing effective method for sequential regeneration of mentioned system and removal of deposited soot are underway.

References

[1] P. Saiyasitpanich, T.C. Keener, S.-J. Khang and M. Lu, Removal of diesel particulate matter (DPM) in a tubular wet electrostatic precipitator, Journal of Electrostatics 65 (2007) 618–624.

[2] U.S. EPA. Health Assessment Document for Diesel Engine Exhaust, Washington, DC, 2002a.

[3] H. Burtscher, Physical characterization of particulate emissions from diesel engines: A review, Aerosol Science 36 (2005) 896–932.

[4] M. Matti Maricq, Chemical characterization of particulate emissions from diesel engines: A review, Aerosol Science 38 (2007) 1079 – 1118.

[5] Anatol Jaworek, Andrzej Krupa and Tadeusz Czech, Modern electrostatic devices and methods for exhaust gas cleaning: A brief review, Journal of Electrostatics 65 (2007) 133–155.

[6] Michele Ambrogio, Guido Saracco, Vito Specchia, Coen van Gulijk, Michiel Makkee and J.A. Moulijn, On the generation of aerosol for diesel particulate filtration studies, Separation and Purification Technology 27 (2002) 195–209.

[7] Takeo Takahashi, Akinori Zukeran, Yoshiyasu Ehara, Haruo Kishida, Tairo Ito and Takeshi Takamatsu, Influence of Gas Velocity and Electric Field Intensity on Particle Deposit and Re-entrainment Phenomena in an Electrostatic Precipitator, Journal of Aerosol Sci. 29 (1998) S481-S482.

[8] J. O. Chae, Non-thermal plasma for diesel exhaust treatment, Journal of Electrostatics 57 (2003) 251-262.
[9] Tomasz Ciach, Tomasz R. Sosnowski and Albert Podgorski, Efficient filtration methods for diesel aerosols, Journal of Aerosol Sci. 26 (1995) S723-S724.
[10] Tomasz Ciach and Tomasz R. Sosnowski, Removal of soot particles from diesel exhaust, Journal of Aerosol Sci. 27 (1996) S705-S706.

Removal of soot particles from diesel exhaust gas using cylindrical plasma system

Mohsen Sadeghi[a,], Abbasali Khodadadi[a], Seyyed Shamsodin Mohajerzadeh[b], Saeed Shokrollahi[a]*

[a] *Catalyst and Reaction Engineering Research Lab, Department of Chemical Engineering, University of Tehran, Tehran, I.R.Iran*

[b] *Thin Film Lab, Department of Electrical and Computer Engineering, University of Tehran, Tehran, I.R.Iran*

Abstract

Developing and testing reliable and efficient method using self-consistent wire-cylinder type plasma system for removing of fine particulate matter from diesel exhaust gas was studied in this work. For the purpose of this research, numerous removal efficiency measurements were performed for various geometric and electric parameter settings under laboratory and real conditions using by Different types of corona electrodes (Ni-Cr smooth and barbed wire) and collection electrodes (Copper simple cylinder, Ni-Cr spiral and porous cylinder).

The results of laboratory experiments, carried out to optimize smooth wire diameter as corona electrode, showed that under constant applied voltage, the removal efficiency increased by diameter decreasing. Other research were performed to select final corona electrode configuration, indicated that the barbed wire (Core Dia: 700 µm) offered better removal efficiency, higher mechanical strength, longer service life and lower physical oscillations in electrical field in respect to the smooth wire (Dia: 150 µm). The data obtained also demonstrated a negligible and satisfactory variation in removal efficiency vs. time in 27 min. Interval for the barbed wire-simple cylinder system.

In the last step, experimental results of fabricated systems consist of barbed wire-simple cylinder and barbed wire- spiral cylinder with sequential regeneration (collected soot incineration by electrical heating) under real condition, indicated that the removal efficiency about 92% and 79% was achieved respectively with Minibus Benz Model O508D exhaust gas operating in steady-state condition.

The data suggest that, the barbed wire-cylinder type plasma system can be considered to apply for efficient elimination of fine particulate matter from diesel exhaust gas.

Keywords: Diesel exhaust; Soot; Plasma system; Barbed wire; Wire-Cylinder;

*Corresponding author.
E-mail address: m_sadeghi@cape.iust.ac.ir

Objectives

Diesel engines emissions (DE) have been the subject of increasing human health concern in recent years. DE emissions contain large quantities of fine particulate matter, which mainly consist of carbonaceous material and soluble organic fraction (SOF). Some constituents of the SOF such as polycyclic aromatic hydrocarbons and nitrated derivatives are mutagenic and/or carcinogenic [1]. Diesel Particulate emissions can be controlled either at their source or by after-treatment technologies. For eliminating of fine particles using second approach, various types of techniques have been explored to find efficient, self-consistent and less costly method [2, 3, 4].

The non-thermal plasma systems is one of the most commonly applied particulate control devices, which are used extensively since give high precipitation efficiency, negligible pressure drop and compact construction [4].

The major objective of this study was evaluating and developing reliable and efficient method using self-consistent wire-cylinder type plasma system for removing of fine particulate matter from diesel exhaust gas.

Experimental

For the purpose of this research, numerous removal efficiency measurements were performed for various geometric and electric parameter settings under laboratory and real conditions using different types of corona and collection electrodes. The carbon black particles (in range of 0.4-1.2 µm) were considered in this study to simulate soot particle behaviors in the system [5]. Loading and average residence time in the system were about 0.28 second and 1.0 g/m^3 respectively and were kept constant for all tests presented here.

The concentration of particles was measured at downstream of the system with a particle counter capable of measuring the number of particles in real time. The net removal efficiency was calculated then as $\varepsilon = (1 - C_{on}/ C_{off}) \times 100$ where C_{on} and C_{off} were the concentrations at the system outlet when the voltage was turn on and off, respectively.

The experiments were carried out with a laboratory scale plasma system consist of an outer copper cylinder (with a length of 500 mm and an inside diameter of 22 mm) and a central discharge electrode wire. The negative voltage (up to a maximum value of 8 kV DC) was supplied to discharge electrode by a high voltage DC supply through a 2.0 MΩ resistor.

To perform investigation on the effect of corona electrode geometry on removal efficiency, two kinds of electrodes consist of Ni-Cr smooth wire (with diameter of 150, 200 and 250 µm) and Ni-Cr barbed wire (with Core diameter of 700 µm) was used.

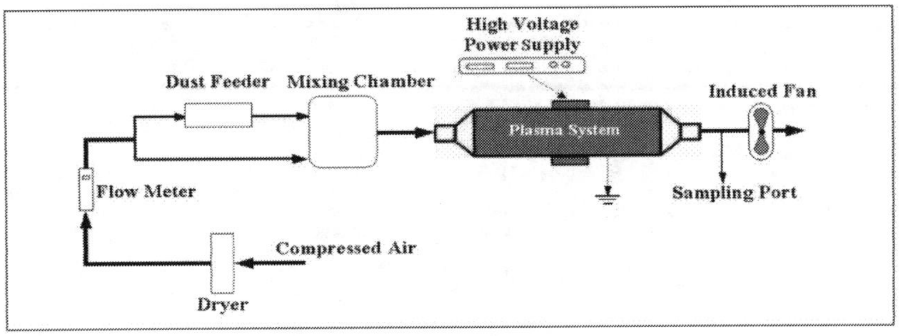

Fig. 1. Schematic diagram of experimental setup.

In order to develop a self-consistent system, sequential incineration of collected particles by applying electrical heating method was proposed for system regeneration. The mentioned method examined under real condition using two kinds of collection electrodes consist of Ni-Cr porous cylinder and Ni-Cr spiral cylinder (made of Ni-Cr narrow strip with a width of 3 mm and a thickness of 500 μm which was wound around the 10 axial supports).

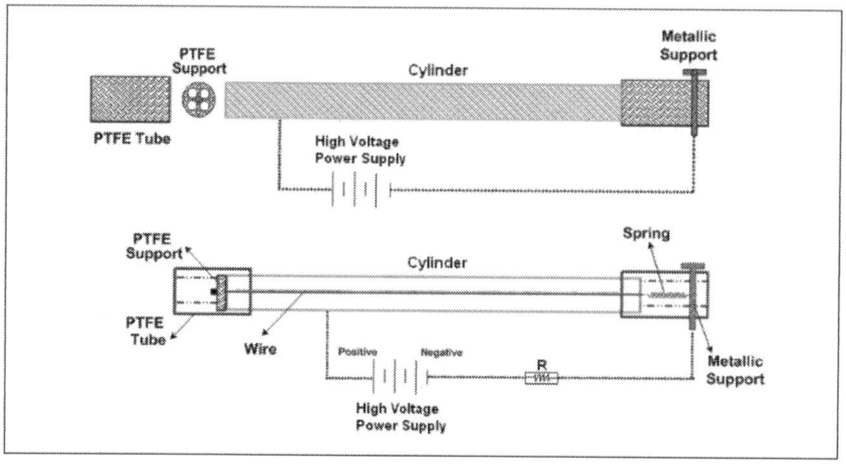

Fig. 2. Structure and Electrical configuration of laboratory scale plasma system.

Results and Discussion

The results of laboratory experiments, carried out to optimize smooth wire diameter as corona electrode, show that under constant applied voltage, the removal efficiency increased by diameter decreasing.

Other research were performed to select final corona electrode configuration, indicate that the barbed wire (Core Dia: 700 μm) offered better removal efficiency, higher mechanical strength and lower physical oscillations in electrical field in respect to the smooth wire (Dia: 150 μm).

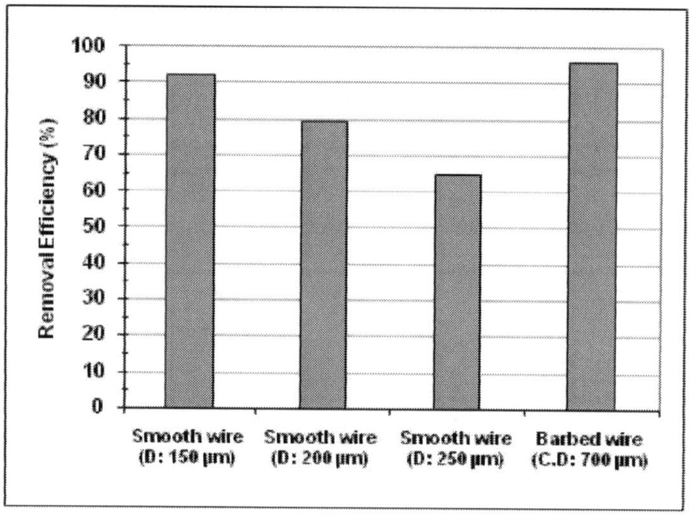

Fig. 3. Removal efficiency of 2 kinds of discharge electrodes used by plasma system. C_{soot}: 1 g/m^3 and Applied voltage: 5.0 kV.

The data obtained also demonstrate a negligible and satisfactory variation in removal efficiency vs. time in 27 min. Interval for the barbed wire-simple cylinder system.

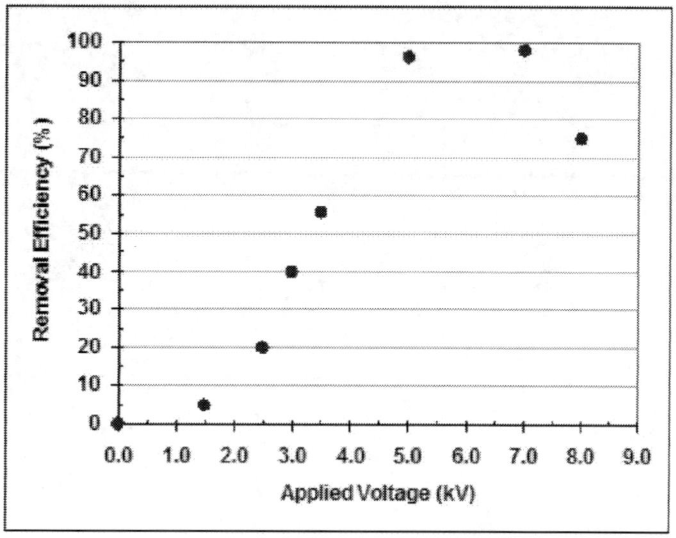

Fig. 4. Removal efficiency of barbed wire-simple cylinder system as a function of applied voltage. C_{soot}: 1 g/m^3.

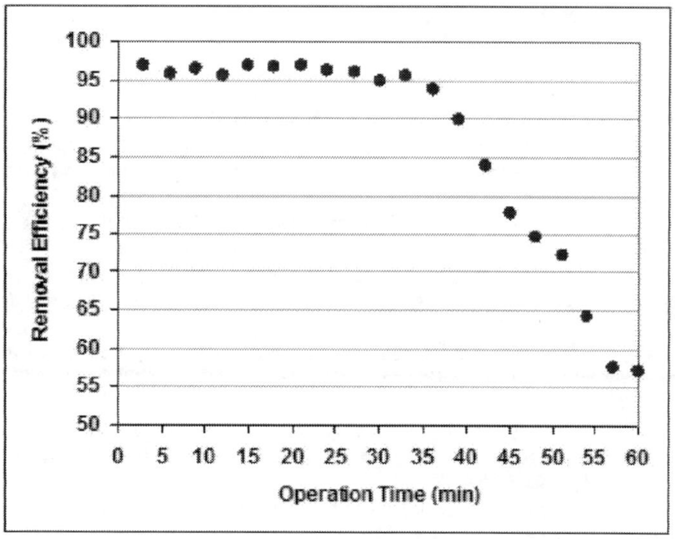

Fig. 5. Removal efficiency of barbed wire-simple cylinder system as a function of operation time.
C_{soot}: 1 g/m^3 and Applied voltage: 5.0 kV.

Under real condition, experimental results of fabricated systems consist of barbed wire- spiral cylinder with sequential regeneration (collected soot incineration by electrical heating) and barbed wire-simple cylinder (both with a length of 700 mm and an inside diameter of 37 mm) indicate that the removal efficiency about 79% and 92% was achieved respectively with Minibus Benz Model O508D exhaust gas operating in steady-state condition.

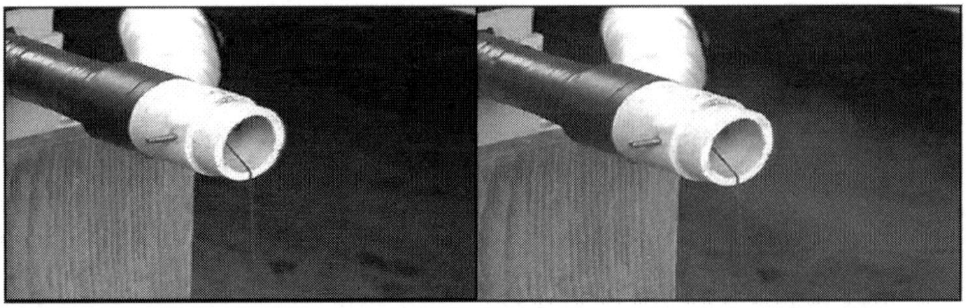

Fig. 6. Evaluating of barbed wire-cylinder type plasma system under real condition.
Left: Plasma is On Right: Plasma is Off

Conclusion

Presented results suggest that, the barbed wire-cylinder type plasma system can be considered to apply for efficient elimination of fine particulate matter from diesel exhaust gas. It should be noted that investigation of fabricated system consist of barbed wire-porous cylinder under real condition is now underway.

References

[1] U.S. EPA. Health Assessment Document for Diesel Engine Exhaust, Washington, DC, 2002a.

[2] P. Saiyasitpanich, T.C. Keener, S.-J. Khang and M. Lu, Removal of diesel particulate matter (DPM) in a tubular wet electrostatic precipitator, Journal of Electrostatics 65 (2007) 618–624.

[3] M. Matti Maricq, Chemical characterization of particulate emissions from diesel engines: A review, Aerosol Science 38 (2007) 1079 – 1118.

[4] Anatol Jaworek, Andrzej Krupa and Tadeusz Czech, Modern electrostatic devices and methods for exhaust gas cleaning: A brief review, Journal of Electrostatics 65 (2007) 133–155.

[5] Michele Ambrogio, Guido Saracco, Vito Specchia, Coen van Gulijk, Michiel Makkee and J.A. Moulijn, On the generation of aerosol for diesel particulate filtration studies, Separation and Purification Technology 27 (2002) 195–209.

[6] Ye Zhuang, Yong Jin Kim, Tai Gyu Lee and Pratim Biswas, Experimental and theoretical studies of ultra-fine particle behavior in electrostatic precipitators, Journal of Electrostatics 48 (2000) 245-260.

[7] Maria Jedrusik, Juliusz B. Gajewski and Arkadiusz J. Swierczok, Effect of the particle diameter and corona electrode geometry on the particle migration velocity in electrostatic precipitators, Journal of Electrostatics 51-52 (2001) 245-251.

[8] Takeo Takahashi, Akinori Zukeran, Yoshiyasu Ehara, Haruo Kishida, Tairo Ito and Takeshi Takamatsu, Influence of Gas Velocity and Electric Field Intensity on Particle Deposit and Re-entrainment Phenomena in an Electrostatic Precipitator, Journal of Aerosol Sci. 29 (1998) S481-S482.

[9] J. O. Chae, Non-thermal plasma for diesel exhaust treatment, Journal of Electrostatics 57 (2003) 251-262.

[10] Tomasz Ciach, Tomasz R. Sosnowski and Albert Podgorski, Efficient filtration methods for diesel aerosols, Journal of Aerosol Sci. 26 (1995) S723-S724.

[11] Tomasz Ciach and Tomasz R. Sosnowski, Removal of soot particles from diesel exhaust, Journal of Aerosol Sci. 27 (1996) S705-S706.

Comment On Coal-Based Chemical Looping Combustion Process

Yan Cao, Department of Chemistry, Institute for Combustion Science & Environmental Technology, Bowling Green, KY

Wei-Ping Pan, Chemistry, Institute for Combustion Science & Environmental Technology, Bowling Green, KY

There is a urgent push focusing on quick ways to curb CO2 emission based on wide-accepted scientific evidences on global warming. Under President Obama's economic stimulus and recovery plan, carbon managements will be the priority in fossil-fuel energy utilization, including coal-fired power plants to reduce CO2 emissions. The United States of America has abundant coal resources. More than 50% of U.S. electricity is derived from coal-based power. Consequently, the U.S. will have to either deal with or lead these upcoming carbon management regulations. However, carbon management in existing coal-fired power plants is burdened with high capital and operational costs based on post combustion capture processes.

This manuscript is to make technical comments on efforts made to develop coal-based chemical looping combustion process, which is an innovative coal-based combustion process dedicated to low cost carbon management and enhanced electricity generation efficiency. Oxygen carriers are tecnical cores of chemical looping combustion process. Selection and preparation of different oxygen carriers directly associated process economics and efficiencies. We are trying to conduct the feasibility study of coal-based chemical looing combustion, and thus point out the opportunity oxygen carriers, that lead to the successful development of coal-based chemical looing combustion process.

Based on selected oxygen carriers, a novel coal-based chemical looping combustion process are proposed. It could be retrofitted to existing coal fired power plants resulting in efficiency enhancement and capacity enlargement as well as 30%-70% CO2 capture under constraint of cost increase. This novel combustion process is the combination of chemical looping combustion (CLC) using novel oxygen carriers and subsequent combustion of residue char in existing, aged, low-efficiency, pulverized coal boilers. The potential impact of this process is the realization of affordable and timely CO2 capture technology that can be introduced into the existing coal-fired power generation fleet. The recovery of coal energy using different combustion technologies will be rationally optimized without energy penalty. Attrition-resisting and thermally stable oxygen carriers will be developed to achieve auto-thermal heat balance of the process, a high purity of CO2, favorable kinetics and participation of free-oxygen. The strategy of controlling multiple major air pollutants (SO2, NOx, and Hg) will be addressed.

See more of this Session: Chemical Looping Processes - II
See more of this Group/Topical: Energy and Transport Processes

Synfuel Production in a Carbon Constrained Scenario Using the Calcium Looping Process

Shwetha Ramkumar, Chemical and Biomolecular Engineering, The Ohio State University, Columbus, OH

Nihar Phalak, Chemical and Biomolecular Engineering, The Ohio State University, Columbus, OH

Liang-Shih Fan, Chemical and Biomolecular Engineering, The Ohio State University, Columbus, OH

The increasing energy demand and the necessity for energy independence have brought coal to liquid (CTL) technologies, which have the capability of near term implementation due to their compatibility with the existing infrastructure, to the forefront. Currently, the indirect production of coal derived liquid fuels is through the coal gasification and Fischer Tropsch (FT) process. It has been estimated that the carbon foot print of the coal to liquids process is at least 150 - 175% higher than petroleum based fuels. By the implementation of carbon capture and sequestration, the life cycle CO2 emissions for the coal to liquids process can be reduced by 20% compared to conventional fuel.

The Calcium Looping Process (CLP) is capable of producing a sequestration ready CO2 stream by capturing all the CO2 emitted during the coal to liquids process. In addition to achieving carbon capture, the CLP improves the efficiency of the coal to liquids process by conversion of the Fischer Tropsch reactor's off gases to hydrogen which is used to adjust the H2:CO ratio of the FT feed and for the product upgrader. The CLP integrates the reforming of hydrocarbons, water gas shift (WGS) reaction and in-situ carbon dioxide (CO2) removal at high temperatures in a single reactor and reduces the overall foot print of the hydrogen production process. The exothermic carbonation and WGSR convert the highly endothermic reforming of hydrocarbons into a heat neutral process thus simplifying the reforming process and reducing the temperature of reforming from >900C to 650C. In this paper, the reaction chemistry, thermodynamics, lab and bench scale experimental studies, process intensification strategies and energy conversion efficiencies will be presented.

See more of this Session: Chemical Looping Processes - II
See more of this Group/Topical: Energy and Transport Processes

Chemical Looping Combustion Process applied to liquid fuels

Ann Forret, Ali Hoteit*, Thierry Gauthier

IFP-Lyon, Solaize, France

Abstract

CLC is a new promising combustion process for CO_2 capture with less or even no energy penalty compared to other processes. Indeed, CLC does not require specific O_2 or CO_2 separation equipment and energy penalty appears very limited. CLC process is an oxy-combustion technology in which the oxygen required for combustion is supplied by metal oxides known as oxygen carriers.

The majority of the work performed to date on CLC has been performed using methane and coal and/or pet coke as the fuel. Up to now, there is no information reported in the literature concerning the use of liquid fuels in CLC. The objective of the present work is therefore to investigate the feasibility of liquid feed injection and contact with oxygen carrier to conduct controlled combustion in CLC conditions. A batch experimental fluidized bed set-up was developed to contact alternatively oxygen carrier with liquid fuels or air. $NiAl_{0.44}O_{1.67}$ was used as oxygen carriers in a fluidized bed reactor and dodecane as the combustible. The fluidized bed experiment results showed that during the reduction period, all the dodecane was converted. A high concentration of CO_2 (> 99%) could be achieved. No formation of carbon was evidenced on the oxygen carrier during the reduction step. Therefore, it was concluded in this study that complete combustion of dodecane is achievable in CLC mode , producing CO_2 and H_2O with stable operation.

Keywords: Chemical-looping combustion; Liquid fuels; CO_2 capture; Fluidization; Oxygen carriers; Clean Combustion

1. Introduction

Over the last 10-15 years, chemical looping routes have started to be investigated in order to provide clean and efficient solutions for several applications such as combustion and hydrogen production. CLC is one of the promising route for CO_2 capture since no specific separation of O_2 from air nor CO_2 from flue gas is required, resulting in much lower energy expected penalties. The chemical looping concept is based on the use of materials with oxidation-reduction cycles that can either oxidize or reduce feedstocks. In combustion, CO_2 rich effluents can be directly produced, enabling subsequent CO_2 transportation and storage. Chemical looping combustion (CLC) is a flameless combustion technology where there is no direct contact between air and fuel. Theoretical analyses of the technology indicate that its use results in increased power efficiency compared to regular combustion [1]. The important feature in the chemical looping combustion is that the carbon dioxide produced from the reduction step is inherently separated from the inert nitrogen and unreacted oxygen from the combustion air. Carbon dioxide is then easily separated from the water by condensation of the water resulting in carbon dioxide available for sequestration. The combustion process does not produce NO_x. The production of a

* Corresponding author: Tel. (+33) 4 78 02 55 71, Email: ali.hoteit@ifp.fr

sequestration ready CO_2 stream does not require any additional separation units and there is no energy penalty or reduction in power plant efficiency. This combustion process is divided into two steps (see Figure 1) : reduction and oxidation.

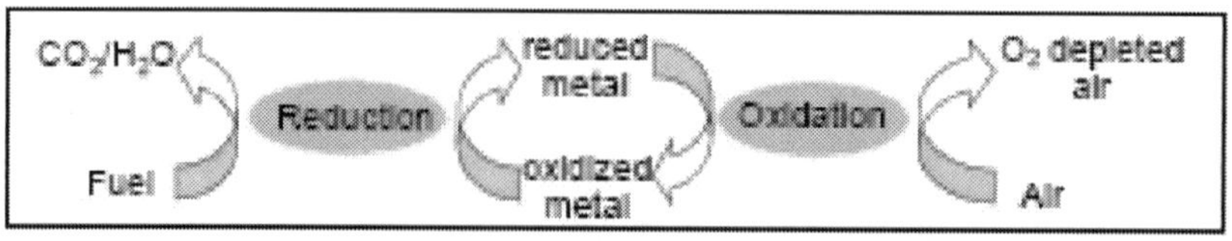

Figure 1: Principle of chemical looping combustion

In the reduction step, fuel gases are oxidized by reduction of a metal oxide via the following reactions:

$$(2n+m)Me_xO_y+C_nH_{2m}\rightarrow(2n+m)Me_xO_{y-1}+mH_2O+nCO_2$$

$$Me_xO_y + CO \rightarrow Me_xO_{y-1} + CO_2$$

$$Me_xO_y + H_2 \rightarrow Me_xO_{y-1} + H_2O$$

(where Me_xO_y is the metal oxide, and $MexOy-_1$ is the reduced metal oxide)

In the oxidation step, the reduced metal oxide is re-oxidized by contact with air via the following reaction:

$$Me_xO_{y-1} + ½\, O_2 \rightarrow Me_xO_y$$

The net result of these two steps is the direct combustion of the fuel gases with oxygen:

$$C_nH_{2m} + ½\,(2n+m)\, O_2 \rightarrow nCO_2 + mH_2O$$

$$CO + ½\, O_2 \rightarrow CO_2$$

$$H_2 + ½\, O_2 \rightarrow H_2O$$

Most of the work around chemical-looping combustion has focused on the use of solid fuel ([2], [3]; [4]; [5]) and on gaseous fuel ([6], [7]). The solid–solid reaction between coal and metal oxide is not very likely to occur at any appreciable rate. Instead the solid fuel needs to be gasified and then the oxygen carrying particles react with the gas produced, making the gasification the time limiting step [8].

2. The aim of the study

Liquid fuels such as heavy fuels resulting from oil distillation or conversion may also be interesting feed stocks to consider. Liquid fuels are challenging feedstock to deal with in fluidized beds. The objective of the present work is therefore to investigate the feasibility of liquid feed injection and contact with oxygen carrier in CLC conditions in order to conduct partial or complete combustion of hydrocarbons in these conditions.

3. Experimental

3.1 Oxygen-carrier materials

The oxygen carrier used in the present study is $NiO/NiAl_2O_4$ which consisted of 60 wt% active phase and 40% support material. This Ni-based Oxygen Carrier was manufactured by Marion Technologies (France). Its average arithmetic and harmonic diameters were 171 and 125 µm, respectively. The bulk density of the particles was

2200 kg/m^3. The BET surface area of the particles is 7 m^2/g. The maximum oxygen-carrying capacity of $NiO/NiAl_2O_4$ was measured as 14.5 wt% with thermogravimatric tests according to literature [9]. Most of the oxygen transfer capacity is accounted for by reduction of NiO to Ni (12.8%), and the remainder by the slower reduction of $NiAl_2O_4$ to Ni and Al_2O_3.

3.2 Reactivity tests in batch fluidized bed reactor

The scheme hereafter (see Figure 2) presents the experimental pilot scale. It consists of a fluidized bed reactor made of quartz with 20-mm inner diameter. The reactor was electrically heated up to 950°C and was operated at atmospheric pressure. K-type chrome/alumel thermocouples was used to measure continuously temperature of the oxygen carrier bed. 45g of oxygen carrier were preheated in an inert atmosphere (N_2) to the desired experimental temperature and then exposed to air to ensure that the particles were fully oxidized before the reduction period.

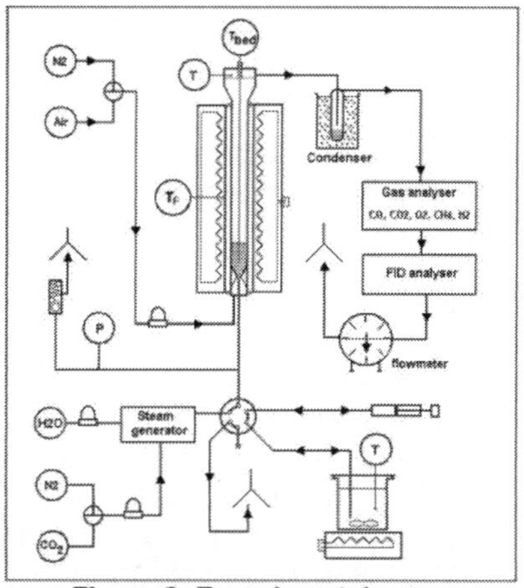

Figure 2: Experimental setup

During reduction step, a pulse of liquid fuel is injected at the bottom of the dense bed. In order to simplify flue gas analyses process of these tests, oxygen carriers are fluidized by nitrogen (instead of CO_2 or steam) during reduction steps. The combustion gas sample was pumped to a water condenser. The H_2, CO, CO_2, O_2, HC and CH_4 gas concentrations were then measured. CH_4, CO and CO_2 were measured by non-dispersive infrared analysis (NDIR); the O_2 concentration was measured in a paramagnetic analyzer and the H_2 concentration by gas conductivity. Hydrocarbons were also measured by Flame Ionization Detector (FID). Data are recorded using a data acquisition system connected to a personal computer running commercial software.

3.3 Evaluation of Data

The conversion of oxygen carrier or the degree of oxidation (X) was defined by the equation (1)

$$X = \frac{m - m_{red}}{m_{ox} - m_{red}} \quad (1)$$

m, m_{ox}, m_{red} are respectively actual mass of oxygen carrier, mass of fully oxidized and mass of fully reduced oxygen carrier respectively. The conversions of oxygen carrier during reduction and oxidation phases of the cycle are evaluated by using the flue gas concentration as given by equations (2) and (3) respectively:

$$X_{i,red} = X_{i-1} - \int_{t_0}^{t_{red}} \frac{Q_{out}}{n_0 P_{tot}} \left(2p_{CO_2,out} + p_{CO,out} + p_{H_2O,out}\right) dt \tag{2}$$

$$X_{i,oxi} = X_{i-1} + \int_{t_0}^{t_{oxi}} \frac{2}{n_0 P_{tot}} \left(Q_{in} p_{O_2,in} - Q_{out}\left(p_{O_2,out} - 1/2\, p_{CO,out} - p_{CO_2,out}\right)\right) dt \tag{3}$$

where, n_0 is the moles of oxygen that can be removed from fully oxidized oxygen carrier. Q_{in}, Q_{out} are the molar flow of the incoming gas and the gas leaving the reactor respectively. P_{tot} is the total pressure. P_{in} is the partial pressure of incoming gas i and $P_{i,out}$ is the partial pressure of gas i leaving the reactor. and t is the time. The complete conversion equation for oxidation phase takes into account the formation of CO and CO_2 during the oxidation period due to the oxidation of carbon formed in the reduction period.

4. Results

In a first experiment, dodecane was injected on inert solids (carborundum SiC) in order to identify specific phenomena linked to vaporization and pyrolysis. In a second step, dodecane was injected on metal oxide particles $NiAl_{0.44}O_{1.67}$.

4.1 Dodecane injection on SiC

Figure 3 shows the conversion of dodecane on inert carborundum SiC at 800°C. 1 ml of dodecane was injected during 2 seconds at t = 1 minute. The oxygen carrier bed is then fluidized by nitrogen. The product stream of dodecane reaction on SiC consists of a mixture of gases, H_2, CH_4, unreacted dodecane (Hydrocarbons) and traces of CO. After 16 minutes of fluidization, nitrogen is then replaced by air. During the oxidation period, after O_2 introduction in the reactor, CO and CO_2 immediately appear at the reactor outlet stream, indicating that some carbon formation occurs during reduction and remains on the oxygen carrier .

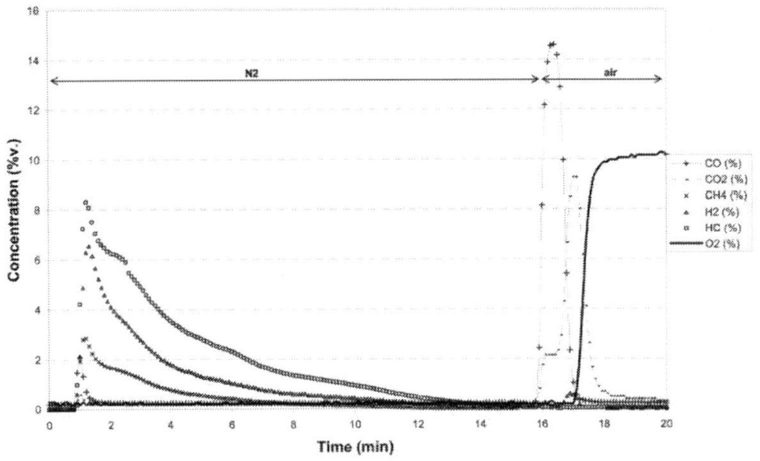

Figure 3: Pyrolysis of dodecane on SiC at 800°C

4.2 Reduction experiments with Dodecane

The oxygen present in 45g of nickel based oxygen carrier ($NiAl_{0.44}O_{1.67}$) totally oxidized is 12.3g. There is therefore in the bed a large amount of oxygen in excess to fully convert 1 mL of dodecane (2.6g) into CO_2 and H_2O.

As shown on Fig.4, after injection of dodecane into the reactor, CO_2 and H_2O are formed and no CH_4, CO nor H_2 can be observed during the carrier reduction time, thus indicating that dodecane conversion is complete .This figure shows also a decrease of the bed temperature due to the endothermic reduction of metal oxide. In the following reoxidation step, upon O_2 introduction to the reactor, CO and CO_2 is not detected at the reactor outlet stream. This result indicate that there is no significant carbon deposit on oxygen carrier at the end of the reduction step. It must be noted that the slight peak of hydrogen measured during the increase of CO_2 concentration ($1<t<1.5$ min) is an artifact: it comes from internal compensation calculation of interference of CO_2 on hydrogen by the analyzer.

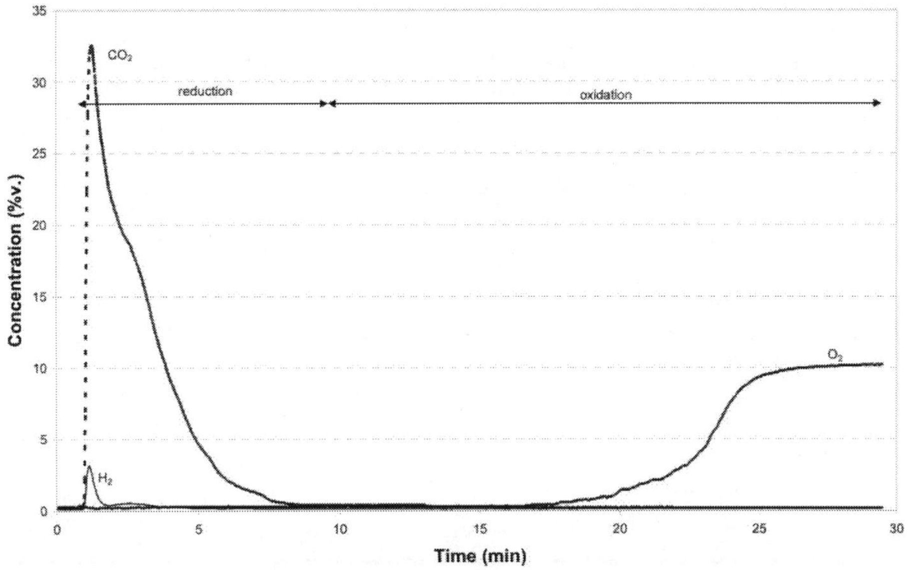

Figure 4: Outlet gas concentration after the injection of dodecane on $NiAl_{0.44}O_{1.67}$ at 800°C

4.3 Series of dodecane injections on $NiAl_{0.44}O_{1.67}$ at 800°C and 900°C

Figure 5 shows flue gas concentration for five consecutive injections of dodecane on a bed of 45g of $NiAl_{0.44}O_{1.67}$ at 800°C. Five injections of 1 mL of dodecane are performed at t = 2.0; 10.7; 21.1; 32.7 et 42.6 min, without intermediate oxidation.

During the two first injections, the conversion of dodecane is complete. This result indicates that, during the second injection time, the amount of available oxygen stored in the oxygen carrier still sufficient to convert dodecane into CO_2 and steam.

Figure 5 shows also the selectivities to synthesis gas. During the third injection, CO and H_2 and CH_4 are detected at the reactor outlet stream. The shift in the reaction is due to the quantity of O_2 present in the oxygen carrier during dodecane injection.

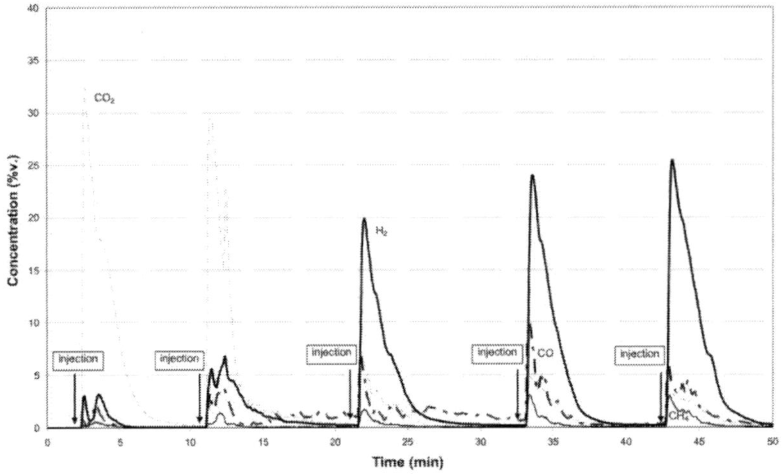

Figure 5: outlet gas concentration for five consecutive injections of 1 mL of dodecane on $NiAl_{0.44}O_{1.67}$ at 800°C (injections at t = 2.0; 10.7; 21.1; 32.7 et 42.6 min).

A similar experiment is performed with the same bed mass at 900°C. Figure 6 shows the flue gas concentrations for eight consecutive injections of dodecane on $NiAl_{0.44}O_{1.67}$. The first four injections of dodecane are converted into CO_2 and steam. That suggests that for the same amount of oxygen stored in the oxygen carriers bed, the quantity of oxygen transferred at 900°C from the oxygen carrier to the fuel is more important than that at 800°C. After 60 minutes of oxidation, an important release of CO is observed, resulting from carbon deposits combustion which was formed during the last metal reduction.

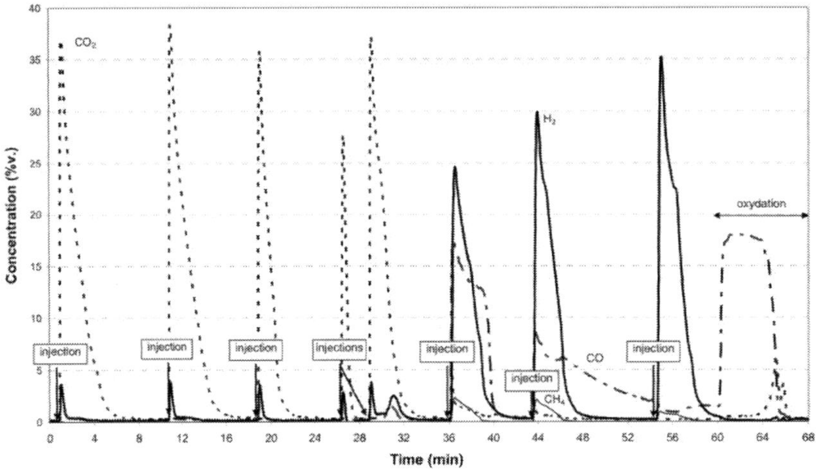

Figure 6: outlet gas concentration for eight consecutive injections of 1 mL of dodecane on $NiAl_{0.44}O_{1.67}$ at 900°C (injections at t= 0.7; 10.6; 18.7; 26.2; 28.7; 36.0; 43.4 et 54.5 min). Oxidation for t>60min

5. Conclusions

Experiments presented in this paper extend the feasibility of the Chemical Looping Combustion (CLC) process to liquid fuels. It was shown that dodecane is totally converted into CO_2 and steam using $NiAl_{0.44}O_{1.67}$ oxygen carrier. No carbon deposits was detected during the reduction step, resulting in an optimal CO_2 capture efficiency. CLC process can then be used to produce energy with a pure flow of CO_2 than can be easy to sequestrate. It is also possible to consider Chemical Looping Reforming for the production of synthesis gas or for cogeneration of H_2 and power.

6. Acknowledgements

The authors are thankful to French National Research Agency for its financial support and to William Pelletant for his co-operation.

References

[1] M. Anheden and G. Svedberg, Energy Convers. Mgmt. 39 (16-18) (1998) 1967-1980.
[2] Cao, Y., Pan, W.-P., 2006. Investigation of chemical looping combustion by solid fuels. 1. Process Analysis. Energy Fuels 20 (5), 1836–1844.
[3] Cao, Y., Casenas, B., Pan, W.-P., 2006. Investigation of chemical looping combustion by solid fuels. 2. Redox reaction kinetics and product characterization with coal, biomass, and solid waste as solid fuels and CuO as an oxygen carrier. Energy Fuels 20 (5), 1845–1854.
[4] Dennis, J.S., Scott, S.A., Hayhurst, A.N., 2006. In situ gasification of coal using steam with chemical looping: a technique for isolating CO2 from burning a solid fuel. J. Energy Inst. 79 (3), 187–190.
[5] Leion, H., Mattisson, T., Lyngfelt, A., 2007. The use of petroleum coke as fuel in chemical-looping combustion. Fuel 86 (12- 13), 1947–1958.
[6] Johansson, M., Mattisson, T., Lyngfelt, A., 2004. Investigation of Fe2O3 with MgAl2O4 for chemical-looping combustion. Indus. Eng. Chem. Res. 43 (22), 6978–6987
[7] Cho, P., Mattisson, T., Lyngfelt, A., 2004. Comparison of iron-, nickel-, copper- and manganese-based oxygen carriers for chemical-looping combustion. Fuel 83 (9), 1215–1225.
[8] Scott, S.A., Dennis, J.S., ., Hayhurst, A.N., Brown T., 2006. In situ gasification of a solid fuel and CO2 separation using chemical looping. AICHE J.52 (9), 3325 – 3328.
[9] Ishida M, Yamamoto M, Ohba T. Experimental results of chemical-looping combustion with NiO/NiAl2O4 particle circulation at 1200 C. Energ Convers Manage 2002;43:1469–78.

Gasification of Biomass and CO_2 Capture Using Chemical-Looping Combustion: Thermodynamic Analysis

Mohammad M. Hossain, Chemical and Biochemical Engineering, The University of Western Ontario, London, ON, Canada

Mohammad R. Quddus, Chemical and Biochemical Engineering, The University of Western Ontario, London, ON, Canada

Hugo de Lasa, Chemical and Biochemical Engineering, The University of Western Ontario, London, ON, Canada

In recent years, biomass including agricultural and municipal waste, has been received increasing attention considering a major source of renewable energy with zero net CO_2 emissions. The use of biomass as fuel, in conjunction with chemical-looping combustion (CLC), offers the additional potential of delivering negative CO_2 emissions. To accomplish this, a possible approach is to gasify the solid fuel producing syngas which can be burned inside the same fuel reactor of a CLC process. The overall gasification and combustion combined process can be described as follows:

Gasification step:

$C_xH_yO_z \rightarrow H_2+CO+CO_2+CH_4+C$

$C+H_2O \rightarrow CO+H_2$

$CO+H_2O \rightarrow CO_2+H_2$

$C+CO_2 \rightarrow 2CO$

Combustion step (with oxygen carrier)

$M_yO_x+H_2 \rightarrow M_yO_{x-1}+ H_2$

$M_yO_x \rightarrow M_yO_{x-1} + CO_2$

The above reactions are heavily influenced by the type of solid fuel, the oxygen carrier, the reaction temperature, and the partial pressure of steam in the fuel reactor (Leion et al., 2007; Cao et al., 2006). This approach has the following main advantages: (i) eliminates extra units for solid gasification and (ii) circumvents the dilution of the syngas with nitrogen (Leion et al. 2007; Scott et al. 2006).

An equilibrium thermodynamic model is reported in this communication able to simulate an integrated Gasifier-Combustor unit using glucose as a feedstock. Glucose is one of the building blocks of cellulose, a main biomass component. Alumina supported nickel based oxygen carrier is considered as the oxygen carrier inthe calculations. The effects of fuel composition, temperature, pressure, and extent of gasification are studied using the Gibbs free energy minimization to determine the product gas composition. This analysis shows that CO_2 can be concentrated and purified up to 99.0% in such a process during the reduction cycle by the gasification products (CO and H_2).

References:

Cao, Y., Pan, W., (2006), Energy & Fuels, 20, 1836-1844.

Scott, S.A., Dennis, J.S., Hayhurst A.N. (2006), AIChE Journal, 52 (9), 3325-3328.

Leion, H., Mattisson, T., Lyngfelt, A. (2007), Fuel 86, 1947-1958.

See more of this Session: Chemical Looping Processes - II
See more of this Group/Topical: Energy and Transport Processes

Insights From Process Engineering Simulations for Envisioning Design of Chemical Looping Combustion Process for Solid Fuels

Asad H. Sahir, Department of Chemical Engineering, University of Utah, Salt Lake City, UT

Adel. F. Sarofim, Dept. of Chemical Engineering, University of Utah, Salt Lake City, UT

JoAnn Slama Lighty, Chemical Engineering, University of Utah, Salt Lake City, UT

Chemical Looping Combustion is currently being investigated as a novel technique for coal combustion with carbon capture. The research community is exploring options to extend the promising results of laboratory studies on chemical looping combustion of coal into an economically viable combustion process.

Process Engineering Simulations for Chemical Looping Combustion of coal have been developed in ASPEN PLUS. A simplified case for chemical looping combustion of coal with copper oxide into gaseous components has been studied using equilibrium reactor models. The fluidized bed fuel and air reactors have been modeled as a series of equilibrium reactors. The temperature of the fluidized bed has been studied as a function of coal, copper oxide flow rates and heat input. It has been observed that the temperature of the conceptualized fuel equilibrium reactor was sensitive to the mass ratio of copper (II) oxide to coal. An inference which could be drawn from the study is that a proper distribution of coal and copper oxide particles is important in maintaining the temperature throughout the length of the fuel fluidized bed reactor. A similar study has been done for analyzing the effect of distribution of copper (I) oxide and air on the temperature throughout the length of the conceptualized air fluidized bed reactor. The analysis suggests possibilities to conceptualize novel process configurations for combustion of coal, taking into consideration that the combustion process for coal occurs over a wide temperature range.

The talk would also feature results of a techno-economic analysis conducted based on the simulations with kinetics of coal combustion incorporated in the process models.

See more of this Session: Chemical Looping Processes - II
See more of this Group/Topical: Energy and Transport Processes

Modeling and Sensitivity Analysis of the Chemical Looping Combustion Process

Balaji Sukumar, Chemical Engineering, Carnegie Mellon University, Pittsburgh, PA

Jovan Ilic, Electrical and Computer Engineering, Carnegie Mellon University, Pittsburgh, PA

Erik Ydstie, Chemical Engineering, Carnegie Mellon University, Pittsburgh, PA

Bruce Krogh, Electrical and Computer Engineering, Carnegie Mellon University, Pittsburgh, PA

Chemical looping is believed to a promising technology in clean combustion of gas or solid fuels for future power generation. In this process, oxygen required for combusting the fuel is provided through metal oxides rather than using air/oxygen. This prevents the production of pollutants like oxides of nitrogen as there is no nitrogen present in the feed. In addition, it has been proved that almost pure carbon dioxide can be produced as a result of this process which makes it easy for carbon sequestration. Noteworthy experimental works have been carried out in the past few years in determining the better metal oxide for these systems. However, the characteristics of metal oxides vary quite significantly with respect to fuel types. Thus, a universal metal oxide which suits for all kinds of fuel is yet to be determined. Therefore, devising a mathematical model to study the effectiveness of various important physical and operating parameters will be beneficial.

The process consists of two fluidized bed reactors connected through loop seals to maintain the pressure drop in the reactors. The inflow and outflow of the solid metal oxide and the fuel should be monitored appropriately for enhanced performance. The process is still at the nascent level and being controlled manually or by simple control algorithms. Once commercialized, there will be a definite need of advanced control strategies for controlled power output. Thus, in this study, we focus on developing a model for the entire CLC system with a view point of determining the optimal operating conditions and the sensitivity of important operating parameters. The model is based on system invariants under thermodynamic equilibrium coupled with the shrinking core mechanism to represent the gas-solid reactions. The model results are validated based on the experimental data available in the literature. To the authors' knowledge, this work is the first in proposing a control oriented model for the CLC system.

Keywords: Modeling, thermodynamic equilibrium, invariants, chemical looping combustion, control.

See more of this Session: Chemical Looping Processes - II
See more of this Group/Topical: Energy and Transport Processes

An Incremental, Multi-Technology Approach to Retrofitting Existing Pulverized Coal Power Plants for Carbon Capture

David C. Miller, National Energy Technology Laboratory (NETL), U.S. Department of Energy, Morgantown, WV

Global climate models indicate that acting rapidly to reduce carbon dioxide emissions will provide greater benefits to the environment than delaying action; however, new technologies, currently under development, may be significantly less costly than those currently available to be deployed. This tension makes it difficult to determine the best course of action. This paper will explore scenarios under which partial capture is achieved with one technology (aqueous amine absorption) in the near term and additional capture capability is achieved through newer technology in the future, such as oxy-combustion, adsorption, membranes or advanced absorbents. In particular, this paper will focus on identifying potential synergies that can be achieved with a multi-technology approach to carbon capture.

See more of this Session: Dynamics, Design and Control of Sustainable Processes and Technologies and Associated Waste Management Principles
See more of this Group/Topical: Environmental Division

Conversion of Solid Fuels in the Iron-Based Chemical Looping Process

Hyung Rae Kim, William. G. Lowrie Dept. of Chemical and Biomolecular Engineering, The Ohio State University, Columbus, OH

Fanxing Li, Chemical and Biomolecular Engineering, The Ohio State University, Columbus, OH

Liang Zeng, William. G. Lowrie Dept. of Chemical and Biomolecular Engineering, The Ohio State University, Columbus, OH

Deepak Sridhar, William G. Lowrie Chemical and Biomolecular Engineering, The Ohio State University, Columbus, OH

Fei Wang, Chemical and Biomolecular Engineering, The Ohio State University, Columbus, OH

Andrew Tong, Chemical and Biomolecular Engineering, The Ohio State University, Columbus, OH

Nobusuke Kobayashi, William G. Lowrie Chemical and Biomolecular Engineering, The Ohio State University, Columbus, OH

L. -S. Fan, Chemical and Biomolecular Engineering, The Ohio State University, Columbus, OH

The chemical looping schemes that directly convert solid fuels such as coal and biomass pose economical and technical attractiveness since it adapts relatively cheap solid fuels, avoiding a capital and energy intensive gasifier. For the solid fuel chemical looping processes to be feasible, the slow kinetics of solid-solid reaction between the solid fuels and oxygen carriers must be overcome. In addition, the effects of contaminants in solid fuels on the oxygen carriers need to be addressed. A two-stage counter-current moving bed is proposed for effective conversion of solid fuels. In the first stage, the oxygen carriers in the upper section convert the gaseous volatiles from the solid fuels. Then, the bottom stage converts char, as the metal oxides are reduced to the lower oxidation states. Prior to the moving bed demonstration, externally-heated fixed bed is used to simulate the reaction between volatiles and iron particles. Additionally, the char conversion enhancement is further studied with steam or CO2 along with the cold model study. These studies show that the oxygen carrier developed at the Ohio State University is capable of converting volatiles from coal and biomass and chars to CO2 with highly recyclable manner. The effects of contaminants such as sulfur and ash on the reactivity and recyclability of the oxygen carrier particle are also studied.

See more of this Session: Dynamics, Design and Control of Sustainable Processes and Technologies and Associated Waste Management Principles
See more of this Group/Topical: Environmental Division

Bio Conversion of Glycerol to Methane and Hydrogen

R. Mark Bricka, Chemical Engineering, Mississippi State University, Mississippi State, MS

Trey Fleming, TreyCo Inc., Vicksburg, MS

Todd French, Chemical Engineering, Mississippi State University, Mississippi State University, MS

Sumesh Arora, Mississippi Alternative Energy Enterprise, Ridgeland, MS

Glycerol is produced as a by-product during the production of biodiesel. Until recently the quantity of biodiesel and subsequently glycerol has been minimal and the market has adsorbed the glycerol and the demand has historically remained high. With the recent increase in biodiesel predication capacity there has been a glut of glycerol production and the demand for glycerol has dropped. In fact, the glycerol produced from biomass is generally substandard and contains many impurities which must be removed to generate glycerol with any value. As a result of such issue the price for crude glycerol has dropped and many biodiesel producers are faced with either paying to dispose of their glycerol or providing it to energy generators at no cost. This has generated the need for alternative uses for crude glycerol.

A study was undertaken to determine if glycerol could be degraded using bacteria to produce methane. The methane from the degradation of the glycerol could then be used as a supplemental energy source. This study consisted of a the fabrication and operation of a pilot digesting unit. The system was seeded with domestic digester sludge and variables such as temperature and pH were adjusted to understand their effects on gas production rate.

Results indicate that both methane and hydrogen can be produced from the glycerol feed. A temperature of approximately 35oC appears to be optimal for gas production. A time lag is experienced prior to gas production but once initiated the system is easy to restart and maintain at a constant gas production rate. This paper will present the result of findings for this study.

See more of this Session: Dynamics, Design and Control of Sustainable Processes and Technologies and Associated Waste Management Principles
See more of this Group/Topical: Environmental Division

Bioremediation Costs Estimation in Lakes and Reservoirs through Dynamic Optimization Based On Hybrid Eutrophication Models

Vanina Estrada, Chemical Engineering, Planta Piloto de Ingenieria Quimica - Universidad Nacional del Sur, Bahia Blanca, Argentina

Maria Soledad Diaz, Chemical Engineering, Planta Piloto de Ingenieria Quimica (PLAPIQUI), Universidad Nacional del Sur - CONICET, Bahia Blanca, Argentina

Long-term studies on Europe and North American eutrophic lakes showed that even when reduction of external nutrient loading has had successful results, eutrophication symptoms, like harmful algal blooms (HAB) development, are retained even for decades after restoration (Jeppensen et al., 2005, Søndergaard et al 2007). Biological and chemical process can delay the recovery of water quality (Gulati and van Donk, 2002; Søndergaard, Jensen and Jeppesen, 2003) mainly due internal phosphorus recycling (the limiting nutrient in freshwater ecosystems) from sediments. In order to accelerate lake recovery, several in-lake restoration methods have been applied (e.g. sediment removal by dredging, hypolimnetic oxygenation, and alum treatment). Bio-manipulation has been one of the most extensive strategies used to prevent algal blooms in European lakes and reservoirs. This method is based on the trophic chain theory and consists of keeping a high grazing pressure on phytoplankton populations performing zooplanktivorous fish removal (Shapiro & Wright, 1984; Jeppesen et al., 1990). Many restoration projects have been reported in the literature in the past decades, some of which were successful (Hansson et al., 1998; Søndergaard et al., 2000; Gulati and Van Donk 2002), whereas others had not obtained the expected results (Hansson et al., 1998).

In this work, we estimate biorestoration costs through simultaneous dynamic optimization based on a hybrid ecological water quality that determines bio-manipulation policies in a currently eutrophic water body, which is the drinking water source for more than 450,000 inhabitants in Argentina. The model considers four biogeochemical cycles, i. e., carbon, nitrogen, phosphorus and dissolved oxygen, three functional groups of phytoplankton (cyanobacteria, diatoms and chlorophyta) and three of zooplankton (cladocerans, copepods and rotifers). The model takes into account different food quality for each phytoplankton group. Previous work (Estrada et al., 2008) showed that in-lake restoration techniques, as well as the reduction of external nutrient load, are needed in order to avoid the appearance of recurrent algal blooms that are occurring for several years.

To solve the model, the partial differential algebraic equations model is transformed into an ordinary differential equations system by spatial discretization into horizontal layers. The objective function is to minimize the offset between total phytoplankton concentration and a desired value below eutrophication limit, and the optimization variable is the zooplanktivorous fish biomass to be removed. The best candidate for bio-manipulation at Paso de las Piedras Reservoir is silverside (Odontesthes bonaeriensis).

The optimal control problem is solved with a simultaneous approach (Raghunathan et al., 2004) applying an Interior Point method with reduced successive quadratic programming techniques, within program IPOPT (Kameswaran and Biegler, 2006).

References

Estrada V., Parodi, M.S. Diaz (2008). A simultaneous dynamic optimization approach for addressing the control problem of algae growth in water reservoirs through biogeochemical models, submitted.

Gulati, R.D. and Van Donk, E. (2002). Lakes in the Netherlands, their origin, eutrophication and restoration: stateof-the-art review. Hydrobiologia, 478, 73-106.

Hansson, L.-A., Annadotter, H., Bergman, E., Hamrin S. F., Jeppesen, E., Kairesalo, T., Luokkanen, E., Nilsson, Søndergaard, M. and Strand, J. (1998) Biomanipulation as an application of food-chain theory: constraints, síntesis, and recomendations for temperate lakes. Ecosystems 1, 558-574.

Kameswaran, S., L.T. Biegler (2006), Simultaneous dynamic optimization strategies: Recent advances and challenges, Comp. Chem. Eng., 30, 1560.

Raghunathan A., S. Diaz, L.T. Biegler, 2004, An MPEC Formulation for Dynamic Optimization of Distillation Operations, Comp. Chem.Eng., 28, 2037.

See more of this Session: Dynamics, Design and Control of Sustainable Processes and Technologies and Associated Waste Management Principles
See more of this Group/Topical: Environmental Division

An Experimental Investigation On Coal Particles Combustion of Subbituminous Coals in Different Oxygen Concentration Conditions

Yong-Gyun Kim, Mechanical Engineering, Pusan National University, Busan, South Korea

Jae-Dong Kim, Mechanical Engineering, Pusan National University, Busan, South Korea

Gyu-Bo Kim, Pusan Clean Coal Center, Pusan National University, Busan, South Korea

Ju-Hun Song, School of Engineering/Pusan Clean Coal Center, Pusan National University, Busan, South Korea

Young-June Chang, School of Engineering / RIMT, Pusan National University, Busan, South Korea

Chung-Hwan Jeon†, School of Engineering/Pusan Clean Coal Center, Pusan National University, Busan, South Korea

Oxy-coal combustion is a promising alternative to achieve carbon dioxide capture and clean coal technology which could be directly applicable to the conventional pulverized coal fired utility power plants. In line with this, an 100MWe oxy-combustion power station project was launched in Korea 2007. For successful installation and operation of such facility, it is essential to understand the coal combustion mechanism under unconventional conditions in which the characteristic of pulverized coal combustion under oxy-coal combustion is quite different from that of air-blowing atmosphere. In this study, the characteristic of coal combustion mainly on devolatilization and char combustion, were investigated through flame lengths. The effects of oxygen mole fraction in N_2 and CO_2 diluent gases respectively, particle sizes for sub-bituminous coals were studied in a laminar flow entrained reactor that produces similar environment of utility furnace in terms of high heating rate and temperature. O_2 mole fraction was varied 0~50%. The flame length becomes shorter as increase of O_2 concentration. It means that the reactivity in both devolatilization and char oxidation process is increased in higher O_2 concentration environment. The char particles' burnout time with increase of O_2 fraction becomes shorter but there is not much difference between those of O_2/N_2 and O_2/CO_2 environments.

See more of this Session: Oxycombustion of Coal - Needs, Opportunities, and Challenges - II
See more of this Group/Topical: Energy and Transport Processes

Ignition of a group of coal particles in oxyfuel combustion with CO_2 recirculation

Alejandro Molina*[,1], Ethan S. Hecht[2], Christopher R. Shaddix[2]

[1] *Grupo de Bioprocesos y Flujos Reactivos, Facultad de Minas, Universidad Nacional de Colombia, Sede Medellin, Colombia*
[2] *Combustion Research Facility, Sandia National Laboratories, Livermore, CA USA*

Recent studies for coal ignition in oxyfuel combustion have been conducted as single-particle experiments or in pilot-scale burners. While single-particle experiments have the advantage of well-controlled atmospheres in which it is possible to infer a good understanding of the phenomena related to pulverized-coal ignition, they raise the question of the effect that burning coal in a multi-particle mode, such as exists in industrial burners, may have on particle ignition. On the other hand, pilot-scale experiments are often difficult to control and may not allow a simple understanding of the ignition phenomenon. This paper presents experimental results obtained under group-combustion regime in a laminar flow environment using coal flows larger than those observed in single-particle experiments but smaller than those of industrial burners. We expect the results exhibit some of the synergistic effects observed in industrial-size coal boiler flames. Furthermore, combustion of groups of coal particles can be used to reasonably infer characteristics of flame stability – something difficult to achieve in single-particle combustion experiments. The experiments were conducted with a U.S. high-volatile bituminous coal under oxygen concentrations ranging from 12% to 48%, with N_2 or CO_2 as balance gas, at two gas temperatures (1130 K and 1650 K). The standoff distance from the coal flame to the burner was used as a metric of ignition delay and the variation in time of the flame location as an indication of flame stability. The results showed that oxygen concentration has a strong effect on ignition delay at 1130 K. The presence of CO_2 as balance gas significantly changed how the standoff distance responded to changes in oxygen concentration, particularly at 1130 K. The theory developed for drop group combustion was reviewed for coal particles combustion. The analysis shows that traditional understanding of group combustion based on the G number needs to be revised to take into account a balance between the fast devolatilization process and the laminar flame speed that control group combustion.

1. Introduction

During coal combustion in oxygen-enriched environments, a CO_2-rich stream is recycled into the furnace to decrease combustion temperature as a suitable strategy to retrofit existing boilers, keeping low cost and generating a nitrogen-free, CO_2-rich stream than can be used for sequestration, enhanced-oil recovery, and ultimately, reduction of greenhouse emissions.

Although the transition from coal combustion in air (~ 21% N_2) to combustion in a CO_2 stream seems as straightforward as recycling a fraction of the exhaust gas to the burner, the different physical and chemical properties of CO_2 when compared to N_2, particularly the higher molar heat capacity of CO_2, cause differences in flame behavior that motivate changes in burner design and boiler operation, some of them motivated by differences in ignition characteristics of the coal particles [1-4].

Single-particle studies have been the first step towards understanding the complex phenomena occurring during coal combustion. Particularly, in oxygen-enhanced coal combustion with CO_2 recirculation, different studies [5-8] showed that higher concentrations of CO_2 decrease the temperature of the volatile cloud flame surrounding the particle and increase ignition delay.

The required step for extension of single particle experiments to boilers would be to characterize coal flames in industrial-size burners. However, industrial-size burners are difficult to control and limit the possible understanding gain from the experiments.

An alternative to single particle and industrial-size experiments are experiments in where multiple particles burn in such a way that interaction among particles occur.

In this research ignition of groups of particles of a U.S. high-volatile bituminous coal is studied in environments containing 12% to 48% O_2, with N_2 or CO_2 as balance gas, at two gas temperatures (1130 K and 1650 K). The results show that the trends observed for single particles are also evident when a group of particles is ignited. The paper also evaluates the theoretical developments on interactive combustion and suggests new theoretical modifications that may be required for oxyfuel combustion.

2. Theoretical considerations

Although at first hand the difference between a single particle-mode combustion and group combustion seems straightforward, several authors have identified different regimes to characterize combustion of more than one particle. While Law [9] considers four different regimes for drop combustion, Annamalai and Puri [10] describe seven different regimes for drops combustion that can be easily extended to the combustion of a group of particles.

a. Isolated-particle combustion (ISPC). Particles burn completely independent one from another.

b. Individual flame combustion (IFC). The number of particles is high enough to decrease the oxygen content, but the distance between particles is large enough that each particle burns independently.

c. Incipient group combustion (IGC). As the number of particles increases, a point is reached where the oxygen concentration within the clouds decreases to a point that the flame radii around the particle is large enough that flames from different particles touch each other. At this point IGC starts.

d. Partial group combustion (PGC). A further increase in the number of particles causes that the inner particles devolatilize without ignition, because of the low oxygen content, and a group flame is established inside the cloud. Particles far from the center keep burning in the IFC mode.

e. Group combustion (GC). A higher particle load increases volatile concentration inside the cloud to a point where oxygen does not penetrate anymore and a sole exterior flame forms outside the cloud. The onset of this regime is known as Critical Group Combustion (CGC)

f. Sheath combustion (SC). For a very large number of particles, oxygen concentration becomes very low and the cloud behaves as a whole that resembles single-particle combustion.

To characterize the extent of group combustion, different authors (see discussion in [9]) defined a G number as:

$$G = Nd_{p0}/\ell \qquad \text{E 1}$$

where N is given by:

$$N = n\ell^3 \qquad \text{E 2}$$

and n is the particle number density, ℓ a characteristic dimension of the cloud of particles and d_{p0} the initial particle diameter.

Although it is clear that group combustion is favored as for large G, the definition of G is strictly geometrical.

3. Experimental methods

The experiments presented in this section are similar to those already described in [11]. The analysis of results is, however, different as it is based on the theoretical considerations presented above.

3.1 Reactor and gas composition and temperature

The optical entrained flow reactor at Sandia National Laboratories was used in the experiments. The system, extensively described in previous papers (see e.g. [6]) was modified so that a higher coal flow (of the order of 0.4 g min^{-1}) than normally used in the single-particle experiments could be introduced into the reactor. Great care was taken to maintain a constant coal flow. To this aim a modified coal feeding system, similar to that described by Graham [12], was developed. The total gas flow into the reactor was also decreased when compares with previous studies [5] to increase the volumetric load of particles into the system and to increase the available residence time for ignition to occur.

Table 1 shows molar composition, calculated adiabatic temperature and total gas flow of the mixtures used in the experiments. Two different characteristic furnace temperatures were investigated: one in which particle ignition is relatively difficult (~ 1130 K) and one in which all but the most unreactive coals readily ignite (~ 1650 K). The 1130 K furnace environment is produced by N_2-diluent mixtures with an adiabatic temperature of 1200 K and by CO_2-diluent mixtures with an adiabatic temperature of 1240 K. For this low temperature furnace environment, the flat-flame burner was operated so that the oxygen concentration downstream from the burner was 24%, 36%, or 48% O_2. The high temperature furnace environment was produced using gas mixtures with an adiabatic temperature of 1800 K for N_2 mixtures and 1830 K for CO_2 mixtures. In this high temperature environment, experiments were conducted at 12%, 24% and 36% O_2 with a N_2 diluent and at 24% O_2 with a CO_2 diluent.

Table 1. Gas composition used in the experiments

Adiabatic temp.			1200 K			1240 K	
O_2	%v	24.0	36.0	48.0	24.0	36.0	48.0
CO_2	%v		1.1		66.1	54.0	42.0
N_2	%v	64.9	52.8	40.7		0.0	
H_2O	%v	10.0	10.1	10.2	10.0	10.0	10.0
Total gas flow	slpm*		40.0			40.0	

Adiabatic temp.			1800 K		1830 K
O_2	%v	12.0	24.0	36.0	24.0
CO_2	%v		3.3		60.0
N_2	%v	68.7	56.6	44.5	0.0
H_2O	%v	16.0	16.1	16.2	16.0
Total gas flow	slpm*		40.0		40.0

*slpm: standard (1 atm, 298 K) liters per minute

3.2 Fuel

A pulverized sample of high-volatile eastern U.S. bituminous coal (Pittsburgh seam, Bailey mine) was sieved to 75-106 μm size fraction. Sieving was conducted for 30-minute periods, several times, until the mass passing the 75 μm sieve was negligible. This procedure was carried out to ensure that there were no fines in the coal, as flame stability and ignition in a group of particles is very sensitive to the particle size, as has been shown in the literature [13] and as was evident in the flow reactor experiments. This coal sample has been analyzed to have 1% moisture, 7% ash, and 35% volatiles (see [11] for a complete analysis).

3.3 Measurement of standoff distance

In the experiments, the standoff distance from the flat flame burner to the coal flame, caused by the simultaneous combustion of multiple coal particles, was used as a metric of ignition delay and the variation in time of the flame location as an indication of flame stability. To measure the standoff distance, the standard setup [14] used in the entrained-particle flow reactor was modified by decreasing the gain in the photomultipliers (PMs) to a minimum value to prevent saturation. The intensity of the PM outputs at a specific distance from the burner were measured during a 10-second time interval. Time-average radiation intensity detected by the PM, as well as the signal standard deviation, were recorded as the optical probe volume height was increased. The experiment was repeated three times to improve the statistical analysis. This protocol was repeated for all the gas compositions described in Table 1.

Because ignition strongly depends on the heat balance surrounding the particles, care was taken to guarantee stable conditions in the experiments before measurements were collected by allowing the combustion gases to heat the quartz chimney for 10 minutes and then having stable coal injection for 5 more minutes before acquiring any data.

4. Results

4.1 Qualitative Analysis

Figure 1 shows digital pictures of coal particle combustion for the optical furnace operating in N_2 atmospheres, at a nominal gas temperature of 1130 K and at four different oxygen concentrations: a. 12%, b. 24%, c. 36%, and d. 48%. For the 12% O_2

case, there was no visible flame production. However, when the oxygen concentration was 24% (Figure 1b), a volatile flame formed at a certain distance from the burner. Based on the description presented above, while most particles burn in the PGC mode, a region at the bottom of the figure clearly burns under a GC. The presence of the volatile flame at the bottom can be deduced from the strong visible light emission associated with soot oxidation (we are currently working on implementing a means of directly detecting the volatiles flame location and size). As oxygen concentration increases, from left to right in Figure 1b, a qualitative observation suggests that the number of particles burning in PCG mode decreases as group combustion becomes more evident. Additionally, the standoff distance of the volatile flame from the burner decreases as the oxygen concentration increases, as expected. Similar pictures were obtained for a CO_2 atmosphere.

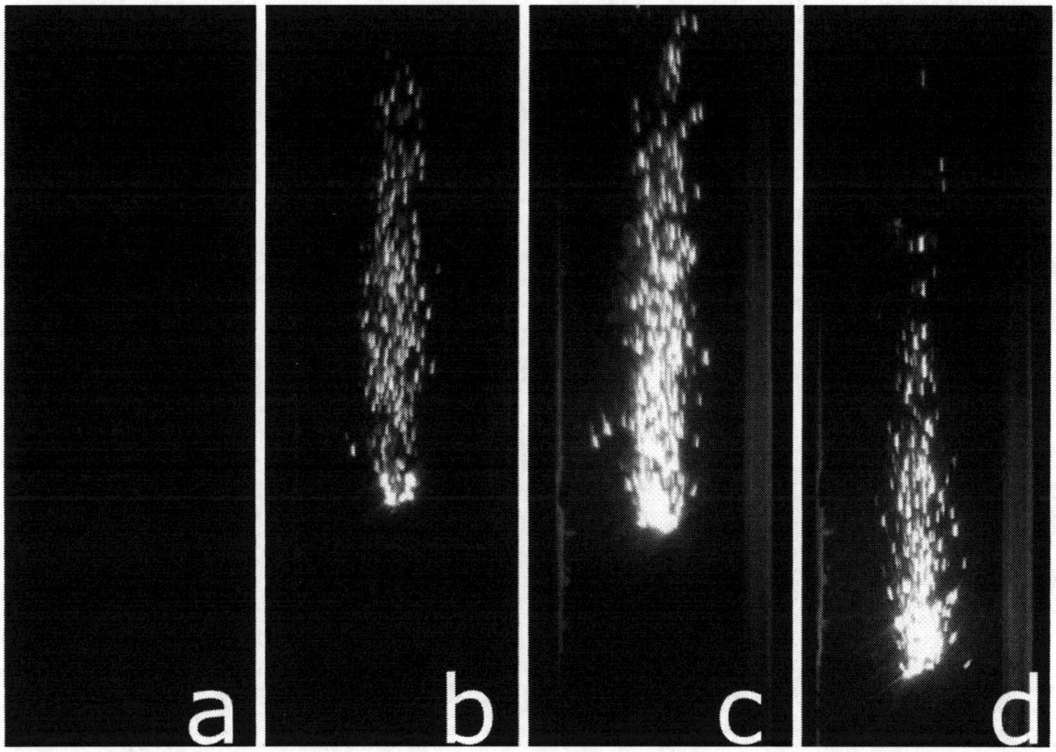

Figure 1. Examples of group of coal particles burning in N_2 at a nominal gas temperature of 1130 K at different oxygen concentrations: a. 12% O_2, 24% O_2,, 36% O_2 and 48% O_2. Particles flow upward from the center of a flat flame burner located at the bottom of the figures.

4.2 Quantitative analysis based on PM data

Figure 2 presents the average, standard deviation, and coefficient of variation (CV defined as standard deviation over average) of the PM signal at different heights (h) above the burner for different oxygen concentrations, for N_2 and CO_2 atmospheres. Figure 2a shows how, as the oxygen concentration increases in a N_2 atmosphere, the distance from the burner where the PM signal becomes detectable, i.e. the standoff distance, decreases. Figure 2b shows the same results for CO_2. While the standoff distances are clearly greater in the CO_2 atmospheres, the effect of the oxygen concentration is not as large as for N_2. It is also clear that the PM signal, an indication of the amount of thermal radiation leaving the flame, is much lower when the background gas is CO_2.

Figure 2c and d show the standard deviation of the PM signal with time. A high value of standard deviation can be associated with an unstable flame that frequently changes with time. However, as the absolute value of standard deviation strongly depends on the magnitude of the PM signal, it might make morasce sense to study the coefficient of variation, which normalizes the standard deviation by the average value. As Figure 2e and f show, the value of CV is higher in the presence of CO_2 than in N_2, and it decreases as the oxygen concentration increases.

Figure 3 presents results equivalent to those of Figure 2 but when the nominal gas temperature was 1650 K and only for a N_2 atmosphere. Figure 3a shows, consistent with the lower temperature results, that as the oxygen concentration increases the standoff distance decreases. However, the difference in standoff distance is minimal and the standoff distances are much shorter than were apparent at 1130 K. The small variation in standoff distance may reflect the stronger effect of particle heat-up (rather than subsequent gas-phase chemical induction time) on ignition at these high temperatures. Perhaps the most interesting result in Figure 3 is the fact that, contrary to what happens at 1130 K, at 1650 K the value of CV decreases as the oxygen concentration increases. This implies that at 1650 K the flame becomes less stable as the oxygen concentration increases, though it should be noted that the CV values are about an order of magnitude lower at 1650 K than at 1130 K.

Figure 4 compares the PM signals for a group of coal particles burning at a nominal gas temperature of 1650 K under a 24% O_2 concentration in N_2 and CO_2. The standoff distance is shorter in N_2 than in CO_2; however, the difference is much smaller than that observed at 1130 K. The lower value of CV for the N_2 environment suggests as well that the particles burn in a more stable flame when in N_2.

The results in Figure 2 to Figure 4 can be explained by results from previous studies [5-8] for single-particle ignition. Higher O_2 concentration decreases the ignition delay time, and, therefore, the standoff distance decreases. This effect is more evident at 1130 K where ignition takes longer, than at 1650 K, where ignition occurs faster. The higher molar heat capacity of CO_2 causes a lower flame temperature for the same oxygen concentration. This is evident from the lower values of PM intensity whenever CO_2 and N_2 atmospheres are compared. The lower flame temperature and, probably, the lower diffusivity of oxygen in CO_2 contribute to longer standoff distances in CO_2. The effect of oxygen concentration on the standoff distance is less pronounced in CO_2 than in N_2, something that needs further analysis to be explained, but that may be related to differences in diffusivity.

4.3 Considerations about G number

For the experiments described above, the G number, as calculated based on E 1, is ~ 3900. This calculation considers a number density of the order of 4.3×10^7 particles m^{-3} and a characteristic cloud length (ℓ) of 2.54×10^{-2} m, defined as half the length of the squared cross sectional area of the chimney where combustion occurs based on Figure 1. This is clearly an oversimplification, as the value of ℓ in the vertical direction may be higher.

This value of G, as computed from E 1 is constant for all oxygen concentrations, as it will only depend on geometrical factors. Actually, ℓ may change with oxygen concentration. However, this change would be minor and difficult to determine solely based on true color images as those in Figure 1. In fact, Figure 1 suggests that ℓ would decrease as oxygen concentration increases, this would render a lower G, which would indicate a lower propensity for group combustion. The fact that G is basically insensitive to oxygen or background composition suggests the need for further theoretical development to understand interactive combustion during oxyfuel combustion.

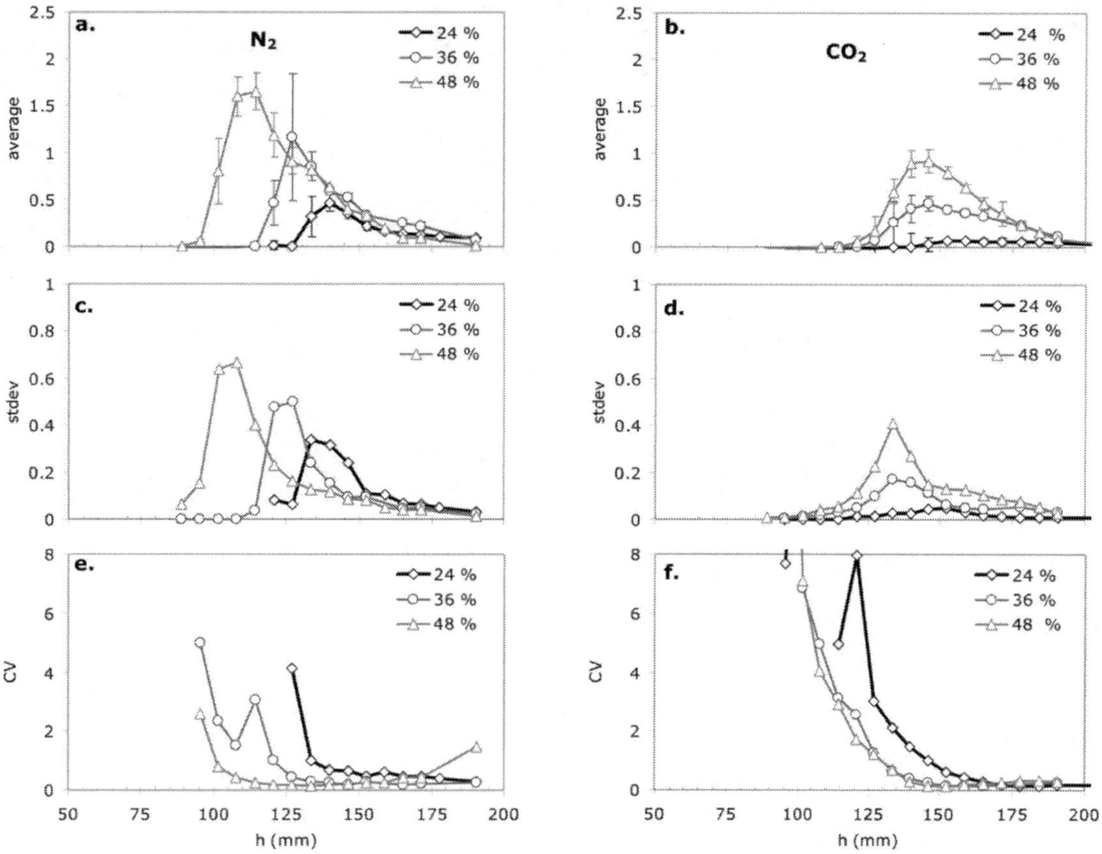

Figure 2. Variation with distance from the burner (h) of the PM signal (a., b.): average, (c., d.) standard deviation and (e., f.) coefficient of variation for different oxygen concentrations for a nominal gas temperature of 1130 K for coal burning in N_2 (a., c., and e.) and CO_2 (b., d and f.) atmospheres . Vertical bars in a. and d. represent the standard deviation from the three tests. Standard deviation in b. and e refers to the standard deviation of the PM signal as it changes in time.

Drop combustion theory [9, 10] suggests that as oxygen concentration increases, particles tend to burn in indifividual flames as a diffusion flame stabilizes given the high amount of oxygen surrounding the particle. Although this concept is valid for drop combustion, its extension to oxy-fuel coal combustion may be limited by the fact that, contrary to drop evaporation, coal devolatilization is an extremely fast process, particularly at high temperatures. Previous studies [6] have shown that as oxygen concentration increases, a volatile flames forms close to the particle surface, for single particle combustion.

For low oxygen concentration, volatiles escape from the particle and will ignite downstream from the particle, at a distance to far from injection at a point where particle density would be too short to grant group combustion. As oxygen concentration increases, volatile ignition occurs sooner in the chimney, when particle density is high enough and group combustion can be sustained.

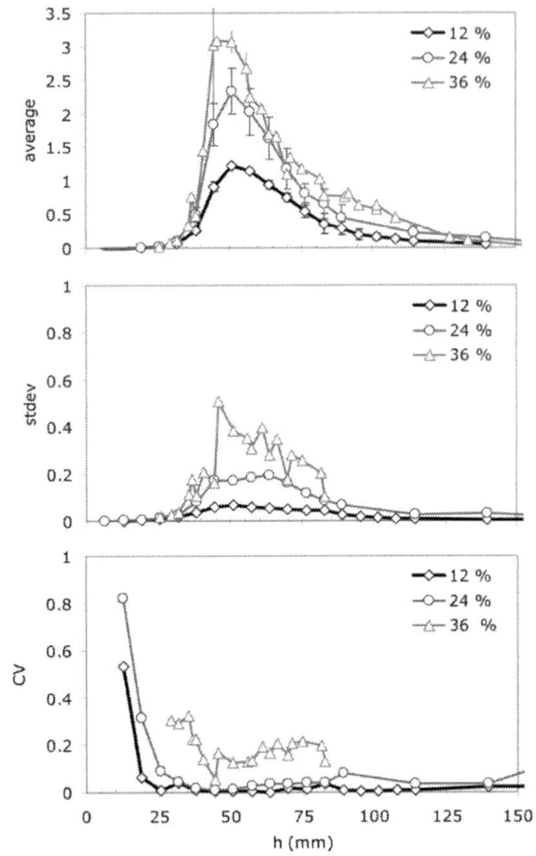

Figure 3. Variation with distance from the burner (h) of the PM signal average, standard deviation, and coefficient of variation for different oxygen concentrations for a nominal gas temperature of 1650 K for coal burning in N_2 atmospheres.

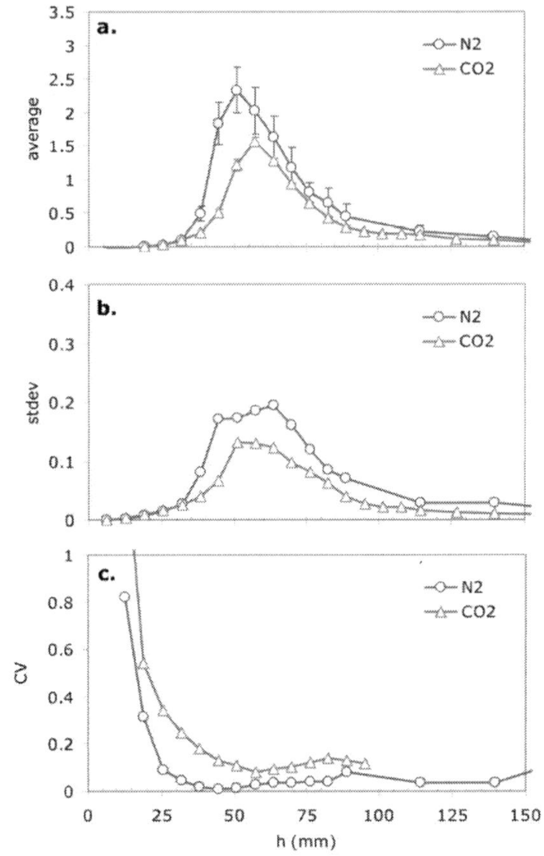

Figure 4. Variation with distance from the burner (h) of the PM signal (a.): average, (b.) standard deviation, and (c.) coefficient of variation for a nominal gas temperature of 1650 K for coal burning under N_2 and CO_2 atmospheres at 24% oxygen concentration. Vertical bars in a. represent the standard deviation from the three tests. Standard deviation in b. refers to the standard deviation of the PM signal as it changes in time.

A thought experiment would suggest that at even higher oxygen concentration (higher than 48% O_2) particles would burn as a single cloud as a volatile flame would form surrounding the cloud. For a higher oxygen concentration to produce single particle combustion, the devolatilization rate should be lower than the laminar flame combustion of the volatiles surrounding the particle.

A simple analysis shows that the laminar flame velocity (S_L) can be approximated as [15]:

$$S_L = \left[-2\alpha(v+1)\frac{\overline{\dot{m}'''_F}}{\rho_{un}} \right]^{\frac{1}{2}} \qquad \text{E 3}$$

where α is the gas thermal diffusivity, v a stoichiometric coefficient, $\overline{\dot{m}'''_F}$ a term that reflects an average combustion rate and ρ_{un} the density of the unburned gases. Combustion with CO_2 recirculation causes a lower laminar flame velocity as the thermal diffusivity is lower than in nitrogen. This analysis suggests that for a constant oxygen concentration, oxyfuel combustion with CO_2 would favor single particle combustion and make the occurrence of group combustion more difficult.

4. Summary

Experiments carried out to determine the effect of gas conditions on the ignition of a group of coal particles in oxyfuel combustion with CO_2 recirculation show that most of the trends can be explained from previous studies using single-particle ignition. In general, the standoff distance from the burner to a volatile flame formed by the group of particles decreases as the oxygen concentration increases and is higher for CO_2 environments. These effects are more noticeable at 1130 K than at 1650 K, where ignition occurs very rapidly.

The group particle experiments give new insight into flame stability. The results show that at a gas temperature of 1130 K, higher oxygen concentration produce more stable flames. An opposite trend was observed when the gas temperature was 1650 K, a surprising result that requires verification with further experiments. At both temperatures, flame stability decreases in CO_2 environments.

At 1130 K and 24% O_2 the particles burn under partial group combustion, however, as the oxygen concentration increases, the system moves to conditions close to total group combustion. The theoretical analysis based on group number theory (G) does not predict clearly the variation of group combustion mode with oxygen concentration or background gas. New theory developments should take into account a balance between the fast coal devolatilization process and laminar flame velocity, which decreases in the presence of CO_2.

Acknowledgment

This research was sponsored by the U.S. Department of Energy through the National Energy Technology Laboratory's Power Systems Advanced Research Program, man-aged by Dr. Robert Romanosky. Susan Maley serves as the NETL technical monitor for this project. Sandia is a multi-program laboratory operated by Sandia Corporation, a Lockheed Martin Company, for the United States Department of Energy's National Nuclear Security Administration under Contract DE-AC04-94AL85000.

References

1. Buhre, B.J.P., Elliott, L.K., Sheng, C.D., Gupta, R.P., and Wall, T.F., *Oxy-Fuel Combustion Technology for Coal-Fired Power Generation.* Prog. Energy Combust. Sci., 2005. **31**(4): p. 283-307.

2. Tan, R., Corragio, G., and Santos, S., *Oxy-Coal Combustion with Flue Gas Recycle for the Power Generation Industry. A Literature Review.* 2005, International Flame Research Foundation: Velsen Noord, Netherlands. p. 1-52.

3. Wall, T.F., *Combustion Processes for Carbon Capture.* Proc. Comb. Inst., 2007. **31**(1): p. 31-47.

4. Wall, T., Liu, Y., Spero, C., Elliott, L., Khare, S., Rathnam, R., Zeenathal, F., Moghtaderi, B., Buhre, B., Sheng, C., Gupta, R., Yamada, T., Makino, K., and Yu, J., *An Overview on Oxyfuel Coal Combustion--State of the Art Research and Technology Development.* Chem. Eng. Res. Des. **In Press, Corrected Proof.**

5. Molina, A. and Shaddix, C.R., *Ignition and Devolatilization of Pulverized Bituminous Coal Particles During Oxygen/Carbon Dioxide Coal Combustion.* Proc. Comb. Inst., 2007. **31**(2): p. 1905-1912.

6. Shaddix, C.R. and Molina, A., *Particle Imaging of Ignition and Devolatilization of Pulverized Coal During Oxy-Fuel Combustion.* Proc. Comb. Inst., 2009. **32**(2): p. 2091-2098.

7. Molina, A. and Shaddix, C.R. *Effect of O_2/Co_2-Firing on Coal Particle Ignition.* in *22nd International Pittsburgh Coal Conference.* 2005. Pittsburgh, PA, USA.

8. Molina, A., Shaddix, C.R., *Efecto Del Co2 En La Velocidad De Combustión De Semicoques De Carbón En Aplicaciones De Oxi-Combustión.* Energética, 2007. **38**: p. 101-106.

9. Law, C.K., *Combustion Physics.* 2006, New York: Cambridge University Press.

10. Annamalai, K. and Puri K., I., *Combustion Science and Engineering.* Crc Series in Computational Mechanics and Applied Analysis, ed. CRC. 2006, Boca Raton, U.S.A.: CRC Press. 1121.

11. Molina, A., Hecht, E.S., and Shaddix, C.R., *Ignition of a Group of Coal Particles in Oxyfuel Combustion with Co2 Recirculation,* in *34th International Technical Conference on Coal Utilization & Fuel Systems.* 2009: Clearwater, Florida, U.S.A.

12. Graham, K., *Submicron Ash Formation and Interaction with Sulfur Oxides During Pulverized Coal Combustion,* 1991, Ph. D. Dissertation. Massachusetts Institute of Technology, Chemcial Engineering: Cambridge.

13. Essenhigh, R.H., Misra, M.K., and Shaw, D.W., *Ignition of Coal Particles: A Review.* Combust. Flame, 1989. **77**(1): p. 3-30.

14. Murphy, J.J. and Shaddix, C.R., *Combustion Kinetics of Coal Chars in Oxygen-Enriched Environments.* Combust. Flame, 2006. **144**(4): p. 710-729.

15. Turns, S.R., *An Introduction to Combustion Concepts and Applications.* 1st ed. 2002: McGraw-Hill, Inc. 565.

Application of Photo-Imaging Techniques to Quantify Flame STABILITY Characteristics DURING Oxy-Coal Combustion

Jingwei Zhang, Department of Chemical Engineering & Institute for Clean and Secure Energy, University of Utah, Salt Lake City, UT

Kerry Kelly, Department of Chemical Engineering & Institute for Clean and Secure Energy, University of Utah, Salt Lake City, UT

Eric G. Eddings, Department of Chemical Engineering & Institute for Clean and Secure Energy, University of Utah, Salt Lake City, UT

Jost O.L. Wendt, Department of Chemical Engineering & Institute for Clean and Secure Energy, University of Utah, Salt Lake City, UT

The purpose of this research is to better understand and to scale the effect of partial pressures of O2 and CO2, two additional degrees of freedom for oxy-coal combustion, on coal jet ignition and flame stability in the near burner zone. A novel methodology is developed to quantify the flame stability and flame length by introducing an image processing technique. Ultimately the experiment is designed to provide data for simulation validation studies that can be used to predict how air fired combustors may be retrofitted to oxy-coal.

Specifically the objective is to explore effects of the partial pressure of O2 and CO2 on coal jet flame stability, which is here quantified by stand-off distance, the distance between burner tip and the base of the detached flame. The experiments are carried on in a 100 kW pilot-scale laboratory furnace outfitted with a coaxial burner (primary jet in the center pipe and secondary stream in the annular sleeve, no swirl, Type 0 turbulent flames) and arrays of electrically-heated panels in the burner zone in order to control variations in near-burner heat loss. Currently fresh, one-through CO2 (instead of flue gas recycle) and O2 streams are introduced to the burner. The furnace also has visual access to the burner zone through quartz windows, which allows for optical measurements. A special CMOS sensor based camera, which is more sensitive to the near infrared wavelength (responsivity: 1.4 V/lux-sec (550nm)), is applied to capture Type 0 axial turbulent diffusion flame shapes for the use of statistical studies of stand-off distance under the influence of changes in various operational parameters, including systematic variations of partial oxygen pressure in both transport and secondary oxidant stream. Statistical analysis is used to scale the effect of oxygen partial pressure in both transport and secondary stream on coal jet ignition and flame stability. Ultimately the statistics from the photo images help to generate probability density functions (PDF's), which can be used to evaluate the experimental accuracy, reproducibility and allow more precise model validation.

Keywords: oxy-coal combustion, flame stability, oxygen partial pressure (PO2), stand-off distance, near burner phenomena, coal jet, type 0 axial turbulent flames, probability density function (PDF), image processing, statistical analysis

See more of this Session: Oxycombustion of Coal - Needs, Opportunities, and Challenges - II
See more of this Group/Topical: Energy and Transport Processes

Utilization of Reburn Reactions for NO_x Control in Oxy-Fuel Combustion

Fredrik Normann[*], Klas Andersson, Bo Leckner, Filip Johnsson
Chalmers University of Technology, Göteborg, Sweden

Introduction

Oxy-fuel combustion implies several new options for NO_x control, but for optimization purposes most of these options require further exploration. The present work focuses on reburning of NO_x. Typically, experiments show that recirculation of NO_x into the flame in oxy-combustion lowers the emission with around 70% per unit of fuel supplied, compared to the emission from air-firing under otherwise similar conditions, see review by Normann et al. [1].

Reburning or fuel staging was proposed by Wendt et al. [2] in 1973 and is thoroughly investigated for air-fired applications (see for example Ref. [3-5]). The objective of reburning is to create hydrocarbon radicals (CH_i), which can break up the NO formed and, then, form N_2 instead. A reburning process consists of a reburn zone and a burnout zone. NO is supplied to the reburn zone. In the air-fired application of reburning, the primary combustion is the source of NO and a secondary fuel stream is supplied to create the reburn zone, while in the oxy-fuel application, discussed in the present work, NO is supplied by the recycle and the primary combustion acts as a reburn zone. The reburn zone should be oxygen lean to promote formation of radicals. In the burnout-zone O_2 is added in sufficient amount to complete the combustion.

The influence of high CO_2 concentration on combustion and gas-phase nitrogen chemistry has recently been investigated in a series of studies by Glarborg and Bentzen [6], and Mendiara and Glarborg [7]. Their work is based on a combination of experiments in a lab-scale flow reactor and detailed chemical kinetic modeling. When substituting N_2 with CO_2 in the oxidizer, they detected an increased formation and similar reduction of NO under oxygen-lean conditions and a decreased formation and higher reduction of NO during stoichiometric and oxygen-rich conditions. A similar investigation was performed by Mackrory [8] who used a combination of experiments with a pulverized coal-fired reactor and a model based on established sub-models. Mackrory concludes that the reduction of NO_x under oxy-fuel conditions is less sensitive to an increased stoichiometric ratio than under air-fired conditions. This was mainly explained by the onset of thermal NO_x formation in air combustion when more oxygen becomes available in the primary combustion zone, which does not occur to the same extent under oxy-fuel conditions. Furthermore, Mackrory concluded that the reduction during oxy-fuel conditions is favored by higher stoichiometric ratio, which causes more NO_x to pass through the primary zone due to the increased recycle of flue gas.

Although the reduction of NO_x by reburning reactions has been proven in several studies, the utilization and optimization of this technique is poorly investigated within the framework of oxy-fuel combustion. The present paper is the first step of a comprehensive work on the utilization of the reburning chemistry in oxy-fuel combustion. The results in this paper present the critical reaction paths in reburning chemistry and the main focus is on key differences between air and oxy-combustion. Differences in radical pool composition between O_2/CO_2 and O_2/N_2 atmospheres are also high-lighted due to its direct coupling to the

[*] Corresponding author. E-mail address: fredrik.normann@chalmers.se

reburning chemistry. An established gas-phase model is applied in the reaction path analysis. Future work will be extended towards experimental studies of the reburning capabilities in oxy-combustion and optimal settings for NO_x control.

Method

The reburning chemistry is investigated with a detailed chemical kinetic mechanism together with the associated thermodynamic library. The mechanism is compiled and validated for oxy-fuel conditions by Mendiara and Glarborg [7]. It includes reactions of importance for the combustion of light hydrocarbons and for the nitrogen chemistry active in reburning of NO_x. The mechanism is implemented in the Chemkin-Pro software [9].

To achieve a similar basis for a comparison between air and oxy-fuel combustion, the replacement of N_2 with CO_2 is the only parameter change, and the inlet oxygen concentration in the oxidizer is set to $21\%_{vol}$ in both cases. In practice, the inlet oxygen concentration in oxy-fuel combustion will probably be higher, and due to the higher heat capacity of CO_2 a similar temperature profile to air combustion is achieved with around $30\%_{vol}$ of O_2. The present simulations are isothermal with the same temperature in both cases to eliminate the influence of temperature. Furthermore, the reburning is simulated as performed during oxy-fuel combustion, i.e. an initial concentration of NO is supplied with the oxidizer also in the air-fired case.

Combustion chemistry

The nitrogen chemistry is a small part of the combustion process compared to the chemistry involved in the CH-oxidation, and small changes to the combustion process can have a large impact on the nitrogen chemistry. Hydrocarbon and nitrogen chemistry are mainly connected through the important radical pool of atomic oxygen (O), hydrogen (H) and the hydroxyl radical (OH). The single most important reaction for a combustion process is the chain branching reaction,

$$H + O_2 \rightarrow OH + O \qquad (1)$$

which consumes one (H) and produces two radicals (O and OH) [10]. Consequently, any combustion reaction increasing the production of H-atoms promotes combustion and any reaction consuming H-atoms inhibits combustion. All hydrocarbon combustion involves H_2 and CO oxidation, Reaction 2, which produces most of the CO_2 in combustion, is also central for the combustion processes [10].

$$CO + OH \leftrightarrow CO_2 + H \qquad (2)$$

The oxidation of CO by Reaction 2 produces H-atoms and is, therefore, of direct importance for Reaction 1.

Figure 1 illustrates the stages of a reburning process. The reburn zone (oxygen lean) is divided in two stages, Stage I and Stage II, and the burnout zone (oxygen rich) constitutes Stage III. In Stage I the fuel is decomposed and oxidized under oxygen-lean conditions and a high level of hydrocarbon radicals is produced. The reaction of hydrocarbon radicals with both H and OH is generally faster than both Reactions 1 and 2 [10]. This leads to a build-up of CO, which starts to oxidize after the oxidation of hydrocarbons is completed. It has been concluded in previous studies [11] that in some parts of an oxy-fuel flame, the elevated concentration of CO_2 may drive Reaction 2 in the reverse direction. The influence of CO_2 on the chain carrying radicals is shown in Figure 2. The reverse Reaction 2 can even compete with the reactions between hydrocarbon and H-radicals and may have an impact on the flame velocity. This is seen in Figure 1 by the slower oxidation of the fuel during oxy-fuel combustion than during air combustion. In Stage II when CO starts to oxidize to CO_2 (the forward Reaction 2) the higher CO concentration in oxy-fuel combustion reduces the relative rate of radical consumption and, eventually, the concentration of radicals in oxy-fuel combustion could exceed that of air combustion, which is seen in the centre diagram of Figure 2. The relative order of the quantity of radicals is changed from H>OH>O in air combustion to OH>O>H in oxy-fuel combustion. In Stage III more oxygen is added to complete the oxidation of the fuel and a similar pattern as in the reburn zone is detected: initially, the concentration of radicals is higher during air-firing but the decrease is slower during oxy-fuel combustion due to the buffering effect by CO_2 in Reaction 2.

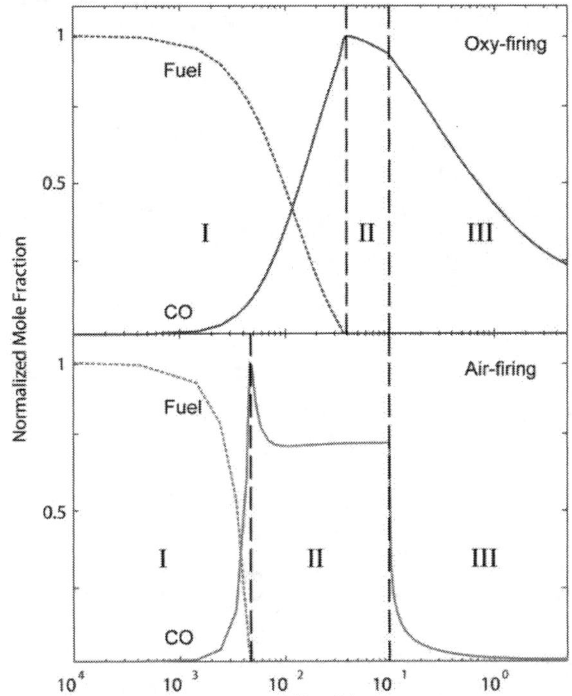

Figure 1. Comparison of the three stages of the reburning process, expressing the normalized mole fractions of fuel and carbon monoxide as functions of resedence time. Dashed verticle lines indicate the break between stages.

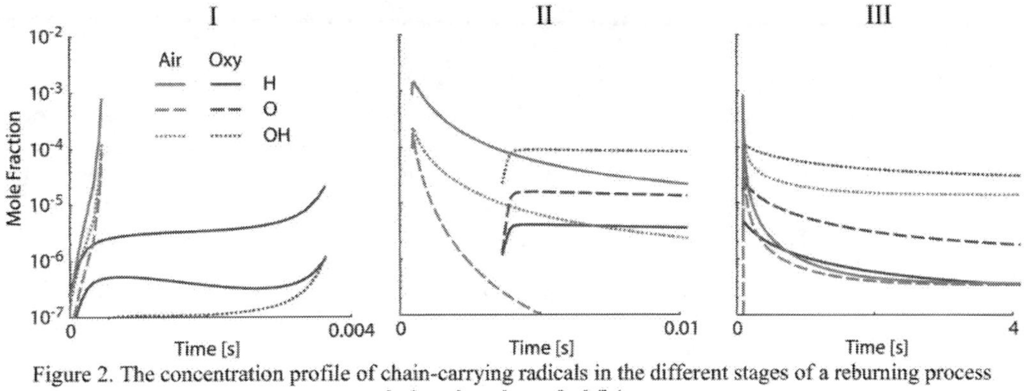

Figure 2. The concentration profile of chain-carrying radicals in the different stages of a reburning process during air and oxy-fuel firing.

Nitrogen chemistry

Figure 3 shows the concentration profiles of NO, HCN and CH_3 in the reburn zone (Stages I and II), which give an indication on how the reduction of NO_x proceeds. The maximal reduction rate is lower during oxy-fuel combustion, due to the lower peak concentrations of important reactants. This is, however, compensated by a longer residence time in the reducing zone (Stage I), due to the slower oxidation of the fuel, which open ups for alternative reaction routes. It can also be seen in Figure 3 that even though there are large differences in combustion conditions and in reduction behavior between air and oxy-fuel combustion in the investigated cases, the overall reduction of NO_x is similar.

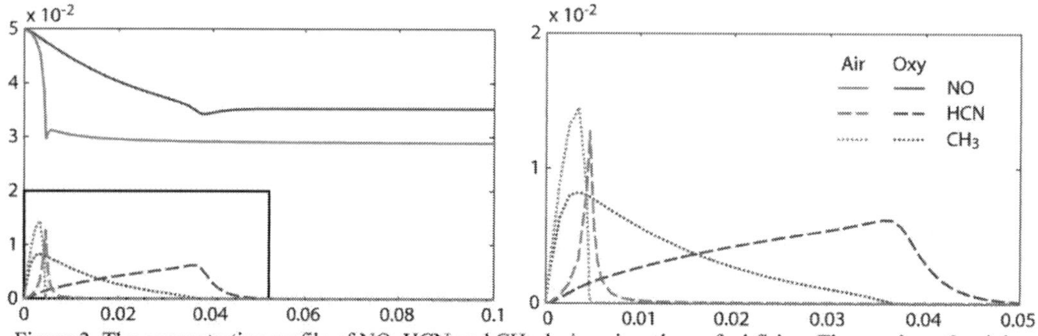

Figure 3. The concentration profile of NO, HCN and CH_3 during air and oxy-fuel firing. The graph on the right hand side is an enlargement of the black square in the figure on the left hand side.

The reaction routes of interest for NO_x reduction in the defined stages of reburning is discussed by means of Figures 4 to 6. Figure 4 shows the governing reaction routes for nitrogen compounds in Stages I and II, while Figure 5 and 6 show the relative importance of each step of these routes during air and oxy-fuel conditions. In Stage I the reduction of NO to HCN dominates nitrogen chemistry. In oxy-fuel combustion less HCN is formed directly from NO, mainly due to lower concentrations of the small hydrocarbon radicals (CH and CH_2). This is, however, compensated by a larger formation of HCN via H_2CN due to higher CH_3 concentration. In oxy-fuel combustion more of the HCN formed reacts to CH_3CN and further to NCO. This is explained by the higher concentration of CH_3 and lower concentration of chain-carrying radicals. In air-firing NCO is also formed directly from the reaction between HCN and O radicals. In Stage II, HCN and CN react to mainly form NCO. In air-firing almost all NCO reacts with an H-radical to form NH and further to N (this route is active already in Stage I). The N atom reacts to form either NO or N_2. In oxy-fuel combustion the concentration of H-radicals are lower, and NCO reacts through alternative paths to form N_2O, NO and N_2. The split between NO and N_2 are dependent on similar reactants in both cases (NO gives N_2, and, O_2 or OH gives NO). The reaction through N_2O to N_2 is more important during oxy-fuel combustion. In summary, the lowered concentration of H-radicals affects the nitrogen chemistry both directly and by changes in the rate of the oxidation of the fuel.

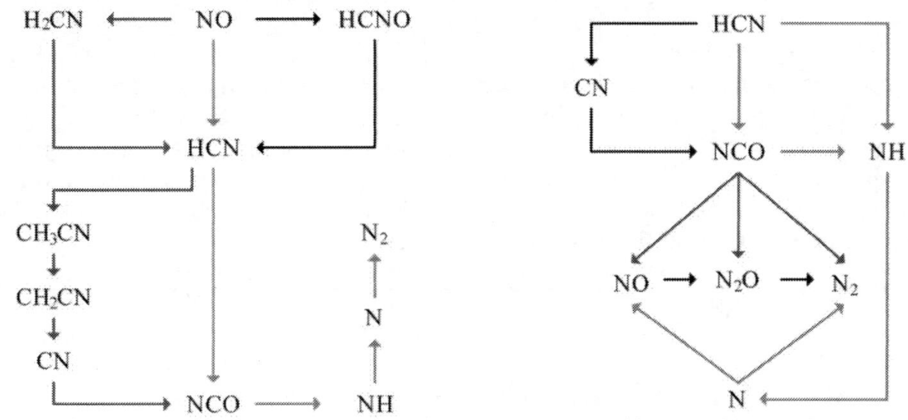

Figure 4. The most important reaction routes in the reburning zone. a) Stage I. b) Stage II. Blue arrows mark routes more important for oxy-fuel combustion; red arrows are the same for air-firing and black arrows mark routes with similar importance for both cases.

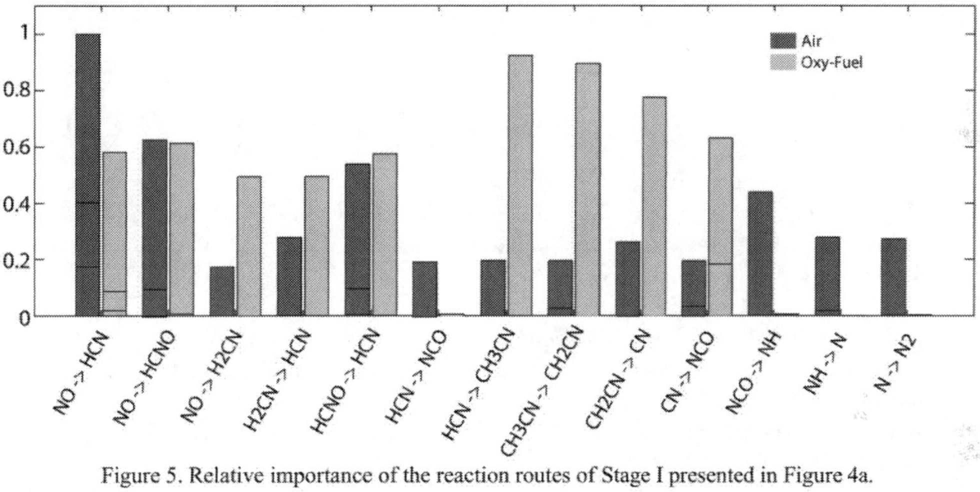

Figure 5. Relative importance of the reaction routes of Stage I presented in Figure 4a.

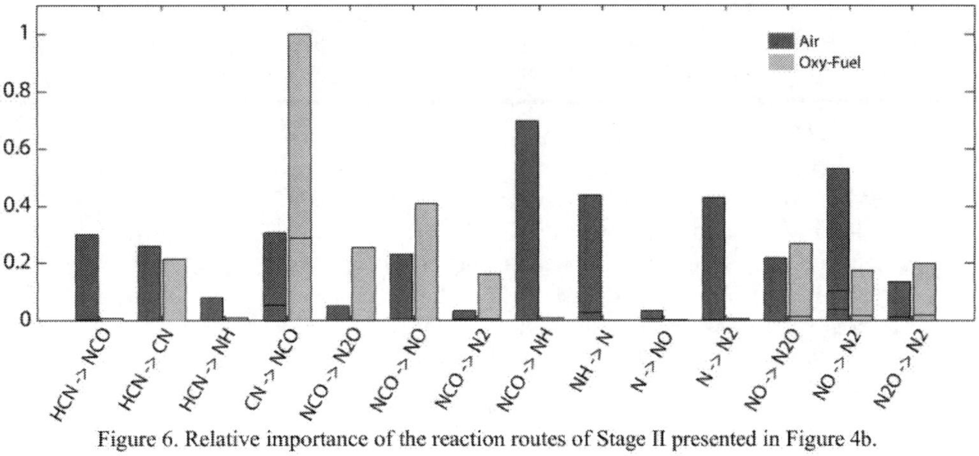

Figure 6. Relative importance of the reaction routes of Stage II presented in Figure 4b.

Conclusions

This work is the first step in an investigation of the reburning process for oxy-fuel conditions. Differences in combustion chemistry and performance of the reburning process during air and oxy-fuel combustion are investigated by means of a detailed reaction mechanism from which it is concluded that there are substantial differences between reaction routes during air and oxy-fuel combustion. The most important factor is the reduced concentration of H-radicals at the elevated concentration of CO_2. The lowered concentration of H-radicals affects the nitrogen chemistry both directly and by changes in the rate of the oxidation of the fuel. Despite the differences in reaction routes the overall reduction of NO_x is similar for the case studied. It is, however, important to further investigate the effect of combustion parameters on the routes of reburning during oxy-fuel conditions to identify eventual differences from the optimizing procedures during air-fired conditions.

Acknowledgements

This work is financed by Vattenfall AB.

References

1. Normann, F., Andersson, K., Leckner, B. and F. Johnsson (2009), "Emission control of nitrogen oxides in the oxy-fuel process", *Progress in Energy and Combustion Science,* 35, pp. 385-397.
2. Wendt, J.O.L, Sternling, C.V. and M.A. Matovich (1973), "Reduction of sulfur trioxide and nitrogen oxides by secondary fuel injection", *Combustion Science and Technology*, 139, pp. 329-363.
3. Kilpinen, P., Glarborg, P. and M. Hupa (1992), "Reburning chemistry: a kinetic modeling study", *Industrial and Engineering Chemistry Research,* 31, pp. 1477-1490.
4. Kristensen, P.G., Glarborg, P. and K. Dam-Johansen (1996), "Nitrogen chemistry during burnout in fuel-staged combustion", *Combustion and Flame*, 107, pp. 211-222.
5. Dagaut, P., Glarborg, P. and M.U. Alzueta (2008), "The oxidation of hydrogen cyanide and related chemistry", *Progress in Energy and Combustion Science,* 34, pp. 1-46.
6. Glarborg, P. and L.L.B. Bentzen (2008), "Chemical effects of a high CO_2 concentration in oxy-fuel combustion of methane", *Energy & Fuels,* 22, pp. 291-296.
7. Mendiara, T. and P. Glarborg (2009) "Reburn Chemistry in oxy-fuel combustion of methane", *Energy & Fuels*, 23, pp. 3565-3572.
8. A.J. Mackrory (2008), "A mechanistic investigation of nitrogen evolution in pulverized coal oxy-fuel combustion", *Dissertation*, Brigham Young University.
9. CHEMKIN-PRO (2008), Reaction design, San Diego.
10. Westbrook, C.K. and F.L. Dryer (1984), "Chemical kinetic modeling of hydrocarbon combustion", *Progress in Energy and Combustion Science*, 10, pp. 1-57.
11. Liu, F., Guo, H., Smallwood, G.J. and Ö.L. Gülder (2001), "The chemical effects of carbon dioxide as an additive in an ethylene diffusion flame: implications for soot and NO_x formation", *Combustion and Flame,* 125, pp. 778-787.

Effects of Gas Residence Time ON NO Reduction IN Oxy-Coal Combustion

Wei-Yin Chen, Chemical Engineering, University of Mississippi, University, MS

Shaolong Wan, Swan@Olemiss.Edu, University of Mississippi, University, MS

We evaluate the potential benefits of long residence time of gas in a recycle system on NO reduction in oxy-coal combustion. In both single-pass and recycle tests, residence time within a range of 0.26 to 1.6 seconds shows only minimal benefits on NO reduction. Thus, the reverse Zeldovich mechanism appears not a governing factor up to 1700 °C, the peak flame temperature in practical boilers. The NO concentrations under equilibrium are much lower than those experimentally observed. Moreover, addition of 500 ppm NO in the feed for simulating the effects of the recycled NO does not change the estimated equilibrium concentrations in burning coal in O2-CO2 environment. Nevertheless, it is not observed in practical operations. Thus, NO reduction in oxy-coal combustion is highly kinetically controlled, not thermodynamically controlled.

See more of this Session: Oxycombustion of Coal - Needs, Opportunities, and Challenges - II
See more of this Group/Topical: Energy and Transport Processes

Cofiring of Coal and Biomass Under Oxycombustion Conditions: NOx Formation and PM Characterization

Benjamin M. Kumfer, Energy, Environmental and Chemical Engineering, Washington University in St. Louis, St. Louis, MO

Scott A. Skeen, Energy, Environmental and Chemical Engineering, Washington University in St. Louis, St. Louis, MO

Richard L. Axelbaum, Energy, Environmental and Chemical Engineering, Washington University in St. Louis, St. Louis, MO

The cofiring of coal with biomass is a promising method under consideration for power generation as a means to reduce net CO_2 emissions and to meet renewable energy standards. Furthermore, if carbon capture and sequestration (CCS) technology is applied to a cofired system, a net removal of atmospheric CO_2 may be realized, since the CO_2 removed by the biomass is not released back into the atmosphere. In this work, the particulate matter and NOx produced from the cofiring of coal and biomass is investigated under both conventional air-fired and oxy-combustion conditions. Oxy-combustion is advantageous since a concentrated stream of CO_2 is inherently produced for CCS. Experiments are performed in a 35 kW, horizontally-fired combustion facility, utilizing pulverized subbituminous coal and various biomass materials. When air is replaced with a mixture of 30% O_2 and 70% CO_2, and no overfire oxygen is utilized, oxy-fuel combustion results in a 20% reduction in NO production. The oxygen concentrations of the primary and secondary streams are systematically varied, while holding constant the total O_2 and CO_2 flow rates, and a minimum emission rate of NO is measured when the primary and secondary O_2 concentrations are both equal to 30% O_2. A further reduction in NO is measured when cofiring with sawdust, primarily due to the reduced nitrogen content of the fuel. The effects of cofiring and oxy-combustion on the fly ash size distribution and the size-fractionated composition are discussed. In addition, the interaction between coal and biomass fly ash in the high-temperature zone and the potential for increased deposition and fouling/slagging is explored.

See more of this Session: Oxycombustion of Coal - Needs, Opportunities, and Challenges - II
See more of this Group/Topical: Energy and Transport Processes

EFFECT OF CARBON DIOXIDE ON SOOT DISTRIBUTION FROM OXY-COAL COMBUSTION

Gyu-Bo Kim, Pusan Clean Coal Center, Pusan National University, Busan, Korea
Do-Wook Choi, Pusan National University, Busan, Korea
Jae-Dong Kim, Pusan National University, Busan, Korea
Ju-Hun Song, Pusan Clean Coal Center, Pusan National University, Busan, Korea
Young-June Chang, RIMT, Pusan National University, Busan, Korea
Chung-Hwan Jeon, Pusan Clean Coal Center, RIMT, Pusan National University, Busan, Korea

Introduction

As an interest in carbon dioxide(CO_2) emission regulation due to climate change and limited energy resource increase, carbon capture sequestration (CCS) technology is being investigated widely. Especially, an oxy-fuel combustion combined with carbon capture sequestration (CCS) technology is known to be one of emerging clean coal technologies for the CO_2 emission reduction. Soot is useful due to an enhanced radiative heat transfer in a coal-fired boiler furnace, while its formation from tar and char should be suppressed to avoid the loss of carbon content in the coal. In this study, soot distribution measurement was carried out with a laser induced incandescence technique in an oxy-fuel diffusion flame burner. A methane-fed diffusion flame was used for ignition of a coal which is independently introduced in the central tube of the burner. The coal is one of coals which are imported to South Korea being used at its coal-fired power plant. An inert gas such as CO_2 was additionally supplied to the oxidant to simulate the N_2-free oxidation condition and to further investigate the effects of CO_2 on coal-derived soot formation and mechanisms. The flame length made solely from coal burning was also observed to provide the oxidation rate of coal as a function of oxygen fractions in a flue gas.

Experimental apparatus and methodology

Laser induced incandescence (LII) is the most powerful method because it can measure soot particle size, soot volume fraction and mass faction etc. simultaneously. It is also preferred in many combustion systems because it is capable of in-situ measurement and high temporal, spatial resolution. Laser-induced incandescence technique is based on the fact that soot particles absorb the laser light and reach, during the laser pulse, well above vaporization temperature of soot.[1] The radiation of the heated particles is then time resolved detected during the cooling phase.

In this experiment, the burner was made for oxy-fuel combustion reactor of coal. The burner consists of three parts: a central tube to supply coal and carrier gas, a middle tube to supply methane and to ignite coal, and an outer tube to supply oxygen and carbon dioxide for forming surrounding gas. A schematic diagram of the burner is shown in Fig. 1. To improve flame stability, sintered metal form is used. Methane was used as the fuel supply for ignition of coal. CO_2 was used as the inert gas and carrier gas. The coal was used one of coals which are imported to South Korea being used at its coal-fired power plant. The optical measurement system for LII is as follows: pulsed Nd:YAG laser (λ_{2nd} = 532 nm, 9 ns FWHM, 10 Hz) as heating source of soot particulate matters, photo multiplier tube for multiply LII signal, 450nm narrow bandpath filter with FWHM=40nm for collecting LII signal, oscilloscope(500Mhz, 2Gs/s) for analyzing LII signal and data acquisition system. A schematic diagram of the system is shown in Fig. 2. Experimental conditions including fuel and oxidant flow rates were listed in the Table 1.

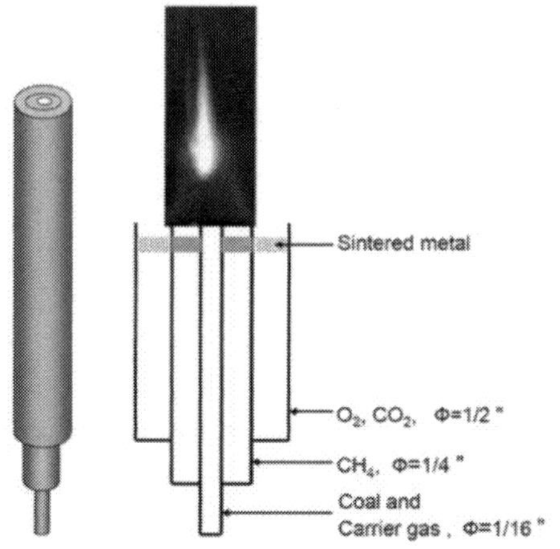

Fig. 1 Schematic diagram of the burner for oxy-fuel combustion

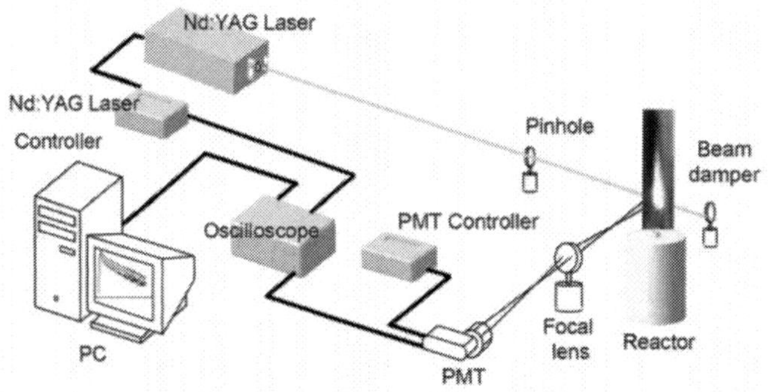

Fig. 2 The optical measurement system for laser induced incandescence

Table 1. Experimental conditions

Item	**Unit**	**Flow rate**
Methane, CH_4	cc/min	36
Oxygen, O_2	cc/min	144
Carrier gas, CO_2	cc/min	63
Inert gas, CO_2	cc/min	210, 350, 490
CO_2 fraction in oxidant stream of O_2-CO_2 mixture	%	65, 75, 80

Soot volume fraction is decided by maximum LII signal.[2] The signal is called prompt LII signal that equal to the LII signal at the maximum temperature of heated particle. Prompt LII signal is proportional to:

$$\text{Prompt LII} \propto N_p\, d_p^{\,x},\ x = 3 + 154\ \text{nm}/\lambda_{det},$$

where d_p is the particle diameter and λ_{det} is the detection wavelength.

In the VIS region, this exponent is about 3. Therefore, the prompt LII signal is proportional to the soot volume fraction.

Experimental results

Fig. 3 shows natural luminosity images of diffusion flame for surrounding gas with air and O_2-CO_2. The flame at far left side image is methane-air diffusion flame. Other images are methane-oxy diffusion flame with carrier gas in the central tube. Flame height of oxy-fuel flame is shorter than one of methane-air flame and the flame shape is symmetrically divided by the effect of carrier gas (CO_2) in the center. It is apparent that there is no indication of soot emitted from all of flame conditions, which would provide a clear background to measure soot amount produced from coal burning under the above oxy-fuel condition.

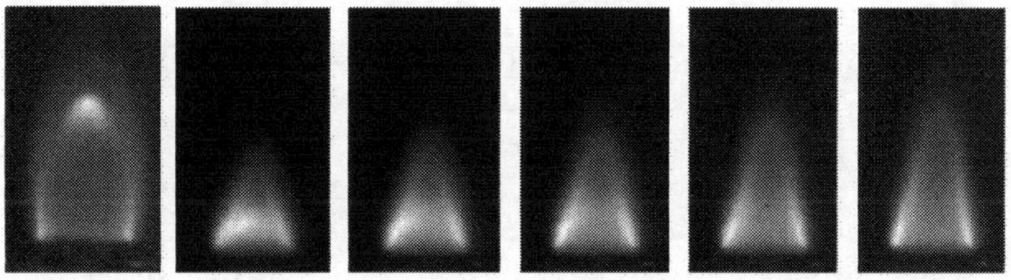

(a) (b) (c) (d) (e) (f)

Fig. 3 Methane diffusion flames in the conditions of air (a) and O_2-CO_2 ; (b) Y_{CO2}=20%, (c) 35%, (d) 45%, (e) 50%, (f) 55%

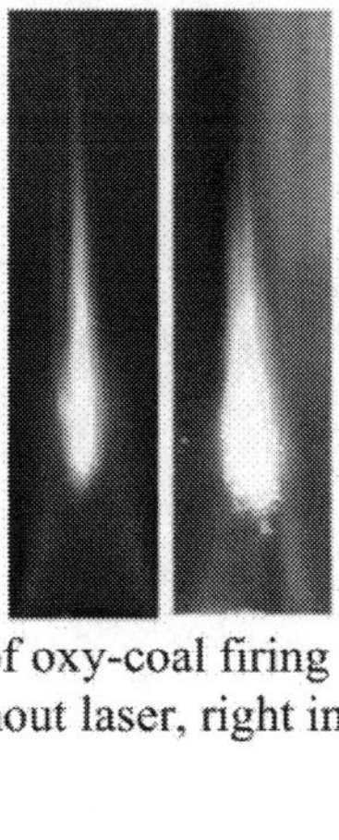

Fig. 4 Comparison of images of oxy-coal firing with 75% CO_2 in the oxidants, left image : without laser, right image : with laser

Fig. 4 shows images of oxy-coal firing and induced by the laser. In the flame image induced by the laser, it will be able to confirm that the scattering light of the laser and incandescence are being mixed. Through the results of the image, it is able to expect that LII signals will include serious noise.

Fig. 5 shows temporal profile of LII signal from methane-air-coal firing as a function of elevation from the flame tip. The signal intensities are decreased as the elevation of coal flame increases. This indicates that soot is formed at early stage in coal combustion which may occur from either the surpression of soot formation due to reduced temperature downstream or a completeness of volatilization process at the height of 10 mm. Fig. 6 shows the comparison with LII signal of methane-air-coal condition and oxy-coal firing. The signal intensity of oxy-coal firing is lower than one of air condition. In the case of oxy-fuel combustion, the soot oxidation is more activated by the effect of the more available oxygen than soot inception and growth is.

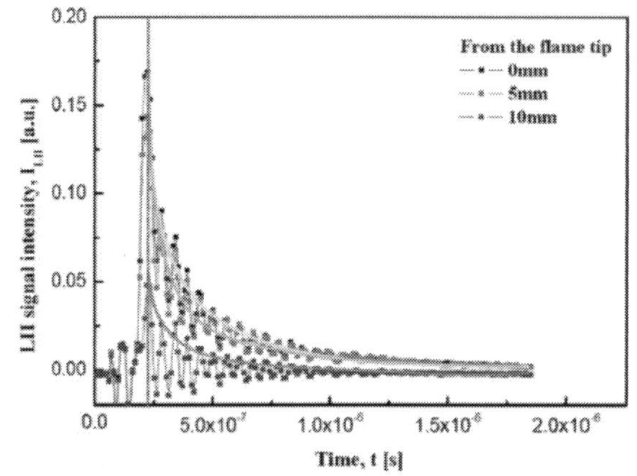

Fig. 5 Temporal profile of LII signal as a function of elevation from the flame tip

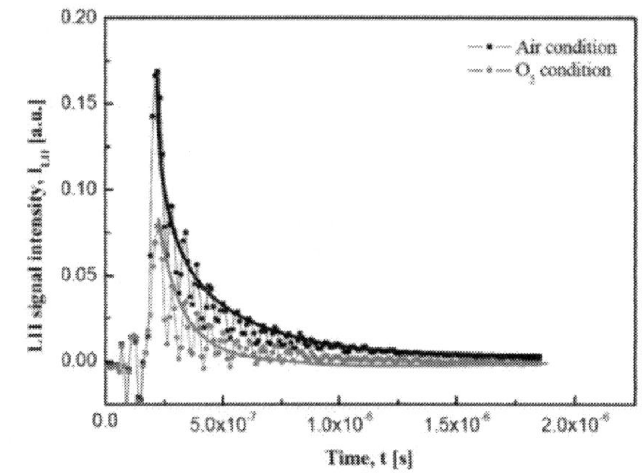

Fig. 6 The comparison of LII signal of methane-air-coal condition with that of oxy-coal firing (Y_{CO2}=75%)

Fig. 7 presents the characteristics of prompt LII signal as CO_2 fraction in oxidants varies over the range between 65 and 80 %. The result shows little dependence of soot formation measured in terms of volume fraction on CO_2 presence in these particular ranges of CO_2 fraction.

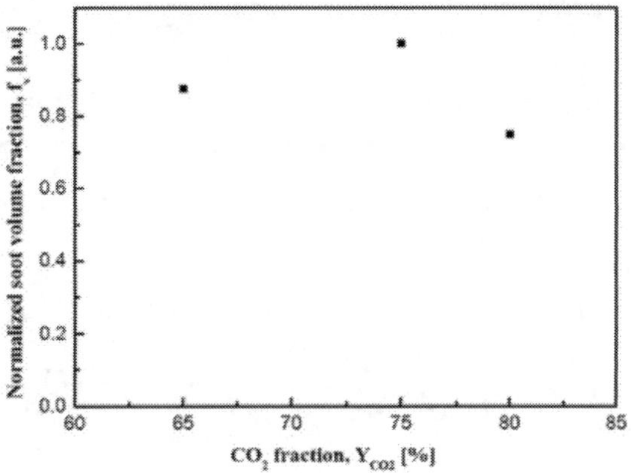

Fig. 7 The characteristics of prompt LII signal as a function of CO_2 fraction during the oxy-fuel combustion

Conclusion

From the results of temporal profile of LII signal obtained from an experiment under oxy-coal firing conditions, the followings are concluded:
1. Soot is observed to be formed at early stage of coal combustion process.
2. Soot volume fraction of oxy-coal firing is lower than one of methane-air-coal condition probably to combined effect of surpressed soot inception and increased oxidation.
3. The effects of CO_2 fraction were not revealed on the soot formation during oxy-coal firing process.

References

1. Bladh H. (2004), "Characteristics of Laser-Induced Incandescence from Soot in Studies of a Time-Dependent Heat and Mass Transfer," *Applied Physics B* 78 pp. 241-248.

2 Speaker, Zizak G. (2000), "Laser Induced Incandescence (LII) of soot," in *the ICS Training course on Diagnostics of Combustion Processes*, Zizak G. NILES, University of Cairo, Egypt

Influence of Biomass On Fine-Particle and Mercury Emissions in Oxy-Coal Combustion

S. Michael Daukoru, Energy, Environmental and Chemical Engineering, Washington University in St. Louis, St. Louis, MO

Achariya Suriyawong, Energy, Environmental and Chemical Engineering, Washington University in St. Louis, St. Louis, MO

Scott A. Skeen, Energy, Environmental and Chemical Engineering, Washington University in St. Louis, St. Louis, MO

Richard L. Axelbaum, Energy, Environmental and Chemical Engineering, Washington University in St. Louis, St. Louis, MO

Pratim Biswas, Energy, Environmental and Chemical Engineering, Washington University in St. Louis, St. Louis, MO

Oxy-coal combustion elicits continued study as a means for obtaining high-concentration carbon-dioxide (CO_2) flue gas that can be readily purified and sequestered. As with most other modalities for addressing the role of CO_2 in global climate change, oxy-coal combustion must contend with the prospect of wide deployment and site-specific access to and preferences for fuels. Biomass co-combustion with coal is one such preference that has the added benefit of net CO_2 reduction and has been widely studied and implemented for existing, conventional coal combustion modalities, in spite of the many technical challenges presented by the introduction of biomass. While the impact of oxy-coal combustion on other emissions such as fine particles and mercury has been examined (Suriyawong et al., 2006), operational demands may result in biomass co-firing in oxy-coal combustion environments.

Therefore, we will compare the effect of sawdust (biomass) on fine particle formation and mercury emissions in oxy-coal combustion to its effect in conventional coal-air combustion for a laboratory-scale self-sustaining combustor. These results will also be compared with those obtained from earlier controlled studies of oxy-coal combustion in drop-tube and self-sustaining combustion environments. It is expected that different temperature-time profiles would result in different vaporization rates, particularly for sawdust, for oxy-coal combustion compared with coal-air combustion. And, increased formation of alkali salts due to the addition of biomass to oxy-coal combustion may increase the potential for mercury oxidation. Practical implications for fine-particle and mercury capture will be suggested.

Suriyawong, A., Gamble M., Lee M.H., Axelbaum R. and Biswas P., "Submicrometer particle formation and mercury speciation under O_2 - CO_2 Coal Combustion", Energy and Fuels, 20 (6), 2357-2363, 2006.

See more of this Session: Oxy-Combustion of Coal - Needs, Opportunities, Challenges - III
See more of this Group/Topical: Energy and Transport Processes

Role of Recycle On Fine-Particle Formation, Mercury Speciation, and Electrostatic Capture During Oxy-Coal Combustion

S. Michael Daukoru, Energy, Environmental and Chemical Engineering, Washington University in St. Louis, St. Louis, MO

Pratim Biswas, Energy, Environmental and Chemical Engineering, Washington University in St. Louis, St. Louis, MO

Oxy-coal combustion is being touted as a potential means for increasing concentration of carbon dioxide in exhaust gases to help in potential remediation and sequestration. The impact of this modality of combustion on fine-particle and trace metal formation and mercury speciation has been examined in both drop-tube and self-sustaining coal combustors (Suriyawong et al., 2006; Suriyawong et al., 2008). Most of these studies have been conducted by combustion of the coal in a mixture of oxygen and carbon dioxide. However, in a full-scale operating scenario, the flue gases will be recycled with only the necessary amount of oxygen added at the inlet with the coal.

In this paper, we will compare the effect of flue gas recycle on fine particle formation and other emissions, to that of the controlled studies performed earlier. Due to the difference in the temperature-time histories and the composition of the flue gases, there are expected to be differences in the combustion characteristics and performance of particle capture devices such as electrostatic precipitators (ESPs). First, the sub-oxide formation pathways are expected to result in different vaporization rates due to the different time-temperature relationships. It is expected that the more volatile elements such as mercury will not show any differences, however, the chemistry in the gas phase maybe different due to the potential for oxidation to increase. The performance of added sorbents - both micrometer and nanostructured, will be different - and the results will be reported. Flue gas recycle is also expected to affect particle penetration and corona inception for the ESP due to the variation of flue gas compositions. Thus, strategies for effective modalities for oxy-coal combustion to minimize emissions of fine particles and mercury will be proposed.

--

Suriyawong, A., Gamble M., Lee M.H., Axelbaum R. and Biswas P., "Submicrometer particle formation and mercury speciation under O_2 - CO_2 Coal Combustion", Energy and Fuels, 20 (6), 2357-2363, 2006.

Suriyawong, A., Hogan C.J., Jiang J., and Biswas P., "Charged fraction and electrostatic collection of ultrafine and submicrometer particles formed during O_2-CO_2 coal combustion", Fuel, 87 (6), 673-682, 2008.

See more of this Session: Oxy-Combustion of Coal - Needs, Opportunities, Challenges - III
See more of this Group/Topical: Energy and Transport Processes

The Behavior of Mercury and Sulfur in An Oxy-Fuel and Integrated Pollutant Removal System

Stephen J. Gerdemann, National Energy Technology Laboratory, U.S. Department of Energy, Albany, OR

Casey Carney, National Energy Technology Laboratory, U.S. Department of Energy, Albany, OR

Thomas Ochs, National Energy Technology Laboratory, U.S. Department of Energy, Albany, OR

Danylo Oryshchyn, National Energy Technology Laboratory, U.S. Department of Energy, Albany, OR

Cathy Summers, National Energy Technology Laboratory, U.S. Department of Energy, Albany, OR

Oxy-fuel combustion uses conventional technology and denitrified air to combust fossil fuels resulting in a flue gas which is primarily CO2. Integrated Pollutant Removal (IPR™) compresses the gas to pipeline pressure and uses the cooling at the IPR inlet, as well as intercooling between compression steps, to remove soluble and entrainable material with the precipitated water. Latent and sensible heat given up by the processed gas at the cooling steps are partially recovered and used to preheat the boiler feed water. Because most of the nitrogen has been removed from the oxidizing gas, the relative concentration of the remaining gasses, such as SO2 and Hg, in the resulting flue gas is increased. The increased concentrations have implications for removal and also corrosion. This paper follows sulfur and Hg from an oxy-fuel fired system through the compression and cooling steps of IPR. A model of the system is compared with actual data from a coal-fired burner test facility.

See more of this Session: Oxy-Combustion of Coal - Needs, Opportunities, Challenges - III
See more of this Group/Topical: Energy and Transport Processes

Formation and Impact of Gases Sulfur Components in an Oxyfuel Process

J. Maier, L. Al-Makhadmeh, G. Scheffknecht
Institute of Process Engineering and Power Plant Technology (IVD)
Pfaffenwaldring 23, University of Stuttgart, 70569 Stuttgart, Germany
Corresponding Author: maier@ivd.uni-stuttgart.de +49 711 685 63396

Abstract

The higher concentrations of sulfur components in the furnace during an oxy-fuel process, compared to conventional combustion, increase the interest of studying sulfur components formation and destruction mechanisms under oxy-fuel conditions and its impacts in power stations. Sulfur components have effects on boiler corrosion, ash deposition, ash precipitation, and acid dew point.

In this investigation the formation and destruction mechanisms of sulfur components under oxy-fuel conditions in the boiler were analyzed during different combustion conditions (staged, un-staged). Experimental investigations were conducted in a once-through 20 kW furnace and in a semi technical 500 kW test facility with flue-gas recirculation. Different coal qualities such as lignites and bituminous coals were used. Besides SO_2, H_2S has been measured in the sub-stoichiometric combustion zone. To study effects on destruction and formation of sulfurous species, the SO_2 concentration in the artificial combustion gas mixture was varied between 500ppm and 6000ppm and detailed inflame measurements of gas components (SO_2, H_2S, NO, CO etc) have been performed. During staged combustion, the effect of burner oxygen ratio on sulfur components speciation has been performed at 0.75, 0.85, and 0.95. In addition, the impact of a wet condition was studied by injecting H_2O steam through the burner. Moreover, the conversion of SO_2 to SO_3 and the behavior of the acid dew point have been investigated. On the other hand, SO_2 capture capacity of coals in the convective section of the furnace has been studied.

Keywords: Sulfur components (SO_2, SO_3, and H_2S); Oxy-fuel combustion; Coal; SO_2 capture

1. Introduction

The fate of sulfur during oxy-coal fired power plant with CO_2 capture is an important issue since it could impact the various aspects of the power plant in terms of the process design, plant construction principles and operations. A general flow sheet of oxy-fuel technology is given in Figure 1, which indicates the locations of potential sulfur impacts.

Santos [1] presented an overview in the understanding of fate of sulfur in coal during oxy-fuel combustion, demonstrating the different factors that could impact the level of SO_2 and SO_3 emissions and the efficiency in the capture of sulfur by the ash. Similar to any air fired furnace, the level of SO_2 emission during oxy-coal fired power plant is proportional to the coal's sulfur content. The in-furnace SO_2 concentration has been reported to increase by a factor of 2-3 during O_2/recycled flue gas (RFG) combustion [2, 3]. The presence of other sulfur components i.e. H_2S, SO_2 and SO_3 is associated with high temperature corrosion at the furnace walls and in the super-heater sections and low temperature corrosion through the condensation of sulfuric acid in the economizer or air heater. In terms of the total sulfur mass output (i.e. specific mass per energy input), the SO_2 emissions in oxy-coal fired cases are about 20-40% lower than in the air-fired cases. The lower specific emissions are primarily attributed to the significant retention of the sulfur in the ash [1] and, therefore, depend on fuel and ash properties. This paper is a summary of IVD activities on the impact of combustion media and recycled SO_2 on the formation of gaseous sulfur components in the flue gas (H_2S, SO_2 and SO_3). In addition, the SO_2 capture capacity of coals (ash) in the convective section of a boiler during different combustion conditions when firing different coal qualities such as lignites and bituminous coals are reported.

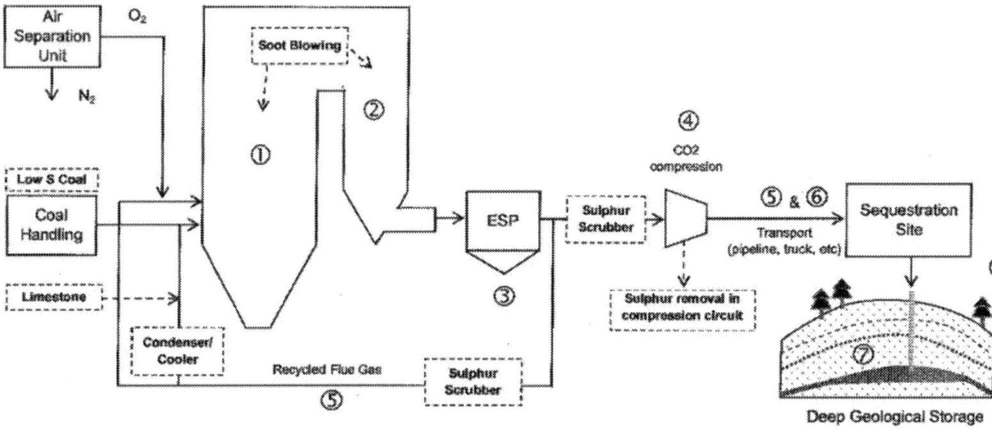

Figure 1: Oxy-fuel flow sheet with locations of the impacts of sulfur and possible mitigation options [4]

2. Experimental Facilities and Approach of investigations

2.1. Coals used

Four coals (Table 1) were used: Klein Kopje (KK), a medium volatile bituminous coal from South Africa, Lausitz brown coal (LA) from Eastern Germany, Rhenish brown coal (RH) from Western Germany and high volatile bituminous coal Ensdorf (EN) from Saar basin in Germany. The brown coals when received from the mine often contains more than 50% moisture but were pre-dried prior to combustion as pre-drying of high moisture lignites is expected to become state-of-the-art for efficiency improvement. The particle size distribution of the coals used is representative of the pulverized coals used in large-scale furnaces.

Table 1: Proximate and Ultimate analyses of the coals used

	LHV [kJ/kg]	Moisture [%, ar[1]]	Ash [%, d]	Volatile [%, daf]	Fixed C [%, daf]	C [%, daf]	H [%, daf]	N [%, daf]	S [%, daf]	O [by diff]
Klein Kopje KK	24932	3.60	19.29	27.76	72.24	83.93	4.50	1.67	0.72	9.18
Lausitz Coal LA	21412	10.20	5.46	57.36	42.64	66.78	5.26	0.65	0.85	26.45
Rhenish Coal RH	20965	11.50	4.07	54.18	45.82	67.96	7.68	0.73	<0.3	23.63
Ensdorf Coal EN	30955	2.42	7.47	37.21	61.90	74.85	5.05	1.59	0.83	17.68

[1] ar: as recieved
daf: dry ash free
by diff: by difference
H, N, S and O: hydrogen, nitrogen, sulfur and oxygen content

2.2. A Once-Through 20 kW Furnace

Figure 2 shows the once-through 20 kW furnace which was used for the major investigations in this summary. It mainly consists of a ceramic tube of 2500 mm length and 200 mm diameter forming the reaction zone. The facility is electrically heated around the reaction zone enabling constant wall temperature up to 1400 °C. The combustion air or O_2/CO_2 mixture is injected through annular clearances, divided into primary and secondary streams. Axial profile measurements of the gaseous and solid samples are carried out by means of an oil-cooled sampling probe which can be moved vertically along the central axis of the reactor from the burnout section to the burner mouth with high flexibility.

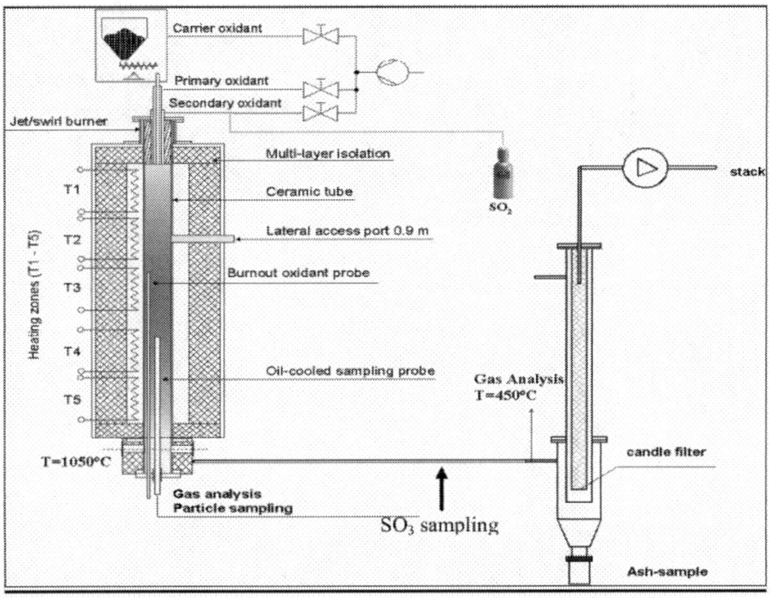

Figure 2: Schematic diagram of the once-through 20 kW furnace

Table 2: Experimental conditions studied in the once-through 20 kW furnace

Staged Combustion in 27%O_2/73%CO_2				
O_2 excess [vol%]		3.52		
Overall O_2 ratio		1.15		
Burner O_2 ratio		0.75	0.85	0.95
Carrier stream	[m³ stp[2]/h]	1.5	1.5	1.5
Primary stream	[m³ stp/h]	2.03	2.03	2.03
Secondary stream	[m³ stp/h]	3.04	3.04	3.04
Staged stream	[m³ stp/h]	3.51	2.32	1.38
SO_2inj	[ppm]	2000, 3000, 6000	2000, 6000	2000, 6000
H_2O	[vol%]	0, 20	0, 20	0

Staged flames for Lausitz coal have been performed in OF27 flame (designating 27%O_2/73%CO_2 flame) at different burner oxygen ratios and to simulate the recycled SO_2 concentration in oxy-fuel process, up to 6000ppm SO_2 was added to the fuel gas through the secondary stream. Moreover, to simulate the recirculated stream in oxy-fuel process where the recycle branch is situated upstream the flue gas condenser (wet recycle), steam injection was also carried out during Lausitz coal combustion. Table 2 summarizes the investigated

[2] stp standard temperature and pressure conditions

experimental parameters for Lausitz coal in OF27. On the other hand, the gaseous concentration measurements were performed at the end of flue gas path and before the candle filter to study SO_2 capture in the convective section of the furnace. For studying the correlation between SO_2 and SO_3 in the 20 kW furnace diluted solution of H_2SO_4 has been injected through the secondary stream to simulate SO_3 concentration in practical oxy-fuel processes. The injection of H_2SO_4 is justified by that the SO_3 in the flue gas forms H_2SO_4 under the temperature window of flue gas path in practical oxy-fuel processes.

2.3. The 500 kW Test Facility

Figure 3 shows the 500 kW test facility. A detailed description can be taken from [3]. As shown in the figure, the SO_3 measurements have been performed upstream the ESP. For SO_3 sampling an adapted controlled condensation procedure, in line with a VDI draft guideline, was applied. In addition information given in the publications [5, 6] where taken into account. Sample gas is extracted from the flue-gas and passed through a condenser helix which is controlled to a temperature of 95 °C. Here, sulfuric acid that has formed in the flue-gas condensates on the helix's inner surface. Afterwards the helix is rinsed with purified water and the sample collected is subjected to analysis by titration. To prevent particles from entering and contaminating the condenser two heated filters with different pore sizes were integrated in the sampling arrangement. The flue-gas temperature during sampling was between 350-380 °C.

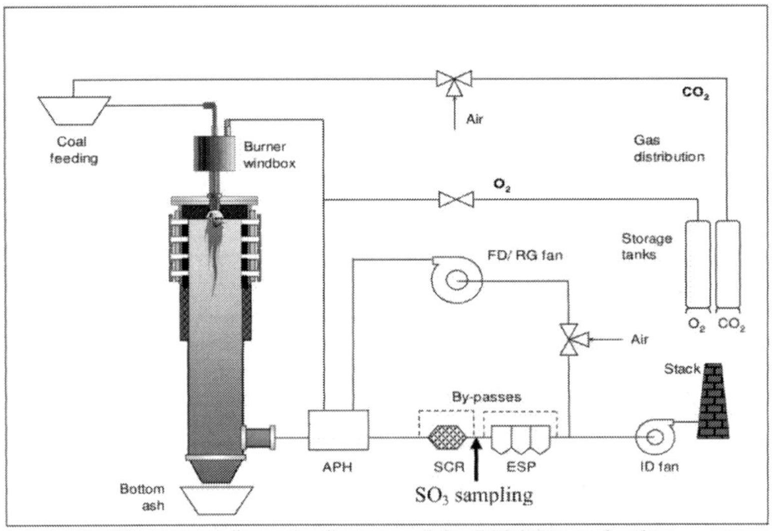

Figure 3: Schematic diagram of the 500 kW test facility

3. Results

3.1. SO_2 and H_2S Formation and Destruction Mechanisms

In conventional combustion, sulfur is only added via the fuel while in oxy-fuel process the recirculated flue gas is another source of sulfur components in the furnace. To study sulfur components speciation in oxy-fuel process, staged flames with SO_2 injection up to 6000ppm in a 20 kW furnace has been performed.

Figure 4 shows the axial concentration profiles of SO_2, H_2S, and O_2 for Lausitz coal combustion in air and OF27 flames with a burner oxygen ratio of 0.75, the SO_2 concentration in the oxidizing agent was set at 3000ppm. For both air-firing and oxy-fuel OF27_3000 cases, it was noticed that the SO_2 concentration increases rapidly up to 0.5 m from the burner followed by a fast reduction to H_2S. The maximum H_2S concentration forms at around 0.9-1.1 m and after the addition of burnout air all sulfur components oxidize to SO_2. On the other hand, the sulfur components speciation is more to the side of SO_2 for oxy-fuel OF27_3000 case, due to the addition of sulfur compound in the form of SO_2 in the oxidizing agent. However the H_2S concentration doubles that of air-firing case due to the reduction of SO_2 to H_2S in the sub-stoichiometric zone. It can be concluded that the H_2S/SO_2 formation and reduction mechanisms in near flame zone are fundamentally similar during air-blown and oxy-coal combustion.

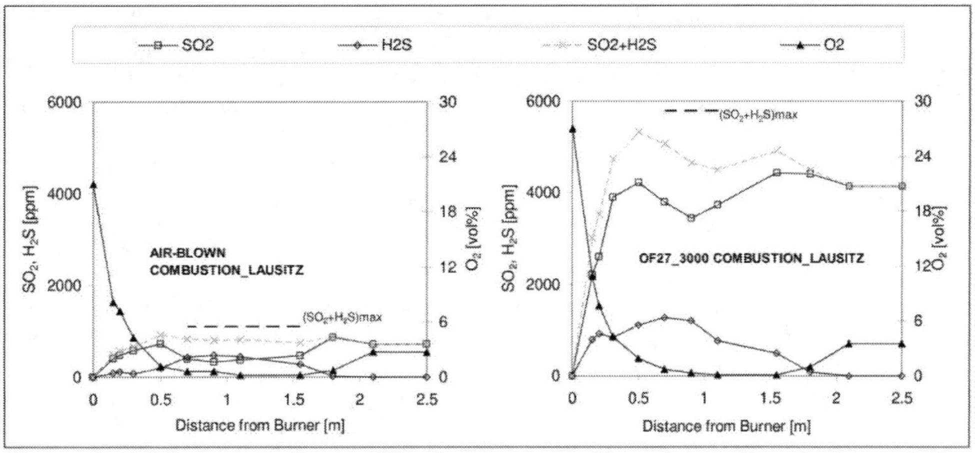

Figure 4: Axial concentration profiles for gaseous sulfur compounds and oxygen during Lausitz coal combustion in air and OF27 with a burner oxygen ratio of 0.75.

Figure 5 shows the absolute and relative H_2S concentration for Lausitz coal combustion between 0.9 and 1.1 m from the burner with different SO_2 concentration in the oxidizing agent, the burner oxygen ratio was 0.75.

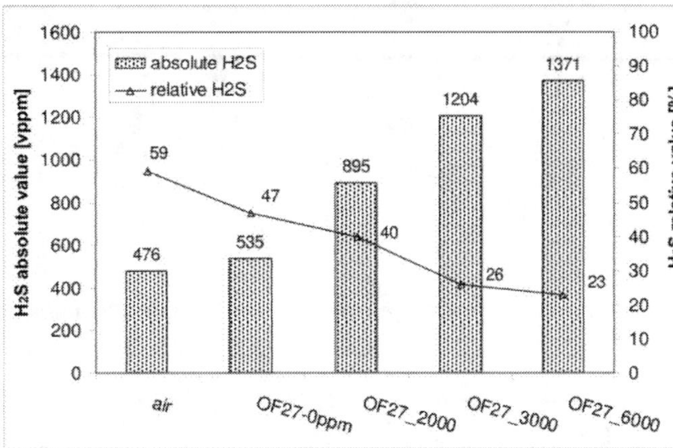

Figure 5: Absolute and relative H_2S concentration between 0.9-1.1 m from the burner during Lausitz coal combustion in air and OF27 combustion

The relative concentration of H_2S (α_{H2S}) or volume fraction at 0.9-1.1 m has been calculated as:

$$\alpha_{H_2S} = \frac{y_{H_2S,d}}{y_{H_2S,d} + y_{SO_2,d}}$$

where $y_{H2S,d}$ and $y_{SO2,d}$ are the volumetric concentration of H_2S and SO_2 in the flue gas (dry), respectively at 0.9-1.1 m.

As can be seen, the fraction of H_2S in the near-flame zone is higher during air-blown combustion compared to oxy-fuel flames, even though the concentration of H_2S is higher and the difference increases as the SO_2 concentration in the oxidizing agent is higher. When the SO_2 concentration in the oxidizing agent was increased from 0ppm to 6000ppm, the H_2S concentration increased from 535ppm to 1371ppm. It can be concluded that the H_2S concentration in the near-flame zone is 2-3 times higher during oxy-coal combustion with SO_2 injection of 2000-6000ppm compared to air-firing case. Therefore, the issue of sulfur induced corrosion in an oxy-coal operated furnace must be considered especially for high sulfur-containing coals. Meanwhile, the relative concentration of H_2S decreases from 47% to 23% since the addition of sulfur compound was in the form of SO_2 in the oxidizing agent as mentioned previously.

Impact of Coal Quality

Besides Lausitz-coal staged flame, staged flames for Klein Kopje and Rhenish coals with burner oxygen ratio of 0.75 have been performed in order to investigate the effect of coal quality on sulfur components speciation. The SO_2 concentration in the oxidizing agent was set at 3000ppm. Generally, it was found that KK and RH coals show the same trend of SO_2 and H_2S concentrations as well as LA coal, even though the chemical speciation of sulfur components differs significantly for the different coals as shown in Figure 6 and Table 3.

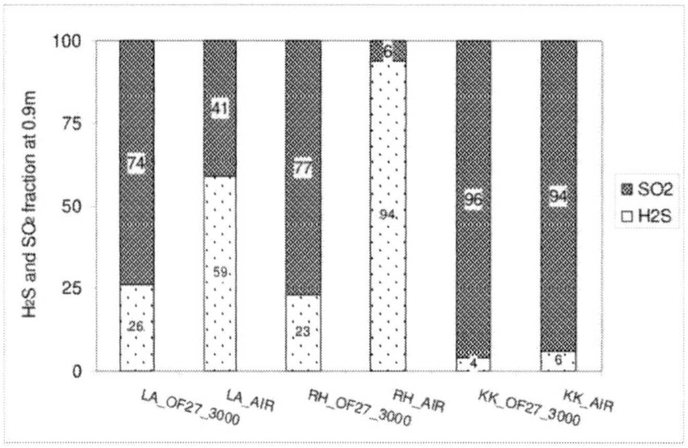

Figure 6: H_2S and SO_2 volumetric fraction for the investigated coals with burner oxygen ratio of 0.75

Table 3: Absolute and relative H_2S values at 0.75 between 0.9-1.1 m for the investigated coals

	LA_OF27_3000	LA_AIR	RH_OF27_3000	RH_AIR	KK_OF27_3000	KK_AIR
H_2S absolute [vppm]	1204	476	910	294	186	37
H_2S relative	26	59	23	94	4	6

[%]						

For example 15% of sulfur (the maximum) was detected as H_2S for Klein Kopje coal while between 25-50% H_2S was detected for Lausitz coal [7]. The quantitative drop in SO_2 level and increase in H_2S level in the sub-stoichiometric region shows a strong relationship to the volatile content of the coal. The high volatile coals, Rhenish and Lausitz lignite showed a strong drop in SO_2 level with high H_2S peak in the sub-stoichiometric region whereas medium volatile coal, Klein Kopje showed only a small drop in SO_2 level and low H_2S peak during air-blown as well as OF27_3000 combustion. On the other hand, from coal pyrolysis investigations in the CO_2 environments at temperature ranges of 700-1150 °C using an entrained-flow reactor, it was found that the S-release for Klein Kopje coal is around 30% while Lausitz coal releases 50% S indicating the importance of sulfur functional groups [8]. It can be concluded that the volatile content and the amount of released sulfur in the sub-stoichiometric zone have a significant effect on sulfur components speciation in the reduction zone.

Impact of Burner Oxygen Ratio

As staging is used increasingly for reduction of nitrogen oxides, the impact of burner oxygen ratio on sulfur components speciation has been studied at 0.75, 0.85, and 0.95 for Lausitz coal. Figure 7 shows the absolute and relative H_2S concentrations at the three investigated burner oxygen ratios without and with SO_2 injection of 2000 and 6000ppm.

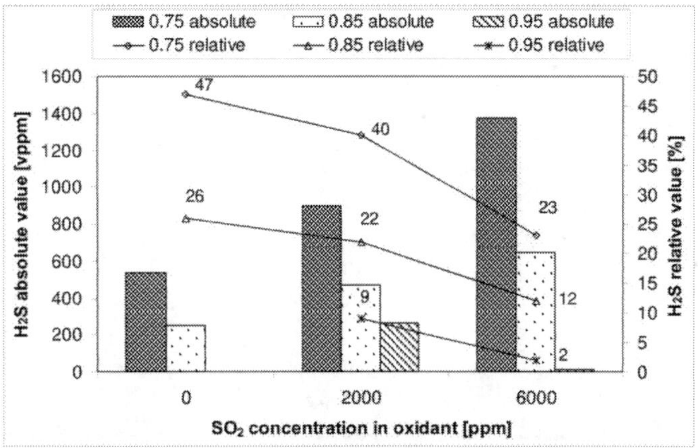

Figure 7: The influence of burner oxygen ratios on the absolute and relative H_2S concentration between 0.9-1.1 m from the burner during OF27 combustion for Lausitz coal

The lower the burner oxygen ratio, the higher the H_2S concentration was found in all flames studied without and with SO_2 injection. It was noticed that the H_2S concentration with burner O_2 ratio of 0.75 doubles that at 0.85 with and without SO_2 injection in the oxidant, e.g. with 6000ppm SO_2 in the oxidant the H_2S concentration increases from 647ppm to 1371ppm as the burner oxygen ratio decreases from 0.85 to 0.75. On the other hand, at burner oxygen ratio of 0.95 the H_2S concentrations are very low. It can be concluded that as the staging is stronger (the burner oxygen ratio decreases) and as the SO_2 concentration in the oxidant increases the reduction of SO_2 and the formation of H_2S increases. In addition, the relative concentration of H_2S decreases as the burner oxygen ratio increases.

Impact of Wet Condition

The effect of the increased water vapor content in the oxidant (27% O_2, 53% CO_2 and 20% H_2O) on the formation of sulfur components has been studied and the results are presented in Figure 8 with burner oxygen ratio of 0.75 and 0.85 and different SO_2 concentrations in the oxidant.

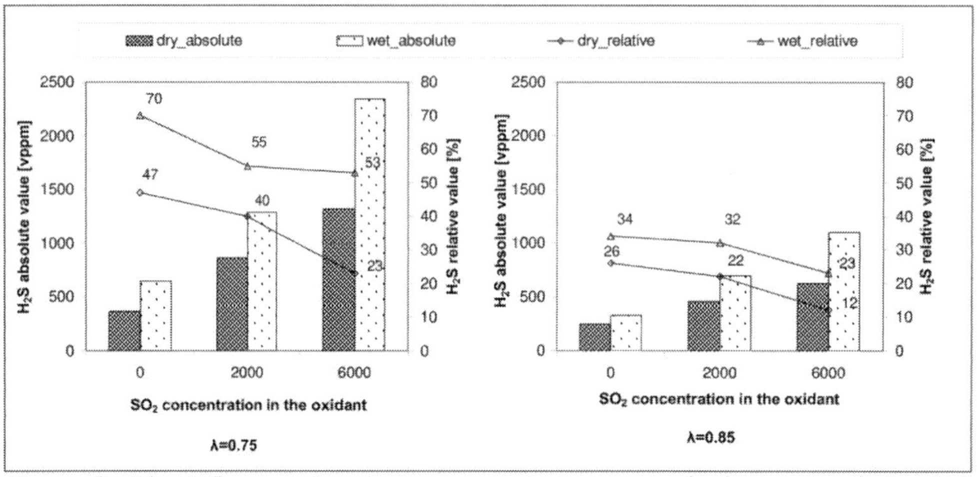

Figure 8: The influence of water vapor content on the absolute and relative H_2S concentrations between 0.9-1.1 m from the burner during OF27 combustion for Lausitz coal with burner oxygen ratio of 0.75 and 0.85(the absolute concentrations on wet basis)

It is clearly seen that the H_2S concentrations under wet condition flames are higher than dry flames at all investigated conditions; the H_2S concentrations are 1.4-1.8 higher compared to dry conditions. The relative H_2S concentration, on the other hand, is 15-30% and 10% higher under wet conditions compared to dry conditions at 0.75 and 0.85, respectively. In addition, as the SO_2 concentration in the oxidant increases with higher water vapor content the higher the H_2S formation will be with both burner oxygen ratios 0.75 and 0.85. The H_2S concentration under wet condition is 2 times higher when SO_2 in the oxidant was 2000ppm with burner oxygen ratios of 0.75 and 0.85 compared to wet condition without SO_2 injection. Moreover, when 6000ppm SO_2 was in the oxidant the H_2S concentration under wet condition was 3-4 times higher than without SO_2 in the oxidant. It can be concluded that the water vapor content intend to increase the H_2S concentration in the reduction zone of the furnace. This can be explained by that even the water vapor retards sulfur conversion to SO_2 in the reduction zone or gives a new hydrogen source which favor H_2S formation. In addition, higher SO_2 concentrations in the oxidant under wet conditions promote higher H_2S formation.

3.2. Reduction of Recycled SO_2 in the Convective Section/Flue Gas Path of the Furnace

Theoretically, the concentration of SO_2 in the furnace in an oxy-fuel process is about 5 times higher than in air firing. Depending on the fuel, however, sulfur is also captured in the ash in the course of flue gas path cooling, so the SO_2 concentration is two-three times greater in oxy-fuel combustion compared to air combustion [2, 3]. In addition, it is reported that the SO_2 concentration did not change significantly with increasing the recycle ratio [9].

For the determination of SO_2 capture potential by the ash in the convective section/flue-gas path, SO_2 measurements were carried out at the end of the flue-gas path just before the candle filter in the 20 kW test facility by gradually increasing the SO_2 injection in the furnace. The flue-gas temperature at the end of radiative section was around 1050 °C and the temperature at the SO_2 measurement point just before candle filter was around 450 °C. Figure 9 shows SO_2 capture in the flue-gas path vs. SO_2 measured at the end of radiative section of the furnace (2.5 m from burner) during air-blown as well as OF27 combustion for all four coals mentioned previously.

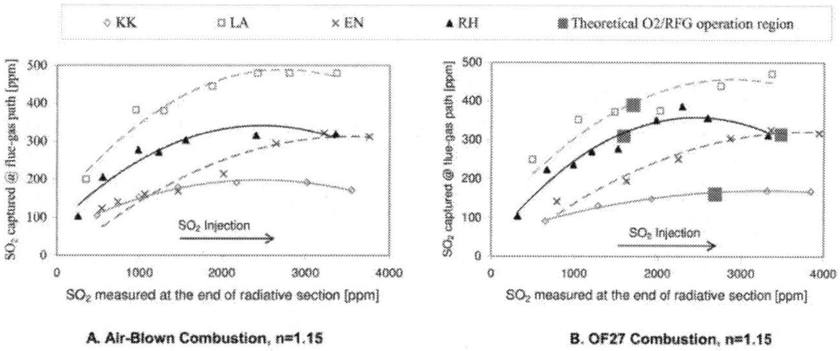

Coal	Ash [wf, mass%]	CaO [wf, mass%]	S in fuel [wf, mass%]	Ca:S molar ratio
Klein Kopje (KK)	19.29	7.8	0.58	2.98
Ensdorf (EN)	7.29	10.50	0.77	1.03
Lausitz (LA)	4.79	23.78	0.35	3.17
Rhenish (RH)	4.07	35.10	0.34	3.96

Figure 9: SO_2 capture in the convective section/flue-gas path during (A) Air-Blown combustion and (B) OF27 combustion (embedded table illustrating coal properties).

As seen from the Figures, SO_2 captured in the flue-gas path increases with increasing SO_2 concentration at the end of the radiative section during air-blown as well as OF27 combustion. However a plateau is reached after a certain concentration of SO_2 at the end of the radiative section for all coals indicating that the SO_2 capture capacity is exhausted and more SO_2 cannot be captured. The capture capacity however shows a weak dependency on the Ca/S molar ratio (see embedded table) as Ca/S molar ratio is highest for Rhenish coal and lowest for Ensdorf coal but SO_2 captured is highest for Lausitz coal and lowest for Klein Kopje coal indicating that the form of calcium present in the ash plays a more important role on SO_2 capture. Klein Kopje char combustion shows that 100% of S is released in the oxidation step while 86-91% is released during Lausitz char combustion indicating again the importance of the form of Ca in the char (ash) [8]. Although the residence time available for interaction between SO_2 in the flue-gas and sulfate forming elements in the ash is not long enough in the flue-gas path, desulfurization efficiency as high as 50% was achieved for Lausitz coal indicating the potential for SO_2 retention in the convective section and flue-gas path if right coal and right temperature gradient is available. The grey symbols indicate the practical SO_2 concentration during combustion for each coal under oxy-fuel condition.

3.3. SO_2 Conversion to SO_3

The conversion from SO_2 to SO_3 has been studied in the 20 kW furnace and the 500 kW test facility (2 campaigns) in air firing and oxy-fuel combustion for Lausitz coal.

Figure 10 shows the correlation between SO_2 and SO_3 concentrations under air firing and oxy-fuel conditions in both test facilities. As can be seen, in oxy-fuel process the SO_2 and SO_3 concentrations are higher than in air combustion. From the 1st campaign results in the 500 kW test facility, the average SO_3 concentration at oxy-fuel conditions has a min/max range of 3.6-12.1ppm while for air conditions the average was 1ppm. In the 2nd campaign, the average SO_3 concentration at oxy-fuel conditions has a min/max range of 2.5-6.0ppm and the conversion rate varies by a factor of 3 to 13 higher than in air firing. Our results are in agreement with others. Wall et al. [9] reported that the SO_3 concentration is approximately two and a half (2.5) to three (3.0) times higher in oxy-fuel combustion than in air combustion. On the other hand, the SO_2 and SO_3 concentrations for Lausitz coal combustion in the 20 kW furnace in air and under oxy-fuel conditions show that as the SO_2 concentration increases the concentration of SO_3 increases.

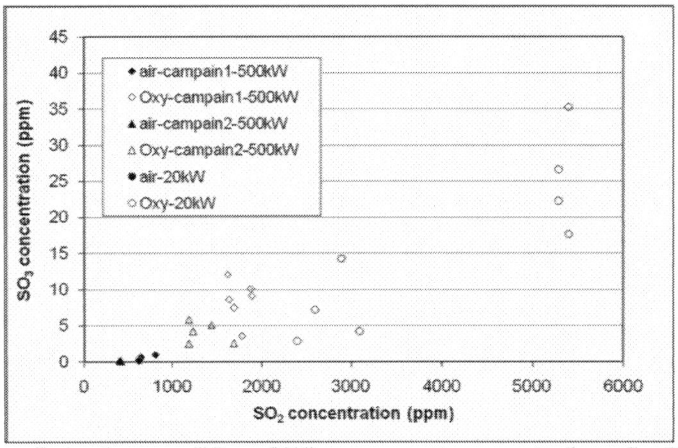

Figure 10: Correlation between SO_2 and SO_3 concentrations results from the 500 kW (1st and 2nd campaign) and 20 kW test facilities

In summary, the trend of higher SO_3 concentration under oxy-fuel conditions and that the SO_3 concentration increases as SO_2 concentration increases have been found from the semi technical facility and the 20 kW furnace despite the variation in the operating conditions for both facilities and the fluctuations during the operation of the 500 kW test facility.

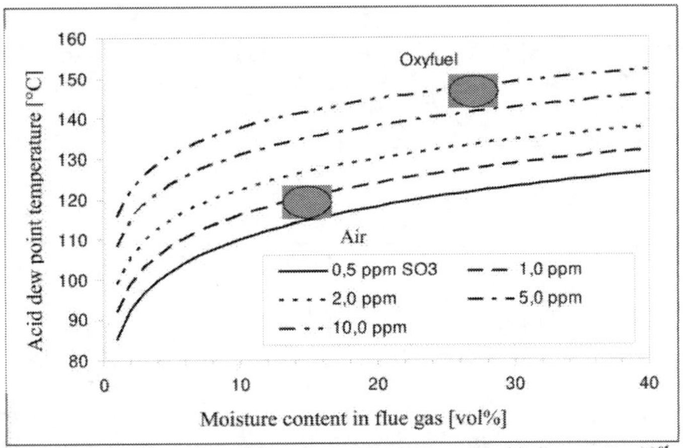

Figure 11: Correlation between SO_3 concentration and acid dew point (1st campaign)

Moreover, the acid dew point temperatures were calculated from the 1st campaign results and presented in Figure 11. It was found that the average acid dew point temperatures are 118 °C for air and 148 °C for oxy-fuel conditions. This difference was to be expected as both the SO_3 concentration and the vapor-phase content of the flue-gas increased significantly at oxy-fuel conditions. The corresponding vapor-phase water contents measured during the SO_3 samplings alternated around the averages by ± 2.3 %-points at air conditions and ± 1.3 %-points at oxy-fuel conditions.

4. Conclusions

Gaseous sulfur components have been investigated in an air and oxy-fuel combustion mode with a 20 kW once-through furnace and 500 kW semi-technical facility with flue-gas recycling. The major conclusions from these results are as follows.

The H_2S/SO_2 formation and reduction mechanisms in near flame zone are fundamentally similar during air-blown and oxy-coal combustion. Although the fraction of H_2S in the near-flame zone or oxygen deficient reduction zone is higher during air-blown combustion, the concentration of H_2S was 2-3 times higher during oxy-coal combustion with SO_2 injection of 2000-6000ppm (representing different O_2/RFG scenarios).

The concentration of H_2S is much higher for high volatile coals compared to medium volatile coals. Therefore, from corrosion point of view, careful attention should be given to coals with high amount of volatiles and sulfur i.e. low rank high sulfur coals.

The burner oxygen ratio, wet condition and the SO_2 concentration in the furnace have a significant effect on H_2S formation and SO_2 reduction mechanisms. Higher H_2S concentration was found when the oxygen ratio was low (< 0.8). In addition, wet condition is a source for higher H_2S concentration in the sub-stoichiometric zone.

Higher SO_3 concentrations and conversions under oxy-fuel conditions compared to air-firing cases have been determined. Considering in addition the increase of the water vapor concentration during wet recycle, the acid dew point could be substantially increased. In addition, good agreement between the results obtained in a semi technical 500 kW test facility and a 20 kW furnace was obtained.

Some of the SO_2 can be reduced in the lower temperature region of the furnace if active sulfate forming elements are available in the coal. Due to high concentration of SO_2 during O_2/RFG combustion, the capture of SO_2 is always higher than in air-blown furnaces by coals containing high shares of sulfate forming elements.

References

[1] Santos, S.:Fate of Sulphur in Coal During Oxyfuel Combustion With Recycled Flue Gas (Review of the Current State of Understanding), CCT2009.

[2] Croiset, E.; Thambimuthu, K.: NO_x and SO_2 emissions from O_2/CO_2 recycle coal combustion, Fuel, 80, pp. 2117-2121, 2001.

[3] Mönckert, P.; Reber, D.; Maier, J.; Scheffknecht, G.: Operation of a Retrofitted 0.5 MW_{th} PF Combustion Facility under Oxyfuel Conditions, 32nd International Technical Conference on Coal Utilization and Fuel System, Florida, 2007.

[4] Wall, T.; Stanger, R.; Maier, J.: Sulfur Impacts in Coal-Fired Oxyfuel Combustion with CCS: Impacts and Control Options. 1st Oxyfuel Combustion Conference.

[5] Fernando, R.: SO_3 issues for Coal_Fired Plant; IEA-COAL, 2003.

[6] Verhoff, Banchero: Predicting Acid Dew Point Temperatures from Flue-Gases; Chem. Eng. Prog., Aug. 1974.

[7] Scheffknecht, G.; and Maier, J.: Firing Issues Related to the Oxyfue Process.VGB PowerTech, 2008. 91-97.

[8] Al-Makhadmeh, L.: Coal Pyroylsis and Char Combustion Under Oxy-Fuel Conditions. Diss. Stuttgart University. Shaker Verlag, 2009.

[9] Wall, T.; Liu, Y.; Spero, C.; Elliott, L.; Khare, S.; Rathnam, R.; Zeenathal, F.; Moghtaderi, B.; Buhre, B.; Sheng, C.; Gupta, R.; Yamada, T.; Makino, K.; Yu, J.: An Overview on Oxyfuel Coal Combustion-State of the Art Research and Technology Development. Chemical Engineering Research and Design, 2009.

Oxy-Coal Combustion in a Pilot-Scale Circulating Fluidized Bed

Lawrence Bool III, Industrial Applications and Healthcare R&D, Praxair Inc, Tonawanda, NY

Eric G. Eddings, Dept. of Chemical Engineering, University of Utah, Salt Lake City, UT

Oxy-coal combustion to facilitate carbon capture and sequestration is the focus of numerous research and development activities. Recently, Praxair collaborated with the University of Utah to study some of the fundamental impacts of oxy-coal combustion in fluidized bed combustion. An existing circulating fluid bed (CFB) combustor was modified to allow firing with oxygen-enriched recirculated flue gas. Experiments with this facility were able to routinely achieve CO2 concentrations greater 85% (dry basis). Bed temperature and pollutant emissions were measured under both air-firing and oxy-coal firing.

INTRODUCTION

Oxy-coal combustion has received a great deal of attention in recent years as a way to facilitate carbon capture and sequestration (CCS). Many groups have demonstrated the feasibility of using oxygen diluted with recirculated flue gas (RFG) as a replacement for air in pulverized coal (PC) fired boilers. Pilot and full-scale demonstrations of oxygen enriched RFG for PC units are planned or ongoing. Others have published limited experimental data and commercial circulating fluid bed (CFB) boiler design concepts based on oxygen enriched RFG. Praxair has teamed with Foster Wheeler, and others, to propose a 50 MW oxy-coal CFB at Jamestown, NY. Even with this work, there are still details that need to be addressed to optimize the use of oxygen in oxy-coal firing. For example, are there mixing strategies to optimize oxygen use and are there trace species that may create problems for the CO2 purification unit (CPU).

Praxair was involved with some of the earliest oxy-coal combustion studies1 and was the first company to inject oxygen into coal-fired utility boilers15. Praxair has teamed with the University of Utah to explore optimized oxygen use for PC and CFB units. Preliminary data has been collected using oxygen-enriched wet RFG in the L1500 furnace, a 1.5 MW pulverized coal combustor, and will be reported elsewhere. Data has also been collected using the laboratory-scale CFB. These data will be discussed in this presentation.

See more of this Session: Oxy-Combustion of Coal - Needs, Opportunities, Challenges - III
See more of this Group/Topical: Energy and Transport Processes

Impact of Autoconversion Schemes On Prediction of the Aerosol Indirect Effect

Athanasios Nenes, Georgia Institute of Technology, Atlanta, GA

Wei-Chun Hsieh, Schools of Earth & Atmospheric Sciences, Georgia Institute of Technology, Atlanta, GA

John H. Seinfeld, California Institute of Technology, Pasadena, CA

Peter J. Adams, Civil and Environmental Engineering and Engineering and Public Policy, Carnegie Mellon University, Pittsburgh, PA

Online GCM (NASA GISS II') simulations of aerosol and cloud droplet number are used to assess the sensitivity of predicted climate and aerosol indirect effects to the treatment of autoconversion for warm clouds. Four autoconversion parameterizations and direct integration of the kinetic collection equation (KCE) with kernels that account for both gravitational and turbulent effects on collection are used to represent autoconversion. Predictions of the combined cloud albedo and lifetime effects on aerosol indirect forcing are found to be quite sensitive to the autoconversion scheme used, with differences between present-day and preindustrial aerosol indirect forcing ranging from -0.89 W/m^2 to -2.05 W/m^2. Autoconversion parameterizations are evaluated by the extent to which the predicted timescale for conversion of cloudwater into drizzle is consistent with observations. Of the schemes considered, the MC, P6, and KK parameterizations tend to predict values within the observed range of autoconversion timescale when clouds are close to the precipitation threshold. Analysis of the link between global annual average changes in LWP, autoconversion, and aerosol indirect forcing shows that the contribution of cloud albedo effects to aerosol indirect forcing is about -0.9 W/m^2, with the remainder being the cloud lifetime contribution. Reduction of autoconversion uncertainty to a factor of 3-5 would result in a 30-50% aerosol indirect forcing uncertainty. Quantifying autoconversion prediction uncertainty using the observational approach outlined here offers a robust method for determining the aerosol indirect forcing uncertainty associated with the combined cloud albedo and lifetime effects.

See more of this Session: Advances in Climate Change Science
See more of this Group/Topical: Environmental Division

Laboratory Measurements of Cloud Condensation Nuclei Activation, Volatility and Growth Kinetics

Gabriella J. Engelhart, Carnegie Mellon University, Pittsburgh, PA

Akua Asa-Awuku, Chemical and Environmental Engineering, UC-Riverside, Riverside, CA

Lea Hildebrandt, Carnegie Mellon University, Pittsburgh, PA

Byong-Hyoek Lee, Carnegie Mellon University, Pittsburgh, PA

Merete Bilde, University of Copenhagen, Copenhagen, Denmark

Athanasios Nenes, Georgia Insititute of Technology, Atlanta, GA

Spyros N. Pandis, Carnegie Mellon University, Pittsburgh, PA

Understanding cloud condensation nuclei (CCN) activation is a vital step towards decreasing the uncertainty in the aerosol indirect effect on climate. In this presentation we will present work characterizing the CCN properties of anthropogenic and biogenic secondary organic aerosols (SOA) produced in the Carnegie Mellon University smog chamber. The smog chamber is a 10 cubic meter temperature-controlled reactor outfitted with ultraviolet light bulbs and instrumentation to monitor gas-phase and particle-phase species. Biogenic precursors included isoprene, the most abundant biogenic non-methane volatile organic carbon, beta-caryophyllene, and a mixure of monoterpenes. Toluene was selected as a model anthropogenic system. SOA was formed from homogeneous nucleation initiated by oxidation. Oxidants included ozone in the dark and hydroxyl radicals from either HONO or HOOH subjected to ultraviolet radiation. CCN concentrations, activation diameter and droplet growth kinetic information were monitored as a function of supersaturation for several hours using a cylindrical continuous-flow streamwise thermal gradient CCN counter developed by Droplet Measurement Technologies (DMT). Some experiments also included activation diameter measurements from a static diffusion CCN counter from DH Associates. We will summarize the findings for CCN activity, aging, and droplet growth kinetics. The hygroscopicity parameter, kappa will allow us to compare across systems while Kohler theory analysis will allow us to highlight some pertinent details of the system CCN activity. All of the CCN systems studied are moderately CCN-active organic compounds.

Filter samples were collected during a number of experiments in order to perform offline CCN analysis by extracting the water soluble organic carbon (WSOC) collected and re-atomizing it into the DMT CCN counter. Köhler Theory Analysis (KTA) is then used to infer the molar mass of the SOA sampled online and offline from atomized filter samples. Assuming complete solubility, the estimated average molar mass of online monoterpene SOA was determined to be 180 plus or minus 55 grams per mole. This value is consistent with existing monoterpene SOA speciation studies. KTA suggests that the aged aerosol (both from alpha-pinene and the mixed monoterpene oxidation) is primarily water-soluble (around 65 percent). CCN activity measurements of the monoterpene SOA mixed with ammonium sulfate suggest that the organic can depress surface tension up to 10 milliNewtons per meter (with respect to pure water). The droplet growth kinetics of monoterpene SOA samples are similar to ammonium sulfate. The CCN activation diameter of alpha-pinene and mixed monoterpene SOA can be modelled to within 10-15 percent of experiments by a simple implementation of Koehler theory, assuming complete dissolution of the particles, no dissociation into ions, a molecular weight of 180 grams per mole, a density of 1.5 grams per cubic centimeter, and the surface tension of water. Isoprene SOA and toluene SOA display similar activation diameters as observed in monoterpene SOA. Finding a small range of activation diameters or a single kappa value for a broad array of precursors would be quite useful for global models and studies of the indirect effect of aerosols on climate.

The model sesquiterpene system, beta-caryophyllene, shows that the CCN contain small amounts of hygroscopic material. Although beta-caryophyllene produces higher yields than monoterpene systems the aerosols are less effective as CCN in the atmosphere. The two CCN counters used in this study have fundamentally different methods for supersaturation generation. The static CCN counter relies upon the nonlinear dependence of water vapor pressure upon temperature to generate supersaturation whereas the DMT CCN counter exploits the higher diffusivity of water than that of heat. The result of these unique methods is that the DMT CCN counter heats the particles along the length of the column causing a small amount of the particle mass to evaporate whereas the static CCN counter cools the particles. A major finding in this work is that the hygroscopic component of beta-caryophyllene SOA is volatile. This leads to discrepancies between CCN instrumentation not observed in monoterpene systems. The finding is potentially quite important for atmospheric measurements impacted by diurnal temperature cycles and thus, potentially, CCN activity diurnal cycles. Evidence of a kinetic barrier to droplet activation was revealed from a comparison of droplet growth kinetics of the CCN and the fraction of WSOC. Filter samples show that the WSOC fraction of the SOA is composed of low molecular weight (less than 200 grams per mole) compounds that are slightly surface active. This is similar to the findings for monoterpene SOA systems indicating that exact speciation knowledge might not be required for predictive understanding of CCN.

See more of this Session: Advances in Climate Change Science
See more of this Group/Topical: Environmental Division

A New Physically-Based Parameterization of Ice Cloud Formation for Large-Scale Simulations

Donifan Barahona[1] and Athanasios Nenes[1,2]

[1] School of Chemical and Biomolecular Engineering, Georgia Institute of Technology
[2] School of Earth and Atmospheric Sciences, Georgia Institute of Technology

Abstract. This study presents a comprehensive ice cloud formation parameterization that computes the ice crystal number, size distribution, and maximum supersaturation from precursor aerosol and ice nuclei with any size distribution and chemical composition. The parameterization provides an analytical solution of the cloud parcel model equations and accounts for the competition effects between homogeneous and heterogeneous freezing, and, between heterogeneous freezing in different modes. The diversity of heterogeneous nuclei is described through a nucleation spectrum function which depends on the surface area, chemical composition, and size distribution of the precursor, insoluble aerosol. The parameterization reproduced the predictions of a detailed numerical parcel model. The average error in ice crystals number concentration was -2.0 ± 8.5% for conditions of pure heterogeneous freezing, and, 4.7 ± 21% when both homogeneous and heterogeneous freezing were active. Apart from its rigor, excellent performance and versatility, the formulation is extremely fast and free from requirements of numerical integration.

Key Words: Heterogeneous nucleation; homogeneous nucleation; Cirrus Clouds.

INTRODUCTION

Representation of ice clouds in climate and weather prediction models remains a challenge due to the limited understanding of ice cloud formation processes and the difficulties associated with their remote sensing (*1*). Ice clouds form by homogeneous freezing of liquid droplets or heterogeneous freezing upon ice nuclei, (IN) (e.g., *2*). Observational data show that the two freezing mechanisms are likely to interact during cloud formation (*3*); their relative contribution is however a strong function of IN, aerosol concentration, and cloud formation conditions (*4, 5*).

Several approaches have been proposed to parameterize ice cloud formation in atmospheric models. Empirical correlations derived from field campaigns are most often employed to express IN concentrations (e.g., *6*). These expressions are simple but only provide the availability of IN over a limited spatial region. Calculation of ice crystal number concentration, N_c, requires the knowledge of cloud supersaturation, s_i, and therefore the usage of a dynamical framework. Recent physically-based parameterizations address this issue (*7, 8*). They however do not allow direct incorporation of experimental freezing data within their framework, and rely on oversimplified assumptions regarding the nature of heterogeneous freezing.

This work presents a new physically-based, analytical scheme to parameterize ice cloud formation in a computationally efficient manner. The new scheme allows the usage of both empirical and theoretical IN data in a simple dynamical framework, and can consider the variability in aerosol and IN composition. The new parameterization builds upon the frameworks of Barahona and Nenes (*4, 9*) that combine homogeneous and heterogeneous mechanisms of ice formation, and explicitly resolves the dependency of N_c on conditions of cloud formation (i.e., *T, p*), updraft velocity, aerosol number and size, and the freezing characteristics of the IN population.

THE PARAMETERIZATION SCHEME

The ice crystal number concentration, N_c, is given by

$$N_c = \begin{cases} N_o e^{-f_c}(1-e^{-f_c}) + N_{het}(s_{\hom}) & f_c > 0 \\ N_{het}(s_{\max}) & f_c \le 0 \end{cases} \qquad (1)$$

where N_o is the deliquescent aerosol (i.e., droplet) concentration, $s_{\max}$ is the maximum supersaturation during cloud formation, f_c is the droplet freezing fraction, and $s_{\hom}$ is the homogeneous freezing supersaturation threshold. The homogeneous contribution to N_c in equation (1) is given by the term $N_o e^{-f_c}(1-e^{-f_c})$. Heterogeneous nucleation is described using a nucleation spectrum function represented as $N_{het}(s_i)$, which can be either experimentally or theoretically derived (*10*).

The parameterization accounts for competition effects between homogeneous and heterogeneous nucleation. For this, the concentration of IN that completely inhibits homogenous nucleation, $N_{\lim}$, is used to scale the fraction of droplets that freezes homogeneously in the presence of IN. $N_{\lim}$ it is given by

$$N_{\lim} = \frac{N^*}{\sqrt{\left.\Delta s^*_{char}\right|_{s_{\hom}}}} \frac{\left(1+s_{\hom}\right)}{s_{\hom}} e^{\frac{2}{\lambda s_{\hom}}} \qquad (2)$$

where N^* and λ are functions of cloud formation conditions (i.e., updraft velocity *T*, and *p*) and of the water vapor to ice deposition coefficient. Δs^*_{char} is given by

$$\left.\Delta s^*_{char}\right|_{s_i} = \left\{ \frac{\Delta s_{char}\left[\frac{4}{3}\Delta s_{char} + 2(s_i - \Delta s_{char})\right]}{(1+s_i-\Delta s_{char})} \right\}^{1/2} \qquad (3)$$

The parameter Δs_{char} is characteristic of the IN population, and is given by,

$$\Delta s_{char} = \min\left[\left(\frac{d\ln N_{het}(s_i)}{ds_i}\right)^{-1}, s_i\right] \tag{4}$$

when applied in Eq. (3), Δs_{char} is calculated at $s_i = s_{\text{hom}}$.

Combined Homogeneous and Heterogeneous Freezing

When both, homogeneous and heterogeneous nucleation take place, heterogeneously frozen crystals tend to deplete water vapor before the onset homogeneous freezing (*11*). Thus, the droplet freezing fraction, f_c, depends on the IN concentration and on the freezing characteristics of the IN population; f_c is given by,

$$f_c = f_{c,\text{hom}}\left\{1-\left(\frac{N_{het}(s_{\text{hom}})}{N_{\text{lim}}}\right)^{3/2}\right\}^{3/2} \tag{5}$$

where $f_{c,\text{hom}} = \frac{\rho_a}{\rho_i}\frac{k_{\text{hom}}^{1/2}}{\beta N_o}\left[\frac{2\alpha V(s_{\text{hom}}+1)}{\pi\Gamma s_{\text{hom}}}\right]^{3/2}$ is the droplet freezing fraction in absence of IN, i.e., pure homogeneous freezing (*9*) which depends only on *T*, *p*, and updraft velocity.

Pure Heterogeneous freezing

If the IN concentration is high, heterogeneously frozen crystals may deplete enough water vapor so that $s_{\max} < s_{\text{hom}}$, inhibiting homogeneous freezing (*4, 5*). In this regime only heterogeneous freezing takes place. N_c and the maximum supersaturation, $s_{\max}$, are given by solving,

$$\frac{N_{het}(s_{\max})}{N^*} = \frac{1}{\sqrt{\Delta s^*_{char}\big|_{s_{\max}}}}\frac{(1+s_{\max})}{s_{\max}}e^{\frac{2}{\lambda s_{\max}}} \tag{6}$$

Δs^*_{char} and Δs_{char} are calculated setting $s_i = s_{\max}$ in Eqs. (3) and (4), respectively.

EVALUATION

The parameterization was compared against the results of a detailed cirrus parcel model (4, 9), for different formulations of $N_{het}(s_i)$ (both theoretical and experimental) and over a wide range of *T*, *p*, aerosol concentration, and updraft velocity values (about 1200 runs). For pure heterogeneous nucleation, the overall error, with respect to parcel model results, for the calculation of s_{max} and N_c, was $-1.68\pm3.42\%$ and -2.0 ± 8.5 %, respectively. For combined homogeneous and heterogeneous freezing the error in the calculation of N_c was $4.7\pm21\%$. The low error indicates no biases in the parameterization. The new parameterization was also compared against the schemes of Liu and Penner (7, LP05) and Kärcher et al. (8, K06) (Figure 1). Results showed that the assumed freezing efficiency of the different IN populations is the main factor determining the effect of heterogeneous freezing on the total ice crystal concentration.

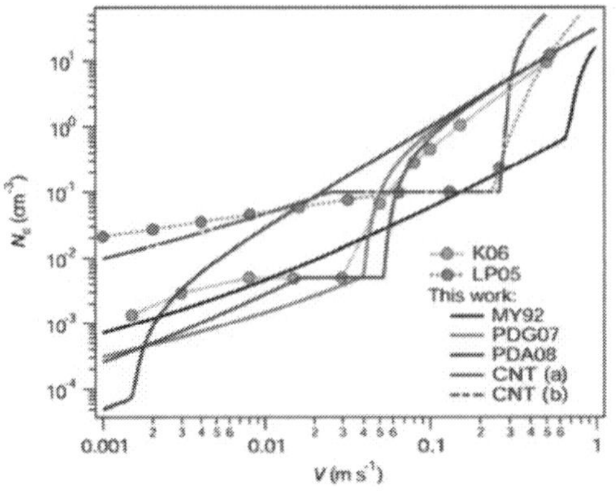

Figure 1. N_c vs. V calculated using the new parameterization using several published heterogeneous nucleation spectra (taken from Barahona and Nenes (*10*)). Also shown are results taken from Kärcher et al. (*8, K06*) and the parameterization of Liu and Penner (*7, LP05*) . Conditions considered were $T = 206$ K, $p = 22000$ Pa, $\alpha_d = 0.5$. $N_{soot} = 0.1$ cm^{-3}, $N_{dust} = 0.1$ cm^{-3}, For CNT (b), $e_{f,soot} = 1.0$, and $\alpha_d = 0.1$.

Acknowledgments

This study was supported by NASA MAP, NASA ACMAP and a NASA New Investigator Award.

References

1. D. E. Waliser *et al., Journal of Geophysical Research* **114**, doi:10.1029/2008JD10015 (2009).
2. H. R. Pruppacher, J. D. Klett, *Microphysics of clouds and precipitation* (Kluwer Academic Publishers, Boston, MA ed. 2nd, 1997), pp. 954.
3. P. J. DeMott *et al., Proceedings of the National Academy of Sciences of the United States of America* **100**, 14655 (2003).
4. D. Barahona, A. Nenes, *Atmospheric Chemistry and Physics* **9**, 369 (2009).
5. K. Gierens, *Atmospheric Chemistry and Physics* **3**, 437 (2003).
6. M. P. Meyers, P. J. DeMott, R. Cotton, *Journal of Applied Meteorology* **31**, 708 (1992).
7. X. Liu, J. E. Penner, *Meteorologische Zeitschrift* **14**, 499 (2005).
8. B. Kärcher, J. Hendricks, U. Lohmann, *Journal of Geophysical Research* **111**, D01205 (2006).
9. D. Barahona, A. Nenes, *Journal of Geophysical Research* **113**, D11211 (2008).
10. D. Barahona, A. Nenes, *Atmospheric Chemistry and Physics* **9**, 5933 (2009).
11. P. J. DeMott, M. P. Meyers, R. W. Cotton, *Journal of the Atmospheric Sciences* **51**, 77 (1994).

Modeled and Measured Carbon Gas Concentrations in Eastern Iowa

Charles Stanier, Department of Chemical & Biochemical Engineering, University of Iowa, Iowa City, IA

A. Jamroensan, Department of Civil and Environmental Engineering, University of Iowa, Iowa City, IA

A. E. Andrews, National Oceanic and Atmospheric Administration, Boulder, CO

J. Schoenfelder, Department of Chemical & Biochemical Engineering, University of Iowa, Iowa City, IA

J. Kofler, National Oceanic and Atmospheric Administration, Boulder, CO

J.C. Williams, National Oceanic and Atmospheric Administration, Boulder, CO

B. Vaughn, Institute of Arctic and Alpine Research (INSTAAR), Boulder, CO

W.D. Neff, National Oceanic and Atmospheric Administration, Boulder, CO

P.P. Tans, National Oceanic and Atmospheric Administration, Boulder, CO

C. Sweeney, National Oceanic and Atmospheric Administration, Boulder, CO

E. Dlugokencky, National Oceanic and Atmospheric Administration, Boulder, CO

S. Montzka, National Oceanic and Atmospheric Administration, Boulder, CO

W. Eichinger, IIHR Hydroscience and Engineering, University of Iowa, Iowa City, IA

P. Lewandowski, IIHR Hydroscience and Engineering, University of Iowa, Iowa City, IA

A. Pettibone, Department of Chemical & Biochemical Engineering, University of Iowa, Iowa City, IA

G. R. Carmichael, Department of Chemical & Biochemical Engineering, University of Iowa, Iowa City, IA

Atmospheric measurements of CO2 concentration and other greenhouse gases have a key role in climate science, climate change detection, and climate policy. It is likely that greenhouse gas (GHG) concentration and flux measurements will grow in importance as GHG management policies are implemented at state, national and international levels. As the chemical engineering community is actively engaged in many aspects of technology development (e.g., energy efficiency, low carbon energy, and carbon sequestration) the GHG monitoring network and its use may be of interest to chemical engineers in the energy and climate fields.

In locations with inaccurate emission inventories, top-down flux estimates are a key source of emissions data. Efforts to narrow the gap between top-down and bottom-up inventories are advancing under the guidance of the interagency North American Carbon Plan. One key activity is the establishment by the National Oceanic and Atmospheric Administration (NOAA) of a high accuracy CO2 monitoring program. CO2 and CO monitoring from a 400 m television transmitter tower located 10 km from the University of Iowa began in July 2007.

The two year measurement record of CO2, CO, 13CO2, COS, and CH4 from this tower will be presented. Corresponding model predictions from University of Iowa air quality models will be analyzed during portions of the 2 year period. Of particular focus will be determining the role of regional air-quality models such as WRF/Chem and STEM can bring to the carbon cycle problems, through their extensive treatment of pollutant gases and atmospheric chemistry processes that are not typically resolved in carbon cycle models.

See more of this Session: Advances in Climate Change Science
See more of this Group/Topical: Environmental Division

Black Carbon and Its Role in Climate

Gregory R. Carmichael, Chemical and Biochemical Engineering, University of Iowa, Iowa City, IA

Chandra Venkataraman, Department of Chemical Engineering, Indian Institute of Technology, Mumbai, India

Bhupesh Adhikary, School of Engineering, Kathmandu University, Dhulikhel, Kavre, Nepal

Sarika Kulkarni, Chemical and Biochemical Engineering, University of Iowa, Iowa City, IA

Black carbon in soot is the dominant absorber of visible solar radiation in the atmosphere. Black carbon is often transported over long distances, mixing with other aerosols along the way. The aerosol mix can form transcontinental plumes of atmospheric brown clouds, with vertical extents of 3 to 5 km. Because of the combination of high absorption, a regional distribution roughly aligned with solar irradiance, and the capacity to form widespread atmospheric brown clouds in a mixture with other aerosols, emissions of black carbon are the second strongest contribution to current global warming, after carbon dioxide emissions. The interception of solar radiation by atmospheric brown clouds leads to dimming at the Earth's surface with important implications for the hydrological cycle, and the deposition of black carbon darkens snow and ice surfaces, which can contribute to melting, in particular of Arctic sea ice. Globally, sources of black carbon emissions include burning of forests and agricultural residues, combustion of fuels for residential energy, brick production, diesel transport and coal-fired thermal power production. In this paper the role of black carbon in climate change and implications for policies directed at reducing short term climate change by targeting BC reductions are discussed.

See more of this Session: Advances in Climate Change Science
See more of this Group/Topical: Environmental Division

Sustainable Bioprocess Synthesis Routes for Tailor-Made Chemicals

Chutima Swangkotchakorn, Chemical and Biochemical Engineering, Technical university of Denmark, Lyngby, Denmark

Rafiqul Gani, Chemical and Biochemical engineering, Technical university of Denmark, Lyngby, Denmark

John Woodley, Chemical and Biochemical engineering, Technical university of Denmark, Lyngby, Denmark

Jan-Dierk Grunwaldt, Chemical and Biochemical engineering, Technical university of Denmark, Lyngby, Denmark

The decrease in the world's crude oil reserve and environmental concerns have continuously driven the current chemical processes which are heavily reliant upon fossil fuel to the use of renewable resources for the production of energy (needed in the transport and industrial sectors), as well as the production of a host of chemicals (needed to sustain chemicals-based consumer products). Since the potential to convert the renewable resources to more valuable products has dramatically increased, it means that alternative strategies can start to compete while minimizing environmental impact and increasing sustainability. Based on the conversion of material, industry would preferentially use biomass, in particular lignocellulose biomass, because it is a non-food feedstock and has low purchase cost, as well as desirable environmental and price attributes. The process for separation of biomass constituents and converting them to high value products is known as biorefining. Two major pathways, hydrolytic and thermochemical, could be considered for biorefineries. Because there are numerous ways, there is a need for a systematic methodology capable of generating, analyzing, and selecting the sustainable processing routes for conversion of a set of raw materials into a set of desired tailor-made products. Tailor-made bio-petrochemicals are those chemicals that are made to meet given specifications, by reducing and/or replacing non-renewable resources (such as petroleum) with appropriate biomaterials or a combination of both. Economic and environment impacts will be minimized through the optimal use of renewable and non-renewable feedstocks.

As a proof of concept, the synthesis routes based on chemical, biocatalytic and fermentation have been generated for two raw material resources (crude oil and lignocellulosic biomass) and a set of products (gasoline, gasohol, ethanol with different water content, etc.). A superstructure-based methodology has been developed to enumerate all possible synthesis routes and their mathematical models to calculate the material, product and residual (waste) flows for each route. Each operation in the corresponding routes is also associated with a cost so that a quick techno-economic analysis of the feasible synthesis routes can be made. Based on these results, the optimal synthesis route is identified for further analysis. The implementation of this superstructure based methodology needs a knowledge based of data for various operating in the synthesis routes and a library of models for each type of operation. The presentation will highlight the application and implementation of the superstructure two-based systematic methodology for a "model" biorefinery process. For this process, a knowledge base has been created with a large amount of collected data. Also, a library of simple mathematical models has been created. The results show the successful identification of a sustainable bio-process synthesis route for a small set of tailor-made chemicals. Future work will be to build on this (superstructure, knowledge base and model library) to develop bigger biorefinery systems.

See more of this Session: Innovation in Sustainable Process Engineering
See more of this Group/Topical: Sustainable Engineering Forum

Synergy of Technological and Ecological Systems for Self-Reliant Process Design

Robert A. Urban, Chemical and Biomolecular Engineering, The Ohio State University, Columbus, OH

Bhavik R. Bakshi, Lowrie Department of Chemical and Biomolecular Engineering, The Ohio State University, Columbus, OH

Traditionally, chemical engineers apply man-made technological principles to design processes that solve practical industrial problems. However, the multitude of goods and services that are extracted as well as those that could be used from nature are usually ignored. This has led to the deterioration of many ecosystems due to industry operating outside the constraints applied by nature and unnecessarily forced engineers to look past the industrial application of nature's unit operations. For example, the natural carbon cycle has been overwhelmed by anthropogenic sources beyond its carrying capacity causing an increase in greenhouse gas concentrations in the atmosphere. Natural goods that were once considered free and infinite such as water and air are now recognized as prized resources and should be considered as such in process design. To responsibly manage these resources and ensure their sustainability, the design and use of natural processes may be required. For example, a beverage plant that depends on large flow rates of water from existing pools may benefit from the design of a wetland system that allows for the on-site generation and filtration of water. This research aims to capture the industrial and ecological connection and design sustainable processes through the Synergy of Technological and Ecological Systems (STEMS). Instead of using technology alone to design processes, STEMS will consider industry and ecology to be one cohesive system with the goal of designing an optimum system that is self-reliant and minimizes ecological impact while maintaining industrial effectiveness. Instead of limiting the design palette to tools ordinarily used by engineers, such as traditional unit operations, STEMS would encourage the usage of natural systems to perform industrial tasks as well as stress the "waste equals food" principle where byproducts from one process are inputs to another. Wetlands can be used for water filtration and storage, trees and turfgrass for carbon sequestration, riparian buffers to prevent nutrient runoff, and various microbial activities for waste conversion into usable fuels. The smart integration of processes can integrate flows of material and energy such that there are minimal wastes of both. The concept of STEMS is demonstrated through the design of two systems: A residential home and a corn ethanol system that uses both technological and ecological systems as design elements. Optimization methods are applied to determine the impact of design alternatives on multiple objectives, such as economic potential and net carbon emissions. The results yield a Pareto optimal solution set that indicates the trade-offs between objectives under the various design variables. These results can extrapolated to process design in general, suggesting that STEMS can yield both economically and ecologically sound designs that minimize waste and promote self-sufficiency.

See more of this Session: Innovation in Sustainable Process Engineering
See more of this Group/Topical: Sustainable Engineering Forum

Effect of Green Roofs on Building Energy Costs

by

Sandra Nunez[1], Robert W. Peters[2], and Matthew Winslett[3]

[1]Graduate Student
[2]Professor of Environmental Engineering
Department of Civil, Construction, and Environmental Engineering
University of Alabama at Birmingham
1075 13th Street South
Birmingham, AL 35294-4440

[3]Manager of Energy Management Division
Facilities Management Department
University of Alabama at Birmingham
Facility Support Building
822 University Blvd.
Birmingham, AL 35294-4564

Introduction

Green roofs (vegetated roofs) are engineered ecosystems that rely on vegetation to provide services such as reduction of roof temperature [Wolf and Lundholm, 2008]. A number of advantages and features have been cited for use of green roofs, including low maintenance planting, improved building aesthetics, controlled rainwater runoff, cleaner air, reduced dust and smog levels, and improved thermal performance by helping to keep the building warm in the winter and cool in the summer [Patel, 2007]. Green roofs help with city greening, mitigating urban heat island effects, and lowering urban stormwater runoff [Berndtsson *et al.*, 2009]. A green roof enhances the thermal performance and helps retain stormwater. Existing buildings can be retrofitted with green roofs to help save energy. Thermal performance indicated a significant reduction (~40%) of a building cooling load during the summer period [Spala *et al.*, 2008]. Similar results were achieved for a nursery school, with reductions ranging from 6% to 49%, and reduction ranging from 12% to 87% on the last floor of the nursery school [Santamouris *et al.*, 2007]. Wong *et al.* [2007] note that green roofs tend to experience lower surface temperatures than the original exposed roof, especially in areas well covered by vegetation. When green roofs are well covered by vegetation, the resulting substrate moisture will tend to keep substrate temperature lower than the original exposed bare roof. They determined that over 60% of the heat gain was mitigated by vegetated roof systems.

Summertime data have indicated significant lower peak roof surface temperature and higher nighttime surface temperature for green roofs as compared to conventional roofs [Sonne, 2006]. The maximum average daily temperature seen for the conventional roof surface was 130°F in his study, while the maximum average day green roof surface temperature was 91°F (~39°F lower than the conventional roof). Wolf and Lundholm [2008] note that performance differences among growth forms and plant species across soil moisture gradients suggest that optimal water loss and roof surface cooling can be enhanced by planting multiple species. Green roofs offer cooling potential (~3.02 kWh/day) to maintain an average room air temperature of 25.7°C [Kumar, 2005]. Green roofs help minimize environmental burdens, conserve energy, and extend the life span of the roofing system in overall sustainability [Cavanaugh, 2008]. Significant research progress has been reported by the City of Chicago [undated], Penn State University [undated, a,b], and Michigan State University [undated].

Reductions in stormwater quantity can be achieved using vegetated roofs which efficiently detain and retain stormwater compared to convention (black) roofs [Carter and Rasmussen, 2006]. Green roofs improve stormwater runoff resulting from roofs [Hathaway *et al.*, 2008]. The green roofs retained a significant proportion of the rainfall (~64%), and peak runoff outflow was significantly reduced (~75%) compared to conventional roofs. Each green roof substantially delayed runoff. From a study in North Carolina's Neuse River Basin, green roofs retained a significant proportion of the rainfall; peak outflows were significantly reduced from the green roofs; and the green roofs had substantial delays in the runoff [Moran *et al.*, 2005]

Urban Heat–Island Reduction and Building Energy Conservation

For millions of Americans living in and around cities, heat islands are of growing concern. This phenomenon describes urban and suburban temperatures that are 2 to 10°F (1 to 6°C) hotter than nearby rural areas. Elevated temperatures can impact communities by increasing peak energy demand, air conditioning costs, air pollution levels, and heat-related illness and mortality. The Environmental Protection Agency (EPA) recommends installing cool or vegetative green roofs, planting trees and vegetation and switching to cool paving materials as a way of reducing the negative effects of urban heat islands. The EPA says green roofs, if installed widely in a city, can contribute to heat island reduction by replacing heat-absorbing surfaces with plants, shrubs, and small trees. The vegetation cools the air through evapotranspiration (or evaporation of water from leaves). Planted rooftops remain significantly cooler than a rooftop constructed from traditional heat-absorbing materials. Further, green roofs reduce summertime air conditioning demand by lowering heat gain to the building.

Results and Discussion

With rising energy costs and reduced state funding, the University of Alabama at Birmingham (UAB) is investigating various techniques by which their utility bills can be reduced and energy can be saved. To facilitate this action, the PI has collaborated with UAB's Facility Management Department for the past 3+ years. Current efforts include having five students perform lighting surveys in various buildings and facilities on campus. These students have been conducting inventories of buildings, noting lighting systems, magnetic versus electronic ballasts, exit signage (incandescent, LED, etc.), potential for motion/occupancy sensors in various rooms, etc. To date this summer, lighting surveys have been conducted on the Business and Engineering Complex (BEC), Hoehn Engineering Building, the 19th Street Parking Deck, and the School of Dentistry. These efforts will help identify where lighting replacement should occur and offer means by which significant energy savings can be achieved at relatively modest cost.

Additionally, the PI has been conducting research (in collaboration with the Facilities Management Department) for the past 2+ years investigating retrofitting of campus building roofs. Fifteen (15) mini-roofs have been set up on top of the Chemistry Building. Each mini-roof is 8-ft long x 4-ft wide x 4-ft deep. A number of different roofing systems are being examined for their energy performance. All roofs are insulated with 2.0-in. of extruded polystyrene which has an *R*-value of 0.10. Each roof is equipped with a temperature sensor which records the temperature inside the roof (located about halfway along the length of the mini-roof). The temperature readings are transmitted electronically to a data logger. The data is sent electronically to the PI each day, with readings recorded every 10 minutes for all 15 mini-roofs. Additionally, a pilot-vegetated roof has been set up on top of Hulsey Center. The min-roofs and pilot green roof are described below.

Mini-Roofs

The roofing systems being studied using the 15 mini-roofs are listed in Table 1 below.

Table 1. Mini-roof descriptions.

Mini-Roof No.	Mini-Roof Description
1 (Sensor B)	White TPO/PVC/Elvaloy fully adhered, FiberTite Membrane.
2 (Sensor C)	Black 60-mil EPDM fully adhered/coated/white urethane, Mule Hide Membrane.
3 (Sensor D)	Black 60-mil EPDM fully adhered, Mule Hide Membrane.
4 (Sensor E)	Beige TPO/PVC/Elvaloy fully adhered, FiberTite Membrane.
5 (Sensor F)	White granular modified, Firestone SBS Modified Membrane.
6 (Sensor G)	Black granular modified, Firestone SBS Modified Membrane.
7 (Sensor H)	Black granular modified coated/white urethane, Firestone SBS Modified Membrane.
8 (Sensor I)	Bituthene IRMA with lightweight "T Clear" pavers.
9 (Sensor J)	Bituthene IRMA with river rock ballast.
10 (Sensor K)	Bituthene IRMA with vegetative green roof, ½-in. drain mat, 350-lbs dry soil.
11 (Sensor L)	Bituthene IRMA with vegetative green roof, 1-in. drain mat, 350-lbs dry soil.
12 (Sensor M)	Black 60-mil EPDM loose, ballasted with river rock, Mule Hide Membrane.
13 (Sensor N)	Black 60-mil EPDM loose, ballasted with #300 marble chips, Mule Hide Membrane.
14 (Sensor O)	White TPO/PVC/Elvaroy loose laid, ballasted with river rock, FiberTite Membrane.
15 (Sensor P)	Bituthene IRMA with vegetative green roof, ½-in. drain mat, 350-lbs dry soil.
Sensor S	Sensor inside mini-roof No.10 (inside the soil of the green roof).

Sensor T	Sensor under the white TPO/PVC/Elvaloy loose laid, ballasted with river rock.

We have been collecting surface temperature measurements during the month of June on the various roofing materials used with our mini-roof systems. During this time period, we have observed roofing surface temperatures ranging from 69°F to 180°F. The lower surface temperature values, 69°F to 119°F, are found in a loose rock or stone roof combination. The higher temperature values occur on the roofing combinations primarily made of a coating or membrane; these roofs exhibit temperatures roughly between 78°F to 180°F. The vegetative mini-roofs exhibit surface temperatures ranging between 71°F to 126°F, while their large counterpart, the pilot roof on top of Hulsey Center, exhibits higher temperatures ranging from 89°F to 142°F. When looking at the specific roofs contained in each subset of Table 2, the lighter the roof, generally the cooler it is. For example, the white Firestone SBS is cooler than the black Firestone SBS on any given day. This thermal property is observed due to the reflectivity of the roof. The darker colored roofs absorb more incoming light radiation than the light colored roofs causing the dark roofs to become hotter. (Since the roofs temperatures observed are taken during the late spring and early summer, it can be inferred that overall roof temperature will increase during late summer).

Table 2. Surface temperature ranges* observed on the mini-roofs.

General Roof Type	Surface Temperature Range
Coating/Membrane (e.g., PVC or Urethane or Firestone SBS) [Colors: Black and White]	78°F –180°F: Black Firestone SBS: 104°F–180°F; White Firestone SBS: 102°F–158°F)
Vegetative (1–2 plant species/dark and light small rock substrate)	71°F–126°F
Loose Rock (e.g., river rock or white marble chips)	69°F –119°F
White Pavers Stone	69°F–119°F
Pilot roof (~20 plant species/dark and light small rock substrate/gray pavers stone)	89°F– 142°F

*Temperature ranges are based on the average values per reading between June 1st, 2009 and July 2nd, 2009.

Photographs of several of these mini-roofs are presented in Figures 1 through 5. Figure 6 presents a typical temperature profile of the 15 mini-roofs during the course of a day. This figure shows a cyclical nature of the temperature readings over a 24 hour period.

Surface temperatures of the roofing materials during hot summer days were ~170–180°F for the black roofs; ~110°F for the clean white roof materials; and ~145°F for the soils of the green roof systems. However, due to transpiration, the temperature of the surface plants was ~115°F. Thus, as the plant systems grew and spread, the surface temperature decreased significantly, resulting in a lower heat load being transmitted into the building. Another noteworthy observation is that as the white roof systems became dirty (accumulating dust, etc.), the surface temperature rose considerably, to ~140°F.

The results from these mini-roof structures have shown the following trends [Nunez and Peters, 2008]:

- Clean white roofs resulted in consistently lower temperatures inside the mini roof than the other roofing materials.
- Black roofs resulted in the highest temperature readings.
- Green roofs resulted in temperatures typically ~2–3 degrees higher than the white roofs; however, they will dampen the drainage of rainfall during a rain storm through the retention of water onto the soil.
- Bituthane (river rock) performs only slightly better than black roofing materials.
- White granular roofing behaved similarly to black granular roofing materials.
- Over time, the reflective (white) roofs become dirty, losing some of their reflectivity, resulting in the roof being less energy efficient.
- White marble chips behaved slightly worse than green roofing materials, but considerably better than black roofing and granular roofing materials.
- While it is too early to come to a definitive conclusion, preliminary evaluations indicate that the "white" and "green" roofs both significantly reduce the roofs surface temperature and therefore, the air temperatures above and around the roof.

Figure 1. Black mini-roof.

Figure 2. White (reflective) mini-roof.

Figure 3. Mini-roof equipped with river rocks.

Figure 4. Mini-roof equipped with crushed marble.

Figure 5. Vegetative mini-roof equipped with sedum plants.

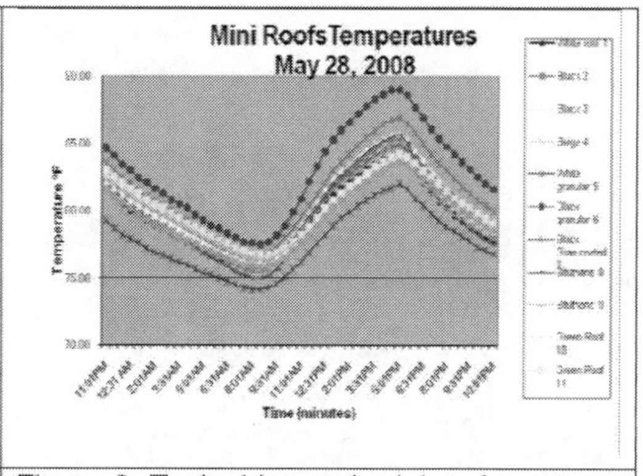

Figure 6. Typical internal mini-roof temperatures during the course of a day.

Based on the information collected on these mini-roof systems, the UAB Facilities Management Department decided to install a vegetated roof on top of Hulsey Center as a roofing retrofit. Under this current proposal, UAB wants to obtain more fundamental information and knowledge for establishing green roofs in the southeastern U.S. In the Birmingham, Alabama area, we receive fairly high rainfalls (approximately 52 inches/year on average). However, during the summer months, it is common to have periods of drought with minimal rainfall and very hot and humid days (with temperatures in the 90+° range). Such climatic conditions require plants that can withstand both significant rainfall events and drought conditions.

Photographs of the construction of the pilot green roof system on Hulsey Center are shown in Figures 7 to 9. Photographs of the system taken in June 2009 are shown in Figures 10 to 12.

Figure 7. Construction phase for installing a pilot vegetative roof on top of Hulsey Center.

Figure 8. Construction phase for installing a pilot vegetative roof on top of Hulsey Center.

Figure 9. Initial vegetative roof immediately after installation on top of Hulsey Center.

Figure 10. Vegetative roof on top of Hulsey Center (June 2009).

Figure 11. Vegetative roof on top of Hulsey Center (June 2009).

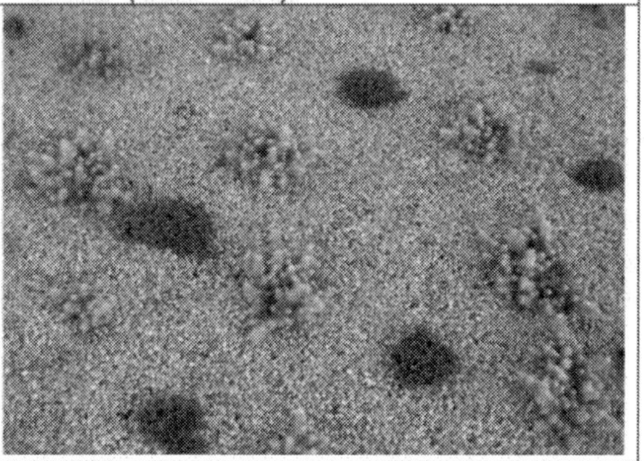

Figure 12. Close-up of vegetative roof on top of Hulsey Center (June 2009).

Of the roofing area of 18,400 ft^2 for Hulsey Center, a pilot green roof has been installed on ~14,940 ft^2 (i.e., occupying ~81.2% of the roofing area, or approximately 0.34 acres. [Prior to installing the pilot green roof, the civil engineering senior design class investigated the loading of a wet vegetative roof on top of Hulsey Center, and found that the current building infrastructure could withstand the loading associated with the vegetative roof. Hulsey Center was originally designed to have two additional floors in the building]. This pilot green roof contains more than 20,000 sedum plants. In the construction of this vegetated roof, the existing roof was removed down to the structural concrete deck. A waterproof roofing membrane was installed directly to the concrete deck. A layer of 3-inches of extruded polystyrene roofing insulation was then applied, to which an 8-oz non-woven geo-textile scrim sheet was applied. Then, 2-ft x 2-ft x 2-in prestressed pavers or brick pavers were installed for design, decoration, and access to the vegetative roof. Then, 3 ½-inches of light weight engineered soil were applied. Sedum plants were planted at the rate of 150 plants per 100-ft^2. The cost of retrofitting the roof and installing a pilot green roof on top of Hulsey Center was ~$159,000.

The current pilot green roof on top of Hulsey Center was installed in July 2008. Research conducted at Penn State University shows that green roofs planted with sedum plants reduce the building energy costs [Penn State Center for Green Roof Research, undated]. Our Facilities Management Department maintains records of the utilities bills (e.g., natural gas, water, and electricity) for each building on campus. The costs of utilities prior to and after implementation of the pilot green roof are shown in Figure 13 through 15 for natural gas, water, and electricity. Since 2006, the costs of natural gas, water, and electricity have increased by 47%, 28%, and 145%, respectively. Due to the increasing rates of these utilities, a more accurate determination is shown with the usage of these utilities, shown in Figure 16 through 18. In the graph symbols, those indicated with a red interior color depict the cost or usage after installation and implementation of the vegetative roof system. The other symbol colors depict the cost or usage prior to installation and implementation of the vegetative roof system. Generally, the usage of natural gas, water, and electricity are lower than that compared to the utility usage prior to installation of the green roof on top of Hulsey Center.

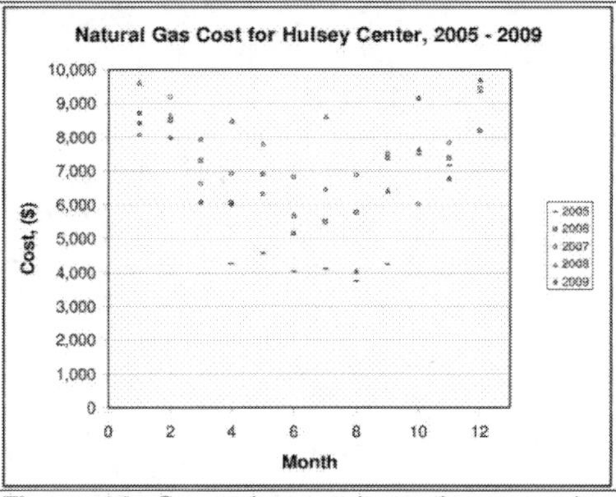

Figure 13. Cost of natural gas for operation of Hulsey Center, 2005 – 2009.

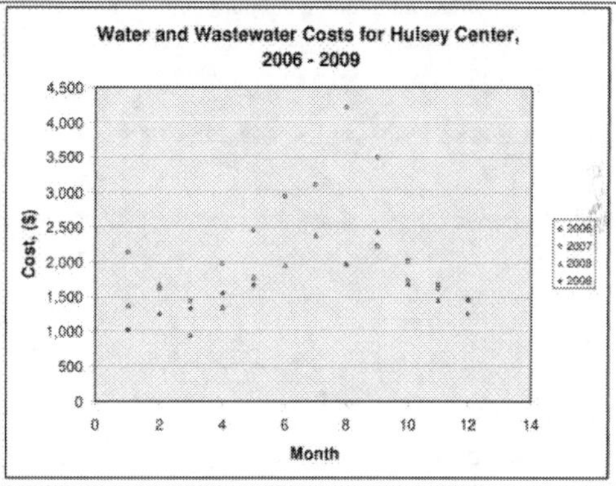

Figure 14. Cost of water/wastewater for operation of Hulsey Center, 2006 – 2009.

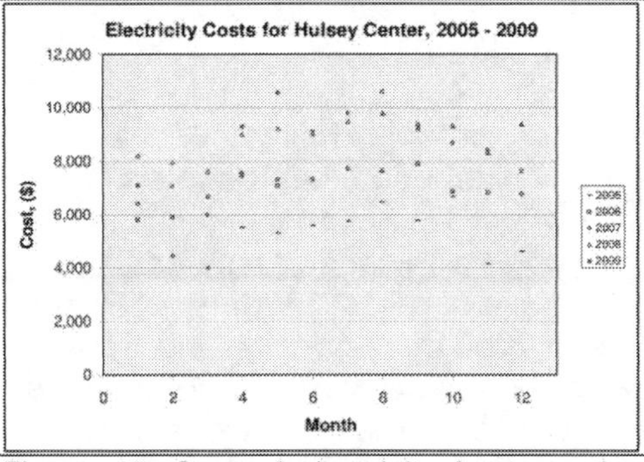

Figure 15. Cost of electricity for operation of Hulsey Center, 2005 – 2009.

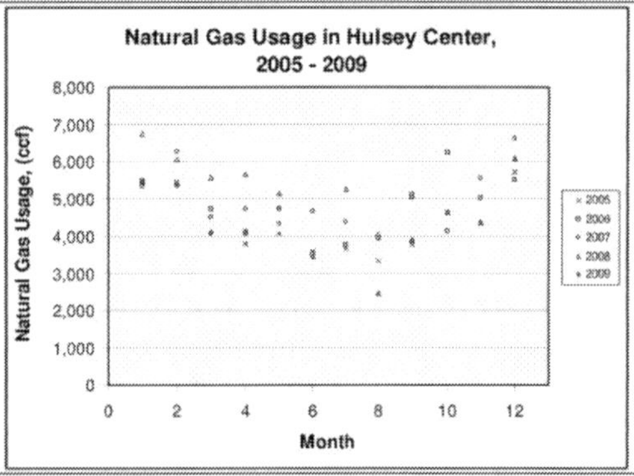

Figure 16. Natural gas usage for operation of Hulsey Center, 2005 – 2009.

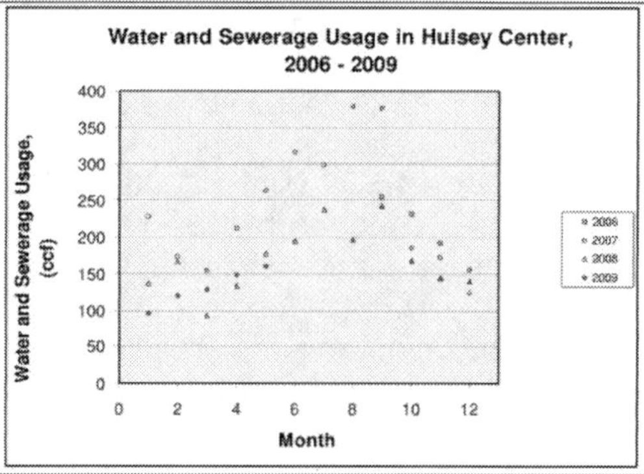

Figure 17. Water/wastewater usage for operation of Hulsey Center, 2006 – 2009.

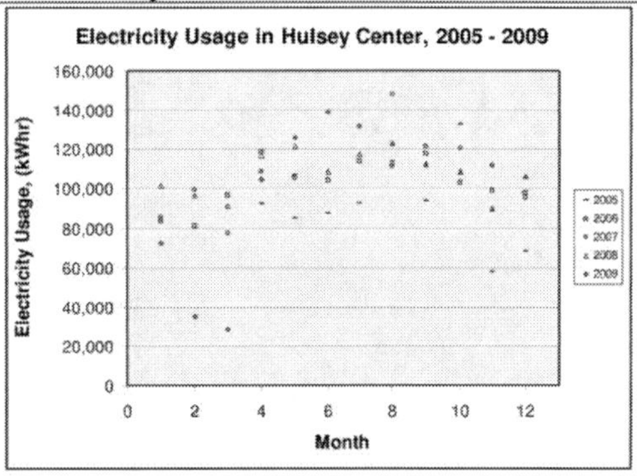

Figure 18. Electricity usage for operation of Hulsey Center, 2005 – 2009.

Summary and Conclusions

Fifteen mini-roofs equipped with different roofing materials have been installed and investigated on the UAB campus. A pilot vegetated roof has been established on top of Hulsey Center. The vegetated roofing systems and the reflective roofing materials result in lower surface temperatures of the roofs (compared to black roofs and concrete materials), thereby lowering the heat load that is delivered to the building. Use of reflective roofs and vegetated roofs result in lower heat loads that must be dissipated through air conditioning operations. From the pilot roof on top of Hulsey Center, we compared the utility bills for three years prior to the installation of the pilot vegetated roof with the utility bills for more than a year since implementation of the pilot vegetated roof. Reductions in natural gas, water, and electricity have averaged 15%, 22%, and 20%, respectively. Retrofitting of building roofs in the Southeastern U.S. offers both environmental and economic benefits.

Acknowledgments

The financial support and technical support of the UAB's Facilities Management Department is appreciated, including the fabrication and installation of the mini-roofs (on top of the Chemistry Building) and pilot roof on top of Hulsey Center. The input and technical assistance provided by Matt Winslett and Kelly Winnett of the Facilities Management Department in supplying the utility bill information is gratefully acknowledged.

References Cited

Berndtsson, J.,C., L. Bengtsson, and K. Jinno, 2009. "Runoff Water Quality from Intensive and Extensive Vegetated Roofs", ***Ecol. Eng.***, ***35***(3): 369–380.

Carter, T.L., and T.C. Rasmussen, 2006. "Hydrologic Behavior of Vegetated Roofs", ***J. Am. Water Resour. Assoc.***, ***42***(5): 1261–1274.

Cavanaugh, L.M., 2008. "Redefining the Green Roof", ***J. Archit. Eng.***, ***14***(1): 4–6.

City of Chicago, Green Roof Test Plots, undated. City of Chicago website: http://www.cityofchicago.org/city/webportal/portalContentItemAction.do?blockName=Environment%2fGreen+Roof+%26+Cool+Roof+Grants+Programs%2fI+Want+To&deptMainCategoryOID=-536887205&channelId=0&programId=0&entityName=Environment&topChannelName=Dept&contentOID=536912453&Failed_Reason=Invalid+timestamp,+engine+has+been+restarted&contenTypeName=COC_EDITORIAL&com.broadvision.session.new=Yes&Failed_Page=%2fwebportal%2fportalContentItemAction.do&context=dept; accessed July 4, 2009.

Hathaway, A.M., W.F. Hunt, and G.D. Jennings, 2008. "A Field Study of Green Roof Hydrologic and Water Quality Performance", ***Trans. American Society of Agricul. Biol. Engrs.***, ***51***(1): 37–44.

Kumar, R., 2005. "Performance Evaluation of Green Roof and Shading for Thermal Protection of Buildings", ***Build. Environ.***, ***40***(11): 1505–1511.

Michigan State University Green Roof Research Program, Department of Horticulture, undated. Michigan State University website. http://www.hrt.msu.edu/greenroof/, accessed July 5, 2009.

Moran, A.C., W.F. Hunt, and J.T. Smith, 2005. "Green Roof Hydrologic and Water Quality Performance from Two Field Sites in North Carolina", ***Proc. 2005 Watershed management Conference – Managing Watersheds for Human and Natural Impacts: Engineering, Ecological, and Economic Challenges***, pp. 1175–1186, Williamsburg, VA, (July 19–22).

Nunez, S., and R.W. Peters, 2008. "Effect of Roofing Materials on Building Energy Efficiency", Paper included in the conference proceedings of the 2008 Annual AIChE Meeting, Philadelphia, PA, (November 16–21).

Patel, A., 2007. "The Green Roof Challenge", ***Struct. Eng.***, ***85***(9): 29–30.

Penn State Center for Green Roof Research, undated. Pennsylvania State University website: http://horticulture.psu.edu/cms/greenroofcenter/about_ctr.html; accessed July 5, 2009.

Penn State Center for Green Roof Research, undated. "Utility Bills Getting You Down? Plant Stuff Up on the Roof", http://aginfo.psu.edu/news/2000/9/roof.html; accessed July 5, 2009.

Santamouris, M., C. Pavlou, P. Doukas, G. Mihalakakou, A. Synneta, A. Hatzibiros, and P. Patargias, 2007. "Investigating and Analysing the Energy and Environmental Performance of an Experimental Green Roof System Installed in a Nursery School Building in Athens, Greece", ***Energy***, ***32***(9): 178–1788.

Sonne, J., 2006. "Evaluating Green Roof Energy Performance", ***ASHRAE J.***, ***48***(2): 59–61.

Spala, A., H.S. Bagiorgas, M.N. Assimakopoulos, J. Kalavrouziotis, D. Matthopoulos, and G. Mihalakakou, 2008. "On the Green Roof System, Selection, State of the Art and Energy Potential Investigation of a System Installed in an Office Building in Athens, Greece", ***Renew. Energy***, ***33***(1): 173–177.

Wolf, D., and J.T. Lundholm, 2008. "Water Uptake in Green Roof Microcosms: Effects of Plant Species and Water Availability", ***Ecol. Eng.***, ***33***(2): 179–186.

Wong, N.H., P.Y. Tan, and C. Yu, 2007. "Study of Thermal Performance of Extensive Rooftop Greenery Systems in the Tropical Climate", ***Build. Environ.***, ***42***(1): 25–54.

Supercritical Carbon Dioxide: A Promising Green Solvent in One-Pot Synthesis of ZrO_2-TiO_2 Nanotubes

Rahima A. Lucky, Chemical and Biochemical Engineering, The University of Western Ontario, London, ON, Canada

Paul A. Charpentier, Department of Chemical and Biochemical Engineering, The University of Western Ontario, London, ON, Canada

Recently, considerable effort has been devoted to synthesizing inorganic one-dimensional (1D) nanostructures (e.g. nanotubes, nanowires) because of their unique physical properties compared to their bulk counterparts. Many exciting potential applications are possible including catalysis, high efficiency solar cells, coatings, and sensors. Amongst the 1D oxide nanomaterials reported, titania (TiO_2) nanomaterials are receiving considerable attention due to titania's high activity, strong oxidation capability, and chemical stability. Modifying TiO_2 with ZrO_2 , one of the most suitable dopants, also enhances its performance by increasing thermal stability, surface area and reducing crystallite size as the performance of titania nanomaterials strongly relies on their crystallinity, crystallite size, crystal structure, and specific surface area.Several different processing techniques have been used for the preparation of TiO_2 based nanostructures, such as anodization, template techniques, hydrothermal processes, and soft chemical processes. However, each of these methods has limitations. Due to these limitations, alternative processes using a green solvent such as supercritical carbon dioxide ($scCO_2$) is attractive to synthesize TiO_2-base nanomaterials.

Here we present a very simple, surfactant free sol-gel synthesis route for producing high surface area, mesoporous, ZrO_2-TiO_2 binary nanotubes, under mild conditions. An acid modified sol-gel technique was employed to synthesize the one-dimensional bimetallic ZrO_2-TiO_2 materials studying several solvents of varying dielectric constants including supercritical carbon dioxide ($scCO_2$),hexane, isopropanal and ethanol. The synthesized materials were characterized by several physicochemical techniques (e.g. electron microscopy (SEM and TEM), BET, XRD, and FTIR) in order to assess the preparation method and the effects of the solvents on the morphology, porous structure, and thermal and crystallization behavior of the nanomaterials.

SEM and TEM analysis show that nanotubes having diameters of 55-140 nm were formed in $scCO_2$,whereas nanorods and/or tubes with diameter ranging from 40-250 nm were seen when the samples were prepared in organic solvent. The BET analysis also show that the surface area, and pore volume followed a linear relationship with dielectric constant of the utilized solvent. When low dielectric constant solvents such as $scCO_2$ were employed, synthesized materials had higher surface area. The presence of such solvents helps in the formation of porous structure with high surface area materials. However, when alcoholic solvents having high dielectric constants were used, dense materials having low surface areas were produced.

These high surface area, mesoporous, ZrO_2-TiO_2 binary nanotubes have lots of potential applications. In addition, this process is very easy to scale up for large scale industrial applications.

See more of this Session: Innovation in Sustainable Process Engineering
See more of this Group/Topical: Sustainable Engineering Forum

Synthesis of Crystal α-Capric Betaine Zwitterionic Surfactant

Yun FANG, Minyun JIANG*

School of Chemical & Material Engineering, Jiangnan University, Wuxi, Jiangsu 214122, China

Abstract

In this paper, *α*-capric betaine (*α*-CB) was prepared through quaterisation of *α*-chlorododecanoic acid (*α*-CDA) dissolved in liquid trimethylamine. Crystal *α*-CB was prepared after evaporating trimethylamine. This novel solvent-free quaterisation technology reduced the byproduct of *α*-hydroxyldodecanoic acid (*α*-HDA) and without solvent contamination. The effect of the molar ratio of raw materials, reaction temperature and reaction period on quaterisation of *α*-CDA with trimethylamine was studied in orthogonal design. Optimization of process condition was approached as follows, n (trimethylamine) : n (*α*-CDA) = 6:1, reacted for 10 h at 80 □ the yield of *α*-CB reached 92.6%.

Keywords: Solvent-free quaterisation, alpha-capric betaine, alpha-chlorododecanoic acid, Synthesis

Introduction

Alkyl betaines and their derivatives represent a class of zwitterionic surfactants with a positive charge on the nitrogen atom and a negative charge on the carboxyl group. They are widely used in personal care products, domestic detergents, industrial detergents and other industrial applies, because of their particular structures and synergistic effect with other types of surfactants[1]. However, the related problems discovered in the synthesis and application of betaine zwitterionic surfactants, such as the complex synthesis route, the expensive raw materials, and the unavoidable side effects brought by byproducts made the development of betaine zwitterionic surfactants suffocated.

In this paper, *α*- chlorododecanoic acid (*α*-CDA), which has been prepared from natural lauric acid by an economic method in our laboratory[2], reacted with liquid trimethylamine in a hermetic pressure reactor that keeps trimethylamine in liquid state, and the product *α*-capric betaine (*α*-CB) was obtained via a solvent-free quaterisation route as Scheme 1.

$$\underset{\text{Cl}}{\text{RCHCOOH}} + 2\text{NH(CH}_3)_3 \longrightarrow \underset{\text{N(CH}_3)_3^{+}}{\text{RCHCOO}^{-}} + (\text{CH}_3)_3\text{NHCl}$$

where $R=C_{10}H_{21}$

Scheme 1 The synthesis route of *α*-CB from *α*-CDA via a solvent-free quaterisation

The previous work in our laboratory[3] discovered that the side reaction followed Scheme 2 was inevitable in aqua system, especially in high pH and high temperature environment. It is proved in this

* Corresponding author: yunfang@126.com. Supported by by National Key Technologies Research and Development Program of China during the 11st Five-year Plan (2007BAE52B01).

paper that quaterisation of α-CDA in liquid trimethylamine successfully prevents from byproduct α-hydroxyldodecanoic acid (α-HDA) occurring.

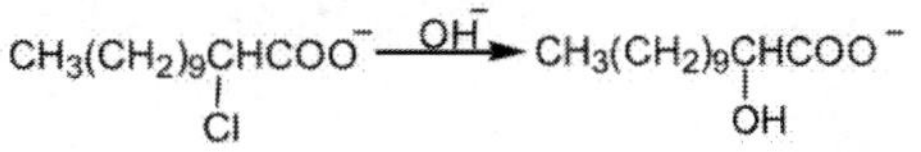

Scheme 2 The main side reaction in the aqua quaterisation of α-CDA

Materials and methods

Materials and equipment α-CDA is prepared in our laboratory[2] with over 96% weight percentage without further treatment. Trimethylamine was purchased from Xinya industrial Co. Ltd., PRC. The solvent-free quaterisation was performed in a 100 ml autoclave (WDF-0.1, Weihai Automatically-controlled Reaction Kettle Co. Ltd., PRC).

Preparation of α- Capric betaine zwitterionic Surfactant The α-chlorododecanoic acid (10 g, 0.04 mol) was highly dispersed in liquid trimethylamine (15 g, 0.25 mol) in a hermetic pressure reactor. The reaction continued for 10 h at 80 □. A yield of 92.6% of α-CB was achieved. After evaporating and recycling the superfluous trimethylamine, pale yellow crystal α-CB was obtained.

Purification of α- Capric betaine Surfactant Purified amphoteric α-CB was obtained according to the following steps: first, adjust the pH of aqueous solution of α-CB to 1.0-2.0 with concentrated hydrochloric acid and then extract the residual fattiness with ethyl acetate; secondly, recrystalize use acetone solvent for three times; thirdly, adjust the aqueous solution of recrystalized α-CB to pH 7.0 with NaOH and then dialysis with semipermeable membrane. Get purified crystal amphoteric α-CB after vacuum drying.

Results and Discussion

Optimizing the Reaction Conditions Orthogonal design was applied to optimize quaterisation condition. The present work was carried out to develop practical reaction conditions. The molar ratio of trimethylamine to α-CDA, temperature and reaction time, which were found to be the key factors influencing the yield of α-CB were investigated. And the results were as shown in Table 1 and Table 2.

Table 1 Factors and levels in orthogonal design

factors	Molar ratio A	Temperature B	Time C
level 1	6:1	70 °C	6 h
level 2	8:1	80 °C	10 h
level 3	10:1	90 °C	14 h

Table 2 The arrangement, result and the analysis of L_9 orthogonal experiment

Run	Factor A	Factor B	Factor C	yield (%)
1	1	1	3	85.0
2	2	1	1	85.2
3	3	1	2	82.6
4	1	2	2	88.1
5	2	2	3	91.2
6	3	2	1	92.3
7	1	3	1	89.3
8	2	3	2	87.5
9	3	3	3	86.9
I /3	87.5	84.3	86.1	
II/3	87.9	90.5	87.7	
III/3	87.3	87.9	88.9	
range	0.7	6.3	2.8	

Table 2 showed that the influence of molar ratio of raw materials was least which means the mole number of trimethylamine is great enough. But the mole number of trimethylamine can not be further cut because trimethylamine serves as both reactant and solvent which facilitates the mass transfer and heat transfer. The results in Table 2 showed that temperature was the main influence factor and reaction at 80 °C was the optimum temperature. The yield of *α*-CB increased along with reaction time. But by considering production cycle and energy saving, we chose 10 h as the optimum reaction period.

The structural characterization was done in details. The infrared spectrum showed a carbonyl band of carboxylate at 1640 cm^{-1} and a weak C–N stretching vibration bond of the quaternary ammonium salt at 960 cm^{-1}while the infrared spectrum of the reactant *α*-CDA showed a carbonyl band of a carboxyl group at 1726 cm^{-1}. The derived ^{1}H-NMR spectrum was analyzed and assigned as follows: 0.73(triplet, 3H, $–CH_3$); 1.23–1.60 [multiplet, 16H, – $(CH_2)_8$–]; 1.72 [multiplet, 2H, $–CH_2–CH–N+(CH)_3)_3$]; 3.08 [single, 9H, $–N+((CH_3)_3)$]; 3.48–3.52 (multiplet, 1H, a-H $–CH_2–CH–N+(CH)_3)_3$.).

In conclusion, *α*-capric betaine (*α*-CB) was obtained via a solvent-free quaterisation route in liquid trimethylamine. This novel solvent-free quaterisation technology reduced the byproduct of *α*-hydroxyldodecanoic acid (*α*-HDA) and without solvent contamination. Optimization of quaterisation process condition was approached as follows, n(trimethylamine): n(α-CDA) = 6:1, reacted for 10 h at 80 °C. The yield of α-CB reached 92.6%.

References

1. Delgado C, Dolores Merch´an M, Mercedes Vel´azquez M, Anaya J. Effect of surfactant structure on the adsorption of carboxybetaines at the air–water interface, Colloids and Surfaces A: Physicochem. Eng Aspects. 2006, 280: 17
2. Qi Liyun, Fang Yun. Synthesis and Physicochemical Investigation of α-Long Alkylchain Betaine Zwitterionic Surfactant. J Surfact. Deterg. 2008, 11: 55-59
3. Jiang Liangyu. Preparation Technology of α-Capric Betaine Derived from α-Chlorododecanoic Acid. Master Dissertation: Jiangnan University, China, 2008

American Clean Energy and Security Act (H.R. 2454) - June 26, 2009

U.S. House of Representatives' American Clean Energy and Security Act ("ACESA," "H.R. 2454," "Waxman-Markey" or the "Bill") will be making its way through the Senate, either in 2009 or 2010. This Bill, if adopted in substantially similar fashion by the Senate, would cut U.S. emissions 17 percent by 2020 from a baseline of 2005 emissions and 83 percent by 2050. The emissions reductions would be achieved through a cap-and-trade approach, rather than a carbon tax. The ACESA would also mandate energy efficiency measures as well as minimum electricity generation from renewable energy by utilities in increasing percentages (6% in 2012, 9.5% in 2014, 13% in 2016, 16.5% in 2018, and 20% in 2021-2039). In support of these mandated efficiency and renewable mandates, ACESA includes provisions issuing, trading, banking, retiring, and verifying renewable electricity credits; and prescribing standards to define and measure electricity savings from energy efficiency and energy conservation measures.

The ACESA as adopted by the House is quite extensive, addressing clean energy (combined efficiency and renewable electricity standards, carbon capture and sequestration, clean transportation, smart grid advancement, transmission planning, nuclear), energy efficiency (building energy efficiency, lighting and appliances, transportation and mobile sources, industrial energy efficiency, public institutions, green resources for energy efficient neighborhoods), reducing global warming pollution (reduction goals, designation and registration of greenhouse gases, program rules, offsets, emissions reductions from reduced deforestation, disposition of allowances, exemptions of GHG from other CAA programs, black carbon, carbon market assurance), transitioning to a clean energy economy (ensuring real reductions in industrial emissions, green jobs and worker transition), energy refund program (exporting clean technology) and adapting to climate change (domestic, public health, natural resources, international and agricultural and forestry related offsets).

Specifically, and as most relevant to AICHE members, ACESA would amend the Clean Air Act (CAA) to require the U.S. EPA ("EPA") to adopt a national strategy to address barriers to the commercial-scale deployment of carbon capture and sequestration; establish an approach to certify and permit geologic sequestration; and promulgate regulations to minimize the risk of escape to the atmosphere of carbon dioxide injected for purposes of geological sequestration.

The Bill would amend the Safe Drinking Water Act to require EPA to promulgate regulations for geologic sequestration, specifically the underground injection wells proposed for use as permanent disposal of carbon dioxide by geologic sequestration.

ACESA includes provisions regarding development of a smart electrical grid to balance peak and baseline electricity demand loading, including electricity grid planning facilitating use of renewable and other zero-carbon energy sources.

ACESA includes provisions requiring amendment of existing rules to require improved energy efficiency in industrial equipment, electric motors and lighting and appliances, while requiring establishment of mechanisms to promote commercial use of alternative energy technologies.

ACESA would require establishment of a program to make monetary awards to encourage owners and operators of electric energy generation facilities or thermal energy production facilities using fossil or nuclear fuel to use innovative means of recovering any thermal energy that is a potentially useful byproduct of their processes to: (1) generate additional electric energy; or (2) make sales of thermal energy not used for electric generation, in the form of steam, hot water, chilled water, or desiccant regeneration, or for other commercially valid purposes.

The Bill designates CO2, CH4, N2O, SF6, HFCs, PFCs and NF3 as GHGs and establishes a carbon dioxide equivalent (CO2e) value for each gas. Then, the bill would amend the CAA to require EPA to promulgate regulations that would cap and reduce GHG emissions, annually, so that GHG emissions from capped sources are reduced to 97% of 2005 levels by 2012, 83% by 2020, 58% by 2030, and 17% by 2050; and track these reductions through a federal GHG registry. The bill would require EPA to establish specific emission allowances (i.e., annual tonnage GHG emission limits) for 2012-2049 and then 2050 and after, while providing compensatory allowances for the destruction and conversionary use of fluorinated gases and the non-emission creating use of petroleum-based or coal-based liquid or gaseous fuel, petroleum coke, natural gas liquid, or natural gas as a feedstock.

The bill would phase in enforceable prohibitions against exceeding these emission allowances (augmented with available offsets) by electricity sources, fuel producers and importers, industrial gas producers and importers, geological sequestration sites, industrial stationary sources, industrial fossil fuel-fired combustion devices, natural gas local distribution companies, nitrogen trifluoride sources, algae-based fuels, and fugitive emissions. Allowances could be traded, banked, borrowed, auctioned, sold, exchanged, transferred, held or retired. EPA would establish a strategic reserve account to hold a certain number of emission allowances for each of calendar years 2012-2050; and auction allowances from the strategic reserve once each quarter.

ACESA would amend the CAA to require EPA to promulgate GHG emission performance standards for specified categories of stationary sources that have uncapped GHG emissions greater than 10,000 tons of CO2e and are responsible for emitting at least 20% of the uncapped GHG gas emission annually; or are responsible for at least 10% of the uncapped methane emissions. The bill would also require EPA to phase down consumption and regulate the production of HFCs and reduce emissions of black carbon.

Significantly, ACESA would prohibit states from implementing a cap and trade program that covers any capped emissions emitted during 2012-2017.

ACESA would require the Federal Energy Regulatory Commission (FERC) to establish, operate, and oversee markets for regulated allowances.

Finally, ACESA includes provisions supporting emerging careers and jobs in renewable energy, energy efficiency, and climate change mitigation and providing climate change adjustment assistance to adversely affected workers.

All 1420 pages of the ACESA as passed by the House are available for review at http://www.govtrack.us/congress/billtext.xpd?bill=h111-2454 . ACESA should be debated in the Senate after the August 2009 recess.

Mary Ellen Ternes, Esq.
Director, AICHE Environmental Division
McAfee & Taft, A Professional Corporation
Maryellen.ternes@mcafeetaft.com

09AIChE
Annual Meeting, Nashville, TN

Greenhouse Gas Emission Regulation Pursuant to the Clean Air Act

November 11, 2009

Bob Martineau, Esq.
Partner, Regulatory & Environmental Law
Waller Lansden
WALLER LANSDEN DORTCH & DAVIS LLP
511 Union Street, Suite 2700
Nashville, TN 37219
615.850.8910
Bob.Martineau@wallerlaw.com

Rick Bolton
Director, Air Quality Services
CTEH
Center for Toxicology & Environmental Health, LLC
124 First Avenue South, Suite 100
Franklin, TN 37064
615.628.3202
rbolton@cteh.com

Overview

09AIChE
Annual Meeting, Nashville, TN

- *Massachusetts v. EPA*
- Statutory Provisions with Endangerment Language
- EPA Proposed Endangerment Finding
- Permitting Issues
- Next Steps for EPA

09AIChE
Annual Meeting, Nashville, TN

Definition of "Air Pollutant" Under Clean Air Act

...Any air pollution agent or combination of such agents, including any physical, chemical, biological, radioactive... substance or matter which is emitted into or otherwise enters the ambient air. Such term includes any precursors to the formation of any air pollutant...

- [§ 302(g)]

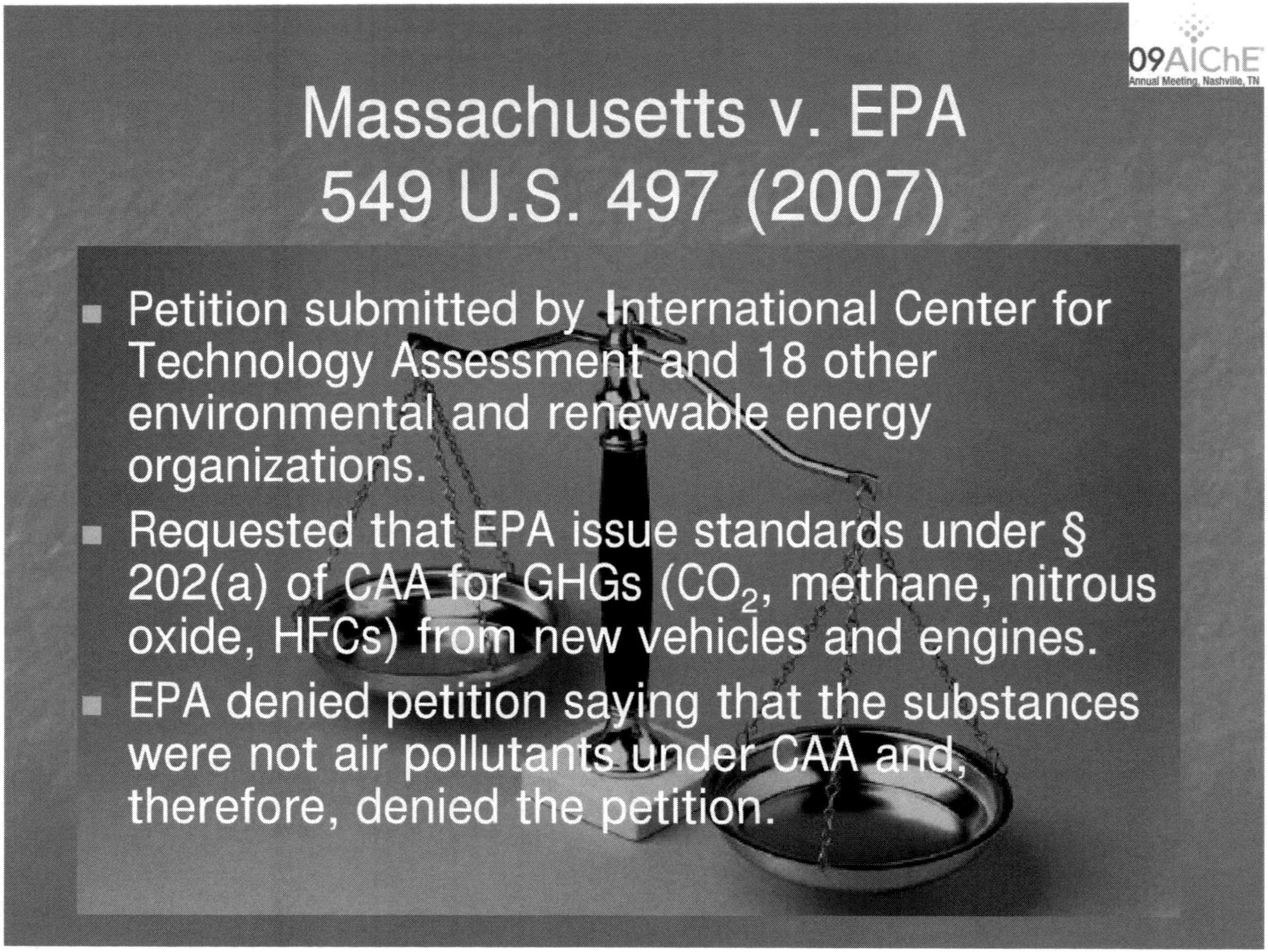
09AIChE
Annual Meeting, Nashville, TN
Massachusetts v. EPA
549 U.S. 497 (2007)
Petition submitted by International Center for Technology Assessment and 18 other environmental and renewable energy organizations.
Requested that EPA issue standards under § 202(a) of CAA for GHGs (CO_2, methane, nitrous oxide, HFCs) from new vehicles and engines.
EPA denied petition saying that the substances were not air pollutants under CAA and, therefore, denied the petition.

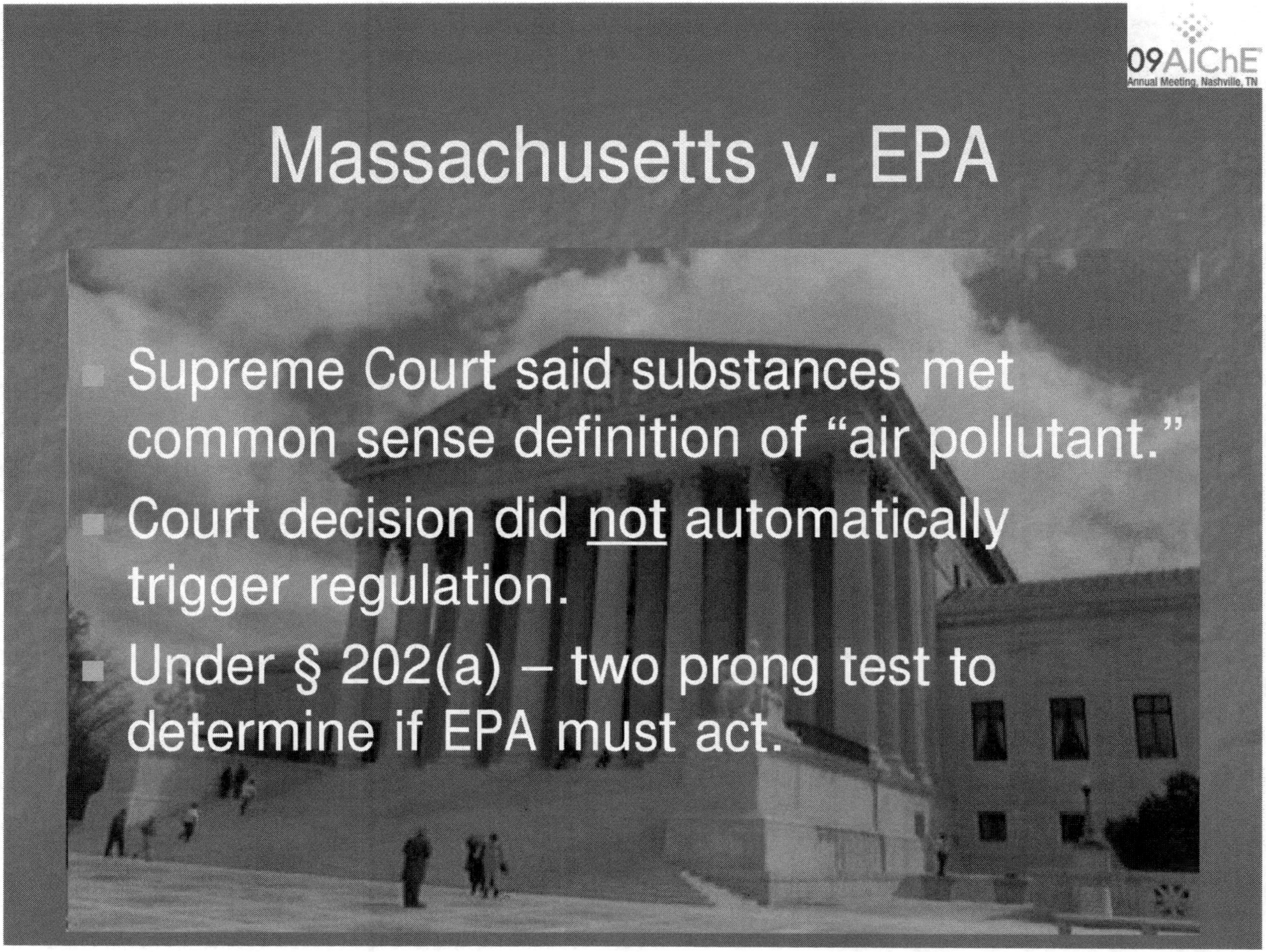
09AIChE
Annual Meeting, Nashville, TN
Massachusetts v. EPA
Supreme Court said substances met common sense definition of "air pollutant."
Court decision did not automatically trigger regulation.
Under § 202(a) – two prong test to determine if EPA must act.

GHG Endangerment Test Under § 202

- CAA Section 202 endangerment language:

  The Administrator shall by regulation prescribe ... standards applicable to the emission of any air pollutant from any class or classes of new motor vehicles ... which in his judgment *cause, or contribute to, air pollution which may reasonably be anticipated to endanger public health or welfare.*

- EPA must make one of the following determinations for motor vehicle GHG emissions based on the available science:
  - there is endangerment, or
  - there is no endangerment, or
  - the science is too uncertain to make a reasoned judgment

Stationary Sources & Potential Regulation of GHGs

- CAA Provisions Potentially Applicable
  - NAAQS (CAA §§ 108-110)
  - NSPS (CAA § 111)
  - Toxics (§ 112)
  - Solid Waste Combustion (§ 129)
  - Ozone Depleting Substances (§ 615)

09AIChE
Annual Meeting, Nashville, TN
Mobile Sources &
Potential GHG Regulations
CAA Provisions Potentially Applicable
§ 211(c) Fuels and Fuel Additives
§ 213 Nonroad Engines
§ 231 Aircraft

09AIChE Annual Meeting, Nashville, TN

Other Endangerment Provisions

- § 108(a)(1):
  *[T]he Administrator shall * * * publish, and shall from time to time thereafter revise, a list which includes each air pollutant (A) emissions of which, in his judgment, cause or contribute to air pollution which may reasonably be anticipated to endanger public health or welfare * * **
- § 111(b)(1)(A):
  [The Administrator] shall include a category of sources in such list if in his judgment it causes, or contributes significantly to, air pollution which may reasonably be anticipated to endanger public health or welfare.
- § 211(c)(1):
  *CAA section 211(c)(1) (regarding regulating fuels and fuel additives) states: "The Administrator may, * * * [regulate fuels or fuel additives] (A) if in the judgment of the Administrator any emission product of such fuel or fuel additive causes, or contributes, to air pollution which may reasonably be anticipated to endanger public health or welfare (B) * * * "*

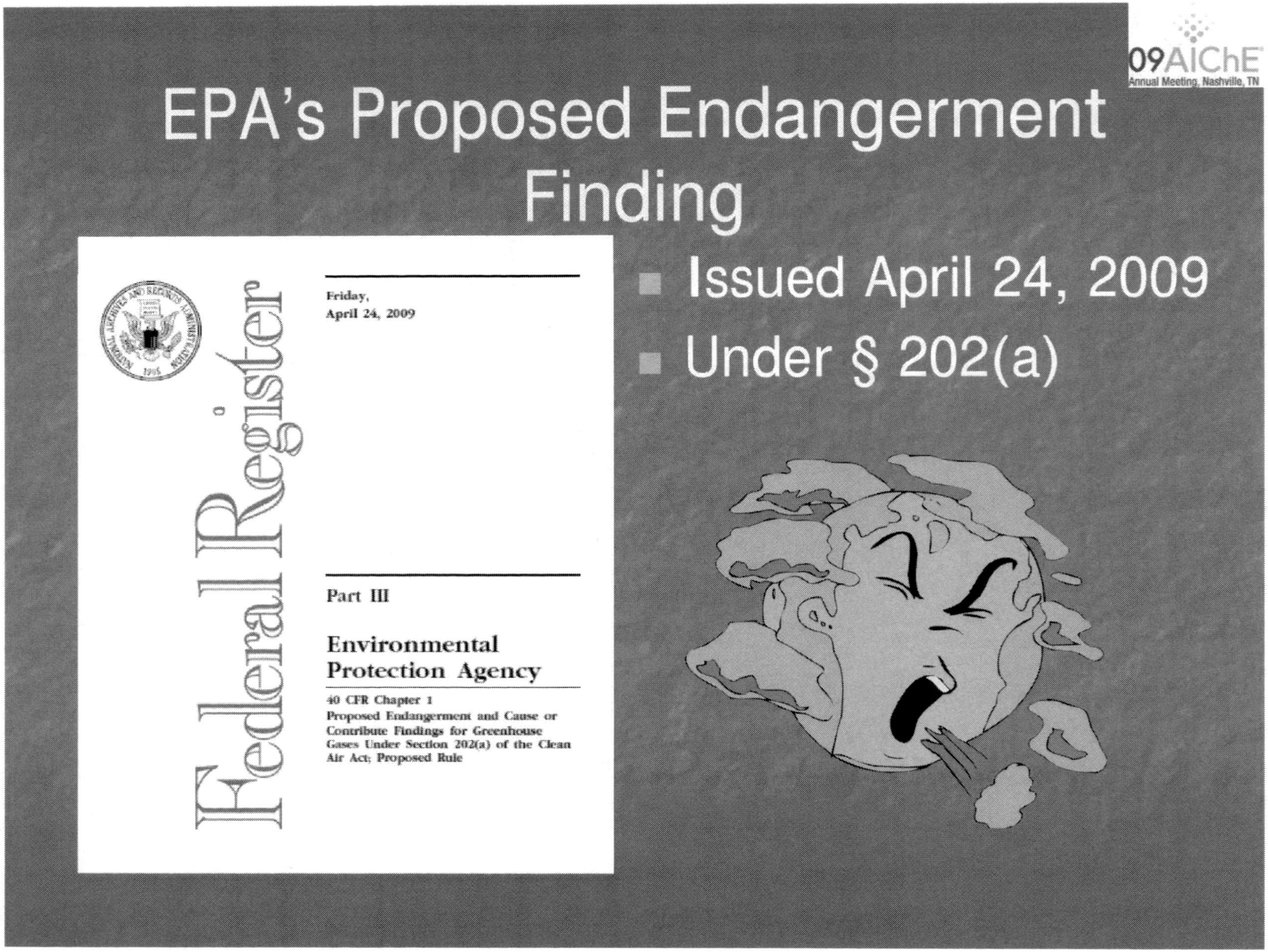
09AIChE
Annual Meeting, Nashville, TN
EPA's Proposed Endangerment Finding
Federal Register
Friday,
April 24, 2009
Part III
Environmental Protection Agency
40 CFR Chapter 1
Proposed Endangerment and Cause or Contribute Findings for Greenhouse Gases Under Section 202(a) of the Clean Air Act; Proposed Rule
Issued April 24, 2009
Under § 202(a)

09AIChE Annual Meeting, Nashville, TN

EPA's Proposed Endangerment Finding

- Endangerment Finding
  - Current and projected [GHG] in the atmosphere threatens public health and welfare of current and future generations
  - GHGs: CO_2, CH_4, N_2O, HFCs, PFCs, SF_6
- Cause or Contribute Finding
  - Combined emissions of CO_2, CH_4, N_2O, and HFCs from new motor vehicles and motor vehicle engines contributes to [GHG] and to the threat of climate change

Implications of Endangerment Finding

- Endangerment finding under one section does not automatically lead to regulation under another section
- Example: compare 202(a) endangerment test vs. fuels and aircraft:
  - Finding on first prong on harmfulness of pollutant similar
  - Second prong finding on cause or contribution from new vehicles would be different than other two categories
  - What constitutes significant enough contribution?

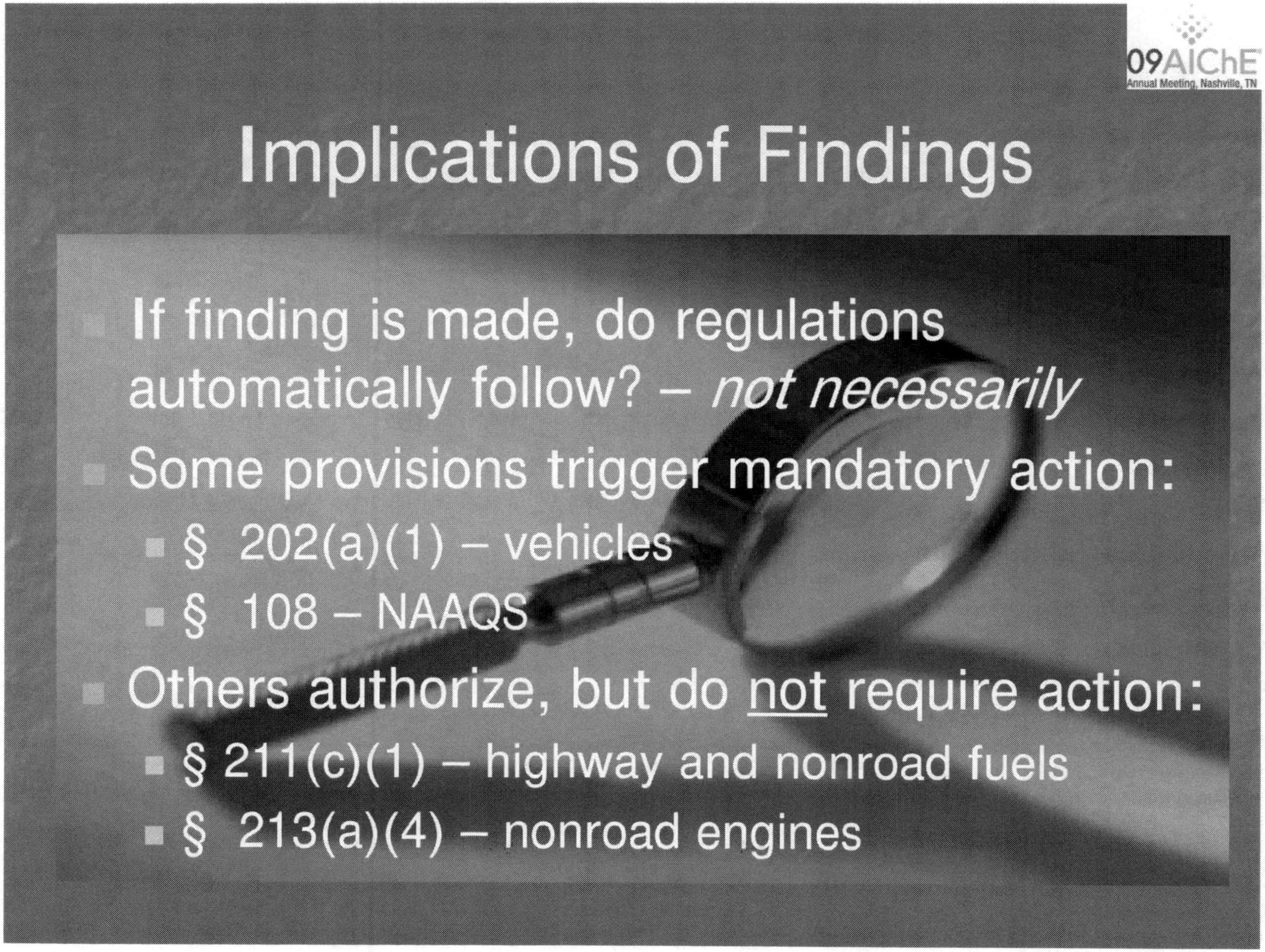
09AIChE
Annual Meeting, Nashville, TN
Implications of Findings
If finding is made, do regulations automatically follow? – not necessarily
Some provisions trigger mandatory action:
§ 202(a)(1) – vehicles
§ 108 – NAAQS
Others authorize, but do not require action:
§ 211(c)(1) – highway and nonroad fuels
§ 213(a)(4) – nonroad engines

Permit Programs

- Title V Permit Program
  - Not supposed to create any new substantive requirements
- PSD/NSR program
  - Covers pollutants regulated under any CAA authority except § 112 or § 211(o)
  - Current Major Source Thresholds Applied to CO_2 would crash NSR system

GHGs and PSD Permits

- Preconstruction permitting program for construction of any major new source or major modification at existing source.
- PSD permits must contain BACT permit limits for "each regulated NSR pollutant." (40 C.F.R. 52.21(j)(2)-(3)
- To date, EPA has said CO_2 or other GHGs are not subject to BACT control.
- Any requirement to control CO_2 or other GHGs under another provision of the Act, would trigger PSD review for that pollutant.

09AIChE Annual Meeting, Nashville, TN

GHGs and PSD Permits

- NSR pollutant equals:
  - Pollutants regulated under NAAQS or NSPS
  - Class I or II substance under CAA Title VI
  - Or any other "pollutant otherwise subsequent to regulation under the Act"
  - Excludes HAPs regulated under § 112(b)(6)
  - Energy Independence and Security Act states that although CAA § 211(o) targets GHG emissions, regulatory status of GHGs not changed yet.

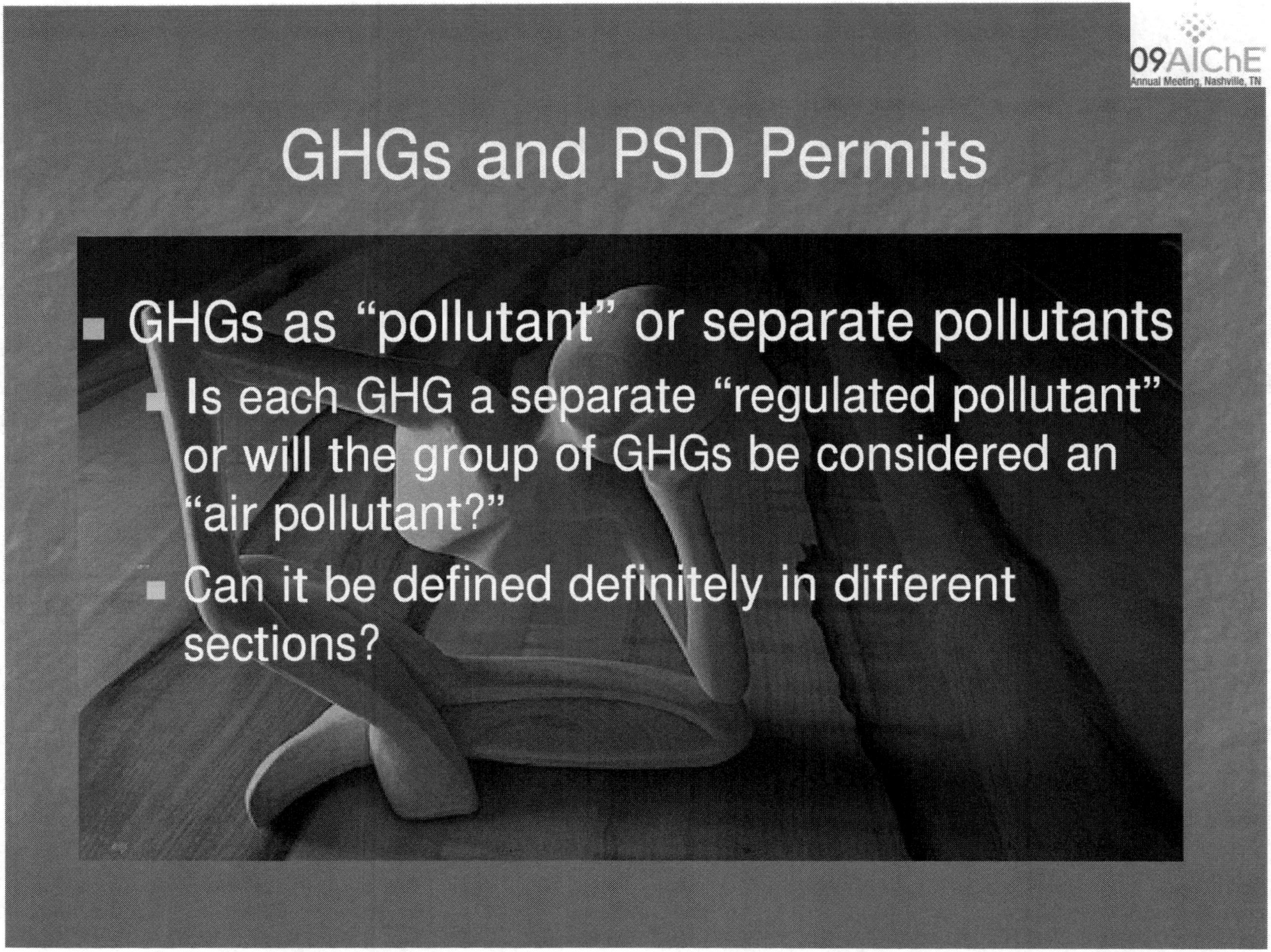
09AIChE
Annual Meeting, Nashville, TN
GHGs and PSD Permits
GHGs as "pollutant" or separate pollutants
Is each GHG a separate "regulated pollutant" or will the group of GHGs be considered an "air pollutant?"
Can it be defined definitely in different sections?

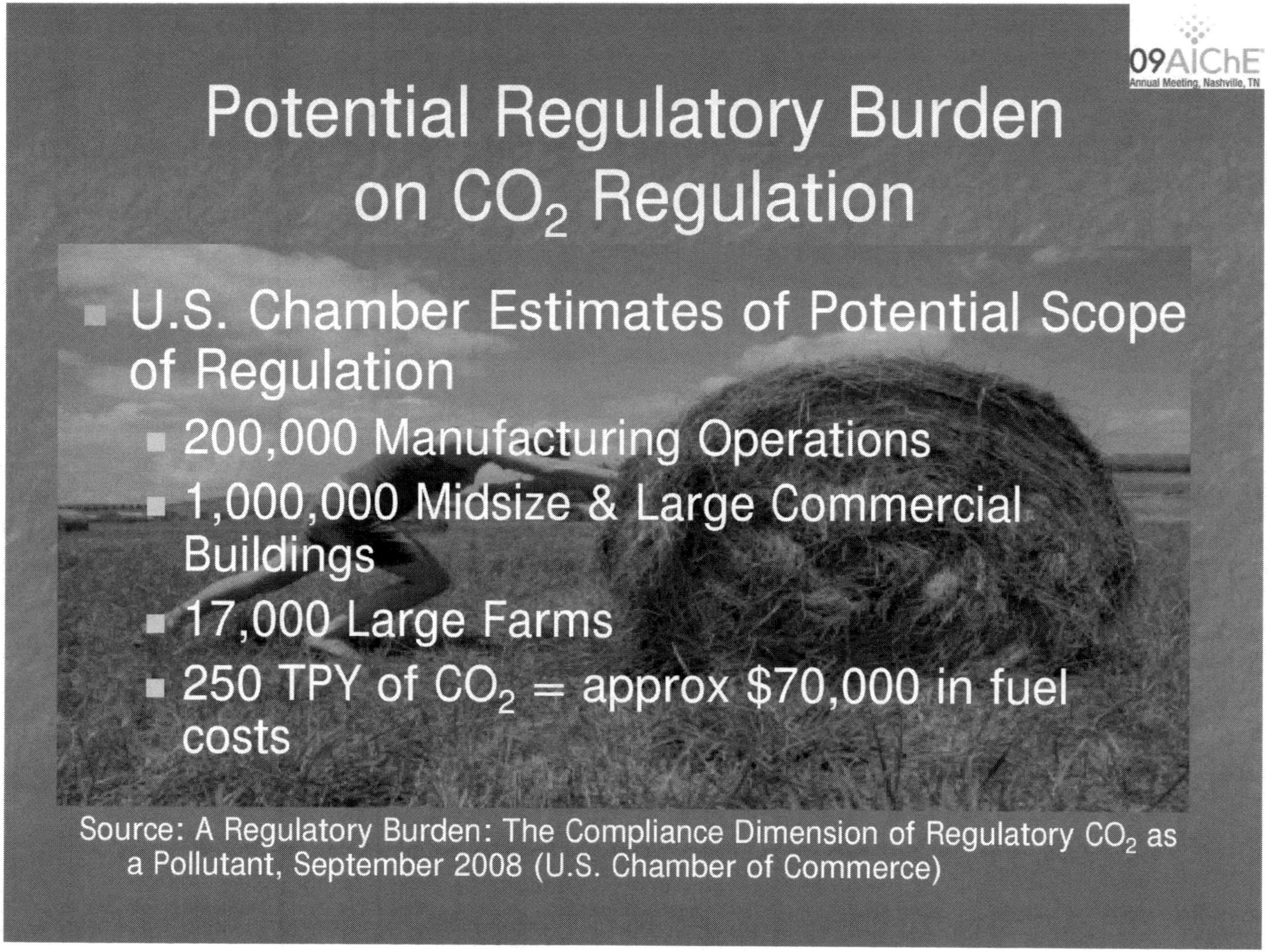
09AIChE
Annual Meeting, Nashville, TN
Potential Regulatory Burden on CO2 Regulation
U.S. Chamber Estimates of Potential Scope of Regulation
200,000 Manufacturing Operations
1,000,000 Midsize & Large Commercial Buildings
17,000 Large Farms
250 TPY of CO2 = approx $70,000 in fuel costs
Source: A Regulatory Burden: The Compliance Dimension of Regulatory CO2 as a Pollutant, September 2008 (U.S. Chamber of Commerce)

Impact of New GHG Legislation on Regional and International GHG Cap and Trade Programs

Author: Chris Moore

Over the past two decades, climate change and associated energy legislation has become a topic of great interest in most businesses and communities across the globe. In recent years, various programs have been introduced with the intent to control the levels of carbon dioxide (CO_2) and other greenhouse gas (GHG) pollutants. Alongside these programs, a myriad of legislative approaches and opinions are being considered that are vast, diverse, ever-changing, and excessively politically-charged. The intent of this abstract is to provide a brief overview of the impact of potential new energy and climate change legislation on implementation of regional and international GHG cap and trade programs.

What is Cap and Trade?

A key component of almost all CO_2 legislation is cap and trade program. Traditionally, in a cap and trade program, the government determines which facilities or type of emissions are covered by the program and then establishes an overall emission target or "cap" for covered facilities. This cap is the sum of all allowed emissions from all facilities included. Once the cap is set, tradable emission allowances are distributed. This distribution can occur via an auction, free allocation or a combination of the two. Allowances are normally defined on a cost per ton basis of the pollutant. The total number of allowances is equal to the level if emissions capped by the program. Facilities subject to a cap and trade program must maintain allocations that cover their allowable levels of emissions. Allowances can be purchased or facilities may choose to reduce emissions through lower-emitting technology and then trade allowances on the open market.

Major International and Regional Cap and Trade Programs

The primary GHG-related cap and trade programs implemented or being developed to date include:

- European Union's Emission Trading Scheme (EU ETS);
- The Northeast Regional Greenhouse Gas Initiative (RGGI);
- The Western Climate Initiative (WCI); and,
- Midwestern Regional Greenhouse Gas Reduction Accord

<u>European Union's Emission Trading Scheme (EU ETS)</u>

The EU ETS was developed over the past decade and is based on meeting the targets established under the Kyoto Protocol by 2012. Formally launched in 2005 and tailored after the US trading schemes for sulfur dioxide and nitrogen oxides under the Clean Air

Act (CAA), implementation of the EU ETS is intended to reduce CO_2 emissions by 20%, increase renewable energy by 20% and improve overall energy efficiency by 20% by 2010. More than 11,000 energy intensive facilities are included in the program across the 27 European Union member states. Allocations under the program are currently free allocations; however, during the third phase (2013-2020), the EU ETS will begin to move toward auctioning of allowances. Also during Phase III, a larger number of facilities will be included in the program as well as adding other GHGs beyond CO_2 such as nitrous oxide and perfluorocarbons.

The Northeast Regional Greenhouse Gas Initiative (RGGI)

RGGI became effective on January 1, 2009 with a goal of reducing CO2 emissions by 10 percent from 2009 levels by 2018. The program encompasses 10 participating northeastern US states:

- Connecticut
- Delaware
- Massachusetts
- Maryland
- Maine
- New Hampshire
- New Jersey
- New York
- Rhode Island
- Vermont

RGGI established a cap of 188 million tons of CO2 for the ten states and only covers fossil fuel-fired electric power plants 25 megawatts or greater. This represents about 225 facilities in the 10-state region. Allowances are auctioned under the program. Key stated goals to be accomplished by RGGI include:

- Reduction of CO2 emissions;
- Support of a green economy by using auction revenues for investment in energy efficiency, renewable energy, and accelerating a regional shift to a clean energy economy;
- Promotion of energy independence; and,
- Provide a model for a national program to reduce CO2 emissions.

The Western Climate Initiative (WCI)

The WCI kicked off in early 2007 and covers the following US states and Canadian provinces:

- Arizona;
- California;
- Montana;
- New Mexico;
- Oregon;
- Utah;
- Washington;
- British Columbia;
- Manitoba;
- Ontario; and,
- Quebec.

Under the WCI, emissions from 90 percent of the facilities in the region will be captured in the program including electricity, industry, transportation and residential/commercial use. Reporting of emissions will begin in 2011 based on 2010 emissions with the first phase of cap and trade beginning in 2012. Beginning in 2015, the program will be expanded to include the remaining 10 percent of facilities not covered in the initial phase.

Midwestern Regional Greenhouse Gas Reduction Accord (Accord)

The Accord was initiated in November 2007 by the Governors of:

- Illinois;
- Iowa;
- Kansas;
- Michigan;
- Minnesota; and,
- Wisconsin.

Additionally the Premier of Manitoba signed the Accord that calls for targets to be established for GHG reductions and implementation of a cap and trade program. Members of the Accord prefer a federal approach to GHG limits and cap and trade; however, they plan to continue to evaluate the prospects of federal plans and act accordingly within the states/territories as necessary.

Federal GHG Legislation

In June 2009, the 112th Congress of the United States passed the first major climate change bill in US history. The bill sponsored by Representatives Henry Waxman (D-Calif) and Ed Markey (D-Mass) establishes CO_2 reduction targets, a federal cap and trade program, a renewable electricity standard (RES), requirements for coal-fired power plants, energy-efficiency standards, worker transition programs, smarter cars and smarter electricity grids.

Emission reduction goals starting in 2012 and based on 2005 emission levels include a:

- 3% reduction in 2012;
- 17% reduction by 2020;
- 42% reduction by 2030; and,
- More than 80% reduction by 2050.

Coverage of the program initially includes approximately 85% of the overall economy including electricity producers, oil refineries, natural gas suppliers, and multiple industry categories. Facilities with CO_2 emissions above 25,000 metric tons per year would be required to acquire a permit for their emissions or, in other words, obtain emission allowances for all CO_2 emissions on an annual basis. Facilities can trade these allowances to other facilities that need allowances or bank them for future usage. Eighty five percent of the allowances would be given away under the current plan to major energy users; however, the free allowances will gradually be decreased over the first 20 years of the program. The remaining 15% of the allowances would be auctioned off. Within 20 years from program implementation, essentially all allowances would become part of an auction program. Current projections of CO_2 allowances expect the per ton cost in 2012 to be around $20-$25 increasing to $50 per ton by 2025.

Revenue from the auctioning of allowances under the cap and trade program would be used to:

- Invest in clean energy technology research and deployment;
- Offset increased energy costs for low-moderate income households;
- Prevent international deforestation;
- Fight climate change in other nations; and,
- Fund worker transition from fossil-fuel dependent industries.

<u>United States Senate GHG Legislation</u>

The US Senate was slated to take action on the Waxman-Markey Bill in September; however, other legislative priorities have caused a delay in consideration. It is expected that the Senate will propose their version of a cap and trade bill early in 2010.

Another key event that will influence US climate policy as it relates to cap and trade programs is the upcoming Copenhagen United Nations Climate Conference in December 2009. Ideally, this conference would result in a new global climate treaty;

however, at a minimum, there is hope that additional clarity can be achieved on key global climate issues. If a new climate treaty is agreed upon, it will replace the Kyoto Protocol which was adopted in Kyoto, Japan, in December 1997.

Impact of Federal Legislation on Current GHG Trading Programs

In comparing the current GHG cap and trade programs with the program included in the Waxman bill, the general tone of the legislation is consistent with current trading programs. A likely outcome assuming eventual passage of a federal cap and trade bill is that existing US programs will be rolled into a national program (i.e., the existing regional programs are super-ceded by the federal program). This approach may provide the most flexibility and assurance for industry and utilities.

Consistent with the Clean Air Act, the Waxman bill protects the individual states rights to be more stringent than federal requirements. This could be an issue if one of the existing groups of states such as RGGI, WCI, or the Accord concludes that the federal rule does not go far enough to reduce CO_2 emissions or have a weak cap and trade program. If a state or group of states chooses to implement a more stringent program, it could result in more expensive state-regional allowances on a per ton basis depending on the quantity of CO_2 emissions and the types of facilities covered by the state/regional program. Depending on the will of the states, as the final federal legislation emerges, there could be an option where regional programs remain in-tact with linkage to the federal program. This hybrid approach may provide opportunities for those states/regions who believe they will have a more effective program if managed within the state/region and yet still have common elements with the federal program.

Regarding how participants in each program may be, required to manage, verify and report GHG reductions and subsequent CO_2 allocations, there is still significant uncertainty. Given that the recently proposed United States Environmental Protection Agency (USEPA) GHG Reporting Rule was designed based on significant review of the existing regional and international programs and other accepted GHG calculation and reporting protocols, this could alleviate many potential problems by not deviating significantly from current efforts.

Conclusion

While many aspects of a federal cap and trade program remain very much in flux, there is growing consistency in the reporting and reduction requirements that will ultimately be included in a final federal cap and trade program. Current federal legislation has drawn upon the significant work completed both internationally and regionally in the US and also input from various stakeholder groups. These efforts should help to minimize the transition to either move from a regional program to a federal program or to enable a regional program to be managed within a corresponding federal program.

Challenges for Chemical Engineers Under California's Global Warming Solutions Act

Roger E. Sowell, Law Office of Roger E. Sowell, Marina del Rey, CA

California's Global Warming Solutions Act of 2006 served as a blueprint for pending Federal legislation. Chemical engineers will have great opportunities and challenges in many industry sectors as a result of Global Warming laws. The presentation will briefly discuss the history and purpose of California's Global Warming law, the timeline and events for implementing the goals in California, the current requirements, then focus on impacts on industry sectors that are of importance to chemical engineers. Among the industries and issues of interest to chemical engineers are: Cap and Trade, Petroleum Refining, Chemicals and Petrochemicals Manufacturing, Hydrogen Production, Power Plants, Cogeneration, Solar Thermal Storage, Ethanol and other Bio-fuels, Cement Plants, Bio-mass Methane Capture, "Green" Building Materials, and Carbon Capture and Sequestration (CCS). The presentation will conclude with what to expect next.

See more of this Session: Climate Change Legislation and Regulation
See more of this Group/Topical: Environmental Division

Challenges for Chemical Engineers Under California's Global Warming Solutions Act

Author: Roger E. Sowell, Esq.*

Law Office of Roger E. Sowell, Marina del Rey, California

*Mr. Sowell is an attorney in Marina del Rey, California, who holds a B.S. in chemical engineering. He practices in the areas of climate change, energy, and engineering law. He worked for 25 years as a chemical process engineer before becoming an attorney.

Abstract

California's Global Warming Solutions Act of 2006, also known as AB 32, requires large reductions in greenhouse gas emissions by 2020. Plants and facilities that consume fossil fuels or produce power will be required to meet the regulations established under AB 32. Achieving the reductions will present many opportunities and challenges for chemical engineers, especially in the areas of oil refining, general industry, cement plants, hydrogen plants, bio-fuels, and many others. Cap and trade issues will present chemical engineers with challenges. More jobs for chemical engineers likely will result, however some plants likely will shut down rather than comply with the burdens imposed by AB 32.

History of AB 32

California's Global Warming Solutions Act, also known as AB 32, was adopted as law in California in 2006, [(1)] after first being proposed as a bill in California's Assembly in 2005. The Democrat-controlled state legislature had little hope of seeing a global warming law at the national level after Republican President George W. Bush was re-elected in November 2004. President Bush had made it clear that he would not ratify the Kyoto Accord that purportedly limits emissions of greenhouse gases (GHG) by signatory countries, nor would he sign any federal bill into law that limited Carbon Dioxide (CO_2) or other GHG emissions. In that legislative environment, California's Assembly Speaker Nunez championed AB 32 into becoming the law in California. The text of AB 32 may be found in California's Health and Safety Code §38500 et. seq. [(2)]

Purpose of AB 32

AB 32 was intended to serve as a blueprint for others to follow in reducing GHGs and increasing economic activity. ARB states that, on average, for each dollar spent on AB 32 requirements, two dollars will be returned. Yet, ARB does not state when those two dollars will be returned, whether in 2 years or 40 years or more. However, many independent economic analyses concluded that the economics of AB 32's Scoping Plan are seriously flawed. ARB states that AB 32 will increase electric power prices. The price of transportation fuels will also increase due to bio-fuels additions. For those segments of the population who are poor or on fixed incomes, who have few or no alternatives to reduce their consumption of electric power and transportation fuel, AB 32 will impose a harsh economic burden. (3)

In addition, the very concept of controlling the earth's temperature by reducing CO2 and other GHGs is fundamentally flawed. The primary goal of AB 32, reducing global warming by reducing GHGs, is very likely unattainable because of the almost non-existent correlation between CO2 and other GHGs and global temperature. A comparison of CO2 concentration in the atmosphere, either from direct measurements from 1959 or by very good estimates from 1900 (see Figure 1), with global temperature anomalies (see Figure 2), shows that CO2 rises continually while global temperature sometimes decreases, sometimes remains constant, and sometimes increases, all for decades at a time. Fundamental process control principles show that, whatever causes global warming and cooling, manipulating CO2 is not an effective means of controlling the earth's average temperature. Dr. Pierre Latour, Ph.D. and P.E. in chemical engineering and expert in process control, also made a compelling argument in the February, 2009, Letter to Editor section of Hydrocarbon Processing on why adjusting CO2 cannot control the earth's temperature. (4) Dr. Latour stated that measuring the earth's temperature is very difficult if not impossible, and under control principles, if you cannot measure it, you cannot control it. Also, Dr. Latour pointed out that there is almost no response, dT/dCO_2, also indicating that CO_2 is not a candidate manipulated variable.

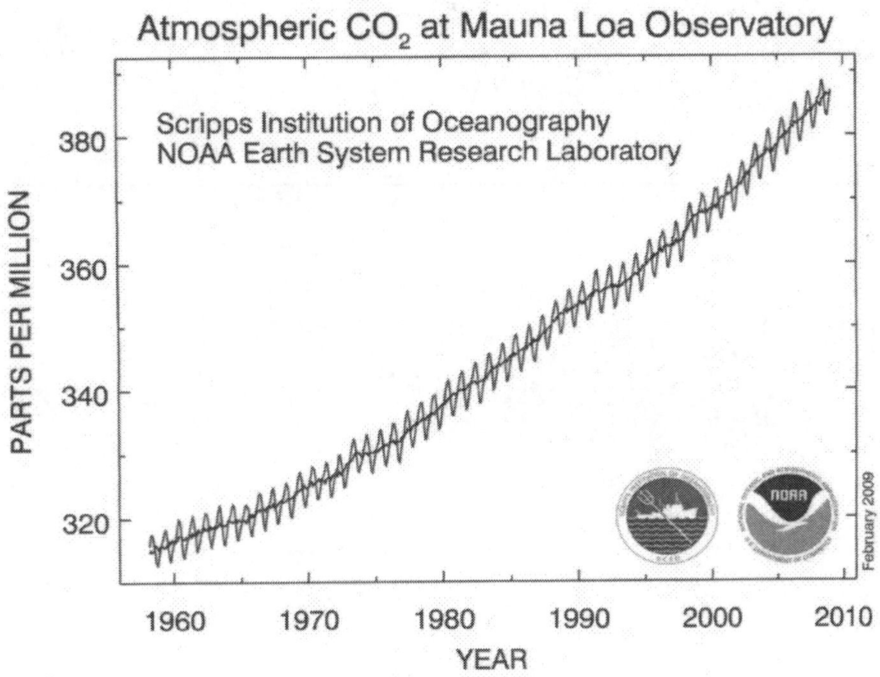
Atmospheric CO_2 at Mauna Loa Observatory
Scripps Institution of Oceanography
NOAA Earth System Research Laboratory
PARTS PER MILLION
380
360
340
320
1960
1970
1980
1990
2000
2010
YEAR
February 2009

Figure 1

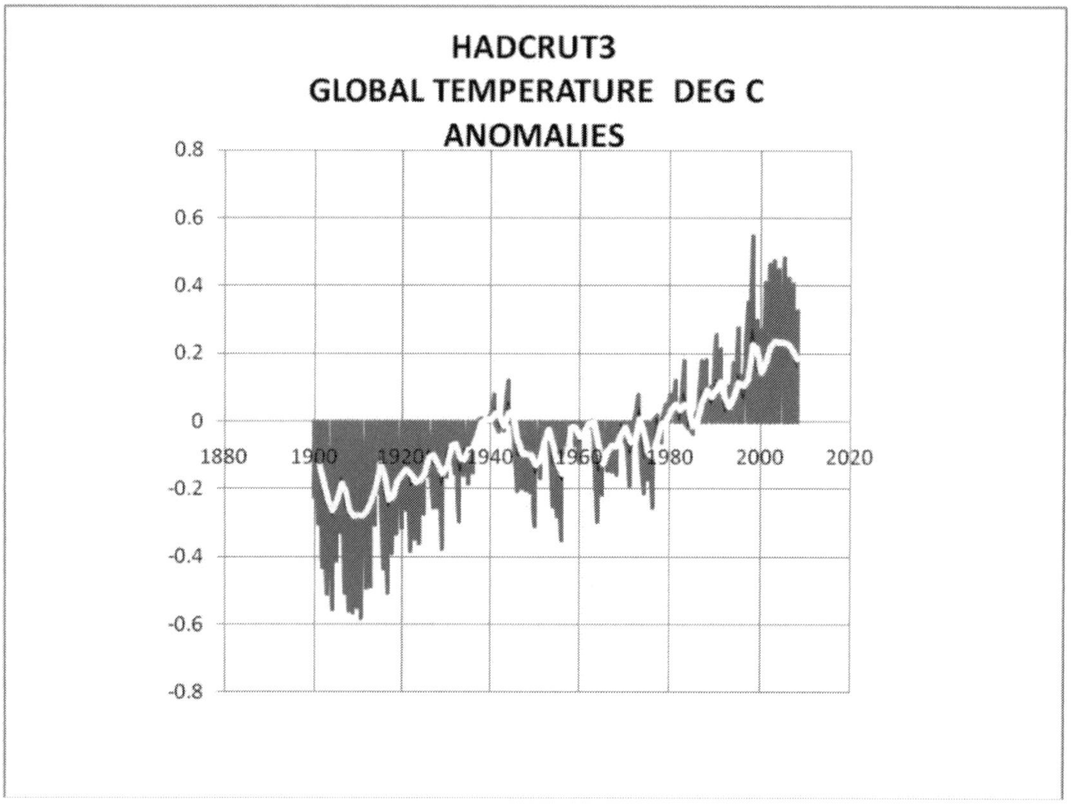

Figure 2

Requirements of AB 32

AB 32 is an enabling act, providing a few specific targets but delegating the creation of the detailed regulations to the California Air Resources Board (ARB) [5]. ARB is directed to coordinate with and consult with appropriate agencies and stakeholders, including the Public Utilities Commission. [6] The specific regulations are to "minimize costs and maximize benefits for California's economy, improve and modernize California's energy infrastructure and maintain electric system reliability, maximize additional environmental and economic co-benefits for California, and complement the state's efforts to improve air quality." [7]

Two dates are critical in AB 32: 1990 and 2020. The total quantity of GHGs emitted in 1990 was to be determined by ARB, and the state's emissions are required to be reduced to the 1990 level by 2020. [8] Another date and requirement is critical, although not part of AB 32, which is 80 percent reduction below 1990 levels by 2050. The "80 by 50" requirement is imposed by the Governor's Executive Order S-21-09 of September 15, 2009. ARB determined that the 1990 emissions were 427 million metric tonnes of CO2 equivalent (MMTCO2e). The expected emissions of GHGs in 2020, absent the requirements of AB 32 (the business as usual case), are approximately 600 MMTCO2e. The GHGs reduction required by 2020 is 173 MMTCO2e, which represents a 28.8 percent reduction from the business as usual case. This reduction by 2020 may be achievable, although at a high economic cost. However, the reduction by 2050 represents a 93 percent reduction over the business as usual case, which is much more problematic (see Figure 3).

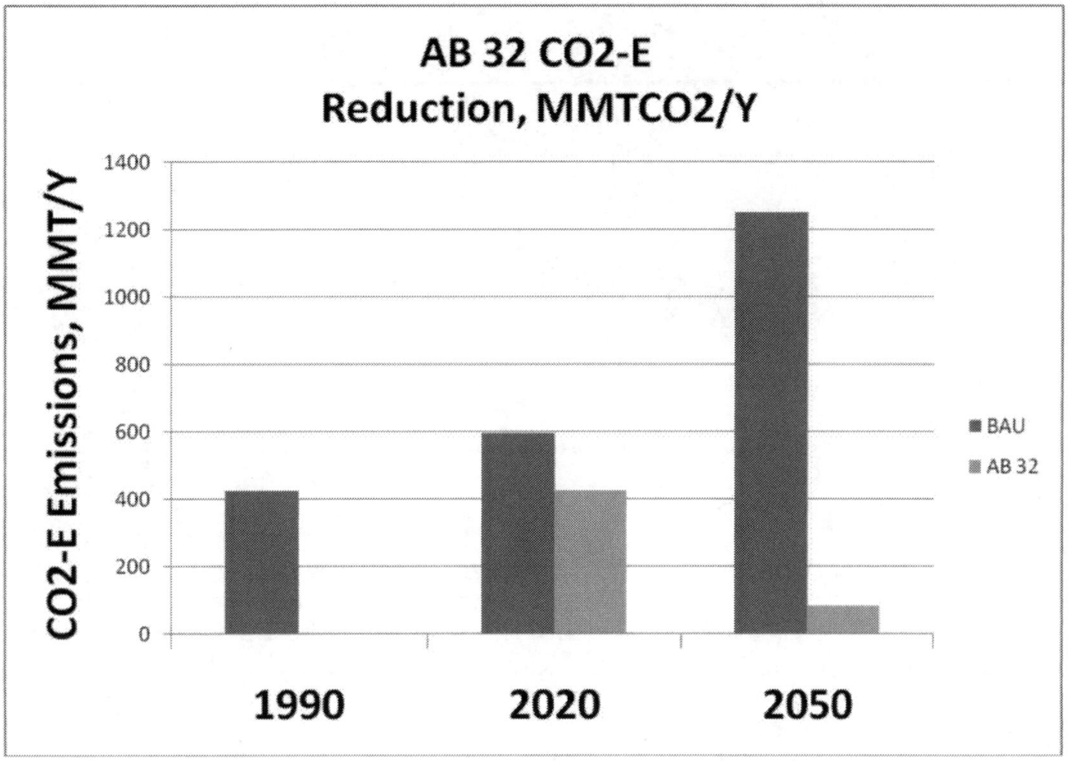

Figure 3

Even if California achieves its goals of GHG reductions, there will be very little impact on the world's energy consumption and GHGs. California consumes approximately 2 percent of the world's total energy (see Figure 4). With a 28 percent reduction by 2020, there will be no noticeable change, and any change will be overwhelmed by increases in growing economies around the world.

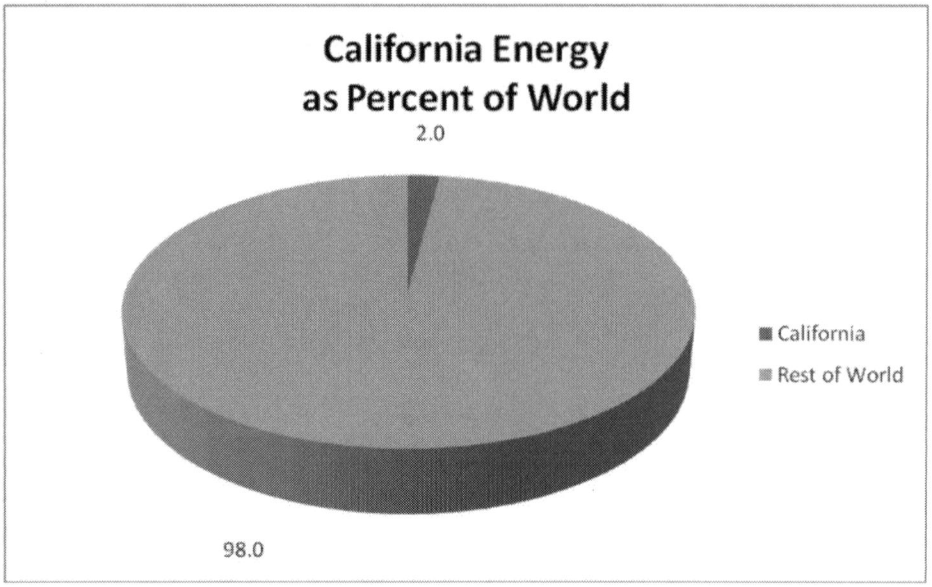

Figure 4

The plan to reduce GHGs to 427 million tonnes per year by 2020 was described by ARB in the Scoping Plan document in late 2008, which divided the California economy into a number of sectors. Each sector has its own requirements for reducing GHGs. The Scoping Plan identified 73 separate reduction items. Opportunities for chemical engineers exist in almost every sector identified in the Scoping Plan. These opportunities are described in the following section.

The Scoping Plan has three basic reduction mechanisms: specific targets or emissions limits, cap and trade, and a fee for emissions. Each economic sector has one or more of the three mechanisms. Some of the 73 items with the largest reductions listed in the Scoping Plan are shown in Table 1 below, along with the anticipated reductions in GHGs. Table 1 also has Scoping Plan items with low emissions reductions that will affect chemical engineers.

Scoping Plan Item	MMTonnes CO2e/Yr
Cap and Trade	34.4
Renewable Portfolio Standard	21.3
Low Carbon Fuel Standard	16.0
Energy Efficiency (Electricity)	15.2
Combined Heat and Power	6.7
Regional Transportation Plan	5.0
Refinery Flare Gas Recovery Systems	0.33
Energy Efficiency and Co-Benefit Audits	N/A

Table 1

The timeline for ARB to write the regulations is 2009 through 2011, with full implementation beginning in 2012. A few regulations have earlier implementation dates. The cap and trade regulation will be finalized in 2011 with implementation starting in 2012.

Reporting of GHGs emissions are required for some facilities, starting with 2009. GHG emission reports will be verified by independent, state-approved, entities beginning in April 2010 for the 2009 reporting year. Chemical engineers will make good candidates for the verification jobs. Mandated reporters include all cement plants, refineries with emissions of at least 25,000 MMTCO2e per year (50 MMBtu/h), power plants of at least 1 MW plus 2500 MMCO2e per year (5 MMBtu/h), and any other facility with the same emissions criteria as refineries. Those entities that are exempt from reporting include hospitals, electric power plants with renewable energy sources, primary and secondary schools, portable equipment, and backup emergency generators. For a comparison of some common fuels and their emission of CO2 compared to MMBtu, see Figure 5.

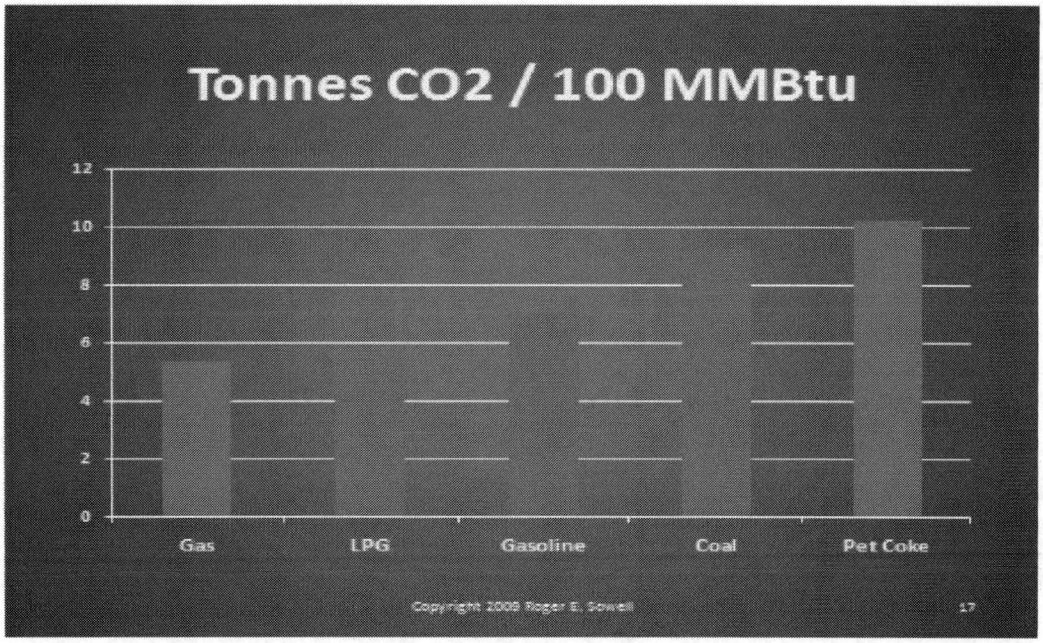

Figure 5

Opportunities for Chemical Engineers

Oil Refining

The oil refining sector will have GHG reporting and verification, cap and trade, energy audits, and specific reduction requirements. The regulations are not yet final for specific reductions, but will likely include flare gas recovery systems, electric heating systems, and low-carbon-forming catalysts for Fluid Catalytic Crackers. Also, the fired boilers will likely be replaced by gas turbine generators with heat recovery steam generators to provide process steam. The cap and trade requirement will provide great challenges and opportunities for chemical engineers as they strive to reduce GHGs by 28 percent by 2020.

One such GHG-reducing project is hydrogen production by partial oxidation of petroleum coke, with CO_2 capture and sequestration. A plant to accomplish this is contemplated for Dow Chemical in Freeport, Texas. [9]

It may be more attractive for oil refiners to cease the refining function, and purchase products from other countries and have the products brought in via ships. Faced with similar legislation in Australia, Caltex announced they would shut down their refineries and operate them as oil terminals. This would entail a loss of property tax, and several hundred high-paying jobs in the community. [10]

<u>Industry and Manufacturing</u>

The Industry and Manufacturing sector includes semiconductor plants, cement plants, hydrogen plants, and other large facilities excluding refineries and oil and gas production facilities. Semiconductor plants are required to reduce their use of perfluorocarbons by 2012 to a level not to exceed 0.2 kg CO2e/ sq cm of wafer. This reduction in perfluorocarbons will reduce GHGs by 0.18 MMTCO2e/yr.

Cement plants are subject to cap and trade, and will likely be required to increase energy efficiency, capture and sequester CO_2, switch fuels from coal and petroleum coke to natural gas, and use alternative cementitious materials rather than limestone-based clinker. Also, the concrete industry will be required to reduce waste, and that in turn will reduce cement production. At least one cement plant in California has announced it cannot remain in business by making the proposed changes under AB 32, and will shut down its operations and lay off the employees.

Hydrogen plants will be required to increase energy efficiency, minimize fuel consumed, and likely will be required to capture and sequester CO_2.

CO_2 Capture and Sequestration

CO_2 capture and sequestration likely will be required for many processes and power plants. Such processes will present many challenges to chemical engineers, in finding appropriate CO_2 removal systems, materials of construction, and low-energy consumption. CO_2 sequestration also may require CO_2 purification, compression, liquefaction, or chemical conversion to another compound.

Low Carbon Fuel Standard

The Low Carbon Fuel Standard refers to bio-fuels, both bio-ethanol and bio-diesel, and likely eventually to bio-jet fuel. ARB in 2009 wrote the regulation for bio-ethanol based on corn fermentation. (11) ARB's analysis shows that corn-based ethanol consumes more energy than is obtained upon combustion in a vehicle engine; this was confirmed a few weeks later by the federal Environmental Protection Agency, EPA. Despite this, corn-based ethanol is mandated not only by California law but also by federal law. Also, there is a shortage of water in California, and corn production requires massive amounts of water. Furthermore, the Governor has issued an Executive Order that requires that 40 percent of all bio-ethanol consumed in California be produced within California. ARB estimates that 25 new ethanol refineries will be required to meet the state's needs. It remains to be seen how the state will grow corn without sufficient water supplies.

Bio-diesel is made from both plants, usually soy beans, and rendered animal fats. Chemical engineers will have opportunities in processing the oils, but also in producing appropriate additives to ensure these bio-diesels meet all the cold temperature properties such as cloud point, cold filter plugging point, and pour point. Also, other diesel properties must be met.

At least one company, Rentech, has announced a plant to manufacture bio-diesel using the Fischer-Tropsch process with municipal solid waste as the feedstock. The bio-diesel will be used as fuel for airport ground service equipment at Los Angeles International Airport, LAX. (12)

Renewables Portfolio Standard

The Renewables Portfolio Standard requires that a large portion of all power sold in California be from renewable sources such as solar, wind, geothermal, and small hydroelectric plants. The AB 32 target and deadline is for 20 percent renewables by 12/31/2010, and 33 percent by 2020 by the Governor's Executive Order. At then end of 2008, the state produced only 13 percent of the electrical power by renewables. Chemical engineers will be challenged to discover and produce high-yielding batteries at an economic price to store electric power until it is needed. Also, solar thermal power plants present challenges to chemical engineers in the heat storage medium, materials of construction, and process control. Furthermore, geothermal power plants require chemical engineers in the design of the various power systems, including binary systems.

Low-Friction Oils

Low-friction oils, or synthetic oils, are required under AB 32 to reduce diesel fuel consumption in heavy-duty trucks. These oils are made from poly-alpha olefins, or PAO. Chemical engineers will be involved in designing and operating PAO plants.

Waste Handling and Recycling

Waste Handling and Recycling activities that involve chemical engineers include capturing landfill gas for use as fuel, anaerobic digestors for organic wastes such as manure, and conversion of municipal solid waste to synthesis gas as fuel for power plants. A U.S. utility patent, 7,452,392, was issued in 2008 to Peter A. Nick and his team of Southern California chemical engineers for a process to consume municipal solid waste and convert the material into a medium-Btu syngas. [13] As this technology gains acceptance, chemical engineers will be required for process design, construction, and operation.

Water

Opportunities for chemical engineers exist in water use, purification, and recycling. Processes that use water provide challenges to develop ways to use less, and purification to the point that water can be recycled instead of discharged. Water used as steam condensing heat sink may be replaced by air coolers or water-spray air coolers, with the corresponding reduction in power production. Municipal waste treatment plants will require upgrades to tertiary treatment so that effluent water may be recycled.

Cap and Trade

Many issues remain unresolved with California's cap and trade regulations under AB 32. Chemical engineers will be affected by the rules no matter what form they take, as the required reductions in GHGs will shape the number and type of projects undertaken by the chemical engineers. The price for a carbon credit will greatly affect economic decisions, whether to build a process revision, or buy the carbon credits.

Federal Laws and AB 32

While there presently is no federal law on global warming, the U.S. House of Representatives early in 2009 passed the Waxman-Markey bill, and a similar bill is pending in the U.S. Senate. Both bills are loosely based on AB 32, with similar goals for GHG reductions by 2020 and 2050. A key difference is that California has no coal-fired power plants, although a small amount of coal is consumed in the state. Also, California is one of the few, if not the only, state with geothermal energy for renewable power production.

The federal law, if passed, will have precedence over the state law, as is the case with all such laws. California's law may be more strict than the federal law, but may not be less strict. Therefore, it is possible that AB 32 will require amendment after a federal law is passed.

Conclusion

Even though reducing CO_2 and other GHGs will be futile in attempting to reduce the earth's temperature, many opportunities exist and will be created for chemical engineers under AB 32. The state's numerous refineries, cement plants, hydrogen plants, power plants, semiconductor plants, and other large energy consumers will all be affected by being forced to reduce GHGs. However, AB 32's requirements may be too burdensome for some, perhaps many, plants. It is anticipated that these plants will shut down rather than be forced to comply with AB 32. A similar closing of nearly half the nation's oil refineries occurred during the past 25 years, largely as a result of environmental and regulatory requirements. Therefore, the good news for chemical engineers is that there will be plenty of work. The bad news is that there may be fewer customers and clients who will need that work.

References

(1) see http://www.leginfo.ca.gov/pub/05-06/bill/asm/ab_0001-0050/ab_32_bill_20060927_chaptered.pdf

(2) see http://leginfo.ca.gov/calaw.html

(3) Sowell, R.E., http://sowellslawblog.blogspot.com/2009/06/ab-32-hypocrisy-vs-health-and-poverty.html

(4) Latour, P. R., Hydrocarbon Processing, February, 2009, Letters to Editor

(5) California Health and Safety Code §38501(f)

(6) California Health and Safety Code §38501(g)

(7) California Health and Safety Code §38501(h)

(8) California Health and Safety Code §38550

(9) see http://www.dow.com/facilities/namerica/texops/news/20071213d.htm

(10) see http://www.bloomberg.com/apps/news?pid=20601081&sid=axlU.9bewCsc

(11) see http://www.arb.ca.gov/regact/2009/lcfs09/lcfsisor1.pdf

and http://www.arb.ca.gov/regact/2009/lcfs09/lcfsisor1.pdf

(12) see http://www.foxbusiness.com/story/markets/industries/energy/rentech-lands-big-biomass-fuel-contract-lax-rtk/

(13) see http://patft.uspto.gov/netacgi/nph-Parser?Sect1=PTO1&Sect2=HITOFF&d=PALL&p=1&u=/netahtml/PTO/srchnum.htm&r=1&f=G&l=50&s1=7,452,392.PN.&OS=PN/7,452,392&RS=PN/7,452,392

MANDATORY GREEN HOUSE GAS REPORTING – 40 CFR 98

BUSINESS IMPACTS OF EPA'S PROPOSED MANDATORY GREEN HOUSE GAS REPORTING RULE REQUIREMENTS

Jeff Twaddle, P.E., Principal Consultant, ERM

7106 Crossroad Blvd, Brentwood, TN 37027

(615) 373-3350; Jeff.Twaddle@erm.com; www.erm.com

Abstract

Mandatory Green House Gas (GHG) Reporting – 40 CFR 98

On April 10, 2009, the USEPA published a proposed rule in the Federal Register to require mandatory reporting of GHG emissions. This rule would require about 13,000 facilities across the U.S. to report CO_2 equivalent metric ton emissions for the year 2010 and following. Explicit methodologies for calculation of emissions are required, as well as a quality assurance plan with calibration requirements for data used in the calculations.

In order to report emissions from 2010, quality assurance for inputs to the emission calculations must be in place by January 2010. Most facilities must make changes to their current practices or put in place new processes and systems in order to comply with this rule by the proposed January 1, 2010 start date. This presentation will provide an overview of the rule and describe the business impacts to industry that will be required to meet its requirements. ERM has held workshops around the country regarding this rule and we will share insights gained from several sectors of industry. Additional topics discussed in this presentation will include:

- Identify the gaps in your current GHG tracking systems
- Map out your next steps for preparing to report
- Shared perspectives from several industries with insights gained through a series of workshops held throughout the United States.
- Case studies from ERM's experienced climate change and air quality teams will relate experiences working with other clients' programs

Preparation and USE of Rare METAL OXIDE Coated ELECTRODE for Electrochemcial Disinfection of Ballast WATER

K.G.Nadeeshani Nanayakkara, Division of Environmental Science & Engineering, National University of Singapore, Singapore, Singapore

Yu-Ming Zheng, Division of Environmental Science and Engineering, Natioanl University of Singapore, Singapore, Singapore

J. Paul Chen, Division of Environmental Science and Engineering, National University of Singapore, Singapore, Singapore

Ballast water is an essential component in shipping industry, providing the balance and structural integrity to the ships in voyage. International maritime organization (IMO) has recognized the ballasting and de-ballasting process as one of the four greatest threats to the ocean environment due to the high possibility of biological invasions through the process. Biological invasions create ecological, economical, and human health issues and thus ballast water has to be treated prior to discharge. IMO adopted a regulation to manage the transfer of invasive species through ballast water. In our early work, electrochemical technology has been evaluated and found to be effective for the disinfection of *E. coli* and *E. faecalis*, two organisms that are regulated by IMO. However, an electrode which can provide enough amounts of disinfectants using a lower amount of energy, while last for a long duration of treatment is essential, if the system is to be introduced in full scale. In this research, a rare metal oxide coated electrode was prepared and used for the generation of disinfectants successfully. Electrode lifetime investigation reveals that the produced electrode is effective in terms of service lifetime.

Key words: Ballast water, Disinfectant, Disinfection, Oxide electrode, Service life

See more of this Session: Novel Membranes and Processes for Water Treatment, Desalination and Purification
See more of this Group/Topical: Environmental Division

Application of Nonwoven Membrane Filtration for the Separation and Recycling of Colloidal Particles

Walter Den, Environmental Science and Engineering, Tunghai University, Taichung, Taiwan

D. W. Yang, Environmental Science and Engineering, Tunghai University, Taichung, Taiwan

This study applies a non-woven membrane filtration system similar to the submerged-type membrane bioreactor (sMBR) for separating submicron inorganic particles from water. Non-woven microporous membranes are isotropic with fibrous structure, and hence are generally less expansive and experience lower trans-membrane pressure as compared with thin-film composite membranes. A number of studies have been reported in applying non-woven membranes in the form of a sMBR to treat wastewater with high BODs or CODs. However, applications of non-woven membranes for separating micrometer- or sub-micrometer-scale inorganic and biological particles have not been extensively studied. Therefore, this study aims to investigate the performance as well as the possible fouling conditions of a non-woven membrane mimicking sMBR to separate inorganic submicron SiO2 and Al2O3 particles from water.

In this study, a tempered glass reactor having a full working volume of 22.4 L was divided into two parts by a baffle wall, namely a mixing zone (16 L) and a membrane separation zone (6.4 L). Non-woven membranes (KNH Enterprise Co., Ltd, Taiwan) with two different nominal pore sizes at 0.4 ìm and 1.0 ìm were used. The membranes were supported by rigid spacers and were fabricated into flat-plate type configuration in which permeate can be drawn through the membrane by suction with a peristaltic pump. The model solutions were prepared by diluting commercialized polishing slurries containing either fumed silica or alumina particles having particle size ranges between 50 and 500 nm. In addition, to avoid biofilm growth during the experiments, an UV-C lamp and sodium hypochloride were been added in the reactor as a disinfectant. This method effectively eliminated biofilm interference that causes biofouling of the membranes. During each test run, the trans-membrane pressure and the permeate flux were recorded, and the permeate quality was determined by turbidity that was well-correlated with the actual solid concentration. Fouled membranes were recovered by a sequence of backwashing, NaOCl soaking, and ultrasonication (10 s). Typical over than 90% of the original specific flux could be attained after each recovery process. Furthermore, membrane autopsy analysis was conducted using scanning electron microscope (SEM) to study the surface blocking and internal fouling of the membrane.

Preliminary study indicated that the optimal filtration parameters between membrane pore size and particle size were related to the specific flux and the permeate quality. The effect of concentration of filtrate on the trans-membrane pressure, and thus the filtration permeability, was inconclusive at this point, as experiments using a wider range of particle concentrations were ongoing. We also observed that biofouling, if not sufficiently controlled, represented a major obstacle in assessing the degree of fouling by the particles in the solutions. The type and the extent of fouling as a function of specific permeate flux will eventually be verified with SEM analyses.

See more of this Session: Novel Membranes and Processes for Water Treatment, Desalination and Purification
See more of this Group/Topical: Environmental Division

Development of Advanced Membranes Based On Molecularly Imprinted Polymeric Particles

Olympia Kotrotsiou, Chemical Process Engineering Research Institute, Thessaloniki, Greece

Chryssoula Gkementzoglou, Department of Chemical Engineering, Aristotle University of Thessaloniki & Chemical Process Engineering Research Institute, Thessaloniki, Greece

Costas Kiparissides, Department of Chemical Engineering, Aristotle University of Thessaloniki & Center for Research & Technology Hellas, Thessaloniki, Greece

Molecular imprinting of synthetic polymers is a process where a functional monomer and a cross-linker are co-polymerized, via a free-radical polymerization mechanism, in the presence of a target molecule (i.e., the imprint molecule) that acts as a molecular template. The functional monomer molecules initially form a complex with the imprint molecules via covalent or non-covalent bonding. This is followed by the polymerization of the functional monomer with a bifunctional or trifunctional cross-linker. The resulting highly cross-linked polymeric structure keeps in position the template molecules through covalent or non-covalent molecular interactions between the monomer and the template functional groups. Subsequent removal of the imprint molecules reveals the specific binding sites that are complementary in size and shape to the template molecule. In the present study, the chemically and mechanically stable molecularly imprinted polymers (MIPs) are employed for the development of an innovative water purification system, exhibiting a highly selective separation mechanism, at molecular level, based on affinity interactions.

Precipitation, suspension and miniemulsion polymerization are employed for the synthesis of MIP micro- and nanoparticles to be used as synthetic receptors selective to specific pesticides. In particular, MIP micro- and nanoparticles were prepared for the selective separation of atrazine from aqueous or organic solutions. Methacrylic acid (MAA) was used as functional monomer and ethylene glycol dimethacrylate (EGDMA) as cross-linker. The molar ratio of the cross-linker to the functional monomer was equal to 4:1.

Precipitation polymerization was carried out in a borosilicate glass tube equipped with a screw cap. The template molecule (i.e., atrazine) was added into the solvent (i.e., dichloromethane) followed by the addition of the functional monomer and the cross-linker. The free-radical co-polymerization of MAA with the cross-linking agent was carried out in the presence of a chemical initiator (i.e., AIBN, 2% w/w relative to the monomers mass). The total monomer mass concentration was typically 2% w/v relative to the solvent volume. The solution was first purged with dry nitrogen for 5 min and the tube was sealed under a nitrogen atmosphere. The precipitation polymerization was then initiated by placing the tube in a preheated water bath at 60°C. The polymerization was carried out for about 16hrs.

Suspension polymerization was carried out in a laboratory scale, water-jacketed glass reactor of 100 ml working volume, equipped with a six-blade impeller, an overhead condenser and a nitrogen purge line. The reaction mixture was thermostated to within $\pm$0.05^{o}C with the aid of a constant temperature bath. Initially, the template molecule, the functional monomer, the cross-linker and the initiator (i.e., AIBN, 2% w/w on the total monomers mass) were dissolved in chloroform. Accordingly, the organic phase (15ml in volume, containing the functional monomer, the cross-linker, the porogen, the template molecule and the initiator) was dispersed into an aqueous PVA solution (35ml in volume, 1% w/w) under the action of a mechanical agitator and a nitrogen atmosphere. Subsequently, the polymerization was carried out, at the specified temperature (i.e., 60°C), for 24hrs.

Finally, miniemulsion polymerization was carried out in a laboratory scale, water-jacketed glass reactor. The organic phase consisted of the template molecule, the functional monomer, the cross-linker, the initiator (i.e., AIBN, 2% w/w on the total monomers mass) and the co-surfactant (i.e., hexadecane) was dispersed into an aqueous SDS solution (50ml, 2% w/w) via vigorous stirring for 1hr at room temperature, followed by sonication for 5 min. Subsequently, the polymerization was carried out, at the specified temperature (i.e., 60°C) for 24hrs.

In all cases the polymer particles were collected by centrifugation. Subsequently, the template molecule was removed from the imprinted polymer particles by means of successive washing cycles with a methanol-acetic acid solution (9:1 v/v). Finally, the washed polymer microparticles were conditioned in methanol. UV spectroscopy was employed to monitor the removal of the print molecules. Non-imprinted polymeric nanoparticles (NIP) were also prepared under the same conditions in order to compare their affinity to the template molecule with that of the MIP particles. Batch-wise guest binding experiments were conducted to evaluate the rebinding capacity of polymers using either organic (i.e., acetonitrile) or aqueous (i.e., water/acetonitrile 9/1 v/v) solutions of atrazine (0.1-1mM). The MIPs were found to exhibit 30-60% higher affinity towards the imprint molecule than that of the non-imprinted polymers (NIPs), prepared in the absence of the template, thus proving their binding specificity and consequently, their capability to be used as highly-selective adsorbing materials.

Additionally, the imprinted polymers synthesized by precipitation polymerization were used for the development of hybrid ceramic membranes. The polymeric nanoparticles were impregnated in the ceramic filters employing vacuum filtration, with the aid of a stainless-steel device, using a small quantity of particles dispersed in distilled water. SEM images proved that the nanoparticles have been successfully impregnated into the filters, while the rebinding experiments in aqueous solutions showed that the hybrid ceramic filters with MIPs adsorb more atrazine than those with NIPs.

See more of this Session: Novel Membranes and Processes for Water Treatment, Desalination and Purification
See more of this Group/Topical: Environmental Division

Novel Filtration Membranes Fabricated by Polyaniline Nanofibers

Nan-Rong Chiou, Chemistry, The Ohio State University, Columbus, OH

Katherine M. Kinstedt, Chemical Engineering, The Ohio State University, Columbus, OH

Arthur J. Epstein, Physics, The Ohio State University, Columbus, OH

Harold W. Walker, Civil, Envir Engr & Geod Sci, The Ohio State University, Columbus, OH

Ly James Lee, Chemical and Biomolecular Engineering, The Ohio State University, Columbus, OH

Polyaniline nanofibers show the superhydrophilic property. When applied as a thin active layer to the carbon based nanostructured membranes, e.g., carbon nanofibers and carbon nanotubes, as well as a traditional support material, these membranes can be created with a novel property such as the superhydrophilicity. This novel property offer a great promise for the field of membrane water purification. The superhydrophilic property of these novel membranes could significantly reduce fouling, while also decreasing the necessary transmembrane pressure, and thus energy expenditure. Herein, we will focus on the synthesis of these membranes with optimized properties for water purification by adjusting the support substrate and the binding agent. The properties of those novel membranes such as pure water flux, pore size distribution and fouling potential will also be discussed.

See more of this Session: Novel Membranes and Processes for Water Treatment, Desalination and Purification
See more of this Group/Topical: Environmental Division

Using Cactus Mucilage to Remove Inorganic Contaminants From Drinking Water

Dawn I. Fox, Chemical and Biomedical Engineering, University of South Florida, Tampa, FL

Thomas Pichler, Geology, University of South Florida, Tampa, FL

Daniel H. Yeh, Civil & Environmental Engineering, University of South Florida, Tampa, FL

Norma Alcantar, Chemical and Biomedical Engineering, University of South Florida, Tampa, FL

Sources of drinking water may become contaminated with several harmful inorganic and metallic species from both natural and man-made pollutant sources. Stricter environmental regulations are fuelling the search for more sustainable and environmentally benign methods to remove contaminants from drinking water. Cactus mucilage, an extract from the pads of the *Opuntia ficus-indica* (also known as Nopal and Prickly pear), was used to remove dissolved arsenic oxides, phosphates and chromium from drinking water.

Packed columns were used to filter test water treated with cactus mucilage. Concentrations of arsenic, phosphates and chromium were measured at entry and exit points of the test columns. Packing media and water chemistry were manipulated for optimization. Results show that simple filters employing cactus mucilage are effective at reducing arsenic, phosphate and chromium concentrations in drinking water. Cactus mucilage filters can potentially provide an environmentally friendly and sustainable alternative for drinking water treatment.

See more of this Session: Novel Membranes and Processes for Water Treatment, Desalination and Purification
See more of this Group/Topical: Environmental Division

Multistage Cross-Flow Filtration Process for Efficient Treatment of Oil-Field Produced Water Using Ceramic Membranes

Peter Czermak, I B P T - B.V.T, University of Applied Sciences Giessen Friedberg, Giessen, Germany

Mehrdad Ebrahimi, I B P T - B.V.T, University of Applied Sciences Giessen Friedberg, Giessen, Germany

Zoltan Kovacs, I B P T - B.V.T, University of Applied Sciences Giessen Friedberg, Giessen, Germany

Peter Mund, Atech Innovations GmbH, Gladbeck, Germany

Peter Bolduan, Atech Innovations GmbH, Gladbeck, Germany

Produced water (PW), formation water or brine as it is sometimes referred to, describes contaminated water that is generated along with the oil and gas in both onshore and offshore production operations. By far the largest by-product or waste stream by volume associated with oil and gas extraction. This water is always cleaned to some extent where the level of cleaning is determined by the intended use and/or current discharge regulations.

The increased volume of produced water handled in petroleum and natural gas production is becoming a major concern, especially with the possibility of further reduction in the oil content allowed in the discharged water (offshore operations), as well as the fact that produced water contains a number of undesirable toxic components. The quantity of water produced per unit volume of hydrocarbons varies greatly, but typically increases with the age of the production field.

This investigation focuses on the characterization of permeate flux using ceramic microfiltration (MF), ultrafiltration (UF) and nanofiltration (NF) membranes for efficient treatment of oilfield produced water. Results for average flux rates, flux degradation, total organic carbon (TOC) removal and oil removal efficiency are shown.

See more of this Session: Novel Membranes and Processes for Water Treatment, Desalination and Purification
See more of this Group/Topical: Environmental Division

Parametric Sensitivity Analysis for Pressure Swing Adsorption Systems

Libin Zhang, Aspen Technology, Inc, Houston, TX

The design of pressure swing adsorption is a challenging task because of the simultaneity of transport, adsorption and desorption phenomena (i.e. fluid flow, mass transfer and/or heat conduction) inherent to the processes, which are mathematically modeled by partial differential equations (PDEs) with spatial temporal variations. The high degree of uncertainty in addition to process operational uncertainties (e.g. daily fluctuations, seasonal variations in production levels) and physical property uncertainties adds complexity to this already challenging design problem. In most of studies, the thermodynamic and transport properties always are assumed to ideal relationship for reducing the complexity of the problem and avoiding the numerical challenges. However, those assumptions give a source of model uncertainty and prevent us to operate and design the PSA system reliable and efficiently. In this presentation, with help of Aspen HYSYS automation, a simple and rigorous thermodynamic and transport properties models are compared to minimize this uncertainty. Parametric sensitivity analysis of thermal, kinetic and material properties will be illustrated for this distributed system in two dimensions with full consideration of parameter and model uncertainty.

We will address the PSA systems with the transient transport phenomena i.e. diffusion, convection, equilibrium, thermodynamic equations of distributed systems in the presence of parameter uncertainties. We will assess the distributed system by solving the transport equations with the help of the discretization scheme i.e. finite volume method. First and second order sensitivity information will be provided to address impacts of variation of the thermal, transport and material properties on the PSA system. In particular we will argue for the cyclic steady state condition and product purity concern, ensuring operability and risk management under uncertainty.

In this presentation we highlight how rigorous thermodynamic model and transport equations with the detailed parameters analysis can help design engineers and plant managers to select the PSA with high performance. Simultaneous consideration of design and uncertainty offers higher flexibility and lower expected cost at the operating stage.

See more of this Session: Environmental Applications of Adsorption - I
See more of this Group/Topical: Environmental Division

Effects of Coal Interaction with Supercritical CO2: Physical Structure

Michael C. McClure, Chemical Engineering, The University of Mississippi, University, MS

Benson Gathitu, Chemical Engineering, University of Mississippi, University, MS

Wei-Yin Chen, Chemical Engineering, University of Mississippi, University, MS

It is known that polar solvents swell coal, break hydrogen-bonds in the macromolecular structure, and enhance coal liquefaction efficiencies. The effects of drying, interaction with supercritical CO2 and degassing on the physical structure of coal have been studied using gas sorption technique and a scanning electron microscope(SEM). Both drying and interaction with supercritical CO2 drastically change the micropore and mesopore: surface area, absolute volume and volume distribution in both bituminous coal and lignite. Degassing removes debris in the pore space which allows for better analysis of the changes in the morphology that were induced by drying and exposure to supercritical CO2. SEM reveals that interaction of bituminous coal with supercritical CO2 results in an abundance of carbon structures similar to the maceral collinite.

See more of this Session: Environmental Applications of Adsorption - I
See more of this Group/Topical: Environmental Division

Effect of the H2 On the CO2 Adsorption Properties of Organic-Inorganic Hybrid Amine-Based Adsorbents

Sunho Choi, School of Chemical & Biomolecular Engineering, Georgia Institute of Technology, Atlanta, GA

Christopher W. Jones, School of Chemical & Biomolecular Engineering, Georgia Institute of Technology, Atlanta, GA

Separation of CO2 from other light gases has been a subject of research focusing on natural gas purification as well as carbon capture from flue gas associated with the combustion of fossil fuels. Among several possible strategies to remove anthropogenic CO2 from flue gas, adsorption by solid sorbents has been widely studied. Current and future coal plants offer two types of CO2 separation challenges. For post-combustion CO2 capture, the separation must be made from a mixture of CO2, nitrogen, water, oxygen, sulfur oxides, and nitrogen oxides. Amine-based sorbents have been widely studied in this application. The second type of CO2 separation arises in an integrated gasification combined cycle (IGCC) power plant, whereby the fuel is gasified, shifted and then the CO2 is removed from a stream comprised mostly of hydrogen, prior to combustion of the purified hydrogen stream. Whereas post-combustion CO2 capture is normally carried out between 45¢ªC and 125¢ªC near ambient pressure, pre-combustion CO2 capture is normally carried out at more elevated temperatures and pressures. Because amine-based organic-inorganic hybrid sorbents operate near ambient conditions, all reported studies of CO2 capture with sorbents of this type have focused on post-combustion capture. Here, we report the first known study of CO2 from a hydrogen stream using amine-based organic-inorganic hybrid sorbents. A series of inorganic-organic hybrid sorbents prepared by incorporating mono-, di-, and tri-aminosilane species on porous silica substrates such as MCM-48 and SBA-15, are studied as model sorbents. The adsorption properties of these materials are characterized using a simulated mixture gas of CO2 and H2, while being compared with the results obtained from a diluted stream of pure CO2. The effect of several experimental variables, including the operating temperature, pore structure of the supports, and presence of moisture in the target gases, is also evaluated, with an overall aim of identifying and quantifying the changes of CO2 adsorption capacities and kinetics in pre-combustion vs. post-combustion applications.

See more of this Session: Environmental Applications of Adsorption - I
See more of this Group/Topical: Environmental Division

Adsorption of Naphthalene and Pyrene On Commercial Activated Carbons

Francisco Murilo T. Luna, Chemical Engineering, Universidade Federal do Ceará, Fortaleza, Brazil

Caio C. B. Araujo, Chemical Engineering, Universidade Federal do Ceará, Fortaleza, Brazil

Carolina B. Veloso, Chemical Engineering Department, Universidade Federal do Ceará, Fortaleza, Brazil

Ivanildo J. Silva Jr., Universidade Federal do Ceará, Fortaleza, Brazil

Diana C. S. Azevedo, Chemical Engineering, Universidade Federal do Ceará, Fortaleza, Brazil

Celio L. Cavalcante Jr., Chemical Engineering, Universidade Federal do Ceará, Fortaleza, CE, Brazil

Polycyclic aromatic hydrocarbons (PAHs) are ubiquitous contaminants and are well known for their toxic, carcinogenic, and mutagenic effects[1]. Several industrial applications use streams with PAHs as inputs or solvents, depending on the type and aromatic concentration. Extensive research has been carried out on the adsorption of organic compounds from dilute solutions by activated carbons. In particular, the adsorption of phenol and related compounds is one of the most studied systems in liquid-phase applications of carbon adsorbents, due to their industrial and environmental importance[2]. However, there are few reports in the open literature for the removal of PAHs from a hydrocarbon mixture. In this study, the adsorption of two model PHAxs, naphthalene and pyrene, on different types of commercial activated carbons was evaluated using batch and fixed bed experiments. The equilibrium isotherm parameters obtained from batch and fixed bed experiments were evaluated and compared. A general model that accounts for both internal and external mass transfer resistances was used to predict the breakthrough curves and estimate also kinetic parameters.

[1] Harvey, R.G., Cambridge University Press, Cambridge, UK, 1991.

[2] Ania, C.O., Cabal, B., Pevida, C., Arenillas, A., Parra, J.B., Rubiera, F., Pis, J.J., Water Research, 41, 333-340 (2007).

See more of this Session: Environmental Applications of Adsorption - I
See more of this Group/Topical: Environmental Division

Utilization of a Formerly Incinerated Sodium Nitrite Bearing Process Stream

Authors: Thomas L Guggenheim, Lioba M Kloppenburg, Sabic Innovative Plastics, 1 Lexan Lane, Mt. Vernon, Indiana, 47620

Abstract: A process to isolate and purify an industrial aqueous sodium nitrite bearing by-product stream was developed and built. This process stream was formerly incinerated at significant cost. The basis for the purification is carbon treatment, and the purified product is of sufficient quality to have found utility in the market place. Aspects of the technology to be discussed are raw material and product specifications, carbon type, carbon utilization, pertinent analytical methods, process safety, and the overall benefits to the environment.

Introduction: Sabic Innovative Plastics currently produces a resin in a multi-step process. One of the steps involves the batch reaction of a bisphenate with a 4-nitro-N-alkylphthalimide in the presence of phase transfer catalyst (PTC) in an organic solvent, equation 1. The bisimide product is soluble in the solvent employed, and the sodium

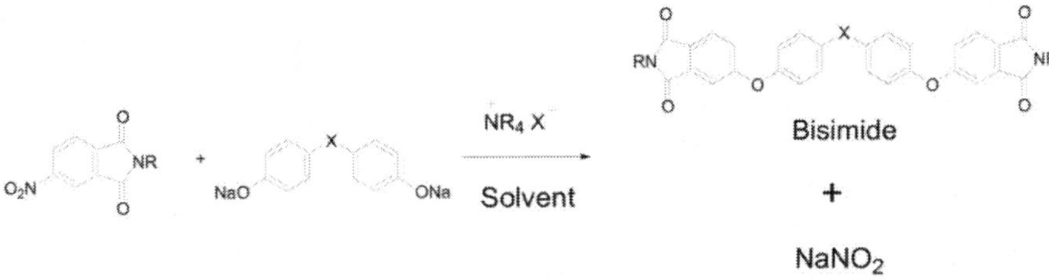

(equ 1)

nitrite by-product is insoluble in the media. The reaction provides a bisimide in >99% yield. The purified bisimide is converted to a dianhydride, which is subsequently used to make a polyetherimide. One dianhydride made by this process is ULTEM® Dianhydride, compound 1. This material is now commercially available.

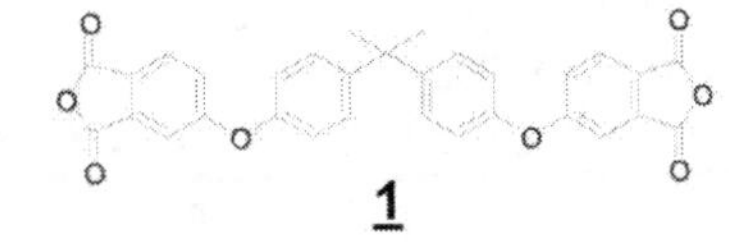

1

Traditionally, the reaction mixture of equation 1 is treated with dilute caustic to remove the sodium nitrite, starting materials and the PTC. The nitrophthalimide is solubilized in the dilute caustic as the corresponding amide-acid salt (2) and nitrophthalic acid salt (3).

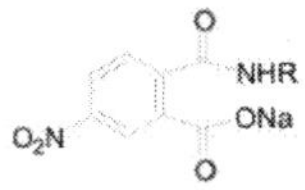

2

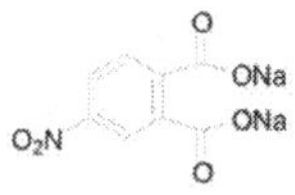

3

The caustic also hydrolyzes and solubilizes some of the bisimide product in the form of bis-amide-acid sodium salts. The wastewater is concentrated and then incinerated at significant cost.

Several other technologies have been investigated to remediate this wastewater. Wet air oxidation was not efficient, supercritical oxidation at high pressure and temperature promised to be expensive to build, traditional biotreatment did not mineralize the organic impurities present, and chemical destruction of the nitrite with urea or sulfamic acid was cost prohibitive. An improved method was developed to recover the sodium nitrite with less organic material present, namely washing the reaction mixture of equation 1 with water instead of dilute caustic.[1] This first wash contains 40 wt% sodium nitrite with about 5000 ppm of total organic carbon (TOC) at a pH of 8 to 12, compared to a traditional caustic wash that was 25% sodium nitrite containing 30,000 to 40,000 ppm TOC and a pH of 12. Isolation of 40 wt% $NaNO_2$ with less organic material opened the possibilities of purifying the wastewater with carbon and marketing the product, or removal or the water to afford a dry material followed by melting at 420 °C to oxidize organic material, or purification via micro-filtration. This paper describes the commercialization of a process to purify the stream with Calgon Reactivated carbon to afford a marketable material.

Results and Discussion:

Commercial aqueous sodium nitrite is typically 40 to 41 wt% $NaNO_2$ with a pH of 8 to 10. Just enough water is added to a batch of bisimide (equation 1) to form an aqueous phase of 40 to 45 wt% sodium nitrite at 50 to 80 °C. The solubility limit of $NaNO_2$ in water at room temperature is ~45 wt%, and ~41% at 0 °C. The aqueous nitrite separated from the organic phase contained 5000 to 6000 ppm TOC, the bulk of the TOC being the PTC. The material has a viscosity of ~3 centipoise.

Three different carbon types were screened to determine the efficiency of TOC removal from the contaminated aqueous $NaNO_2$. One hundred gram portions of the impure aqueous $NaNO_2$ were charged to 250-mL glass bottles and treated with crushed carbon of a certain type at increasing amounts (0.5, 1.0, 2.0, 5.0, 10.0, and 15 grams). The bottles were shaken for 12 hours, the carbon filtered off, and the TOC of the supernatant determined. The filtrate was also analyzed by high pressure liquid chromatography (HPLC) to quantify the amount of each of the known organic impurities originally present. It was found that 1 pound of impure 40% $NaNO_2$ required ~0.1 pound of Calgon Reactivated carbon (DSR-A, 40 x 5) carbon to provide a material with <50 ppm TOC and <1 ppm of any of the know organic materials. Calgon CPG carbon performed better, and wood based Colorsorb G carbon performed worse, in terms of carbon capacity. DSR-A is the most affordable of the carbon materials tested, and it can be regenerated at Calgon. Lignite type carbon cannot be regenerated and must be land filled or burned.

[1] See US Patent, 6,251,354, 6/26/01 to GE Plastics (now Sabic IP).

The process was piloted using DSR-A Calgon Carbon to produce truckload quantities (4000 gallons) of purified product to develop a manufacturing process, to determine the quality of the material produced, to develop a market for the product, and to determine the economic viability of the overall process.

Two vessels that held 10,000 pounds of carbon were plumbed in series, Figure 1.

Figure 1. Simplified Process Diagram for the Carbon Purification of Raw 40 Wt% Aqueous $NaNO_2$.

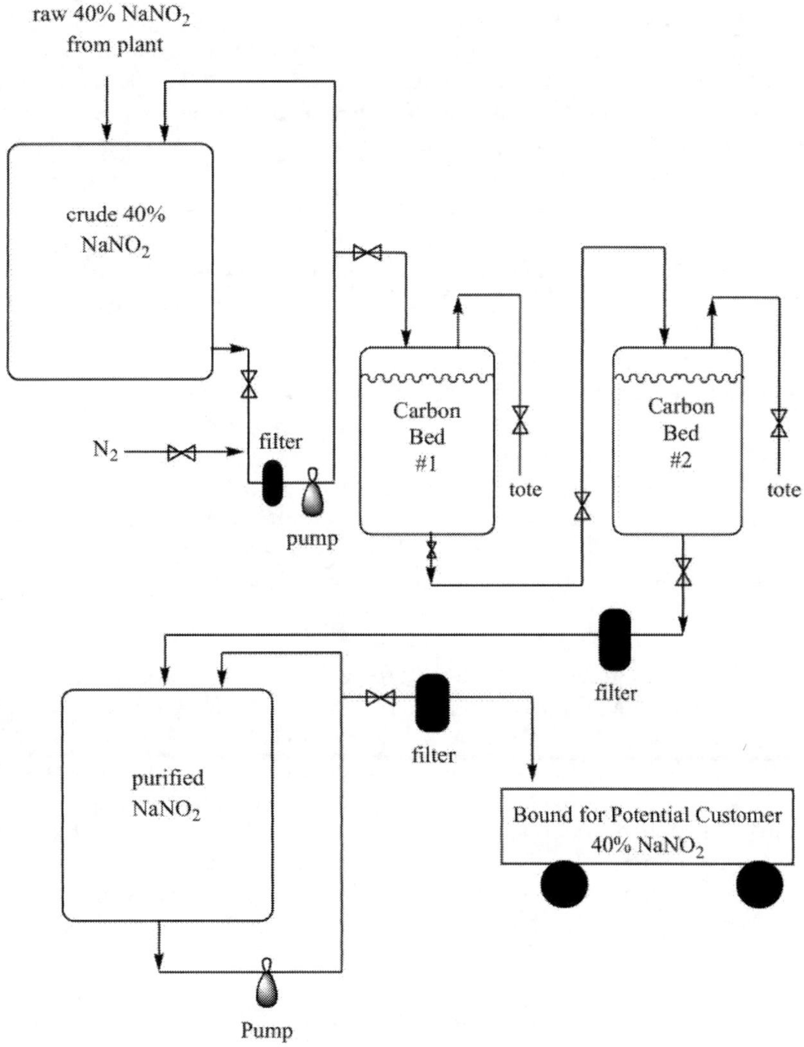

DSR-A 40 x 5 carbon was slurry loaded into the carbon beds, and the clean water was drained from the carbon bed. Raw material was collected from the plant (~40% aqueous

$NaNO_2$, ~5000 ppm TOC, pH = 11) in tank trucks. The material was filtered with a low efficiency 10-micron bag filter to remove any precipitated organic material into a raw feed tank.

Each carbon bed was ~4000 gallons in volume. The carbon bed level in each bed was approximately 2 feet from the top tangent line of the tank. Each bed contained 10,000 pounds of carbon, 345 ft3 of carbon, or about 2580 'gallons' of carbon. Approximately 20% of the space that the carbon occupied was filled with the carbon particles, 40% of the space was the volume of the pores in the carbon particles, and the last 40% of space was that between the particles (a Calgon rule of thumb). The valve at the bottom of the lead bed was closed. The raw feed was pumped to the top of the first carbon bed at ~50 gpm. The lead bed was filled with material; a small amount of material was allowed to exit the vent of the bed into a tote to insure the bed was full. Once the first bed was full, the valve was opened at the bottom of the first bed to allow flow into the top of the second bed; the valve below the lag bed was closed. The second bed was allowed to fill, and a small amount of material allowed to exit the top of the lag bed into a tote. The beds were then allowed to sit (soak) for about 2 hours. The bottom valve on the lag bed was opened and MFW was pumped to the beds and thence to an awaiting purified product hold tank at 60 gallons per minute (gpm). The product from the first and second beds were taken at set intervals and analyzed for TOC, wt% $NaNO_2$, and signature organic compounds (HPLC).

The initial material from the beds was diluted by the water originally on the carbon and was concentrated to 40 wt% nitrite. After the lag bed had broken-through, the material in the beds was pressed out using nitrogen pressure and was recycled to the feed tank. The beds were then filled with water and allowed to soak. The beds were then drained to a separate tank. This procedure was repeated. The combined soak material (~15 wt% $NaNO_2$) was concentrated to 40 wt% $NaNO_2$ and sent to the raw carbon bed feed tank. The carbon was then transferred out of the bed with water and sent to Calgon for regeneration.

The water in the pores of the carbon diluted the first amount of product. (Future campaigns will use 45 wt% feed.) Five trucks were fed to the beds and 5 trucks of product were collected. Approximately 209,000 pounds (~82,300 pounds of $NaNO_2$) of raw material was fed to the beds and 200,800 pounds (72,828 pounds of $NaNO_2$) were collected. The beds were run until known organics broke-through the lag bed. The trial showed that ~0.1 pounds of carbon was necessary to purify ~1.0 pound of the feed, in line with the laboratory carbon isotherm studies. Analyses of the five trucks collected are shown in Table 1. The last truck of material did not contain any known organic constituent above its detection limit (~1 ppm). The pH of the feed was ~10 to 11, yet the product had a bulk pH of 7.

The material was analyzed for trace metals and common anionic impurities, Table 2. Commercial material was also analyzed for comparison.

Table 1. Analysis of Carbon Purified 40 wt% Aqueous $NaNO_2$.

Purified Truck Order	Truck #	TOC Monomer (ppm)	pH	alkalinity as wt% Na2CO3	Color	lbs aqueous NaNO2 solution	wt% NaNO2	~lbs NaNO2
1	824	0	7	ND	light yellow	39760	28.8	11451
2	823	1	7	ND	straw yellow	45600	36.5	16644
3	829	7	7	ND	straw yellow	39440	38.2	15066
4	808	19	7	ND	yellow	41660	38.9	16206
5	171	80	7	ND	brownish yellow	34340	39.2	13461
						200800		72828

Table 2. Anion and Metal Analyses of Carbon Purified 40 wt% Aqueous $NaNO_2$.

	ICP (ppm)										IC (ppm)		
Product Truck #	Ca	Cu	Fe	K	Mg	Ni	Si	Zn	Mo	Cr	Cl	NO3	SO4
1	276	0.4	3.3	163	14.3	0.8	1.2	1	3.8	nd	1110	141	426
2	177	nd	3.3	75	8.8	1.2	0.2	0.5	1.6	0.1	804	78	160
3	96.5	nd	2.9	43	6	1.1	0.2	0.8	0.8	0.1	889	69	78
4	61.6	nd	2.8	36	4.5	1	0.2	1.4	0.4	nd	777	71	45
5	33.2	nd	2.8	33	3.2	0.8	nd	1.7	nd	nd	724	71	49
Rapauno Lot 18812	4.2	nd	0.5	123	1.1	0.2	19.7	0.2	nd	0.1	722	19	110
Commercial	2.5	1.4	nd	4.4	0.8	0.3	14.6	nd	nd	nd	<100	2651	26

The process was scaled up several more times with Calgon React Carbon as described above (except that the concentrated carbon bed soaks from the previous trial was processed with the raw feed from the plant), and a couple of interesting observations were made. In the second 5 truck campaign the pH of the product, collected in two tanks, was higher and residual calcium was lower, Table 3. It was speculated that the contact

Table 3. Product Analyses from the Second Carbon Purification of 40 wt% $NaNO_2$.

	ICP (ppm)													IC (ppm)					
Tank #	Al	Ca	Cu	Fe	K	Mg	Ni	Si	Zn	Mo	Cr	Ti	P	Cl	NO3	SO4	TOC	wt% Na2CO3 equiv	wt% NaNO2	pH
Tank #1	7.6	18.5	nd	36	122	nd	9	4.4	nd	nd	nd	0.5	nd	730	10	180	40	0.03	40.31	10.5
Tank #2	0.6	13.5	nd	nd	84	0.3	nd	5.9	0.1	nd	nd	nd	nd	722	19	110	39	0.03	40.17	10.5

pH of the carbon used in the initial trial was lower than the carbon used in the 2nd trial. Contact pH is a measure of acid groups bound to the carbon, and high acid groups apparently resulted in a lower pH product. The headspace in beds was analyzed for oxides of nitrogen (NOx) and none was detected ($NaNO_2$ will generate NOx under acidic conditions). The bulk of Calgon regenerated carbon comes from carbon used at municipal water plants. The carbon picks up a considerable amount of calcium and magnesium at these plants and is then returned to Calgon for regeneration. The regeneration process involves heating the carbon at up to 900 °C under a controlled amount of oxygen and steam. The regeneration process does not remove the calcium and other metals, and these metals can leach from the bed in the next life of the carbon. It was speculated that this is what happened in the trials described thus far.

The most noticeable difference in the product from the 1st trial and the 2nd trial was the color of the product. The first trial gave material that was a light straw yellow in color. The second trial afforded material that was deeper yellow in color, and material coming off the beds just before major TOC break-though had an orange tint. Commercial 40 wt% aqueous $NaNO_2$ is very light yellow in color.

The carbon utilization for the 2nd trial was better than the first; ~0.085 pounds of carbon was necessary to purify 1.0 pounds of feed, a 15% improvement in carbon utilization.

A third large-scale trial was conducted as described above to afford material with the characteristics shown in Table 4. The color of the product again was a darker yellow

Table 4. Product Analyses from the Third Carbon Purification of 40 wt% $NaNO_2$.

	ICP (ppm)													IC (ppm)						
Tank #	Al	Ca	Cu	Fe	K	Mg	Ni	Si	Zn	Mo	Cr	Ti	P	CL	NO3	SO4	TOC	pH	wt% Na2CO3 equiv	wt% NaNO2
Tank #506	12.0	21	0.3	35.0	61.7	nd	13.2	nd	nd	nd	nd	0.6	6.2	700	< 10	120	46	10.5	0.04	38.4
Tank #509	9.9	37	0.2	34.9	43.1	nd	13.3	nd	nd	nd	nd	0.6	1.8	650	< 10	21	67	10.5	0.02	40.3

with an orange tint, compared to the light straw yellow color of commercial grade material. The carbon utilization was again ~0.085 pounds of carbon to purify 1.0 pounds of feed.

Fines particles of carbon were periodically a problem. A multi-layer bag filter provided by Filtration Systems was employed to effectively remove the fines from the product. This bag was of Nylon construction and consisted of one 5μ layer and two 1μ layers of material. It removed the vast majority of carbon, but pressure build was periodically an issue.

Based on these trials, the draft specifications for the product was established, Table 5.

The purified material has been evaluated in several applications. Uses that can tolerate the presence of significant amounts of chloride, sulfate, calcium, and minor amounts of magnesium, aluminum, potassium, and iron are currently under study.

Table 5. Draft Specifications of Sabic Innovative Plastics 40 wt% Aqueous $NaNO_2$.

Specification	UOM	Lower Limit	Upper Limit	Analytical Method	US Commerical Grade Quality
sodium nitrite	% by wt	40	41	pan solids	40-42
total organic carbon	ppm by wt	0	100	UV persulfate TOC analyzer	<10
pH	pH units	7.5	12	pH electrode	9-10
chloride anion	ppm by wt	500	2000	dilute then ion chromatography	<100
sulfate anion	ppm by wt	0	400	dilute then ion chromatography, not refined yet	<30
nitrate anion	ppm by wt	0	200	dilute then ion chromatography	1000 to 10000
calcium	ppm by wt	0	400	ISE	<2
magnesium	ppm by wt	0	20	tbd	<2
sodium carbonate equivalence	% by wt	0	0.4	titration	<0.2
sodium chloride	ppm by wt	0.46	0.8	dilute then ion chromatography	<150
color	visual	light yellow	orange	UV	light yellow
carbon fines	visual		pass	1u filtration	none

Summary and Conclusions:

An aqueous waste stream composed of 25 wt% sodium nitrite and 5 wt% organic was formerly incinerated at significant cost. A process was developed to isolate this waste stream as a 40 wt% $NaNO_2$ stream containing ~5000 ppm total organic carbon. It was discovered that removal of the sodium nitrite from the reaction mixture resulted in an increased yield of the bisimide product (equation 1), by virtue of the fact that there was significantly less hydrolytic yield loss through the reaction of the bisimide with the caustic wash. The raw 40 wt% $NaNO_2$ steam was then purified using Calgon reactivated carbon (Calgon React DSR-A 40 x 5 mesh). The purified product contained <1 ppm of the known organics present in the raw waste stream. Commercial outlets for the material have been identified. The economics of the process were verified in process trials.

Experimental:

Total organic carbon of 40 wt% aqueous $NaNO_2$ were measured on an OI Analytical 1010 or 1030 Automated Wet Oxidation Analyzer. Samples were diluted 1 to 100 with water before analysis.

The amount of sodium carbonate equivalents was determined by titration with 0.1N HCL using phenolphthalein as an indicator. An ethanolic solution of phenolphthalein was prepared (0.5 wt% in ethanol). A 25 gram sample of 40 wt% aqueous NaNO2 was diluted with 150 mL of water and two drops of indicator solution was added. This was then titrated to the endpoint with the acid.

Metals in the product were determined by an inductively couple plasma technique using a Varian MPX (Axial view) ICP system equipped with a Spray Chamber.

Anions in the product were determined using ion chromatography using a Dionex ICS-2000 Ion Chromatography system, with an AS11 Analytical column with a AG11-HC Guard column and a CR-A Trap column, equipped with an ASRS 300 4mm Suppressor, column temperature of 35 °C, and DS6 heated Conductivity detector, with a gradient eluent (0-9 minutes ramps from 2.5 mM KOH to 4.0 mM KOH; 9-13 minutes ramps from 4.0 mM KOH to 25.0 mM KOH; 13-27 minutes 25.0 mM KOH), at a flow rate of 1.5 mL/min.

HPLC analysis used a reversed phase column. The method is available upon request.

The amount of sodium nitrite present in an aqueous sample was determined by diluting an appropriate amount of ~40 wt% material with water and adding a known excess amount of ceric ion to completely oxidize the nitrite, and then back-titrating the residual ceric ion with ferrous ion.

Acknowledgements: Bringing this process to commercial reality involved many technologists. Chief among them were Chris Poirier, Tom Sommerfield, Michael Buchanan, Joshua Croll and Jacob Oberholtzer. Other contributors were Randy Northrop, Ihab Odeh (formerly Sabic IP), Robert E. Greenwell, Byantha Paul, Bill Greathouse, Al McMahon, Kirk Blackburn, Jeannie Duckworth, Chris Leitsch, Mark DeLong, Annette Bell (formerly GEP), Darlene Nance, Peter Frederick (Microbac Labs), Stephen Dimond, Tammy Moor, John Wood, John Pace, ULTEM Operators, Rick Saunders (Calgon Carbon), Mick Greenbank (Calgon Carbon).

Utilization of Microfibrous-Supported Sorbent Materials and Novel Packaging Designs for Improved Indoor Air Quality at Reduced Energy Consumption

Ryan A. Sothen, Chemical Engineering, Auburn University, Auburn University, AL

Yanli Chen, Chemical Engineering, Auburn University, Auburn, AL

Bruce J. Tatarchuk, Chemical Engineering, Auburn University, Auburn, AL

The demand for improved air filtration due to poor indoor air quality (IAQ), potential threat of biological outbreaks, and increasing control standards in industrial applications has created a need for medium efficiency particulate air filters combined with a gas phase sorbent/catalyst scrubber. These dual-functioning materials excel at removing airborne pollutants such as dust, formaldehyde, and volatile organic compounds (VOC's) that are commonly sited to cause and aggravate asthma, allergic reaction, cancer and other respiratory diseases. The increased filtration performance, however, comes at a price. Dual-functioning materials have substantially higher flow resistances than traditional filtration media due to the decreased fiber diameters, tighter packing densities, and presence of adsorbent/catalyst particles. The utilization of these filters results in a significantly higher energy consumption; thus, these materials can become price prohibitive due to their considerable pressure drops.

Improvements to current filter designs must be made in order to successfully employ a dual-functioning media at a reasonable cost. High resistance media are commonly housed in a pleated filter. By folding the media into a corrugated arrangement, the face velocity and subsequently the pressure drop across the material is reduced. Multi-Element Structured Arrays (MESA's) represent a new approach that expanses on the pleated filter's premise. MESA's consist of numerous pleated filters mounted into a single, corrugated arrangement to further reduce the face velocity and pressure drop across the system. This unique arrangement has been demonstrated to incorporate four times the available media area into a filtration unit while operating at one-third the initial flow resistance of a traditional pleated filter.

The current study investigates the utilization of microfibrous-supported sorbent materials (MSSM) for use as a dual-functioning media. MSSM achieve high adsorptive rates and sorbent utilization by entrapping small (< 200 microns) activated carbon particles within a matrix of polymeric fibers. The fibers double as a particle filter capable of removing 50% of airborne particles as small as 0.5 microns. The filtration performance of the MSSM media were evaluated and compared to traditional filtration materials by investigating the following parameters: initial permeability, removal efficiency, aging characteristics, and breakthrough time. Based on these measurements, a MESA filter unit composed of MSSM media could be modeled for lifetime energy consumption and useful working time.

See more of this Session: Environmental Applications of Adsorption - I
See more of this Group/Topical: Environmental Division

Carbon-Based Chemical Filters Enhanced with Photocatalysis for Removing VOCs From Indoor Air

Hui-Lan Chang, Environmental Science and Engineering, Tunghai University, Taichung, Taiwan

Walter Den, Environmental Science and Engineering, Tunghai University, Taichung, Taiwan

Chemical filters are common designed for the purpose of purifying indoor air, especially for controlled environments demanding low tolerance of contamination, exemplified by their extended use in cleanrooms for microelectronics and optoelectronics devices, as well as for pharmaceutical products. In the ventilation control system in these critical environments, chemical filters must be installed for the removal of gaseous contaminants from various possible sources, including the outside air, the recirculation air, and the point-of-use sub-environments which require purification of specific contaminants.

Activated carbons in granular (GAC) or fibrous (ACF) structure are typically used as the base materials and adsorbent for the removal of organic contaminants in air. Activated carbons can also be impregnated with either phosphoric (or citric) acid for the absorption of basic gases, or potassium permanganate (or hydroxide) for the absorption of acidic gases. The common problem with chemical filtration is that the sorbent materials will eventually be saturated beyond a critical point at which filter replacement becomes necessary to maintain the stringent requirement of air cleanliness in the manufacturing environment. One of the possible methods to prolong the useful life of a chemical filter is to incorporate photocatalysts in the filter structure. This study therefore investigates the effect of GAC-based chemical filter combining with photocatalysis in three different forms,namely (i) chemical filter containing GAC-loaded TiO2 powders; (ii) chemical filter preceded with a sheet of TiO2-loaded non-woven filter; and (iii) chemical filter pressed with a TiO2-coated non-woven filter.

The test reactor consisting of a cylindrical pyrex column (i.d. 80 mm) having two removal segments, namely a photocatalysis section (length 150 mm) and an outlet section (length 55 mm). The filter testing pieces were grooved between the two sections and airtight secured by an O-ring. A quartz tube (OD 45 mm, length 140 mm) in concentric with the column contained a 13-W ultraviolet lamp (Sankyo, Japan) irradiating at a primary wavelength of 365 nm. As it will be discussed in this paper, the test piece was deemed "inactive" in area which was constricted by the flow path around the quartz tube in the center of the reactor. The estimated active area of the test piece was 34 cm2. The irradiation intensity on the perimeter of the testing piece was 0.20 mW/cm2. Toluene and isopropanol were targeted as the test contaminant and their vapors were generated from a syringed pump in a heated three-way glass branch. The air was humidified by passing through a water vessel before mixing with the toluene vapor and metered into the test reactor. The standard air flow rate into the test reactor was 1.0 L/min, corresponding to a face velocity of 0.005 m/s.

Experimental results showed that the chemical filter preceded with a separate sheet of TiO2-coated non-woven fabrics performed with superior results than directly loading TiO2 powders on carbon granules. Also, we have demonstrated that, by pressing the two filter sheets together, a possible synergistic effect occurred to further enhance toluene photocatalytic degradation to extend the adsorption breakthrough. The photocatalytic sheet serves to achieve pre-decomposition of organic contaminants into lower concentrations and/or smaller molecules for the subsequent adsorption by chemical filter, and oxidization of the adsorbed organic pollutants to re-open adsorption sites on the chemical filter. However, high concentration of toluene would lead to catalyst "poisoning" as a results of accumulation of intermediate products. These intermediate products may also promote competitive oxidation and adsorption with their parent contaminants, resulting in reduced photocatalyst oxidation efficiency, and thus shortening the expected breakthrough times.

See more of this Session: Environmental Applications of Adsorption - I

See more of this Group/Topical: Environmental Division

Competitive Adsorption of Ca^{2+}, K^{+}, Cd^{2+} and Pb^{2+}

Wang Tao[1]*, Zhou Yun[2]

[1]*The Key Lab of Hollow Fiber Membrane Materials and Membrane Process of Ministry of Education , Tianjin Polytechnic University , Tianjin 300160 , China*[1]

[2]*Tianjin Lonwin Network Science & Technology Development Corp, Tianjin 300181, China*

Abstract: Active sludge was acclimation by Ca^{2+}, K^{+}, Cd^{2+} and Pb^{2+} together, then it be used to study single and competitive isothermal adsorption of the four kinds of metal ions. It show that the regulations of Ca^{2+}, K^{+}, Cd^{2+} and Pb^{2+}'s single isothermal adsorption accord with Langmuir model. Competitive adsorption between hard-hard ions, soft-soft ions and hard-soft ions was also studied. These competitive adsorption processes were proved accord with Freundlich model. It was found that competitive adsorption obviously happened between the same kinds of hard or soft metal ions while happened not obviously between different kinds of metal ions. Soft ions could be better absorbed than hard ions in biofilm. We explain this phenomenon by coordination radicle bond theory.

Keywords: competitive adsorption, hard ions, soft ions, Freundlich model

Introduction

There has been increasing concern and more stringent regulation standards pertaining to the discharge of heavy metals to the aquatic environment, due to their toxicity and detriment to living species including humans. Heavy metals have a range of adverse impacts on the body. Exposure to nickel can cause dermatitis and allergic sensitization[1]. The harmful effects of cadmium include a number of acute and chronic disorders, such as renal damage, emphysema, hypertension, and testicular atrophy[2].

Many different materials have been examined for use as adsorbent for heavy metals. The proposed adsorbends are either microbial[3], naturally occurring[4], industrial waste[5] or agricultural by-products or waste[6]. Al-Degs[7] investigated the technical feasibility of using diatomite and Mn- diatomite in removal of Pb^{2+}, Cu^{2+}, and Cd^{2+}. Kandah[8] studied the removal of Zn^{2+} using disposal sheep manure waste. Since most water contaminated with heavy metals contain more than one heavy metal ion, it is necessary to determine the binding ability of the adsorbent with multi-metal solutions. Therefore, some reseachers studied the effect of competition for multi-metal binding by different adsorbent[9,10].

Pearson classified metal ions into hard ions (such as K^{+}, Mg^{2+}, Ca^{2+}), soft ions (such as Hg^{2+}, Cd^{2+}, Pb^{2+}) and fringe ions (such as OH^{-}, HPO_4^{2-}, $RCOO^{-}$) according to their capability of forming coordination key. In this work, the single and competitive adsorption between hard-hard ions, soft-soft ions and hard-soft ions was studied.

[1] Correspondence author*. Tel: +86-22-24528104, Email address: wangtao_20000@hotmail.com
Co-author Email address: Zhou Yun: zhouy@lonwin.com.cn

The research was supported and by Nature Science Fund of Tianjin, China(07JCZDJC01400)and Nature Science Fund of China (50808130)

Materials

In experiment of single kind of metal ions adsorption, solution of metal ions had been prepared at certain concentration in series. Each solution was added into conical flask together with biofilm weight about 7g. Keep pH value of solution at 7.0. Put conical flasks into shaker then. Under the experiment conditions of 25℃ and 150 rpm, The adsorption reaction time would be 30 min. Mixture in flasks would be centrifugal separate for 10 min at 3000 rpm. Balance Concentration of metal ions in solution was analysed by atomic absorption spectrophotometer (Type HITACHI 180-80, Japan) .

Theories

It will be influence each other when there is more than one kind of ions exist in solution adsorption process. Isothermal adsorption model become more complex under disturb and competitive adsorption. So did the experiment data were used to according with isothermal adsorption model and express influence between ions. Many reports had introduced competitive adsorption model such as Langmuir model.

$$q_{eq,i} = \frac{Q_i^{\ 0} b_i C_{eq,i}}{1+\sum_{i=1}^{N} b_i C_{eq,i}} \qquad (1)$$

$Q_i^{\ 0}$ and b_i were obtained from single ion isothermal adsorption model, and more parameters competitive adsorption model was developed on the base. [11,12]。

$$q_{eq,i} = \frac{Q_i^{\ 0} b_i C_{eq,i} / \eta_i}{1+\sum_{i=1}^{N} b_i C_{eq,i} / \eta_i} \qquad (2)$$

η_i came from competitive adsorption data of i, it is called interplay parameter. If $\eta_i \approx 1$, model (2) turn into model (1) . It means that there isn't any competitive adsorption between ions. If $\eta_i \leq 1$, it is means competitive adsorption exist.

Duality experiential Freundlich Isothermal adsorption model could be showed as

$$q_{1,eq} = \frac{a_1^{\ 0} C_{1,eq}^{\ b_1^{\ 0}+b_{11}}}{C_{1,eq}^{\ b_{11}} + a_{12} C_{2,eq}^{\ b_{12}}} \qquad (3)$$

$$q_{2,eq} = \frac{a_2^{\ 0} C_{2,eq}^{\ b_2^{\ 0}+b_{22}}}{C_{2,eq}^{\ b_{22}} + a_{21} C_{1,eq}^{\ b_{21}}} \qquad (4)$$

$a_i^{\ 0}$ and $b_i^{\ 0}$ were obtained from single ion isothermal adsorption Freundlich model , other parameters such as b_{11} , a_{12} , b_{12} , b_{22} , a_{21} , b_{21} were come from duality experiential Freundlich Isothermal adsorption model.

Results and Discussions

Langmuir and Freundlich isothermal adsorption equation were used to fitting experiment data. Equation parameters and result can be seen in Tab 1 and Fig 1 to 4

Table 1 Adsorption model parameters of Ca^{2+}, K^{+}, Cd^{2+} and Pb^{2+}

	Langmuir model			Freundlich model		
	Q_0	b	R^2	a^0	b^0	R^2
Ca^{2+}	1.6435	0.0246	0.99773	0.06669	0.68502	0.99037
K^{+}	0.04919	0.01412	0.99971	0.001135	0.74314	0.99552
Cd^{2+}	2.0975	0.5658	0.99994	0.8679	0.31632	0.97088
Pb^{2+}	16.897	2.61905	0.99106	14.72313	0.58902	0.99617

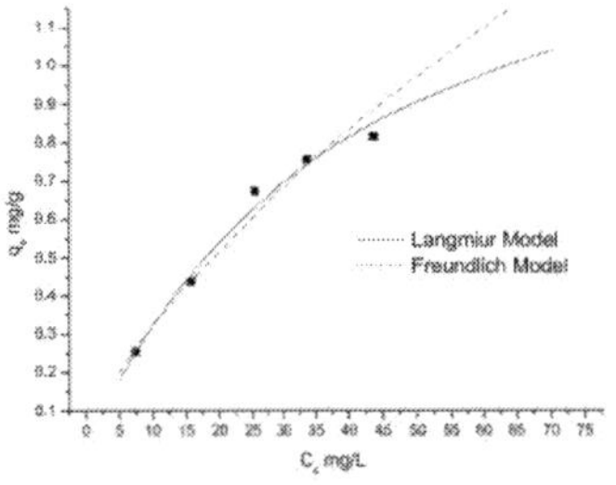

Fig.1 The adsorption isotherm of Ca^{2+}

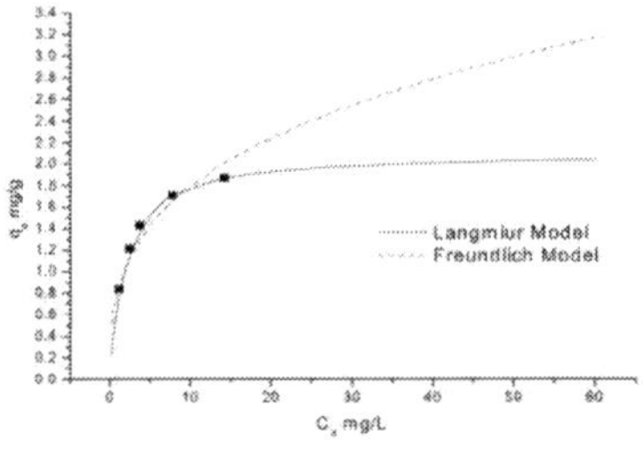

Fig.2 The adsorption isotherm of Cd^{2+}

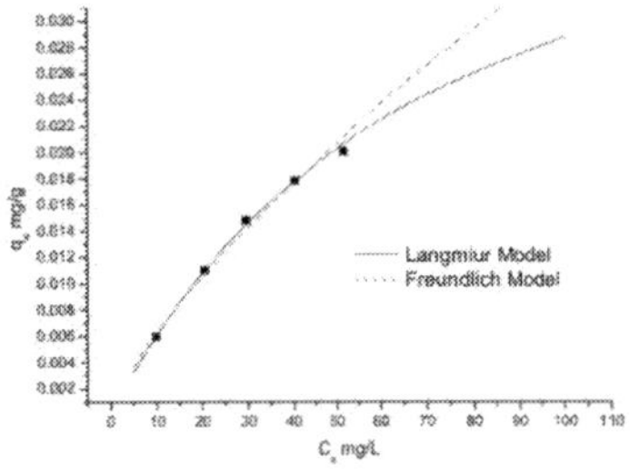

Fig. 3 The adsorption isotherm of K^{+}

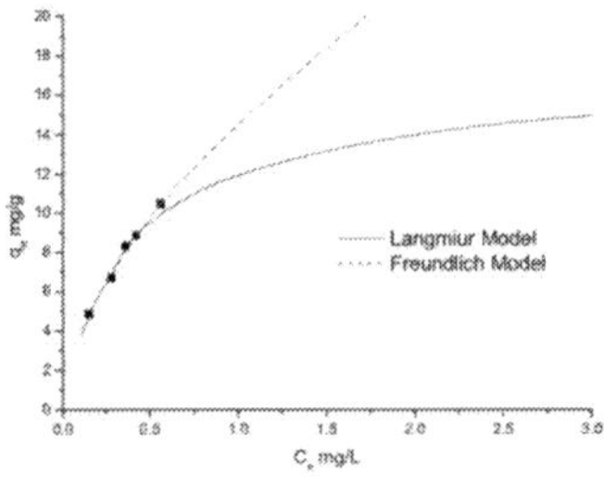

Fig.4 The adsorption isotherm of Pb^{2+}

Balance adsorption amount in biofilm of Ca^{2+}, K^{+}, Cd^{2+} and Pb^{2+} is of 1.6425mg·g^{-1}, 0.04919 mg·g^{-1}, 2.0975 mg·g^{-1} and 16.897 mg·g^{-1}, so $Pb^{2+} > Cd^{2+} > Ca^{2+} > K^{+}$

In experiment of competitive adsorption, we studied 3 kinds of reaction.

1) Hard-hard: Study adsorption amount of Ca^{2+} on biofilm with different concentration of K^{+}

2) Soft-soft: Study adsorption amount of Cd^{2+} on biofilm with different concentration of Pb^{2+}

3) Hard-soft: Study adsorption amount of Ca^{2+} on biofilm with different concentration of Pb^{2+}

Solution of every two kinds of metal ions had been prepared at certain concentration in series. Other conditions are as same as experiment of single metal ions except adsorption reaction time would be 20 min. Freundlich competition adsorption model was used to fitting experiment data. Equation parameters and result can be seen in Tab 2 to 4 and Fig 5 to 7

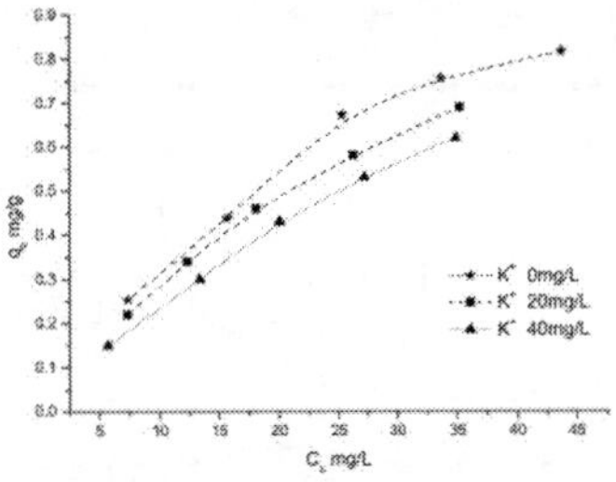

Fig 5 Competition adsorption isotherm of Ca^{2+} with K^+

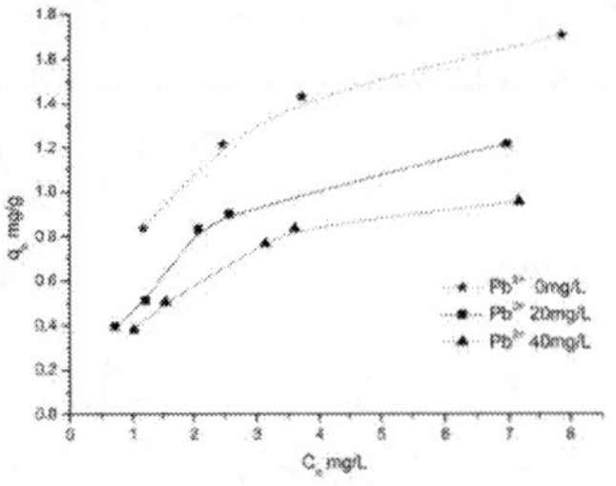

Fig 6 Competition adsorption isotherm of Cd^{2+} with Pb^2

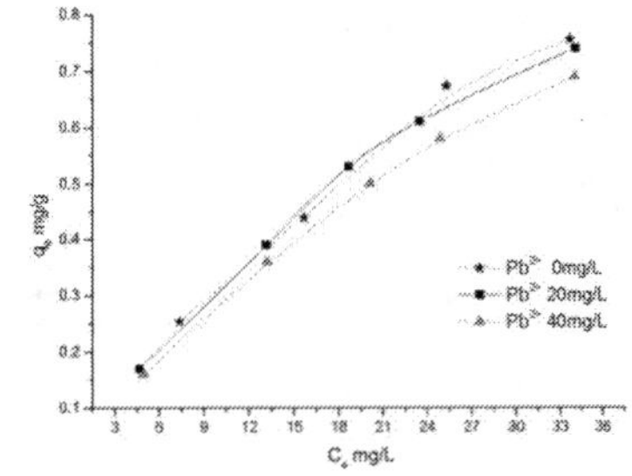

Fig 7 Competition adsorption isotherm of Ca^{2+} with Pb^{2+}

Table 2 The binary Freundlich adsorption isotherm model parameters of Ca^{2+} and K^+

Binary Freundlich model				
Ca^{2+}	b_{11}	a_{12}	b_{12}	R^2
1-5#	0.66855	1897.96	-2.88406	0.99541
6-10#	0.66007	6383.61	-2.31729	0.99758

Table 3 The binary Freundlich adsorption isotherm model parameters of Cd^{2+} and Pb^{2+}

Binary Freundlich model				
Cd^{2+}	b_{11}	a_{12}	b_{12}	R^2
1-5#	0.36405	0.19718	-1.6693	0.96317
6-10#	0.93985	1.87212	1.67962	0.99993

Table 4 The binary Freundlich adsorption isotherm model parameters of Ca^{2+} and Pb^{2+}

Binary Freundlich model

Ca^{2+}	b_{11}	a_{12}	b_{12}	R^2
1-5#	0.69271	477.43535	6.82804	0.98886
6-10#	0.69753	0.10694	-1.46138	0.99869

We draw conclusion that competitive adsorption obviously happened between the same kinds of hard or soft metal ions. To the metal ions which initial concentration was fixed, its adsorption amount in biofilm would be decreased with increasing of concentration of another kind of ions because same kind of ions find same coordination radicle, so the high concentration kinds of ions would gained more active coordination radicle when adsorption happened in biofilm. Competitive adsorption happened not obviously between different kinds of metal ions because different kind of ions needs different coordination radicle and different coordination method.

Soft ions could be better absorbed than hard ions in biofilm because what the hard ions need coordination radicle is always contain oxygen atom such as OH^-, HPO_4^{2-}, $RCOO^-$ etc which generally exist in anaerobic biofilm layer and what the soft ions need such as CN^-, RS^-, $-SH^-$ is always exist in aerobic biofilm layer. When liquid flow slowly (it means shear force is weak) and substrate enough, anaerobic biofilm layer always thicker than aerobic biofilm layer, so the coordination radicle which soft ions needed is more than coordination radicle hard ions needed. Another reason is that soft ions bond its coordination radicle by covalent bond. It is a far more firm bond method than hard ions join with its coordination radicle.

Acknowledgments The research was supported and by Nature Science Fund of Tianjin, China (07JCZDJC01400) and Nature Science Fund of China (50808130)

Reference

[1] I. P. Suhasini, G. Sriram, S. R. Asolekara, et al, Sep. Sci. Technol. 1999, 34, 239-247
[2] C. P. Huang, F. B. Ostovic, J.Environ, Eng. 1978, 104, 863
[3] U. Yetis, A. Dolek, F.B. Dilck, et al, Wat. Res. 2000, 34, 4090-4100
[4] A. G. Sanchez, E. A. Ayuso, J. D. Blas, Clay Min. 1999, 34, 469
[5] A. L. Delgado, F. A. Peterz, F. A. Lopez, Wat. Res. 1998, 34, 4989
[6] M. A. Ferro-Garacia, et al, Carbon, 1988, 26 ,263
[7] Y. S. Al-Degs, M. F. Tuttunju, R. A. Shawabkeh, Sep. Sci. Technol. 2000, 35, 2299
[8] M. Kandah, Chem. Eng. J. 2001, 84, 543
[9] J. L. Gardea-Torresdey, et al, Haz. Mat. 1999, 69, 41
[10] Y. S. Ho, G. McKay, Chem. Eng. Sci. 2000, 55, 817
[11] F.Gritti, G. Guiochon, J Chroma A, 2004, 1028,121
[12] Z. Aksu, Sep & Puri Technol, 2001, 21, 285

Due to unavailable paper there is a gap in pagination.

Pages 435-436

Optimization of the Economic and Environmental Performance of An Ammonia-Water Absorption Cooling Cycle

Berhane H. Gebreslassie, Department of Mechanical Engineering, University Rovira i Virgili, 43007 Tarragona, Spain

Gonzalo Guillén-Gosálbez, Department of Chemical Engineering, University Rovira i Virgili, 43007 Tarragona, Spain

Laureano Jiménez, Department of Chemical Engineering, University Rovira i Virgili, 43007 Tarragona, Spain

Dieter Boer, Department of Mechanical Engineering, University Rovira i Virgili, 43007 Tarragona, Spain

Actually, it seems clear that a drastic change in the energy structure should to be made in the developed countries, and energy efficient technologies and renewable energies have to be more promoted. Cooling and refrigeration represents an important contribution to the total energy consumption. Most of the installed units are based on the mechanical vapour compression systems driven by electrical motors. As a consequence, the electrical demand has been rising, which worsens the trouble for both the generation and distribution networks, along with the associated environmental issues.

A more sustainable concept, known as thermal solar refrigeration, is based on the use of an absorption unit driven by low grade heat from industrial processes as waste heat or cogenerations or solar thermal panels. The idea of producing cold from solar thermal energy is not new at all. However, despite of their benefits, both environmental and economical (for their low operating cost), their deployment in the market is null. The reason for such a low market impact could be attributed to high purchase cost and low integration of the devices involved in such installation, and making a custom design for each application necessary.

Among the absorption systems, mainly two different working fluid mixtures are used: ammonia-water of water-Lithiumbromide (LiBr). The water-LiBr working pair is suitable for cooling applications above 0ºC and shows in general a higher coefficient of performance (COP) than ammonia-water. Ammonia-water can work at subzero temperatures and the heat dissipation temperature is not limited by crystallization. Thus air-cooling can be considered and no wet cooling tower is required.

The objective of the present work is to provide a quantitative decision-support tool for the optimal design of environmentally conscious absorption cycles. The approach presented relies on the development of a multi-objective formulation that simultaneously accounts for the minimization of cost and environmental impact at the design stage. The latter criterion is quantified by the Eco-indicator 99, which follows the principles of life cycle assessment (LCA). The design task is formulated as a bicriteria nonlinear programming (NLP) problem, the solution of which is defined by a set of Pareto points that represent the optimal trade-off between the economic and environmental concerns considered in the analysis. These Pareto solutions can be obtained via standard techniques for multi-objective optimization. The main advantage of this approach is that it offers a set of alternative options for system design rather than a single solution. From these alternatives, the decision-maker can choose the best one according to his/her preferences and the applicable legislation.

The capabilities of the proposed method are illustrated in a case study problem that addresses the design of an ammonia-water absorption cooling system. As driving energy both steam from fossil fuel and solar thermal energy will be considered.

See more of this Session: Novel Membranes and Processes for Water Treatment, Desalination and Purification II

See more of this Group/Topical: Environmental Division

Selective Magnetic Filtration of Noble Metals From Dilute, Acidic Streams Using Carbon-Coated Nanomagnets

Michael Rossier, Department of Chemistry and Applied Biosciences, ETH Zurich, 8093 Zurich, Switzerland

Fabian M. Koehler, Department of Chemistry and Applied Biosciences, ETH Zurich, 8093 Zurich, Switzerland

Evagelos K. Athanassiou, Department of Chemistry and Applied Biosciences, ETH Zurich, 8093 Zurich, Switzerland

Robert N. Grass, Department of Chemistry and Applied Biosciences, ETH Zurich, 8093 Zurich, Switzerland

Detlef Gunther, Department of Chemistry and Applied Biosciences, ETH Zurich, 8093 Zurich, Switzerland

Wendelin J. Stark, Department of Chemistry and Applied Biosciences, ETH Zurich, 8093 Zurich, Switzerland

Magnetic filtration provides an efficient way to control large liquid volumes with a small amount of a specific reagent. At present, the lack of sufficiently stable nanomagnets has limited the magnetic filtration to routine processing in biochemical separations, particularly in diagnostics. Unfortunately these high-price applications are limited to water and neutral pH as the present oxide-based nanomagnets have a poor stability and low binding capacities. Recently we prepared metallic nanoparticles with a graphene-like carbon coating of 1-2 nm by reducing flame spray synthesis [1,2]. This coating protects the metal nanoparticles from oxidation up to temperatures of 190°C and against dissolution in acids. On the other hand it allows the introduction of covalently bound functional groups via diazonium chemistry [3] or physical adsorption, giving access to organic chemistry as a method to design the particle surface. The functionalized magnetic particles represent a promising modular platform for removing a wide range of contaminants. Attaching a thiourea-like chelating agent allows the efficient and selective removal of precious metal ions in highly acidic solutions. The low costs and high particle stability favor this processing method and associated material for large-scale separation application where metal ions are present at ultra low concentrations. This contribution will show how noble metal can be rapidly collected from acidic mining streams down to the microgram per liter level [4]. [1] R. N. Grass, W. J. Stark, J. Mater. Chem. 2006, 16, 1825. [2] R. N. Grass, M. Dietiker, R. Spolenak, W. J. Stark, Nanotechnology 2007, 18, 035703. [3] R. N. Grass, E. K. Athanassiou, W. J. Stark, Angewandte Chemie 2007, 26, 4909 [4] M. Rossier, F. M. Koehler, E. K. Athanassiou, R. N. Grass, B. Aschlimann, D. Guenther, W. J. Stark , manuscript in preparation (2009) Figure 1: Selective gold and platinum extraction in a solution composed of 1g/l Fe, Cu and Zn and 1mg/l Au and Pt.

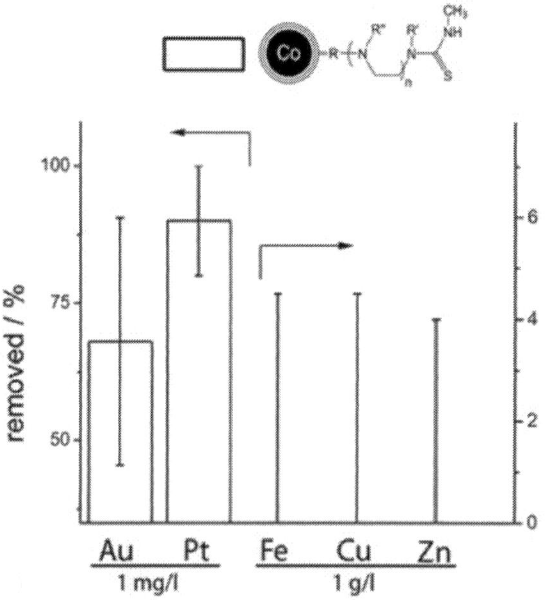

See more of this Session: Novel Membranes and Processes for Water Treatment, Desalination and Purification II
See more of this Group/Topical: Environmental Division

Water Transport Mechanisms in Polylactide by Quartz Crystal Microbalance/Heat Conduction Calorimetry

An Du, Department of Chemical and Biological Engineering, Drexel University, Philadelphia, PA

Donghun Koo, Department of Chemical and Biological Engineering, Drexel University, Philadelphia, PA

Richard Cairncross, Department of Chemical and Biological Engineering, Drexel University, Philadelphia, PA

Amongst all bio-based polymers, polylactide (PLA) is well-known for its renewability and being able to biodegraded. PLA therefore is favored in packaging and medical applications. Water transport in PLA has been studied on several commercial and synthesized films of PLA prepared by spin coating and heat treatment to control crystallinity. Morphology of PLA was investigated by SEM and DSC methods. Water sorption in all PLA samples was studied by the quartz crystal microbalance/heat conduction calorimetry (QCM/HCC) technique. Sorption isotherms were constructed at 35°C for different water activity levels. Despite the difference in morphology, all PLA samples showed a negligible affect of thermal treatment on the amount of water sorption.

Amorphous PLA was used for the evaluation of mathematical models predicting water sorption mechanisms in PLA. Experimental results were fitted to Henry's law, Flory-Huggins theory, dual mode sorption model, and the engaged species induced clustering (ENSIC) model. ENSIC model accounting for the probability of addition of water in water/polymer system provided the best fit amongst the four models. By DSC experiment, loosely bound water was also found by a melting peak at near -0.62°C. Therefore, both DSC and curve fitting results showed a potential of water clustering existence in PLA.

Additional evidence of water clustering in PLA comes from sorption enthalpy which was measured with the QCM/HCC instrument by two different methods. The first method is to apply the Van't Hoff's expression relating sorption enthalpy to the solubility measured at a series of different temperatures. The second method uses the thermal power signals from the HCC. By integrating the area under a sharp peak in thermal power caused by a change in moisture content, the sorption enthalpy can be estimated. Sorption enthalpy of water in commercial PLA was found to be close to -40 kJ/mol, which is very similar to the heat of condensation of water (-44 kJ/mol).

See more of this Session: Novel Membranes and Processes for Water Treatment, Desalination and Purification II
See more of this Group/Topical: Environmental Division

Rapid Reverse Osmosis (RO) Membrane Performance Analysis Using An Autonomous RO Desalination System

Han Gu, Chemical & Biomolecular Engineering, University of California, Los Angeles, Los Angeles, CA

Michal Uchymiak, Chemical and Biomolecular Engineering, University of California, Los Angeles, Los Angeles, CA

Alex Bartman, Chemical and Biomolecular Engineering, University of California Los Angeles, Los Angles, CA

Andi Rahardianto, Chemical and Biomolecular Engineering, University of California, Los Angeles, Los Angeles, CA

Panagiotis Christofides, Department of Chemical and Biomolecular Engineering and Department of Electrical Engineering, University of California, Los Angeles, Los Angeles, CA

William J. Kaiser, Department of Electrical Engineering, University of California, Los Angeles, Los Angeles, CA

Yoram Cohen, Chemical & Biomolecular Engineering, University of California, Los Angeles, Los Angeles, CA

For many communities around the world, seawater and brackish water desalination is increasingly becoming a necessity for augmenting dwindling fresh water supplies. At present, reverse osmosis desalination is the leading technology for brackish and seawater desalination. Widespread implementation of RO technology, however, remains a challenge due to the wide variety of feed water quality, membrane scaling and fouling problems, and suboptimal energy use with the current construction and operational strategy of Ro systems. Given the multitude of challenges of RO desalting, a systematic and robust approach is required for assessing process feasibility, evaluating process strategies (e.g., to minimize fouling/scaling), and optimizing process performance (e.g., with respect to energy use). In the present study, a mini-modular-mobile (M3) RO system capable of autonomous operation and with remote control capability was developed by the UCLA Water Technology Research Center for rapid reverse osmosis membrane performance analysis. The M3 RO system, equipped with readily configurable spiral wound membrane elements, is capable of real time remote monitoring of process stream conditions, automated data acquisition, web-based data storage, and advanced remote control schemes. The system can be used in a once-through or total recycle operational modes . The system enable rapid membrane performance analysis, which in the present work was accomplished in a total recycle mode. Pure water membrane permeability, intrinsic salt rejection, and concentration polarization in the M3 system were characterized over a wide range of operating conditions (i.e., RO water recovery and feed flow rates). Based on the laboratory experiments and data analysis, the range of safe operating conditions for the M3 RO system was characterized for field deployment of desalting brackish water of high mineral scaling propensity. The use of the M3 system was subsequently demonstrated in the field for desalting agricultural drainage source water in California's San Joaquin Valley with high gypsum scaling potential. Rapid membrane performance analysis was demonstrated over a variety of field conditions. The present study introduces a new paradigm RO desalting evaluation using an autonomous system, enabling rapid assessment of RO membrane performance both in controlled laboratory and field settings. -->

See more of this Session: Novel Membranes and Processes for Water Treatment, Desalination and Purification II
See more of this Group/Topical: Environmental Division

Field Demonstration of Brine Demineralization for High Recovery RO Desalination

Brian C. McCool, Chemical Engineering Department, University of California, Los Angeles, Los Angeles, CA

Anditya Rahardianto, Chemical Engineering, University of California, Los Angeles, Los Angeles, CA

Yoram Cohen, Chemical Engineering, University of California, Los Angeles, Los Angeles, CA

Reverse osmosis desalination of brackish water in the San Joaquin Valley may be feasible only at high recoveries (>85%) because of the high cost of brine disposal. However, because brackish waters in this region tend to be at or near saturation with respect to gypsum, high recoveries cannot generally be achieved with a single stage RO process. Recent studies at UCLA have suggested that high recoveries may be achievable with a two-stage RO process with induced interstage precipitation to demineralize the primary RO concentrate, enabling a secondary RO stage to achieve the desired overall recovery level. In the present study a single-stage pilot-plant operating near its gypsum scaling imposed recovery limit (62% recovery) produced a supersaturated concentrate stream that was then demineralized by seeded gypsum precipitation to produce a desupersaturated stream (gypsum saturation index ≈ 1). On-line monitoring of pH and calcium concentrations during the induced gypsum precipitation batch process allowed the degree of gypsum precipitation and the secondary RO recovery limit to be quantified. The ability to desupersaturate the primary RO concentrate operating at 62% recovery was successfully demonstrated and would enable the two-stage RO process to operate at the desired high recoveries (>85%).

See more of this Session: Novel Membranes and Processes for Water Treatment, Desalination and Purification II
See more of this Group/Topical: Environmental Division

ELEMENTAL MERCURY OXIDATION AND ITS RESULTANT OXIDIZED MERCURY ADSORPTION BY CUPRIC CHLORIDE-IMPREGNATED CARBON SORBENT

Xin Li[1], Joo-Youp Lee[1], Sang-Sup Lee[2], and Tim C. Keener[3]

[1] *Department of Chemical and Materials Engineering, University of Cincinnati, Cincinnati, OH*
[2] *Department of Energy Resources Engineering, Stanford University, Stanford, CA*
[3] *Department of Civil and Environmental Engineering, University of Cincinnati, Cincinnati, OH*

Introduction

Many mercury sorbents have been tested at various levels to reduce mercury emissions from coal-fired power plants. Among the sorbents, raw activated carbon has been extensively studied and has shown the capability to adsorb elemental mercury (Hg^0) primarily in the presence of hydrogen chloride (HCl). [1-3] While the performance of chemically treated activated carbons in Hg adsorption has been examined in laboratory and field studies[1-4], very few studies are found to investigate the kinetics and mechanisms for Hg^0 oxidation and adsorption by chemically treated carbon sorbents.

In our previous studies, [9-15] cupric chloride impregnated carbonaceous substrates demonstrated excellent Hg(0) oxidation. The results indicated that Hg(0) vapor reacts with $CuCl_2$ and the desorbed resultant oxidized mercury is re-adsorbed onto substrate surfaces, suggesting different sites for Hg(0) oxidation and re-adsorption of the resultant oxidized mercury. In this study, X-ray absorption fine structure (XAFS) spectroscopy was used to identify the resultant oxidized mercury. Several previous studies[5-7,19] have been conducted on the speciation of mercury captured on the sorbent surfaces and mercury-sorbent surface interactions in order to understand mercury capture mechanisms. Huggins et al.[5] conducted an XAFS study on the sorption of Hg(0) and $HgCl_2$ vapors onto sulfurized, iodated, and raw activated carbons under simulated flue gas conditions. The study results suggested that $HgCl_2$ vapor appeared to be captured by physisorption and Hg(0) vapor appeared to be captured by chemisorption by forming sorption complexes with the promoted chemical elements (S and I) as well as the acidic gases of HCl and SO_2 in the simulated flue gas. Hutson et al.[7] recently studied the speciation of bromine and mercury and their binding mechanisms on the carbon surface using XAFS and XPS with brominated ACs under simulated flue gas conditions. The study results indicated that bromine appears to be primarily covalently bonded to sp^2 and sp^3 carbon atoms with some physisorbed HBr. Hg(0) vapor then reacts with Br sites and the resultant oxidized mercury is most likely to be present as $HgBr_2$.

XAFS is typically divided into two regimes: x-ray absorption near-edge spectroscopy (XANES) and extended x-ray absorption fine-structure spectroscopy (EXAFS). XANES is strongly sensitive to formal oxidation state and coordination chemistry of the absorbing atom, while the EXAFS is used to determine the species of the neighbors of the absorbing atom, distances, and coordination number. In this study, XAFS is used to probe the speciation of mercury from various spent $CuCl_2$-impregnated sorbents, and their results are presented.

Experimental Section

Sorbent Preparation

Three sorbents were prepared to study mercury speciation: (1) Norit's DARCO FGD raw AC

(Raw-AC); (2) $CuCl_2$-impregnated DARCO FGD AC ($CuCl_2$-AC); and (3) HCl-impregnated DARCO FGD AC (Cl-AC). The detailed synthesis techniques used for the second sorbents are described in our previous studies.[12,15] The HCl-impregnated sorbent followed the same procedures with a 0.3N HCl solution.

Fixed-Bed Reactor System for Mercury Loading onto Sorbents

Elemental mercury vapor was supplied by an Hg(0) permeation tube (VICI Metronics, Inc., Poulsbo, WA) in the VICI Metronics Dynacalibrator permeation oven held at 90 °C in either a nitrogen or an oxygen carrier gas into a fixed-bed reactor system. A high inlet Hg(0) concentration of ~3μg/L was introduced to the system in order to saturate the sorbent in a short period of time. The total gas flow rate was 500 mL/min. Approximately 100 mg of a sorbent was exposed to the Hg(0)-laden gas stream until the sorbent reached an equilibrium capacity. The fixed-bed reactor was placed in an oven maintained at 140 °C, which is a typical sorbent injection temperature between an air preheater and a particulate control device. An online Hg analyzer (UV-1201S with mercury analysis kit, Shimadzu Corp., Columbia, MD) was used to observe breakthrough curves and determine an equilibrium adsorption point. The XAFS spectroscopy used in this study is known to have a detection limit of ~300 ppmw, and is expected to detect mercury captured in most of the sorbents used in this study. Four different carrier gases were used in this study: (1) N_2 (Nitrogen ≥99.999% UHP, Wright Brothers, Inc., Cincinnati, OH); (2) O_2 (Oxygen ≥99.6%, Wright Brothers, Inc., Cincinnati, OH); (3) 10 ppmv HCl in N_2; and (4) 10 ppmv HCl in O_2. The gases were selected in order to investigate the capability of HCl and O_2 in Hg(0) oxidation.

XAFS Experiments and Analysis

Hg L_{III}(12 284 eV) edge X-ray absorption experiments were performed using the 20-BM-B (Sector 20 ID) beamline at the Advanced Photon Source (APS) at Argonne National Laboratory (ANL, Argonne, IL). A Si (111) monochromator was used, and energy calibration was performed using Au metal foil. XAS data were collected in both transmission and fluorescence mode, and the latter used a 13-element Ge detector. The fluorescence data had better signal-to-noise ratio, and was thus applied to data processing. At least five scans were used to improve the signal-to-noise ratio. The powdered samples were loaded into thin Teflon sample holders between layers of Kapton tape. XAFS data was processed using Athena and Artemis interfaces[17]. The XAFS data analysis consisted of converting $\mu(E)$ to $\chi(k)$, Fourier transform $\chi(k)$ to $\chi(R)$, and shell fitting calculation using FEFF and IFEFFIT software.

Results and Discussion

XAFS Spectra Results

The major features of oxidized mercury that occur on the Hg L_{III}-edge XANES spectra are two inflection points, which give rise to two distinct peaks in the first derivative spectra whereas elemental mercury exhibits a single peak. The Inflection Points Difference (IPD) and relative intensities of these two peaks reflect the chemical bonding of the nearest neighbor coordination sphere around the mercury ion.

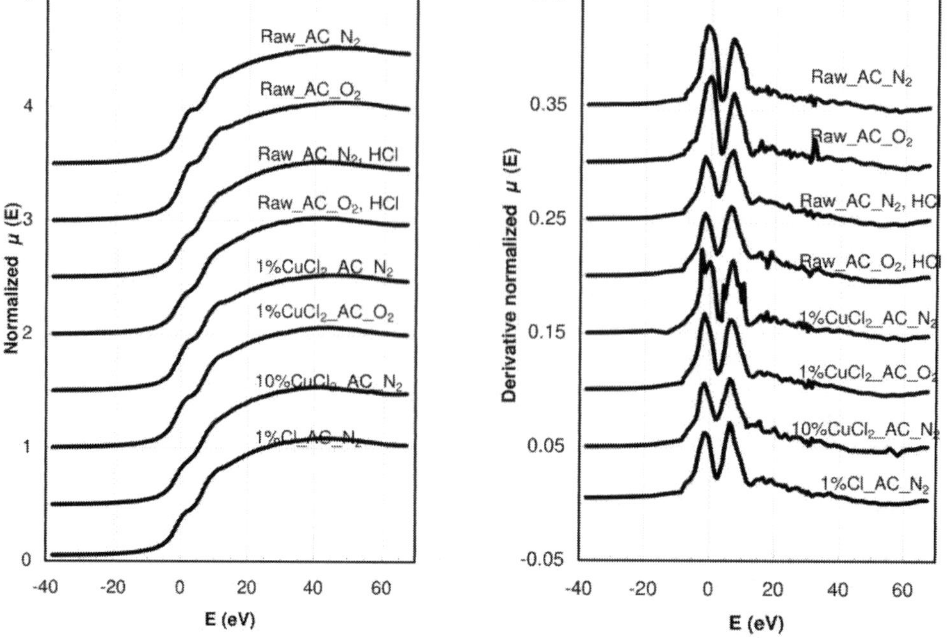

Figure 1. The XANES portion μ(E) of Hg L_{III}-edge XAFS spectra and first derivative of the spectra

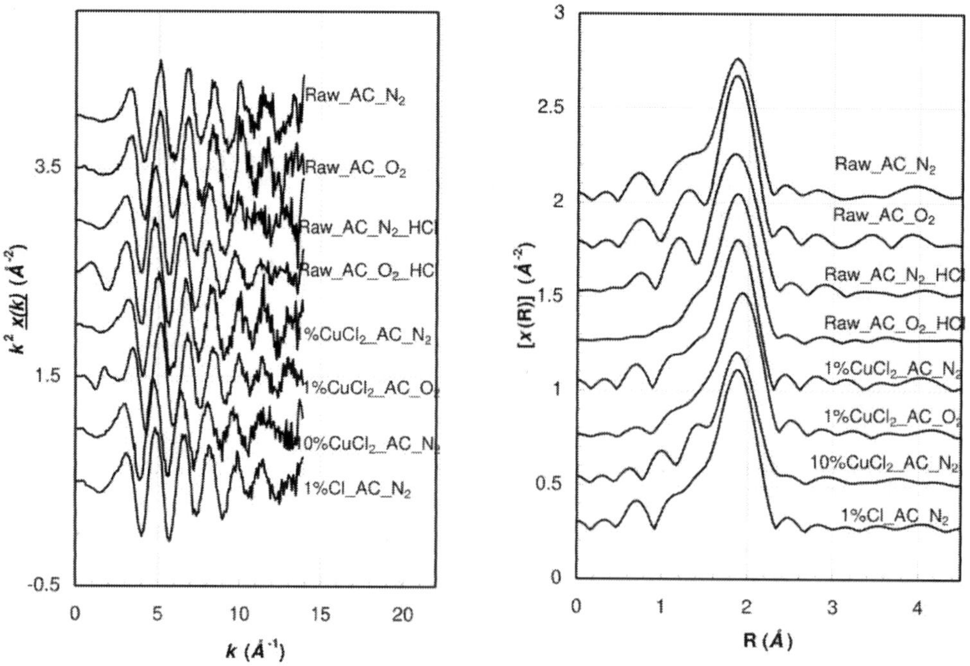

Figure 2. XAFS spectra χ(k) and Fourier transformed EXAFS spectra χ(R) (shown in magnitude |χ(R)|)

The Hg L_{III}-edge XANES spectra and the first derivative of the spectra are shown in Figure 1. The EXAFS spectra and its Fourier transformed spectra are shown in Figure 2. The IPD, peak height ratio, and radial structure function (RSF) values based on the XAFS spectra are summarized in Table 1.

Table 1. XAFS features for Hg reference compounds and samples

		First Derivative XANES Separation (eV) (IPD)	Peak Height Ratio	RSF Peak Position(Å)
Reference compounds	HgO (red)	13.2[a](12.8[b])	1.75 [b]	1.69 [a] (1.68 [b])
	$HgSO_4$	11.4[a](9.6[b])	1.10 [b]	1.75 [a]
	$HgNO_3$	10.7 [a]		1.75 [a]
	Hg_2Cl_2	8.1 [a] (8.5 [b])	0.98 [b]	2.00 [a]
	$HgCl_2$	8.0 [a] (8.4 [b])	1.10 [b]	1.85 [a] (1.92 [b])
	HgS (red)	7.4 [a] (7.6 [b])	1.40 [b]	1.97 [a] (2.03 [b])
	$HgBr_2$	7.7 [a]		2.11 [a]
	HgI_2	6.5 [b]	0.90 [b]	2.61 [b]
Spent sorbent samples[c]	Raw-AC-N_2	7.8	1.22	1.86
	Raw-AC-O_2	7.4	1.32	1.86
	Raw-AC-N_2, HCl	7.9	0.97	1.84
	Raw-AC-O_2, HCl	7.8	1.03	1.87
	1%(wt) $CuCl_2$-AC -N_2	7.6	1.00	1.87
	1%(wt) $CuCl_2$-AC -O_2	7.6	1.09	1.94
	10%(wt) $CuCl_2$-AC -N_2	8.0	0.99	1.87
	1%(wt) Cl-AC -N_2	8.0	0.98	1.87

(Notes): a: these values are from reference 7;
b: these values are from reference 6;
c: spent sorbent is labeled as 'sorbent-carrier gas', for instance, 'Raw-AC-N_2, HCl' means sorbent Raw-AC that has been tested under O_2 with 10 ppm HCl.

As shown in Table 1, oxygen-containing mercury compounds (e.g., HgO, $HgNO_3$, $HgSO_4$, etc.) show the largest IPD values, and mercury halides fall into the next smaller IPD group. It is interesting to note that all the sorbent samples showed similar shapes and values of the XANES and XAFS spectra although these samples were prepared with different sorbent and tested under different carrier gases (i.e., N_2 and O_2). When only IPD values are compared, the samples could be HgS, $HgCl_2$, and/or Hg_2Cl_2. However, when peak height ratio and RSF values are also compared, sorbents tested in a HCl-laden gas and $CuCl_2$-impregnated sorbents are most likely to be $HgCl_2$. It is interesting to note that HCl- and $CuCl_2$-AC samples seem to form $HgCl_2$ regardless of a carrier gas (i.e., N_2 or O_2).

In a previous study, DARCO FGD activated carbon was analyzed using XPS survey scan and found the activated carbon contains 0.4% sulfur, but no chlorine was detected. [8] However, the values of peak height ratio and RSF peak position for HgS reported in the literature are 1.40 and 1.97 (or 2.03), and are not close enough to the values obtained for Raw-AC-N_2 and Raw-AC-O_2. Thus HgS is not likely to be formed.

EXAFS Fitting Results

EXAFS fitting has been used to obtain more detailed quantitative information on mercury bonding.

EXAFS data include oscillation information due to photoelectrons bouncing off nearest neighbor atoms, which can be modeled and interpreted using EXAFS equation:

$$\chi(k) = \sum_{Shells} \frac{NS_0^2 e^{-2k^2\sigma^2} e^{-2R/\lambda(k)} f(k)}{kR^2} \sin[2kR + \delta(k)]$$

By fitting the experimental data $\chi(k)$ with the equation, the species of the neighbors of the absorbing atom, distances(R), and coordination number(N) can be determined. The process is also called fitting. Figure 3 shows Hg EXAFS spectra $|\chi(R)|$ and their theoretical fits with the Raw-AC-O_2 sample (others give similar spectra). The curve fittings with S or Cl atom match well with the experimental profile (shown as Raw_AC-O2), whereas the curve fitting with O atom does not match with the experimental profile. It suggests that the resultant oxidized mercury could have a bonding with S or Cl atom, not with O atom.

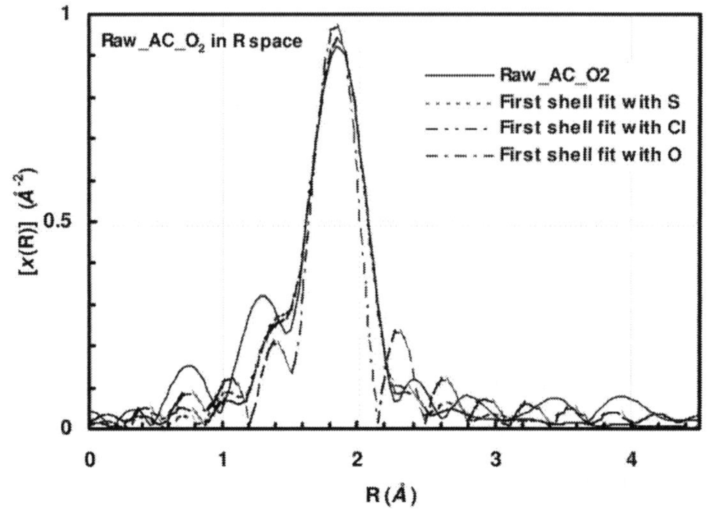

Figure 3. Experimental Hg L_{III}-edge EXAFS spectrum and theoretical fits

Potential Hg(0) oxidation and adsorption mechanisms on $CuCl_2$ impregnated activated carbon

It is currently hypothesized that the reaction of Hg(0) with $CuCl_2$ is a semi-catalytic reaction such that chloride is consumed to form oxidized mercury (e.g., $HgCl_2$) and copper lowers an activation energy barrier for the oxidation reaction. Then, the resultant oxidized mercury is desorbed from reaction sites and is re-adsorbed onto nearby $CuCl_2$-free carbon sites inside the pores if available as shown in Figure 4. It is also further hypothesized that mercury binding on the surface of activated carbon can be described by two generalized types of sites: catalytic sites with a low binding energy (for catalytic reaction and desorption) and stable sites with a high binding energy (for re-adsorption).

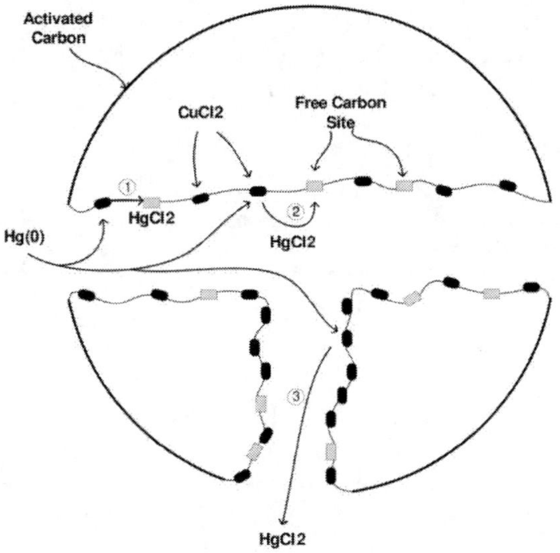

Figure 4. Suggested Hg(0) oxidation and adsorption mechanism on the $CuCl_2$ impregnated activated carbon

Conclusions

Mercury XAFS of spent sorbents provided the valuable information on the oxidation state of mercury and the nearest coordination shell around mercury atom. This XANES study confirmed that $HgCl_2$ is very likely to be a major oxidation reaction product when $CuCl_2$ and HCl are impregnated onto raw activated carbon regardless of a different carrier gas (i.e., N_2 or O_2). Additional EXAFS fitting calculations supported that the resultant oxidized mercury could be either HgS or $HgCl_2$, and ruled out a possible formation of HgO. Based on these results, Hg(0) vapor is very likely to form $HgCl_2$ as a result of the reaction with $CuCl_2$. The oxidized mercury was highly dispersed and disordered, and a crystalline structure was not observed.

References

1. Pavlish, J.; Sondreal, E.; Mann, M.; Olson, E.; Galbreath, K.; Laudal, D.; Benson, S. Status review of mercury control options for coal-fired power plants. *Fuel Process. Technol.* 2003, 82, 89.
2. Srivastava, R.; Hutson, N.; Martin, B.; Princiotta, F.; Staudt, J. Control of Mercury Emissions from Cola-Fired Electric Utility Boilers. *Environ. Sci. Technol.* 2006, 40, 1385.
3. Jones, A. P.; Hoffmann, J. W.; Smith, D. N.; Feeley, T. J.; Murphy, J. T., DOE/NETL's Phase II Mercury Control Technology Field Testing Program: Preliminary Economic Analysis of Activated Carbon Injection. *Environ. Sci. Technol.* 2007, 41 (4), 1365-1371.
4. Granite, E. J. Halogenated Sorbents for Mercury Control. Presented at the DOE/NETL Mercury Control Technology R&D Program Review, Pittsburgh, PA. July 12-14, 2005, http://www.netl.doe.gov/publications/proceedings/05/Mercury/panel/Granite_Panel-071205-pm .pdf.

5. Huggins, F. E.; Huffman, G. P.; Dunham, G. E.; Senior, C. L. XAFS Examination of Mercury Sorption on Three Activated Carbons. *Energ. Fuel.* 1999, 13, 114.
6. Huggins, F. E.; Yap, N.; Huffman, G. P.; Senior, C. L. XAFS characterization of mercury captured from combustion gases on sorbents at low temperatures. *Fuel Process. Technol.* 2003, 82, 167-196.
7. Hutson, N. D.; Attwood, B. C.; Scheckel, K. G. XAS and XPS Characterization of Mercury Binding on Brominated Activated Carbon. Environ. *Sci. Technol.* 2007, 41, 1747.
8. Laumb, J. D.; Benson, S. A.; Olson, E. A., X-ray photoelectron spectroscopy analysis of mercury sorbent surface chemistry. *Fuel Process. Technol.* 2004, 85 (6-7), 577-585.
9. Lee, J.-Y.; Ju, Y.; Lee, S.-S.; Keener, T. C.; Varma, R. S. Novel Mercury Oxidant and Sorbent for Mercury Emissions Control from Coal-Fired Power Plants. *Water Air Soil Poll: Focus.* 2008, 8, 333.
10. Lee, S.-S.; Lee, J.-Y.; Keener, T. C. Elemental Mercury Control by Novel Oxidant and Sorbent in an Entrained-Flow System. *Water Air Soil Poll: Focus.* 2008, 8, 343.
11. Lee, S.-S.; Lee, J.-Y.; Keener, T. C. Novel Sorbent for Mercury Emissions Control from Coal-Fired Power Plants. J. *Chin. Inst. Chem. Eng.* 2008, 39, 137.
12. Lee, S.-S.; Lee, J.-Y.; Keener, T. C. Performance of Copper Chloride-Impregnated Sorbents on Mercury Vapor Control in an Entrained-Flow Reactor System. *J. Air Waste Manage.* 2008, 58, 1458.
13. Lee, S.-S.; Lee, J.-Y.; Keener, T. C. Mercury Oxidation and Adsorption Characteristics of Chemically Promoted Activated Carbon Sorbents. *Fuel Process Technol.* 2009, doi:10.1016/j.fuproc.2009.06.020.
14. Lee, S.-S.; Lee, J.-Y.; Keener, T. C. The Effect of Methods of Preparation on the Performance of Cupric Chloride-Impregnated Sorbents for the Removal of Mercury from Flue Gases. *Fuel,* 2009, 88, 2053.
15. Lee, S.-S.; Lee, J.-Y.; Keener, T. C. In-Duct Mercury Capture Using Cupric Chloride-Impregnated Carbons in a Pilot-Scale Entrained-Flow System. *Environ. Sci. Technol.* 2009, 43, 2957.
16. Newwille, M. Foundamentals of XAFS, Revision 1.7 July 23, 2004. Retrieved from http://www.xafs.org/Tutorials?action=AttachFile&do=get&target=Newville_xas_fundamentals.pdf
17. Ravel, B. ATHENA User's Guide.Version 1.5 for Athena version 0.8.56, 2009 Retrieved from http://cars9.uchicago.edu/~ravel/software/doc/Athena/html/athena.pdf
18. Lennie, Alistair R.; Charnocka, John M.; Pattrick, Richard A. D. Structure of mercury(II)–sulfur complexes by EXAFS spectroscopic measurements *Chemical Geology*. 2003, 199, 199.
19. Wells, A.F. *Structural Inorganic Chemistry*, Clarendon Press, Oxford, 1984

Activated Carbon and Mercury Interactions

Erdem Sasmaz, Department of Energy Resources Engineering, Stanford University, Stanford, CA

Keith Mosher, Deparment of Energy Resources Engineering, Stanford University, Stanford, CA

Shela J. Aboud, Department of Energy Resources Engineering, Stanford University, Stanford, CA

Jennifer L. Wilcox, Department of Energy Resources Engineering, Stanford University, Stanford, CA

Mercury is reported as a hazardous air pollutant (HAP) by The Clean Air Act (CAA) of 1990. Currently within the United States there are over five hundred 500-megawatt coal-fired power plants and 50 tons mercury/year are emitted. To predict the level of mercury emissions from coal-fired power plants it is important to understand its heterogeneous pathway. Many studies indicate that the mechanism of mercury adsorption on activated carbon surfaces is complicated and depends on mercury speciation, flue-gas temperature, active sites on the sorbent surface and effect of other flue gases. Density functional theory calculations are carried out to investigate the binding of mercury on simulated activated carbon surfaces. Activated carbon is modeled as graphene repeated in the x-y planes. The active sites are formed either by removing the carbon atoms on the graphene edges or creating unsaturated carbon atoms at the edge sites. The edge sites are functionalized with halogens such as chlorine or bromine. The mechanism of mercury binding in the presence of active sites is investigated by considering both heterogeneous oxidation of mercury on the surface and the possible oxidized forms of mercury. Density of states (DOS) analysis is carried out to interpret the binding energies calculated on active sites of activated carbon to explain the mechanism of mercury binding. Experimental studies are also conducted using a bench-scale packed-bed system to further understand the interaction of mercury and activated carbon-based sorbents. Flue gas is simulated by burning methane in a laminar flow tubular burner. An electron ionization quadrupole mass spectrometer is used to analyze the outlet gas stream coming from the reactor and directly speciate between the oxidized forms of mercury in the gas phase.

See more of this Session: Environmental Applications of Adsorption - II
See more of this Group/Topical: Environmental Division

Analysis of the Adsorption Mechanism of Hg in SCR Catalyst: V2O5

Ana Suarez Negreira, Department of Chemical Engineering, Stanford University, Stanford, CA

Shela Aboud, Department Energy Resources Engineering, Stanford University, Stanford, CA

Jennifer Wilcox, Department Energy Resources Engineering, Stanford University, Stanford, CA

Over 32% of anthropogenic mercury emissions in the United States are from the flue gas emitted by coal plants. In 2004, 44.2 tons of mercury was emitted to the atmosphere just from U.S. coal-fired power plants. The environment and health effects associated to the emission of Hg have pushed the power plants to use control technologies for removing mercury from the flue gas. It is possible to rely in part on existing control devices in place for NOx, particulate matter, and SOx to assist in the oxidization and subsequent capture of mercury throughout the quench zone of combustion flue gases. One such co-benefit is the vanadium-based selective catalyst reduction (SCR) unit, which is used primarily for the reduction of NO and NO2 to N2 and water vapor by passing ammonia across the catalyst. This catalyst may also act as an oxidizing catalyst for Hg0 to Hg2+. In addition, it has been suggested that the catalyst is also responsible for oxidizing SO2 to SO3.

The primary focus of this work is on the potential oxidation of mercury over the SCR unit. The mechanism for mercury oxidation across the vanadium oxide catalyst is currently unclear with multiple reaction pathways being possible. Density functional theory calculations have been carried out to model the kinetic and thermodynamic behavior of the Hg on the V2O5(001)

Thermodynamic stability characterization of the different V2O5 surfaces is carried out to determine which surface is more abundant at the conditions of the SCR, which further provides indication to which adsorption sites are more active to initiate the mercury oxidation process. The different oxygen atoms on the V2O5 (001) surface are distinguished by the coordination number (number of V atoms that coordinated with oxygen). It has been hypothesized that the most active sites for mercury oxidation are the one-coordinated oxygen atoms in the V2O5 (001) surface. However, the binding energies and density of states (DOS) calculations of Hg on the other surface oxygen atoms, two- and three-coordinated sites, are also carried out to fully understand the mechanism of mercury binding. Additionally, water vapor is present in the oxidizing environment of combustion flue gases so that hydrated surfaces are also investigated for mercury reactivity.

Finally, complementing the theoretical analysis of Hg adsorption and oxidation on V2O5, samples of crystalline vanadium oxide is characterized and tested for mercury reactivity in a packed-bed reactor within a simulated flue gas. An electron ionization quadrupole mass spectrometer is used to directly measure the extent of oxidized mercury in the gas stream after the packed-bed. The catalyst was characterized using XPS before and after exposure to flue gas to determine the nature of the surface.

See more of this Session: Environmental Applications of Adsorption - II
See more of this Group/Topical: Environmental Division

ON-Site PRODUCTION of Activated CARBON for Removal of Mercury FROM Coal Flue Gas

Massoud Rostam-Abadi, Institute of Natural Resource Sustainability, University of Illinois at Urbana-Champaign, Champaign, IL

Yongqi Lu, Institute of Natural Resource Sustainability, University of Illinois at Urbana-Champaign, Champaign, IL

Ramsay Chang, EPRI, Palo Alto, CA

Cassie Shaban, EPRI, Palo Alto, CA

Many coal-fired power plants are considering activated carbon injection (ACI) as an option for controlling mercury emissions. A major operating cost of the ACI is the cost of the activated carbon (AC), estimated to be in the range of \$0.75 to \$1.5/lb. For a 500 MW base-loaded plant with a 90% capacity factor, this would result in an annual cost in the range of \$0.5 to 5 million/year. If AC could be produced on site for less than half the cost of the purchased commercially available ACs, a 500 MW plant could save \$0.25 to 2.5 million/year, making ACI for mercury compliance a more cost-effective technology.

Electric Power Research Institute (EPRI) and the University of Illinois at Urbana-Champaign (UIUC) have developed a patented technology (U.S. Patents 6,451,094 and 6,558,454 B1 issued in 2002 and 2003, respectively) for on-site production of AC to help reduce the operating cost of AC injection processes for mercury control. This technology, Sorbent Activation Process (SAP), will help reduce the operating cost of activated carbon injection processes for removal of mercury emissions from coal combustion flue gas. SAP involves processing coal to AC in an entrained-bed reactor (Figure 1). The feedstock coal is the same coal as that burned by the utility. SAP can be used to prepare ACs with various surface areas, pore structures, and surface chemistry (halogenated AC) from bituminous, subbitumious, and lignite coals. The product gas from SAP unit containing the AC is then directly injected into the flue gas duct upstream of a particulate control device, such as an electrostatic precipitator (ESP), or downstream of an ESP and collected in a baghouse (not shown) to avoid affecting fly ash properties. Alternatively, although not shown in Figure 1, a cyclone could be use to separate the AC from activation product gas before injection to the flue gas duct. The product gas from the activation unit is then recycled to the boiler.

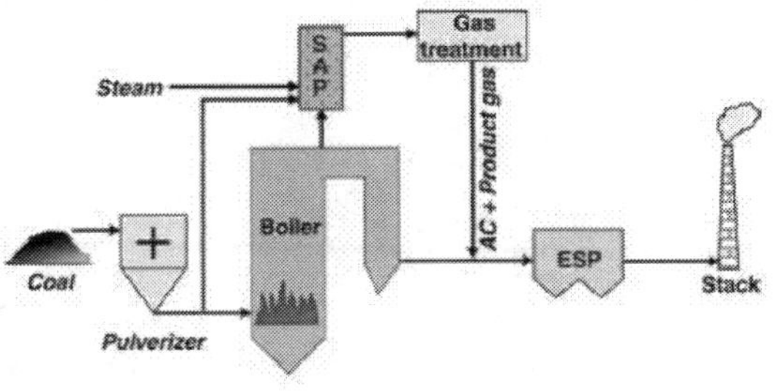

Figure 1. On-site activated carbon (AC) production process.

A study was initiated to evaluate the proof-of-concept of the SAP using both a bench-scale and a pilot-scale unit. The bench SAP, Figure 2, was used to produce 1 to 5 pounds AC per hour from different types of coals. More than 50 tests were performed with the bench SAP using an Illinois coal, a Powder River Basin (PRB) coal, and a lignite coal. The purpose of these tests was to compare quality of the AC products, as measured by their surface area and mercury adsorption capacity, and to obtain process scale up data. Equilibrium mercury capacities of the ACs were measured at URS Corporation (Austin, TX) using a fixed-bed reactor. The relationship between the mercury capacity and BET surface area of bench SAP ACs, with both values normalized with the carbon content in the AC (Hg/C and BET/C), revealed that: 1) PRB-derived ACs had higher surface areas and higher mercury adsorption capacities than their Illinois coal- and lignite-derived ACs, 2) mercury adsorption capacity increases linearly with carbon surface area in the range of 100 to 600 m^2/g-C, 3) data for Illinois coal-, PRB-, and lignite-derived carbons follow a linear plot indicating that mercury adsorption performances of activated carbons derived from the three coals are comparable at a given value of surface area, and 4) capacities of SAP ACs were >80% of the capacity of Darco Hg, one of the most commonly used commercial AC for mercury removal.

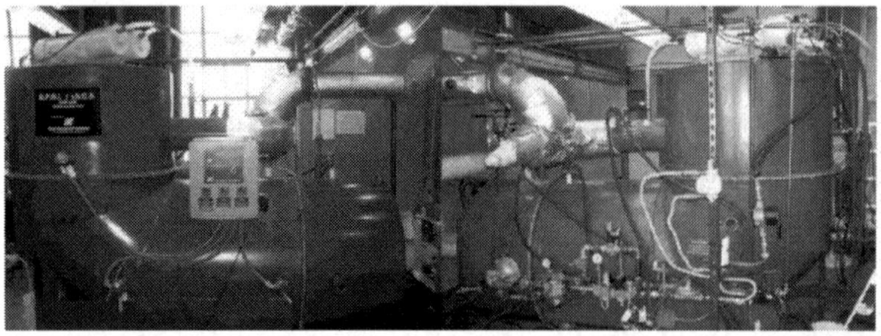

Figure 2. **The bench scale SAP reactor (front view and back view).**

Carbon monoxide (CO) gas is generated during pyrolysis and oxidation of coal as well as gasification of coal char with steam under SAP process conditions. A CO concentration >15 vol% was observed in SAP flue gas in some of the bench tests. It is very likely that the concentration of CO in the SAP flue gas must be reduced in order to meet the required power plant's air permit limits on CO. It should be noted that the volume of the SAP flue gas generated for an injection rate of 5 lb AC/Macf is about 500 times lower than the volume of flue gas of a 500 MW boiler. Therefore, a 500/1 volume dilution will be achieved when SAP flue gas containing AC particles is injected into the boiler's flue gas.

A theoretical kinetic study was initiated to explore the conditions under which CO in the SAP flue gas could be oxidized without burning the AC particles. The results from this study revealed that at a certain temperature range, CO in SAP flue gas can be oxidized to CO2 with a very small amount of carbon being oxidized. Bench-scale tests were performed to investigate the competitive oxidation of CO and coal char when carbon particles existed in the CO containing flue gas, Figure 3. In the presence of the carbon particles, an immediate increase in CO concentration (>4 vol%) in the flue gas was observed, indicating that the CO was produced by either the thermal release of oxygen functional groups from the carbon surface and/or partial oxidation of the carbon particles. Air injection at a location downstream of the coal injection port reduced the CO concentration in the flue gas from about 3.8% to below 1%. The CO concentration decreased with increasing the air injection rate. The results from these tests show the effectiveness of air addition as an option to reduce CO concentration in SAP flue gas.

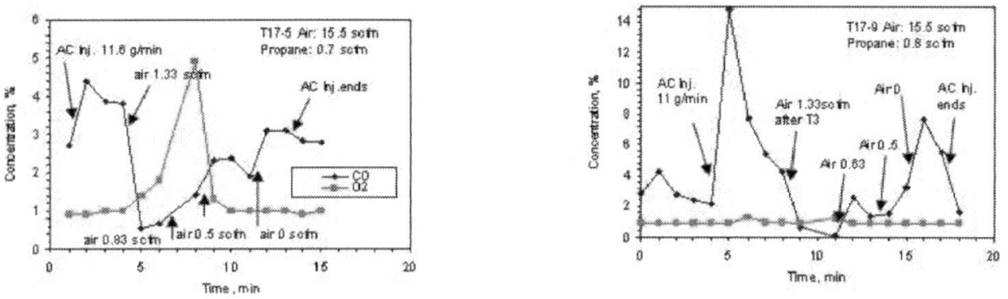

Figure 3. CO concentration profiles with air and carbon/coal injection (O_2 measurement was drifted to ~0.9% during these tests).

The process engineering data collected during the bench SAP tests were used to design and fabricate a pilot SAP unit capable of producing 50 pounds AC per hour. Pilot SAP was integrated with a coal-fired boiler rated at approximately 325,000 acfm (half of 220 MW) at a power plant. Fourteen days of start up and shake down tests have been performed to date to evaluate the performance of the pilot SAP, which is the first-of-its-kind in the world, and its deviation from the design specifications.

A cost analysis study was performed for a 600 lb/hr AC production scale. This amount of AC is equivalent to injecting 5 lb/Macf into the flue gas generated from a 500 MW (net) boiler burning an Illinois coal. The results from the cost study revealed that, for an Illinois coal-to-AC SAP plant, the total capital requirement is estimated $1.2M, and the annual O&M cost amounts to about $1.0M. The cost of AC production and injection using the SAP unit at the 600 lb/hr scale is estimated at $0.28/lb carbon produced.

See more of this Session: Environmental Applications of Adsorption - II
See more of this Group/Topical: Environmental Division

MODELING OF ADSORPTION DATA ON THE REMOVAL OF HEAVY METALS

1. Mrs.G.Anusha,Lecturer,Departmnet of Civil Engineering,
***Bannari Amman Institute of Technology,Sathyamangalam**
2.Dr.Suneeth Kumar ,Professor &Head, Departmnet of Civil Engineering,
***Bannari Amman Institute of Technology,Sathyamangalam**
3. Dr.J.Jeyanthi, Assistant Professor, Departmnet of Civil Engineering ,
***Government College of Technology,Coimbatore**
***Autonomous Institution affiliated to Anna University**

1. INTRODUCTION

1.1 GENERAL

Heavy metals are natural constituents of the Earth's crust and are present in varying concentrations in all ecosystems. Heavy metals are stable and persistent environmental contaminants since they cannot be degraded or destroyed.. The main anthropogenic sources of heavy metals are various industrial sources, including present and former mining activities, foundries and smelters, and diffused sources such as piping, constituents of products, combustion by - products, traffic, etc. Heavy metals are also known to produce adverse effects on the environment and human health.

1.2 POLLUTION DUE TO HEAVY METALS

Heavy metals include all metals with atomic number greater than 23 and specific gravity more than 5. Heavy metals and their compounds, unlike most of the other pollutants occur naturally in the environment also.

Heavy metal pollution in the environment can originate from natural as well as from man made sources. The natural sources of heavy metal pollution are geological weathering and volcanic activities. Electroplating industries, battery manufacturing units are some of the man made sources of heavy metal pollution. Heavy metals are toxic even at low concentration to humans and other living beings.

2. LITERATURE REVIEW

2.1 ADSORPTION WITH ACTIVATED CARBON

In organic compounds in wastewater are resistant to biological degradation and many are toxic or nuisances (odour, taste, colour forming) even at low concentration. Low concentration metals are not readily removed by conventional treatment methods. Activated carbon has an affinity for inorganics and its use for inorganic contaminant removal from wastewater is wide spread.

The larger surface area, a critical factor in the adsorption process, enhances the effectiveness of the activated carbon for the removal of inorganic compounds from wastewater by adsorption. The surface area of activated carbon typically can range **from 500 - 1400 m^2**/g.

These are some important literatures based on the studies related to activated carbon.

- **Orhan Altin et.al,** have studied the effect of pH, flow rate and concentration on the sorption of Pb and Cd on montmorillonite. Continuous column adsorption of lead (Pb) and cadmium (Cd) was studied using pH adjustment and calcium-saturated montmorillonite in a short stainless steel column. At intermediate pH (4-6), ionic size played the major role in adsorption and ion exchange. At low flow rates, sorption of both lead and cadmium increased due to the long retention time in the column. It was found that, when both lead and cadmium ions were present in the feed, adsorption remained the same while that of cadmium decreased compared with single ion experiments.
- **Karuppanna Periasamy and Chinnaiya Namasivayam** studied the removal and Recovery of Cadmium from Wastewater by a Low-Cost Adsorbent. They have reported that carbon derived from peanut hull, a waste agricultural by-product, removed Cr (V1)) and Hg (I1) efficiently from aqueous solutions. The investigation reported, deals with a comparative study of peanut hull carbon and commercially activated carbon for the removal of Cd (I1) from aqueous solutions.
- **S. Jeyaseelan and x.g. Chen** made their investigation on the study of leaching and adsorption of heavy metals by the activated carbon derived from sewage sludge. Six types of activated carbons were produced from digested sewage sludge by different manufacture routes and characterized by the surface area and porosity analysis. The results showed that all these activated carbons had a notable surface area and porosity

• **R. Donat et.al,** made their investigation on the removal of heavy metal cations by natural zeolites In this study, the adsorption pattern of natural zeolites with respect to Co^{2+}, Cu^{2+}, Zn^{2+}, and Mn^{2+} has been studied. It was found that the adsorption phenomena depend on charge density and hydrated ion diameter. According to the equilibrium studies, the selectivity sequence was given as $Co^{2+} > Cu^{2+} > Zn^{2+} > Mn^{2+}$. These results showed that natural zeolites hold great potential to remove cationic heavy metal species from industrial wastewater.

3. MATERIALS AND METHODS

3.1.EXPERIMENTAL PROCEDURE

3.1.1 PREPARATION OF SUGAR MILL SLUDGE BASED ACTIVATED CARBON

Sugar mill sludge was collected and dried in an oven at a temperature of 170°C for 24 hrs. It was then packed in an air tight cylindrical iron container with a completely sealed lid to prevent the entry of air during the process of charring. The sealed iron container was heated in a muffle furnace by slowly raising the temperature upto 600°C and maintained the same for 1 hour. During this process the sugar mill sludge was converted into charred coal. The activated carbon thus prepared was subsequently washed with distilled water, oven dried and packed in a polythene cover and kept in air tight container for further experimentation.

3.1.2 ANALYSIS OF NICKEL USING ATOMIC ABSORPTION SPECTROPHOTOMETER

Atomic absorption spectrophotometer was used to analyse the initial and residual metal concentration after each and every experiment. This instrument is more accurate in analysing the heavy metals (Model GBC 902). Flame method was used for the analysis of metal and the flame used was air acetylene flame. To analyse, first the instrument was set at a wave length of 351.5nm, slit width at 0.5 and lamp current at 4 mA. It was standardized with 20 ppm, 40 ppm, 60 ppm , 80 ppm concentration of metal solution. After calibration, the instrument was put in concentration mode. Then by aspirating the sample, the metal concentration will directly be evaluated by the instrument and was shown as digital display and the value could be printed by the printer.

3.2. ADSORPTION STUDIES

The evaluation of sugar mill sludge based activated carbon chosen for Nickel removal in this study has been done in batch reactor to assess the parameters influencing the adsorption process such as contact time, optimum dosage, adsorption isotherm, pH , initial concentration and to predict the approximate residual Nickel attainable.

3.2.1 BATCH ADSORPTION EXPERIMENTS

Batch reactors have been employed in these experiments where in predetermined quantities (150 ml) of desired concentration were agitated continuously with known weight of the Sugar Mill Sludge based activated carbon.The mixing arrangement was facilitated by wrist action shaker containing eight holders (reactors). After the contact time (5 min) , 2 ml of sample was drawn from each of the eight reactors, filtered using a whatman filter paper and was analysed for its residual Nickel concentration. The same procedure was adopted for all batch studies to determine of optimum pH, optimum time adsorption, optimum sorbent dosage, optimum concentration and isotherm studies.

3.2.2. COLUMN STUDIES

3.2.2.1 Fixed bed Column adsorption system

This type of system is most often used to treat large quantities of wastewater in fixed bed

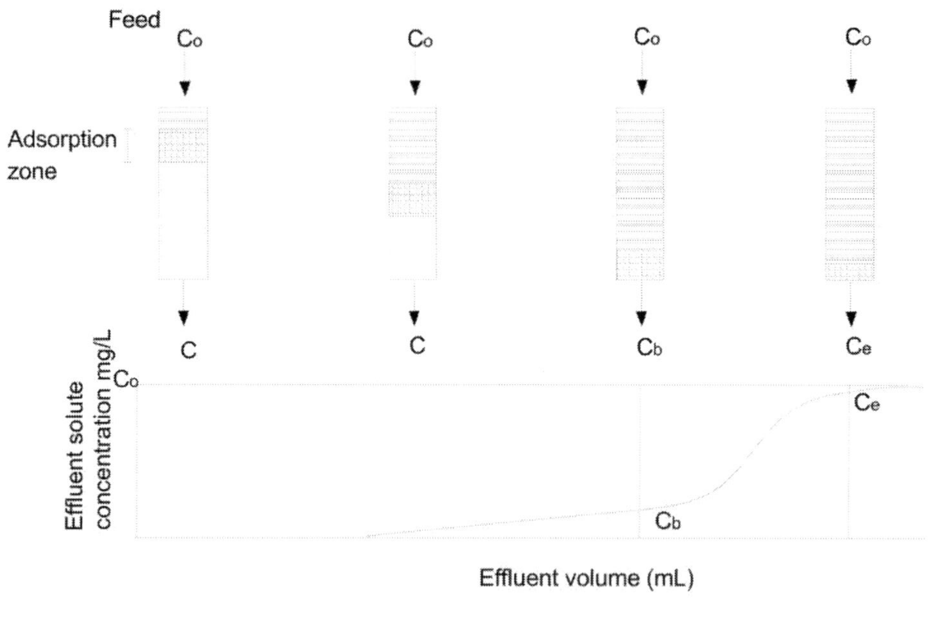

adsorption system. The continuous flow could be either in upward or down ward directions. Packed bed up flow carbon columns for full counter current operation are suitable only for low turbidity water (i.e.) water having a turbidity of 2.5 JTU. Carbons finer than 8 X 30 mesh should not be used for up flow fixed beds because of clogging and high head loss problems. In liquid up flow operations, a portion of the adsorbent is usually removed periodically from the bottom and equal amount of regenerated solid is added to the top.

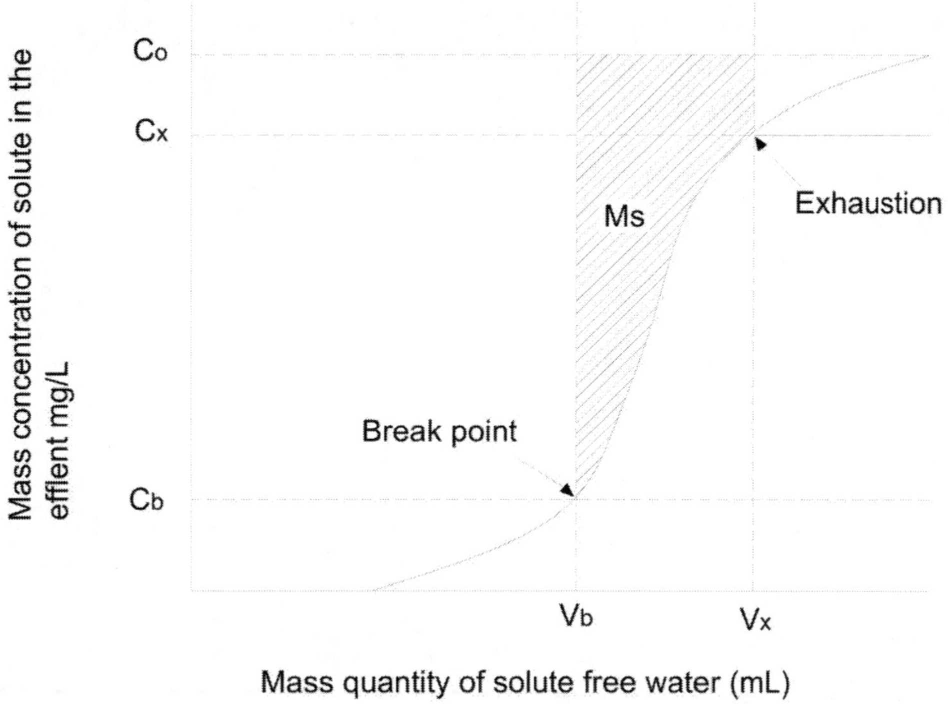

Fig 3.2 Ideal break through curve in a fixed bed adsorption system

3.3 MATLAB

MATLAB is a high-performance language for technical computing. It integrates computation, visualization, and programming in an easy-to-use environment where problems and solutions are expressed in familiar mathematical notation. Typical uses include Math and computation Algorithm development Data acquisition Modeling, simulation, and prototyping Data analysis, exploration and visualization of Scientific and engineering graphics, Application development, including graphical user interface building MATLAB is an interactive system whose basic data element is an array that doesn't require dimensioning.

3.4 TREND ANALYSIS OF PARAMETERS

To study the pattern of trend of the parameters during the study period, the polynomial trend equation namely, cubic trend equation of the form, $Y= b_o + b_1 t + b_2 t^2 + b_3 t^3 + b_4 t^4 \ldots$, where b_i's (i=1,2 ,3,4) are trend coefficients and b_o = constant, , t_i = i th time (i=1,2,.......,) is fitted and the results are presented below. The F-values indicate the overall significance of the trend equation fitted. The R^2 the coefficient of determination indicates that to what extent the trend coefficients are able to explain the variations of the dependent variables under study.

4. RESULTS AND DISCUSSION

4.1 COMPARISON CHARTS FOR THE RESULTS OBTAINED MANUALLY WITH THOSE OF MATLAB AND REGRESSION

The following charts are the results of the percentage removal of various metals for various bed depths of carbon.

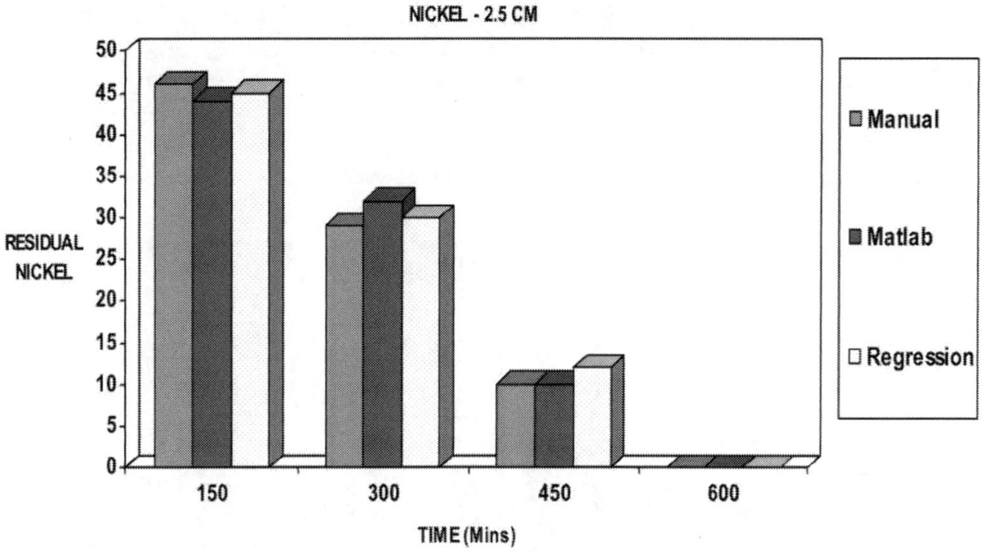

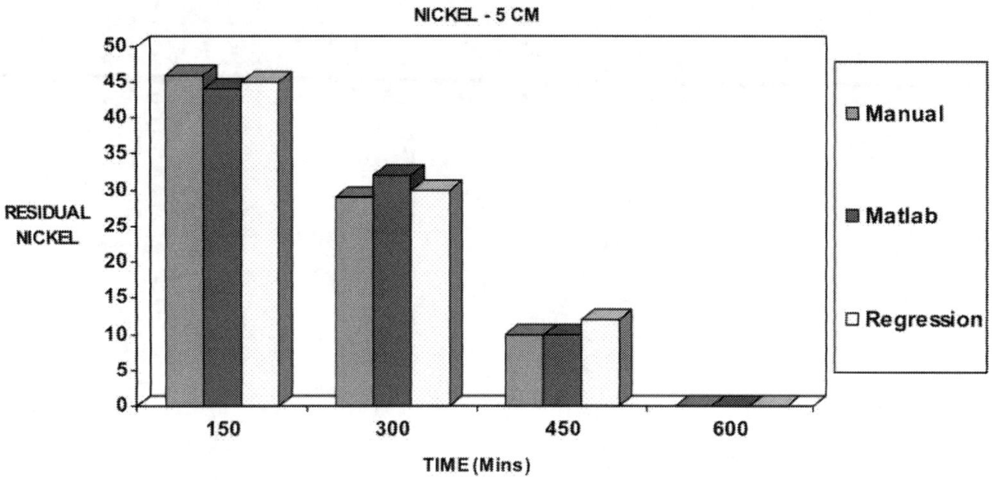

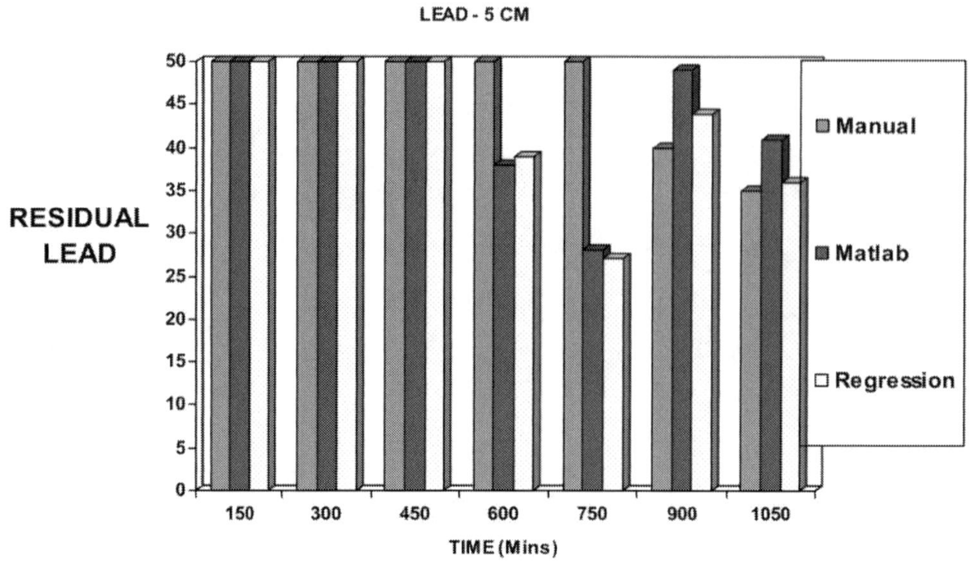
LEAD - 5 CM
RESIDUAL LEAD
50
45
40
35
30
25
20
15
10
5
0
150
300
450
600
750
900
1050
TIME (Mins)
Manual
Matlab
Regression

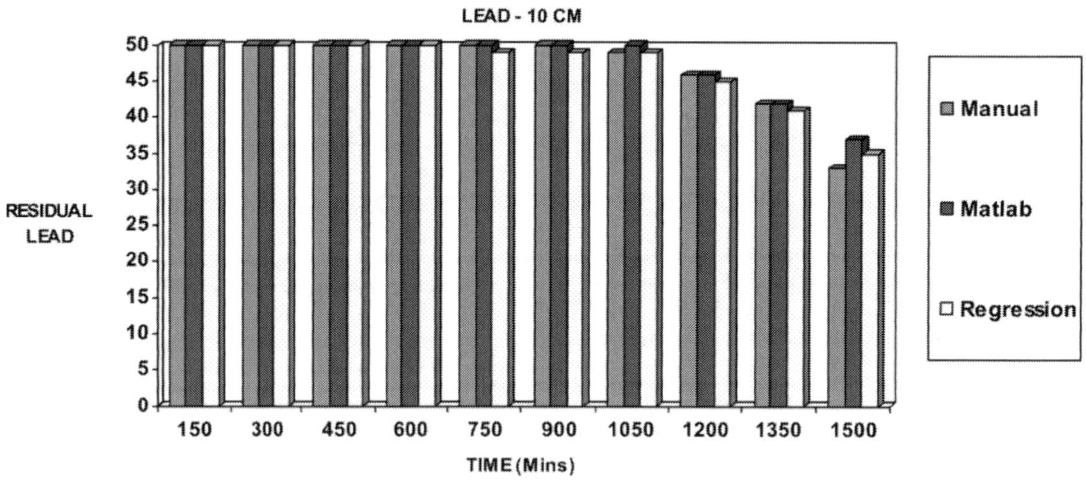
LEAD - 10 CM
RESIDUAL LEAD
50
45
40
35
30
25
20
15
10
5
0
150
300
450
600
750
900
1050
1200
1350
1500
TIME (Mins)
Manual
Matlab
Regression

5. CONCLUSION

To determine the removal efficiency of heavy metals such as Nickel, Lead & Cadmium from the aqueous solution the activated carbon prepared from sugar mill sludge was used as the adsorbent and the removal efficiency was noted at various time interval and at bed depths of 2.5cm, 5cm and 10cm.The data generated using the experiments were utilized in developing a program in MATLAB (Version7).Using the program for a specific metal at any bed depth and at any time can be determined.

6. REFERENCES

1. Amir Hossein Mahvi, Dariush Naghipour, Forugh Vaezi and Shahrokh Nazmara 2005 "Teawaste as An Adsorbent for Heavy Metal Removal from Industrial Wastewaters", Americal Journal of Applied Sciences 2 (1) : 372-375, ISSN 1546 – 9239
2. N. Ayawei, M.Horsfall Jnr and A.I.Spiff, 2005 Rhizophora Mangle Waste As Adsorbent For Metal Ions Removal From Aqueous Solution,"European Journal of Scientific Research, Vol 9, ppNo. 1,
3. Corapcicoglu, etal, "The absorption of Heavy Metals on to hydrous Activated Carbon", Water Research, pp 1031-1044.
4. Gajghaje and Saxena, (1991) "Removal of lead from acqueous solution by active carbon", Indian Journal of Environmental Heath, Vol No.33, pp 374-379.
5. Mohamed Nageeb Rashed, (2001), "Fruit Stones as Absorbents for the Removal of Lead Ion from Polluted Water", Chemistry Department, Faculty of Science, 81528 Aswan, Egypt.
6. Jadon S.P.S. (1999) "User soil as filtrate of the removal of toxic metals from polluted plating wastewater", Proceedings of the International Conference on Chemistry and Thirty – Sixth Annual Convention of Chemists, pp19.

Selection of CO_2 Chemisorbent for Fuel-Cell Grade H_2 Production by Sorption Enhanced Water Gas Shift Reaction

Michael G. Beaver, Chemical Engineering, Lehigh University, Bethlehem, PA

S Sircar, Chemical Engineering, Lehigh University, Bethlehem, PA

H.S. Caram, Chemical Engineering, Lehigh University, Bethlehem, PA

New experimental data are reported to demonstrate that high purity H_2 can be directly produced by sorption-enhanced water gas shift (WGS) reaction using synthesis gas ($CO + H_2O$) as sorber-reactor feed gas. An admixture of a commercial WGS catalyst and a proprietary CO_2 chemisorbent (K_2CO_3 promoted hydrotalcite or Na_2O promoted alumina) was used in the sorber-reactor for removal of CO_2, the WGS reaction by-product, from the reaction zone. The promoted alumina was found to be a superior CO_2 chemisorbent for this application because (a) it could directly produce a fuel-cell grade H_2 product (<10-20 ppm CO) at reaction temperatures of 200 and 400 oC, and (b) it produced ~ 45.6% more high purity H_2 product per unit amount of sorbent than the promoted hydrotalcite at 400 oC. Furthermore, the specific fuel-cell grade H_2 productivity by the promoted alumina at a reaction temperature of 200 oC was ~ 3.6 times larger than that at 400 oC. These striking differences in the performance of the two CO_2 chemisorbents were caused by the differences in their CO_2 sorption equilibria and kinetics. The Sorption Enhanced Water Gas Shift reaction allows for the simultaneous production of H_2 and its separation from CO_2 in a single reactor unit. This concept can be applied to both coal and biomass gasification processes for the direct production of high purity H_2 while also capturing the CO_2 produced by the process.

See more of this Session: Hydrogen Production and Carbon Redirection Technologies
See more of this Group/Topical: Environmental Division

System Design, Development, and Integration of a Solar-Energy Driven Process for Microalgae Fixation of Carbon Dioxide

Walter Den, Environmental Science and Engineering, Tunghai University, Taichung, Taiwan

Sherwin Yang, Byotec Inc., Miao Li, Taiwan

Due to the strong evidence of human-induced climate change, the international community has taken a collaborative approach in an effort to reduce the emission of global-warming gases, most notable carbon dioxide generated from energy providers, especially coal-burning plants. The strategetic policy proposed by the ROC government is also concurrently promoting energy-saving and/or renewable energy technologies as part of the adaptation and mitigation options. In these technologies, biological mitigation have been regarded as one of most environmentally benign processes that can meet the foresight of sustainable socio-economical development, and will be the core of the "green technologies" for the future development.

This study involves an integrated carbon reduction system that encompasses biological process and renewable energy technology. The system will use a specific microalgae incubated in a close-loop photobioreactor to achieve carbon dioxide fixation, whereas the spent microalgae will become a rich source of added-value products such as biodiesel, feedstock or compliment food. The energy that drives the system will be provided by an III-V compound semiconductor, high-efficiency concentration photovoltaic cell module that illuminates low-consumption LED as the light source. This system thus forms a self-sufficient energy loop with continuous reduction of carbon dioxide by the high-density microalgal culture. The conceptual framework of this system takes advantage of the maturing industries in both the photovoltaic solar cell, LED, and microalgal cultivation in Taiwan.

This study is still in a prelimnary stage, with separate studies involving microalgae growth and carbon fixation capability, solid/liquid separation processes, as well as light utilization rate of the photobioreactor.

See more of this Session: Hydrogen Production and Carbon Redirection Technologies
See more of this Group/Topical: Environmental Division

Hydrogen Production Plant with Urea Co-Production

Paul Ghougassian, Chemical Engineering, UCLA, Los Angeles, CA

Vasilios Manousiouthakis, Chemical Engineering Department, University of California at Los Angeles, Los Angeles, CA

While fossil fuels have been very useful in shaping the industrial world as we know it today, their use has been very detrimental to the environment. Nowadays, there is a very strong outcry to move away from this conventional source of energy and revenue to cost ratio remains competitive with respect to gasoline and entice people into purchasing hydrogen vehicles.

In this work we present a hydrogen production process in which hydrogen and urea are generated, instead of the conventional hydrogen and carbon dioxide. The feed to the steam methane reformer (SMR) consists of water and natural gas mixed simultaneously in the appropriate ratio. The outlet of the SMR consists of synthesis gases which are used in the production of urea. In order to address and fulfill the energy requirements of the entire plant, heat and power integration techniques are applied into the process allowing us to maintain a self-powered process plant, reducing the carbon emissions to near zero and being economically feasible.

See more of this Session: Hydrogen Production and Carbon Redirection Technologies
See more of this Group/Topical: Environmental Division

Global Optimization of Ideal Gibbs Reactor for Carbon Dioxide Capturing

Jorge Pena Lopez, Chemical Engineering Department, University of California at Los Angeles, Los Angeles, CA

Vasilios I. Manousiouthakis, Chemical Engineering Department, UCLA, Los Angeles, CA

An increment of carbon dioxide concentration in the atmosphere has been related to increases in the average world temperature, supporting norms that regulate atmospheric carbon dioxide emissions from various industrial sources. One of the most common operations in industrial processes is the burning of a fuel to produce energy. This suggests high temperature streams containing carbon dioxide exit the combustion process. In this work we propose a methodology for the removal of carbon dioxide from water-air mixtures via calcium oxide adsoportion, based on Gibbs free energy minimization in a multiphase system.

The case study presented in this work consists of a first step in which a gaseous mixture of oxygen, nitrogen, water and carbon dioxide reacts at constant temperature and pressure with solid carbon oxide fixed in a batch reactor. The possible reacting products are calcium carbonate and calcium hydroxide. In a second step the bed is regenerated using steam at a higher temperature. The total Gibbs minimization problem is non-linear due to the presence of condensed species, which ultimately yield a total Gibbs free energy equation that is a combination of a linear equation, a negative entropy and a positive entropy functions. The problem is reduced into variables that bound the feasible region of the problem and by the Weirstrass theorem we establish the existence of a minimum. After evaluating the bounds of the variables we establish 9 possible cases for the position of the global minimum and inspection of the cases yields the equilibrium composition of the system.

See more of this Session: Hydrogen Production and Carbon Redirection Technologies
See more of this Group/Topical: Environmental Division

Occurrence of Fluoride in Alluvial Shallow Aquifers of Semiarid Region of Punjab, North-West India

H. S. Hundal, Soils, Punjab Agricultural University, Ludhiana, India
Kuldip Singh, Soils, Punjab Agricultural University, Ludhiana, India
Dhanwinder Singh, Soils, Punjab Agricultural University, Ludhiana, India
Mandeep Hundal, Cheyenne regional medical center, Cheyenne, WY

INTRODUCTION

Fluoride (F^-) has considerable physiological significance for man and animal. Its level in environment, especially in drinking water is closely associated with human health (Zhang et al. 2003 and Yadav et al. 2008). Low concentration of fluoride in drinking water (<0.50 mgl^{-1}, therefore, low intake dose) lead to dental caries (Edmunds and Smedley 1996). Generally, fluoridation is recommended where water supply contained F^- less than 1.50 mgl^{-1} (WHO 2004). Concentration higher than 2 to 4 mgl^{-1} can result in dental fluorosis and skeletal fluorosis, respectively. The effects are permanent and incurable. The fluoride concentration in groundwater of India varied from 0.12 to 24.17 mgl^{-1} (Kumar and Gopal 2000). The manual of water supply (Ministry of Health, Government of India 1962) lists several places in India including semi-arid region of Punjab where people are reported to suffer from fluorosis through consumption of underground waters containing relatively higher concentrations of F^- anion. High concentration of F^- in drinking waters abstracted from shallow aquifers in semi-arid region of Punjab may be the primary cause for yellowing of teeth of local population. The extent of fluorosis has been related to the concentration of fluoride in drinking water, period of consumption as well as the age of the consumer (children are more prone to its toxicity). It also caused joint pains, severe dental and skeletal fluorosis in some people living in those areas. High fluoride concentration in drinking water has also been linked with cancer, which is not uncommon in semi-arid region of Punjab. Extensive survey had been conducted to measure fluoride in India (Madhavan and Subramanian 2004; Choubisa1997; Suthar et al 2008; Khaiwal and Garg 2006; Singh et al. 2007; and Subba Rao, 2008) and abroad (Herlinger and Viero 2007; Gomez et al. 2009) to catalog different aquifer above or below the safe limits of fluoride in drinking water, but no or little attention have been made to determine the effect of oxidation- reduction reactions (redox status) in alluvial aquifers and its impact on Fluoride content in underground waters. The present investigation aims to determine the concentration of F^- in alluvial shallow aquifers of semiarid region of Punjab, north-west India. An attempt was also made to examine the dynamic of redox potential (oxidation- reduction reactions) involved in the shallow alluvial aquifers due to exposure of water to air oxygen and its ultimate impact on fluoride content in underground waters.

MATERIALS AND METHODS

Sample collection and pre-treatment of samples

Thirty six water samples were collected in October 2007 from shallow aquifers located in the semi-arid region of Punjab, north-west India. Fifty six water samples were also collected in

October 2008. The location of sampled aquifers during both the field works are shown in Figure 1. The soils in this region can be classified on physiographic units of sand dunes, inter-dunal areas and alluvial terraces (Sidhu et al 1995). Geographically, the area of sampling lies between $29^{0}\ 32'$ to $30^{0}\ 36'$ N latitudes and $74^{0}\ 38'$ to $75^{0}\ 10'$ E longitudes. The distribution of rainfall (< 400mm) show that the amount of water received is completely utilized as actual transpiration and there is no rainwater surplus throughout the year. However, during the

Figure1. Water sampling from alluvial aquifers in October 2007 (▲) and 2008 (×) from different locations of Punjab, North-west India

last four decades, river water supplied through canal for irrigations of agricultural crops. Before sampling the groundwater was drawn for half an hour to empty the hand pump/ tube well pipes in order to collect the fresh water from the aquifers. At the time of sampling, bottles were thoroughly rinsed 3-4 times with groundwater to be sampled. Polyethylene bottles of 250 ml capacity were used for collection of water samples. Bottles were fully filled with water and closed to avoid in any air bubble. Water samples were analyzed for pH, total dissolved solids, bicarbonate, and chloride as per the standard procedure given in American Public Health Association (APHA 1998).

Redox meter

Field measurement of redox potential (Eh) were also made at the time of sampling. The Eutech make Cyber Scan pH 1100 Bench pH/mV RS232 meter was used to measure the redox potential in milli Volt (mV). Its mV range is -1,850 to +1,850 mV and resolution accuracy is 0.1 mV and ±0.2 mV. The redox potential meter was adjusted to +226 mV at 25^0C with Zobell's solution (Nordstrom 1977) and checked after each observation, during the investigation period.

Duplicate water samples collected from various locations were brought back to laboratory in cooled ice box during transportation and stored in refrigerator (at 5^0) until analysis. The analyses were performed within 24 hour after collections of water samples. The F^- concentration in water was determined by potentiometer method, using the F1052F solid-state membrane electrode in conjunction with K401 calomel electrode. The primary standard solution of fluoride ion at a concentration of 0.1 Mole per liter was obtained by dissolving 4.2 gm of Sodium Fluoride GR (Merck) in a 100 ml flask. Secondary standard solutions with pF value of 2 to 6 in 100 ml volumetric flasks were prepared by serial dilution from primary standard solution (0.1MF). The ionic strength of the secondary standard solutions was kept constant to 0.01M by addition of 10ml of 1M KNO_3. The Blank solution was double distilled water containing potassium nitrate (0.01M) in the concentration as in fluoride containing solutions. The regression curve obtained for fluoride standards by potentiometer method was pF =129.5 X (mV)-431.5 (R^2=0.9956). Five ml of 1M KNO_3 was added in 50 ml water samples to maintain same (0.01M) ionic strength for determination of fluoride in test samples. After measurement of fluoride from test samples, dilution factor of 1.1 was applied to compute fluoride concentration in water samples.

Inductively coupled argon plasma (ICAP-OES)

The acidified water samples were analyzed for Na, K, Ca, Mg, S, Fe and Mn on Thermo Electron Corporation's Emission Spectrometer, iCAP 6300. It is manufactured in United Kingdom and is world's most compact bench top ICP system with an Echelle spectrometer wavelength range of 166-847 nm. It has solid state RF generator working at a frequency 27.12 MHz and charge injection device (CID) detector. Plasma ignition and operation are fully automated and controlled from PC using iTEVA software. The aqueous sample is introduced with inbuilt 3 channel peristaltic pump through glass concentric nebulizer and glass cyclonic type spray chamber to demountable quartz torch, iCAP 6300 duo model allows for simultaneous analysis of all elements from ppb to % level. The detection limits for Na, K, Ca, Mg, S, Fe and Mn on the iCAP duo emission spectrometer are 1.80, 0.60, 0.003, 0.010, 2.22, 0.25, and 0.07 ppb at 589, 766, 393, 279, 181, 260 and 258 nm, respectively.

RESULTS AND DISCUSSION

Fluoride Occurrence in Ground Water

The fluoride concentration in thirty six water samples collected from different sites during October 2007 varied from 0 to 1.17 mg $F^- l^{-1}$ with a mean value of 0.39 mg $F^- l^{-1}$. The median, 75th and 90th percentile F^- concentration were 0.32, 0.64 and 0.95 mgl^{-1}. The fluoride concentration in 56 water samples collected from various locations in October 2008 ranged from 0.88 to 14.04 mg F^{-} l^{-1} with a mean value of 8.22 mg F^{-} l^{-1}. The median, 75th and 90th percentile F^{-} concentration were 8.46, 9.84 and 10.95 mgl^{-1}.

Fluoride occurrence in alluvial aquifers of semi-arid region of Punjab, north-west, India recorded during both years (2007-08) varied from 0 to 14.04 mgl^{-1} with mean value of 5.16mgl^{-1}. The median, 75th and 90th percentile F^{-} concentrations were 6.54, 9 and 10.46 mgl^{-1}, respectively. World health organization (WHO 2004) has laid down a limit of F^{-} (1.5 mgl^{-1}) in drinking water. According to the World health organization (WHO 2004) all the water samples collected in October 2007 contain F^{-} below the safe limit of 1.5 mg F^{-} l^{-1}, whereas 60 percent of the total water samples collected in 2008 contains F^{-} well above the safe limit.

The high pH and dissolved-solid slightly enhanced F^{-} content in ground water. The concentration of F^{-} declined with increasing concentration of Ca but no or little effect was found with increasing concentration of Mg on F^{-} content in ground water. Fluoride concentration enhanced with increasing alkali metals like Na and K but the effect was more pronounced with Na than K ion. Ground water with elevated concentration of Cl^{-}, HCO_3^{-} and SO_4^{2-} tend to have high fluoride concentration. These relationships suggest that competitive adsorption of these anions with fluoride leads to its moderate to high concentration in ground water. The lower concentration of both iron and manganese are exhibited in alkaline aquifers due to precipitation of amorphous oxy-hydroxides of iron and manganese and this newly formed adsorbent are more efficient for adsorption of fluoride and other oxy-anions than crystalline materials (Schwartzman and Cornell 2000). Therefore no consistent effect of dissolved iron and manganese content on fluoride concentration in groundwater was recorded in alluvial aquifers of Punjab.

The specific adsorption of F^{-} anion on Fe oxy- hydroxides is not dependent on the surface charge of the adsorbent materials and can change the charge characteristics of the minerals. The pH of the alluvial aquifers in the semiarid areas varied from 7.14 to 9.10 and thus it's always more than the point of zero charge (pH>6.5) of Fe oxy-hydroxides found in the aquitards. Therefore, specific adsorption of F^{-} involved ligand exchange between hydroxyl and fluoride ions according to the following equation:

$$\equiv FeOH + F^{-} + OH^{-} = FeF + OH^{-}$$

However, there is less likelihood of formation of ferric hydroxides [Fe $(OH)_3$] under reduced condition of aquifers due to soil pores saturated with water under alluvial soils. In oxidized aquifers, hydrous oxides of iron is not present as discreet mineral but deposited in the form of coatings on sand, silt and aluminosilicate minerals likewise, kaolonite, muscovite and biotite. Iron hydrous oxides in particular, have high surface areas and are strongly reactive (Schwartzman and Cornell 2000). Specific adsorption of fluoride on the surface of hydrous oxides of iron in the aquitards of alluvial aquifers could be the concentration-limiting process in ground water.

Relationship between Redox potential (oxidation-reductions) and fluoride content of alluvial aquifers systems

Iron hydroxide minerals are ubiquitous in aquifer sediments throughout the world (Zachara et al. 2001). In most natural groundwater system iron is the dominant (by 1 to 3 orders of magnitude) buffering species involved in redox reactions (Vance 1996). The role of aqueous iron chemistry (oxidation-reduction reaction) in homing or release of heavy metals and arsenic and in ground and surface water has been well established (Stumm and Lee 1960; Jones-Lee and Lee 2005). The oxidation state in which iron exists in particular aquatic system depend upon its redox status, which in turns govern by the presence and absence of dissolved oxygen (DO). Iron can exists in natural waters in metallic and ferrous iron (Fe^{2+}) under reduced condition and ferric iron (Fe^{3+}) under oxidized condition. In the oxic (oxygen-containing) water, iron in the oxidized ferric (Fe III) form exists as particulate or colloidal ferric hydroxide. This form of iron could acts as scavenger for fluoride anion. In an anoxic sediment environment, the ferric hydroxide is reduced to ferrous iron which could precipitates as a black iron sulfide mineral. Some of the chemicals that had been associated with the ferric hydroxide are mobilized in the aqueous medium, while some chemicals, especially metals and arsenic can be precipitates as metal sulfides.

Redox potential determined of thirty six water samples collected in October 2007 varied from +98 to +390 mV with a mean value of +197 mV. The redox values of median, 75th and 90th percentile were +167, +236 and +350 mV, respectively. As the redox potential enhanced from +98 to +390 mV in water samples collected from different alluvial aquifers during 2007, the F^- concentration declined from 1.17 to 0 mg l^{-1} (Fig. 2). The results presented in Figure 2 illustrate that the redox potential is predominantly controlling the concentration of F^- in water pumped out from alluvial aquifers of semi-arid region of Punjab.

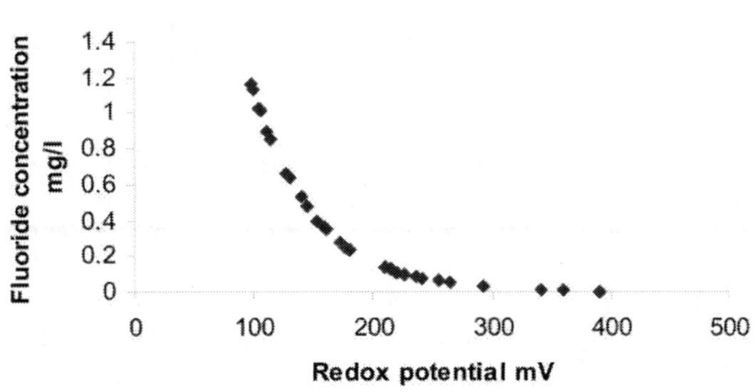

Figure 2. Effect of redox potential on fluoride concentration in groundwater

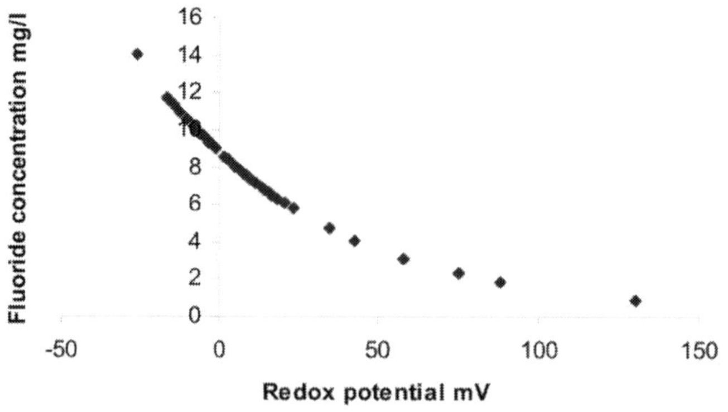

Figure 3. Effect of redox potential on fluoride concentration in groundwater

The redox potential of alluvial aquifers determined during October 2008 varied from -26.0 to +130 mV with a mean value of +9.31 mV. The redox potential values for median, 75th and 90th percentiles were +3 mV, +14.75 and +37.4 mV, respectively for shallow alluvial aquifers. The results presented in Figure 3 elucidate that as the redox status of water sampled for alluvial aquifer declined from +130 mV to -26 mV, the fluoride content in these ground water systems increased from 0.88 to 14.04 mgl^{-1}.

The influence of redox potential on the F^- concentration in all the 92 water samples collected during both years of 2007 and 2008 are depicted in figure 4. The redox potential of alluvial aquifers measured during both years varied from -26 to 390 mV with mean value 81 mV. The median, 75th and 90th percentile redox potential were 17, 152 and 232 mV, respectively. The F^- concentration in groundwater's declined from 14.0 to 0 mg F^- l^{-1} with increase of redox status from -26 to +390 mV. The redox potential of alluvial aquifers determined during October 2007 (+98 to +390 mV) showed oxic conditions. However, redox potential measured during October 2008 (-26 to +130 mV) relatively oscillate towards reduced status in alluvial aquifers of semi-arid region of Punjab. Nevertheless, the oxidation-reduction reaction in the present investigation emphasized strong affect on F^- content in water samples collected from different aquifers in semi-arid region of Punjab. Thus it is conclusively evident that redox potential determined during both the years strongly governs the F^- concentration in the shallow aquifers. Fluoridation of water abstracted from alluvial aquifers of semi-arid region of Punjab is required to maintain F^- 1.5mgl^{-1}, when the redox potential of water is above +95 mV and remediation is recommended when redox potential of water is below +95 mV.

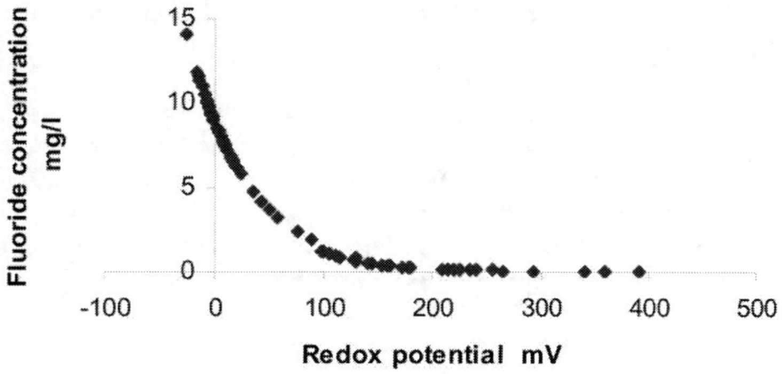

Figure 4 Effect of redox potential on fluoride concentration in groundwater

The redox status of alluvial aquifers in semi-arid region could be affected by recharge of surface canal, tank, and pound water and through surface flooded irrigation waters to agricultural fields. Groundwater level fluctuation in semi-arid region also leads to oscillate reduction-oxidation conditions, which could affect homing or mobilization of F^- in the shallow alluvial aquifers. Some farmer's revealed that in the growing season, daily ground-water level fluctuation as well as annual fluctuation was significant level due to the intensive ground-water pumping in the growing season. The ground-water levels changes also depend upon the water requirements by different crops, type of water management's practices for irrigation or domestic use and may be recharge of aquifers by canal or river water as well as through rainfall. Ground-water level fluctuations in semi-arid region could also lead to fluctuations in oxidation- reduction conditions, which could eventually affect mobilization-immobilization of fluoride in the aquifer sediments.

Remediation of fluoride in drinking water

The use of oxidation-reduction reaction mechanism is also one of the most convenient and inexpensive technology to remediate the groundwater containing high concentration of fluoride. This technology is simple to adopt at rural household level. It needs to collect underground water from hand pumps or tube wells in a big container or cemented open tank having large surface area and stand for few days to get it aerated or to interact with atmospheric oxygen in order to achieved elevated redox potential. Most groundwater in shallow aquifer contain native iron, which readily oxidized upon aeration and form ferric hydro oxides precipitates with very large and highly reactive surface sites for specific adsorption of F^- anion. The basic purpose is to elevate the redox status of the aquifers or collect water to promote or facilitate formation of oxides/hydroxides of iron minerals having large surface area with strong sorption sites to remove fluoride from water below the safe limit for drinking and domestic use. Generally, spring and well water tend to contain higher concentration of fluoride than surface waters from lakes and streams (Weinstein and Davison 2004). In fluoridated areas, the drinking water, obtained from surface water (oxic water) with an average fluoride concentration of 0.1- 0.2 mgl^-

^{1} is raised to optimal level of 0.7 to 1.2 mgl^{-1} by addition of sodium fluoride, hydrofluosilicic acid, or sodium silicofluoride.

In India ground water recharge has also been used to decrease the fluoride content of ground water. Since surface water, i.e. canal water has got higher redox status, which is usually above +240 mV (Hundal et al. 2009). The mixing of canal or surface oxic water elevate the redox status of ground water and thus induce adsorption on freshly precipitated iron oxyhydroxides particulates. If the high fluoride concentrations in groundwater could be mobilized from an adsorbed pool through reduction of ferric compounds to dissolve ferrous iron, then this would give the possibility of using ground water recharge to remove at least a substantial part of the fluoride.

Groundwater (hand pumps and tube wells) is the primary source of drinking water for these village populations. Local residents are reluctant to use low fluoride canal and surface (pound) waters as potable due to pollution of the sources from disposal of industrial, domestic waste, decaying dead animals and their excreta by some irresponsible people. However, problems can also arise with bacterial contaminations and adequate sanitary (chlorination of water) is required. The elevation of redox (oxidation-reduction mechanism) potential technique is most reliable, cheap and simple to adopt for remediation of fluoride in drinking groundwater at rural household level. It needs to collect underground water from hand pumps or tube wells in a big container, earthen pots or cemented open tank having large surface area and stand for few days to get it aerated or to interact with atmospheric oxygen in order to achieved elevated redox potential. Higher redox potential reduces fluoride content in stored water, which can be used after decantation or filtration to another vessel.

REFERENCES

APHA (1998) *Standard methods for the examination of water and waste waters.* 20th edn. Washington DC, USA.

Choubisa, S. L. (1997) Fluoride distribution and fluorosis in some villages of Banswara district of Rajasthan. *Indian Journal of Environmental Health*, 39: 281-288.

Edmunds, W. M., and Smedley, P.L. (1996) Ground water geochemistry and health: In: *Environmental Geochemistry and Health with special reference to developing countries;* Appleton, J.D., Fuge, R., McCall G.J.H. (eds.); Geological Society, Special Publication, 113: 91-105.

Gomez, M.L., Blarasin, M.T. and Martinez, D.E. (2009) Arsenic and fluoride in a loess aquifer in the central area of Argentina. *Environmental Geology*, 57:143-155.

Herlinger Jr R., and Viero, A.P. (2007) Hydrogeochemistry of the Coxihla das Lombas aquifer, Brazil. *Environmental Chemistry Letters*, 5:91-94 DOI10.1007/s10311-006-0083-9 .

Hundal, H. S., Kuldip- Singh, and Dhanwinder-Singh. (2009) Arsenic content in ground and canal waters of Punjab, North-West India. *Environment Monitoring and Assessment,* DOI 10.1007/s10661-008-0406-3.

Jones-Lee, A., and Lee, G.F. (2005) Role of iron chemistry in controlling the release of pollutants from resuspended sediments. *Remediation Journal,* 16:33-41.

Khaiwal, R., and Garg, V. K. (2006) Distribution of fluoride in groundwater and its suitability assessment for drinking purposes. *International Journal of Environmental Health & Research,* 16: 166–169.

Kumar, S., and Gopal, K. (2000) A review on fluorosis and its preventive strategies. *Indian Journal of Environmental Protection*, 20: 430-440.

Madhavan, N., and Subramanian, V. (2004) Fluoride and arsenic content in contaminated groundwater of Rajasthan and West Bengal. In national seminar on arsenic and fluoride contamination in groundwater at NERIWALM, Tezpur, India,: 268-274.

Ministry of Health, Government of India Manual, 1962.

Nordstrom, D.K. (1977) Thermochemical redox equilibrium of Zobell's solution. *Geochemica et Cosmochimica Acta*, 41: 1835-1840.

Schwartzman U, Cornell U. (2000) Iron oxides in the laboratory preparation and characterization; Welli-VCH

Sidhu, G., Walia, C., Lal, T., Rana, K., and sehgal, J. (1995) Soils of Punjab for optimizing land use. Solis of India Series 4, National Bureau of Soil Survey and Land Use Planning, Nagpur, India, pp. 1-75.

Singh, B., Gaur, S., and Garg, V. K. (2007) Fluoride in drinking water and human urine in southern Haryana. *India Journal of Hazardous Material,* 144: 147–151.

Stumm, W., and Lee, G. F. (1960) *The chemistry of aqueous iron. Schweizerische Zeitschruft fur Hydrologie*, 22:95-139.

Suthar, S., Garg, V.K., Jangir,S., Kaur , S., Goswam, N., and Singh, S. (2008) Fluoride contamination in drinking water in rural habitations of Northern Rajasthan, India. *Environment Monitoring and Assessment*, 145:1–6 DOI 10.1007/s10661-007-0011-x.

Vance, D. B. (1996) Redox reactions in Remediation. 2. The 4 Technology Solutions. *Environmental Technology*, 6:24-25.

Weinstein, L. H., and Davison, A. W. (2004) *Fluorides in the Environment.* CABI Publisher, CAB International, Oxon, UK.

WHO (2004) Guidelines for drinking water quality (p. 515) Geneva: World Health Organization.

Yadav, J.P., Lata, S., Sudhir, J., and Kataria, S.K. (2008) Fluoride distribution in groundwater and survey of dental fluorosis among school children in the villages of the Jhajjar District of Haryana, India. *Environment Geochemistry and Health*, DOI 10.10007/s10653-008-9196-3.

Zachara, J. M., Fredrickson, J.K., Smith, S. C., and Gassman, P. L (2001) Solubility of Fe(III) oxide-bound trace metals by a dissimulator Fe(III) reducing bacterium. *Geochemica et Cosmochimica Acta,* 65: 75-93.

Zhang, B.O., Hong, M., Zhao, Y., Lin, X., Zhang, X., and Don, J.(2003) Distribution and risk assessment of fluoride in drinking water in the west plain region of Jilin province, China. *Environmental Geochemistry and Health*, 25: 421-431. doi:10.1023/B:EGAH.0000004560.47697.91.

Continuous Column Modeling of Copper Biosorption Using Pretreated *Aspergillus Niger* Biomass

Mausumi Mukhopadhyay, Tejinder Kaur and Richa Khanna,*
Chemical Engineering Department, S.V.N.I.T, Surat, India

Abstract

In this study, fixed bed model and reduced lumped diffusion model was used to explain biosorption behavior of Cu (II) in a continuous column of pretreated *Aspergillus niger* biomass. The breakthrough was measured for both the model as a function of influent flow rate and bed height for a feed solution at 10 mg/l metal ion concentration and compared with experimental results. MATLAB was used to solve the partial differential equations and breakthrough was plotted. The breakthrough data obtained experimentally was comparable with data evaluated by these models.

Keywords: Continuous column; Copper; *Aspergillus niger*; Biosorption; MATLAB

1. Introduction

Biosorption can be used as an alternative method for removal of heavy metals from wastewater [1]. The removal of traces of toxic copper metal ions by biosorption from wastewater has gained interest in recent years [2]. Recent batch studies have identified the potential for copper biosorption by pretreated *A. niger* biomass [3]. However, the continuous operation of a reactor is a more efficient mode of operation in comparison to a batch process. Hence there is the need to perform column studies.

Different types of reactors can be used for column studies. Packed bed sorption has a number of advantages including the possibility of achieving a reasonably high sorption capacity with very low effluent concentrations. Adsorption takes place from the inlet of the column and proceeds to the exit and a high degree of purification can be achieved in a single process step [4]. Relatively few reports exist of column studies for biosorption of metals. Kapoor and Viraraghavan [5] studied column operation for the removal of copper by immobilized *A. niger.* Volesky et al [6] reported sorption of copper by a packed bed of *Sargassum filipendula* biomass. Vijayaraghavan et al. [7] reported on the operation of a column for sorption of copper by *Turbinaria ornate* biomass.

Removal of copper by pretreated *A. niger* has been studied in a packed bed column and detailed data reported elsewhere [8]. Important design parameters such as column bed height and flow rate of metal solution into the column have been identified. In this study the breakthrough profile for the sorption of copper is analyzed using fixed bed model and reduced lumped parameter diffusion model as a function of influent flow rate and bed height for a feed solution at 10 mg/l metal ion concentration and compared with experimental results. MATLAB is used to solve the partial differential equations and breakthroughs are plotted.

2. Material and Methods

2.1. Growth of Culture

The fungus *Aspergillus niger* strain NCIM 618 (ATCC 10594), was obtained from NCIM, Pune, India and used for the sorption studies. The laboratory strain of *A. niger* was propagated in potato dextrose agar (PDA) consisting of 39g/l PDA and 0.1 g/l of yeast extract. The culture was maintained on slants and was incubated for 5-7 days at 33 ± 20°C. The growth medium used had the following composition (g/l): dextrose, 30; peptone 10; $(NH_4)_2HPO_4$, 0.4; KH_2PO_4, 0.2; $MgSO_4$, $7H_2O$, 0.2. All reagents were from Himedia or Merck (Mumbai, India). The pH of the growth medium was adjusted to 5.8-6.0 by the addition of 0.1 N HCl prior to autoclaving. Once inoculated, flasks were shaken at 200 rpm for five days at 33 ± 20°C. After harvest, the biomass was transferred to a 1% formalin solution for 24 hrs. The biomass was then dried at 600°C for 20-24 hrs and powdered in a mortar. The powdered biomass residue was sieved and in the range of 0.75 to1.0 mm particle size was used for further studies.

2.2. Column experiment

The sorption studies were carried out in a glass column of 1.5 cm outside diameter with 0.3 cm wall thickness and 11 cm (total height 22 cm) length. Packed bed experiments were carried out at 330°C. A perforated polymeric plate was fitted at the top and bottom of the column for better flow distribution of the metal solution and to support the biomass during the studies. Experiments were conducted to study the effect of bed height, flow rate of the influent metal solution at a fixed initial concentration of metal in influent.

Pretreated biomass of 1, 2 and 3 g was packed in the glass column and the sorption experiment was performed corresponding to 2.1 cm, 3.1 cm and 4.1 cm of initial height of the column. The sorption was conducted at flow rates between 1.6-9.8 ml/min. Cu (II) solution of concentration 10 mg/l, was fed from the top of the column. Effluent samples were collected at different times from the bottom of the column. Column effluent samples were analyzed by atomic absorption spectroscopy (AAS-SL 173, EIL, India).

2.3. Modeling

The breakthrough profile for the sorption of copper is analyzed using fixed bed model and reduced lumped parameter diffusion model as a function of influent flow rate and bed height for a feed solution at 10 mg/l metal ion concentration.

2.3.1 Fixed simulated bed for adsorption

The fixed bed model [9] is a semi-empirical one. The mathematics of adsorption in beds i.e. the adsorption of solute from liquids on the surface of solids is either diffusion control or kinetic control. There is a fundamental difference in the mathematical analysis, depending upon whether one assumes that equilibrium attains or whether one assumes a kinetic approach. Here nonlinear adsorption equilibrium is considered and the model is developed for the simulated fixed bed operation. In this case first order partial differential equations need to be solved. For the fixed bed model the data used are as follows: longitudinal dispersion coefficient, $D_L = 298$ cm^2/min; superficial velocity, $v = 1.1415$ cm/min; bed porosity, $\varepsilon_b = 0.58$; density, $\rho = 0.999$ gm/cm^3. The following initial and boundary conditions have been considered for bulk concentration, $C_b = 100 < z <= L$, $t=0$.

2.3.1 Reduced lumped diffusion model for adsorption

Reduced lumped diffusion model [10] is an ideal model. It assumes no axial and mass transfer kinetics. It focuses on the influence of nonlinear thermodynamics of phase equilibrium. It is mainly used for large samples with highly efficient columns and there is good agreement with the experimental chromatograms. For the reduced lump diffusion model the data used are as follows: superficial velocity, $v = 1.1415$ cm/min; bed porosity, $\varepsilon_b = 0.58$; density, $\rho = 0.999$ gm/cm^3.The following initial and boundary conditions has been considered as bulk concentration, $C_b = 100 < z <= L$, $t=0$.

3. Results and discussion

Sorption of copper by pretreated *A. niger* biomass is presented in the form of a breakthrough curve where effluent concentration C is plotted against time. Fig.1 shows the breakthrough profile of copper biosorption at different bed heights for a given flow rate. The metal uptake capacity increased with increasing bed height, from 2.1 cm to 4.1 cm. In the column experiment, the increase in the uptake capacity with the bed height is due to the increase in the available binding sites for the adsorption. The effect of flow rate on breakthrough capacity is indicated in Fig. 2. As the flow rate increases, the breakthrough time decreases. This is due to the low residence times that result at a high flow rate, and also possibly due to leaching of copper from the biomass at such flow rates [8].

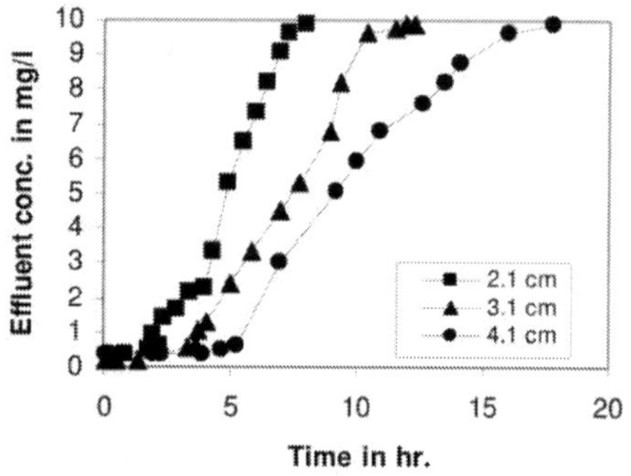

Fig.1 Effect of bed height on the breakthrough profile for sorption of copper (F=5.2 ml/min, C_0 = 10 mg/l)

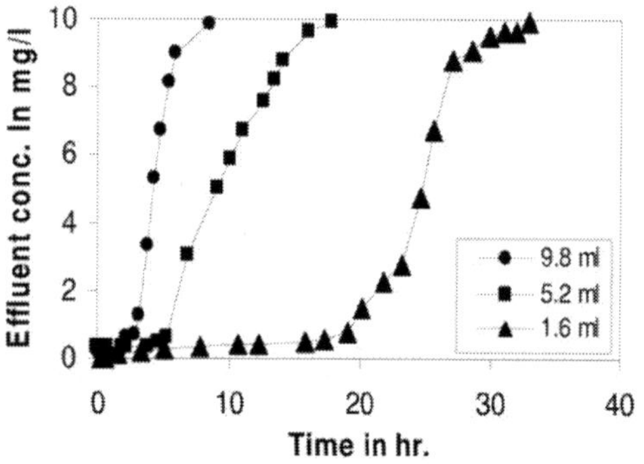

Fig.2 Effect of flow rate on Breakthrough capacity at different flow rates for sorption of Copper ion (Bed height: 4.1 cm)

3.1. Simulated fixed bed model

Fig 3-5 shows the simulated fixed bed model plots as concentration, time and volume of effluent (using MATLAB) for bed height 2.1, 3.1 and 4.1 cm respectively for 5.2 ml/min inflow rate. The amount of solute adsorbed in the bed decreases as the time of adsorption increases due to decrease in the driving force as the time proceeds.

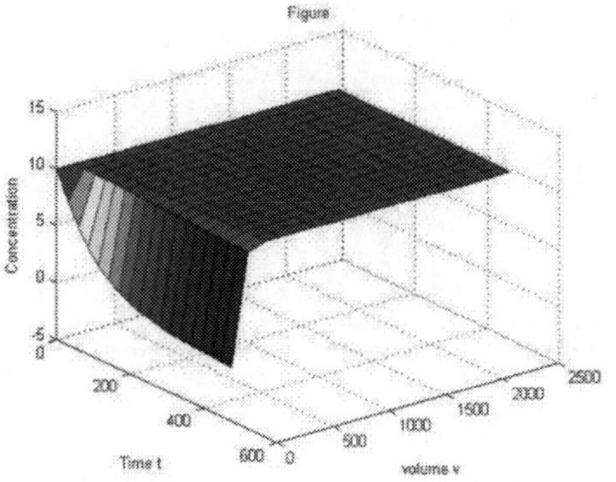

Fig. 3 Plot of concentration vs. time vs. volume of the effluent for bed height 2.1cm and flow rate 5.2 ml/min for fixed bed model

The reason for a constant decrease in the concentration of the bed towards the end of the column is that as the fluid reaches the end of the column the concentration of the fluid is decreased to a very low value and so towards the end the driving force for adsorption decreases. The plot also shows that the concentration of the solute in the effluent increases as volume of the liquid passing through the column increases. A breakthrough point is reached after which the degree of adsorption gets decreased rapidly as the bed saturates. After some time the adsorption ceases completely and the concentration of the elutant becomes equal to the concentration of the effluent.

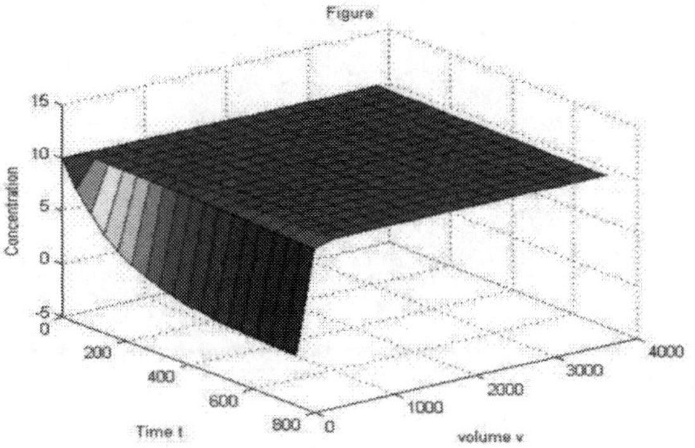

Fig. 4 Plot of concentration vs. time vs. volume of the effluent for bed height 3.1cm and flow rate 5.2ml/min for fixed bed model

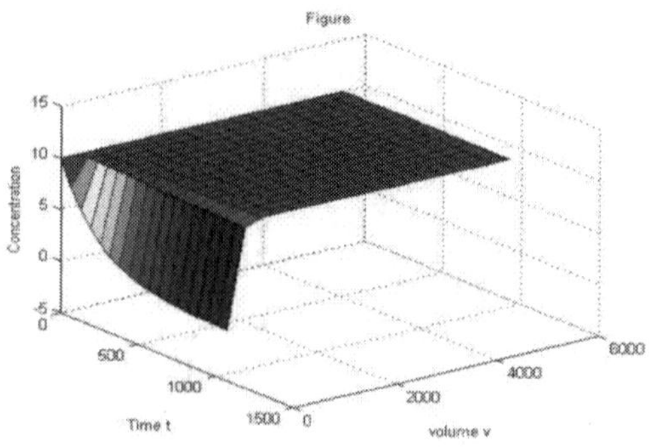

Fig. 5 Plot of concentration vs. time vs. volume of the effluent for bed height 4.1cm and flow rate 5.2 ml/min for fixed bed model

3.2. Reduced lumped diffusion model

Fig 6-8 shows the reduced lumped diffusion model plots as concentration, time and volume of effluent (using MATLAB) for bed height 4.1 cm at 1.6, 5.2 and 9.8 ml/min inflow rate respectively. It shows that the amount of solute adsorbed in the bed decreases as the time of adsorption increases due to decrease in the driving force as the time proceeds.

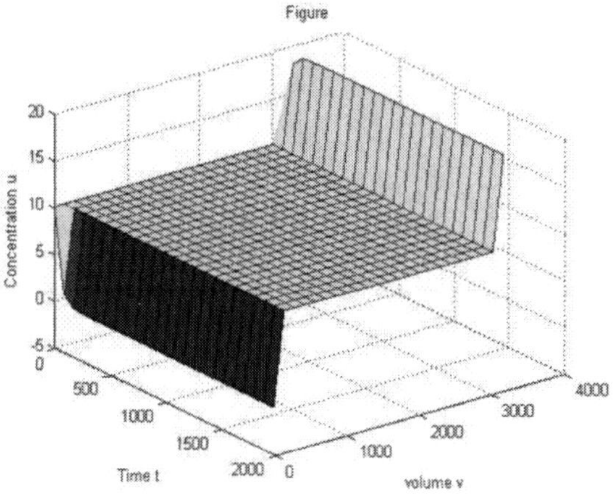

Fig. 6 Plot of concentration vs. time vs. volume of the effluent for bed height 4.1cm and flow rate 1.6 ml/min for reduced lumped diffusion model

The plot also shows that the concentration of the solute in the effluent increases as volume of the liquid passing through the column increases. A breakthrough point is reached after which the degree of adsorption gets decreased rapidly as the bed saturates. After some time the adsorption ceases completely and the concentration of the elutant becomes equal to the concentration of the effluent.

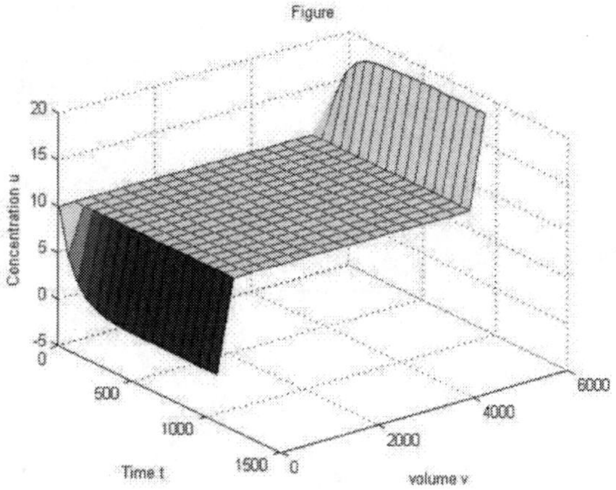

Fig. 7 Plot of concentration vs. time vs. volume of the effluent for bed height 4.1cm and flow rate 5.2 ml/min for reduced lumped diffusion model

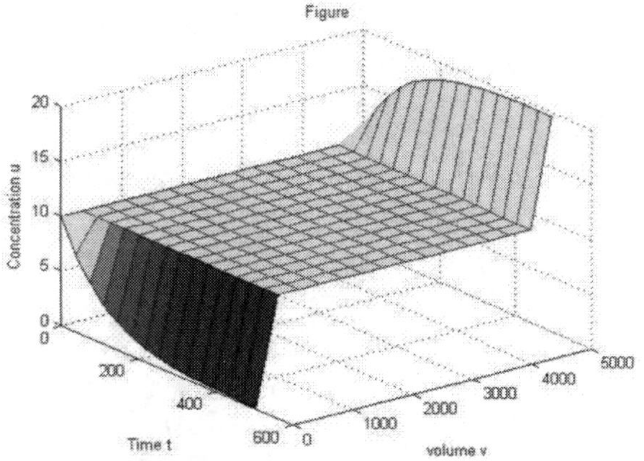

Fig. 8 Plot of concentration vs. time vs. volume of the effluent for bed height 4.1cm and flow rate 9.8 ml/min for reduced lumped diffusion model

4. Conclusions

The fixed bed model and reduced lumped diffusion model are both successfully applied to adsorption phenomenon in continuous columns for copper removal by pretreated *Aspergillus niger*. The variation in adsorption rate and the breakthrough curves are predicted for each system of the study involving various bed heights and flow rates. The amount of the solute removed is the maximum in the initial period but as time passes the bed gets saturated and the amount removed decreases as the driving force decreases. The breakthrough data obtained experimentally was comparable with data evaluated by these models.

Reference

[1] J.M. Tobin, D.G. Copper, R.J. Neufeld, Uptake of Metal Ions by *Rhizopus arrhizus*, *Appl. Environ. Microbiol.* **47** (1984) p. 821.

[2] M. Mukhopadhyay, S.B. Noronha, G.K. Suraishkumar, Copper Removal from Industrial Wastes, *Proc. of Int. Workshop on RD Frontiers in Water and Wastewater Management*, NEERI, Nagpur, India, 2006, p. 593.

[3] M. Mukhopadhyay, S.B. Noronha, G.K. Suraishkumar, Kinetic modeling for the biosorption of copper by pretreated *A. niger* biomass. *Bioresour. Technol.* **98** (2007) p.1781.

[4] M. Suzuki, Adsorption Engineering. Elsevier, Amsterdam, 1990, p.151.

[5] A. Kapoor, T. Viraraghavan, Removal of Heavy Metals from Aqueous Solutions using Immobilized Fungal Biomass in Continuous Mode, *Water Res.* **32** (1998) p.1968.

[6] B. Volesky, J. Weber, J.M. Park, Continuous-flow Metal Biosorption in a Regenerable *Sargassum* Column, *Water Res.* **37** (2003) p. 297.

[7] K. Vijayaraghavan, J. Jegan, K. Palanivelu, M. Velan, Batch and Column Removal of Copper from Aqueous Solution using a Brown Marine Alga *Turbinaria ornate, Chem. Eng. J.* **106** (2005) p.117.

[8] M. Mukhopadhyay, S. B. Noronha and G. K. Suraishkumar. Copper Biosorption in a Column of Pretreated *A. niger* Biomass. *Chem. Eng. J.* **144** (2008) p. 386.

[9] M.O. Corapcioglu, C.P. Huang, The Adsorption of Heavy Metals onto Hydrous Activated Carbon, Water Res. **21**(1987) p.1031.

[10] A.K. Akulov, S.F. Grebenschikov and Yu.I Pakhomov, Modeling the Dynamics of Adsorption on Activated Carbon-fibers, J. Appl. Chem. USSR **59** (1986) p. 1142.

Low aqueous solubility measurement (down to 10^{-12} in molar fraction) of a 5 arenes polycondensed hydrocarbons using dynamic saturation method

J. Saab[1], R. Abou Naccoul[2], I. Mokbel[2], J. Stephan[3], G. Bassil[2], N. Ouaini[1], J. Jose[2]

[1]Equipe thermodynamique de l'environnement et analyses avancées, Holy Spirit university, Science and engineering faculty, Kaslik, Lebanon ;
[2]Equipe thermodynamique et analyse en ligne, Analytical sciences laboratory, UMR5180, Claude Bernard Lyon1 university, Villeurbanne, France;
[3]Lebanese university, Science Faculty, Chemistry department, section II, Fanar, Lebanon

Abstract

Aqueous solubility of a 5 arenes polycondensed hydrocarbon was studied. Thus a dynamic saturation apparatus was realized based on the saturation of a heated fluid passing through a saturation cell. The transported solute is subsequently trapped in a specific extraction column. Back flashing method is then applied to dissolve the compound. The analysis that follows, allow the determination of solubility at the consign temperature. The validation of our apparatus was achieved by measuring the aqueous solubility of the ethyl benzene at 313.15 K. A standard deviation about 0.2 % was obtained, knowing that the literature value was done by means of a static method. Solubility of polycyclic aromatic hydrocarbons (PAH) ranged from $1.42.10^{-10}$ to $9.21.10^{-12}$ in molar fraction. The data were fitted to a 3 parameters Heidman's equation in order to calculate the thermodynamic properties of dissolution.

Introduction

In our environment, polycyclic aromatic hydrocarbons are generally present in a mixture form. Their sources are mainly anthropogenic, although episodically some natural processes such as combustion (forest fires, volcano) can be at the origin of a great production of HAPs. The most important sources come from the processes of pyrolysis and in particular from the incomplete combustion of the organic matters like cooking, waste incineration, combustion of coals and oils. The theoretical number of HAPs likely to be met is higher than 1000. According to the number of cycles, HAPs are classified into light fraction (up to 3 arenes) and heavy fraction (4 arenes or more) having different physicochemical and toxicological characteristics **[1]**.

Generally, the majority of HAPs have very low volatility, their vapor pressure decreases with the molecular mass increase. Their solubility varies between 30 mg/L for the light compounds and 10^{-4} mg/L for heaviest; this variation is mainly due to the complexity of the molecular structure**[2-3]**.

In addition, the knowledge of water solubility grants access to the estimation of the octanol- water partition coefficient. The latter explains the migration of HAPs towards lipids in order to study their bioconcentration factor**[4-5]**.

In this study we present the aqueous solubility determination of a 1,2,3,4 Dibezanthracene in ultra pure water and drinking water. The solubility was determined in the range of temperature (298.15 K - 333.15 K) at atmospheric pressure.

Experimental

Chemical

1,2,3,4-Dibenzanthracene (table 1), Benzo(a)pyrene, Ethylbenzene, used as solutes were commercials samples of the highest purity available. Their manufacturers and purities were as follow :1,2,3,4-DBA (Acros, 97%), Benzo(a)pyrene (Acros, 96%), Ethybenzene (Jansen chimica, 99%), Packing material used for the saturation cell was Chromosorb W 60/80 mesh. The Gas chromatograph used was an *Agilent 6890N* type, with a split/splitless injector and a 30 m capillary column (AT^{TM}-5, 0.25mm i.d., 0.25µm film thickness) along with an inert mass selective detector *Agilent 5975* type. Water used for aqueous solubility determination was ultra-pure. Dichloromethane and methanol used as mobile phase was supplied by Aldrich with a purity of 99.9%. All compounds were utilized without any further purification.

Table 1 : physical chemical properties of 1,2,3,4-Dibenzanthracene

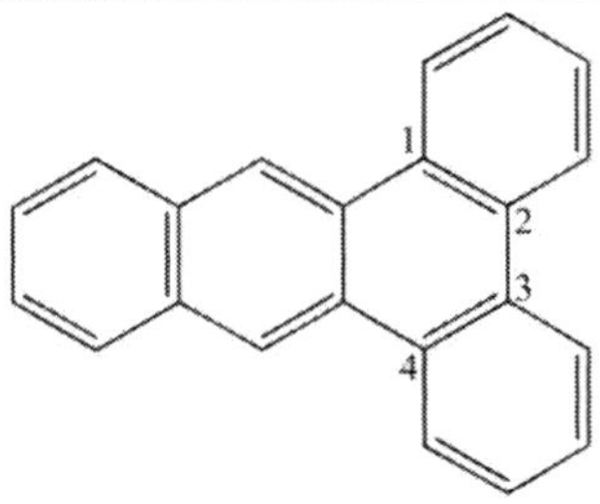	Molecular formula $C_{22}H_{14}$
	Molecular weight 278.1 g/mol
	Boiling temperature 518°C
	Melting temperature 205-207°C
	Predicted solubility in molar fraction 5.56E-11 at 25°C

Apparatus and operation

The dynamic flow saturation apparatus is shown in figure 1. The apparatus is mainly divided into three compartments. First part is the generator, where solvents are pumped. Second is the saturation compartment, where fluid is saturated. Third is the analysis compartment, where final composition is determined.

The solvent delivery system is a high pressure low stream pump able to deliver a flow of 0.1 ml/min. The pump is equipped with a double entry in order to deliver aqueous and organic solvents to the system. The aqueous solution delivered by the pump, enters the preheating coil where it acquires the consign temperature. The preheating coil and the saturation cell are kept in a chromatographic oven where temperature varies within ± 0.01 K by means of PID controller. The used oven can insure accurate temperature from 273.15 K to 673.15 K. Thermocouple placed on the saturation cell allow us to verify the read value on the oven panel. Water exiting the preheating coil get to the cell that contains a stationary phase (Chromosorb W 60/80 Mesh) pre-impregnated by the studied compound. The cell is equipped by sintered stainless steel of low porosity in order to prevent the transport of the compound and even emulsion. The sintered materials are chosen with precaution. A very low porosity causes the raise of the pressure in the generator column. In contrast, a high porosity will not prevent the micelle from passing, and leading to an error by excess in the final result.

Exiting the generator column, the primary circuit, linked to V_1, passes through the refrigerator. The latter is used in order to minimize the solute affinity to water, to make the trapping system more efficient. Then, the solute is trapped subsequently in the inversed phase column, and the solute free water is recuperated and weighed. Water's mass is used in solubility determination.

The elution is insured by means of adequate solvent such as methanol, dichloromethane, toluene etc. The choice is widely dependent on the compound's structure and hydrophobicity. The solvent passes through the T switch, and directly to the inversed phase column by means of the V_2 valve. Finally, an internal standardization quantification method was performed using a gas chromatograph coupled to inert mass spectrometer. Chromatographic separations were accomplished with a 30 m capillary column (AT^{TM}-5, 0.25mm i.d., 0.25μm film thickness).

Additional precaution measures were applied when solutions containing ions were used. The salt aggregation in the presence of organic solvent, used for elution, made impossible the direct elution after the trapping process. Therefore, all internal connection had to be rinsed to eliminate all residual salt **[6]**.

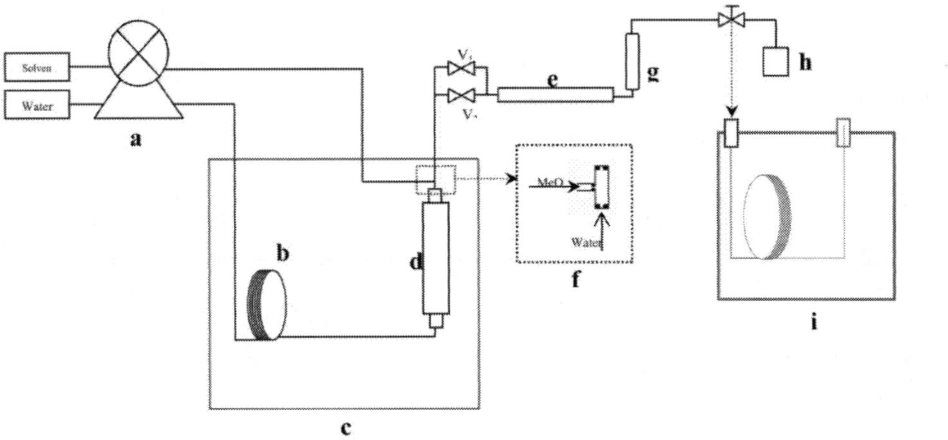

a-Solvent Delivery System ; b- Preheating Coil ; c-Oven ; d-Saturation Cell ; e-Refrigerator ; f-T switch ; g- Inversed phase column ; h- solvent recuperation ; i- Gas chromatograph coupled to a mass spectrometer detector

Validation Process

The validation of our apparatus was achieved by measuring the aqueous solubility of the ethyl benzene at 313.15 K. The latter's solubility was done by batch method. Thus, it was chosen in order to validate the process as well as the procedure **[7]**.

Table 2 : Solubility of ethylbenzene at ~ 313 K in water by dynamic method

Temperature in Kelvin	Values of solubility in mole fraction X^{sol} (dynamic method)
313.85	3.38E-05
314.05	3.26E-05
313.15	3.56E-05
313.25	3.45E-05
312.95	3.38E-05
Mean value	3.40E-05
Standard deviation (STD)	1.11E-06
(RSD) (%)	3.3

The experiment was repeated five times and the values are listed in Table 2. The experimental value (X^{sol} = 3.40E-5) was in good agreement with the ones in the literature review (X^{sol} = 3.41E-5) with a relative deviation better than 0.2%.

Solubility determination

The aqueous solubility of the 1,2,3,4 DBA was studied in multiple systems. First, the solubility was determined in pure water, second drinking water (bottled mineral water). The solubility data were expressed in mole fraction values via equation 1.

$$X = \frac{n_{DBA}}{n_{water} + n_{DBA} + n_{salt}} \quad \textbf{(1)}$$

Where X is the mole fraction and n is the mole of each species participating in the equilibrium. The presence of salts in aqueous solutions make the introduction of the n_{salt} a necessity.

The solubility of 1,2,3,4 DBA varied between 1E-10 and 9E-12 upon temperature and water composition. The evolution showed a regular ascending trend with temperature for both waters studied. Nevertheless, the solubility showed a slight variation when moving from pure to drinking water (figure 1). This effect is called salting out, and is caused by the presence of inorganic salts in the drinking water.

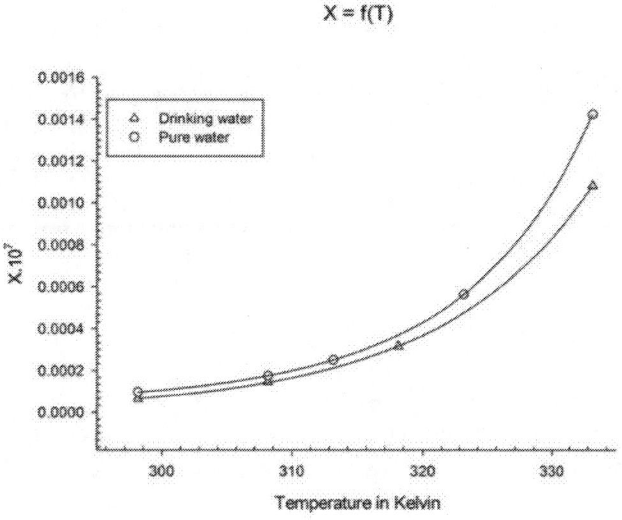

Figure 2 : Solubility variation of 1,2,3,4 DBA function of temperature and aqueous medium composition

For further thermodynamic dissolution properties determination, the solubility data were fitted to a three parameter equation called Heidman equation 2 **[8]**.

$$\ln \chi = A + \frac{B}{T_K} + C \ln(T_K) \tag{2}$$

Derivation of this equation grants determination of enthalpy of dissolution and transfer, thus entropy and Gibbs free energy evaluation.

Conclusion

Experimental technique for mass transfer evaluation was set. This technique is a reliable experimental method for thermodynamic of dissolution thus transfer study. The apparatus will allow the study of heavy organic molecules such PAHs, pesticides and many other contaminants in different mediums. Upon solubility determination, the 1,2,3,4DBA could be classified as highly hydrophobic molecule; It's vapour pressure determination will allow the evaluation of the escaping tendency from an aqueous medium through Henry's law constant.

References

[1] TEMIME-ROUSSEL, B. Contribution a l'étude de la partition des HAP entre les phases gazeuse et particulaire : validation de la technique de prélèvement par tube dénuder annulaire. **2002.** Université de Provence - Aix-Marseille I. UFR des sciences de la vie, de la terre et de l'environnement. AIX-MARSEILLE I.

[2] Altschuh, J., Bruggemann, R., Santl, H., Elchmger, G. et Piringer, O. G. Henry's law constants for a diverse set of organic chemicals: experimental determination and comparison of estimation methods. **1999.** *Chemosphere.* Vol.39 p.1871-1887.

[3] SOLEILLE, S. et BRIGNON, J. M. *HAP.* **2006.** INERIS.

[4] VIJAYALAKSHMI, K. P. et SURESH, C. H. Theoretical Studies on the Carcinogenicity of Polycyclic Aromatic Hydrocarbons **2008.** *Journal of Computational Chemistry.* Vol.29 p.1808-1817.

[5] GABET, S. Remobilisation d'Hydrocarbures Aromatiques Polycycliques (HAP) présents dans les sols contaminés à l'aide d'un tensioactif d'origine biologique. **2004.** Universite de Limoges. Limoges.

[6] Saab, J., Ouaini, N., AbouNaccoul, R., Mokbel, I., Stephan, J. and Jose, J. Thermodynamics' of dissolution for 1-naphtol N-methylcarbamate in pure, mineral and Mediterranean sea water by dynamic saturation method. **2008.** *Fluid Phase Equilibria.* Vol.264 p.156.

[7] Freire, M., Razzouk, A., Mokbel, I., Jose, J., Marrucho, I. and Coutinho, J. Solubility of Hexafluobenzene in aqueous salt solutions from (280 to 340) K. **2005.** *J. Chem. Eng. Data.* Vol.50 p.237-242.

[8] Heidman, J. L., Tsonopoulos, C., Brady, C. J. and Wilson, G. M. High-temperature mutual solubilities of hydrocarbones and water. Part II: Ethylbenzene, Ethylcyclohexane, and n-Octane. **1985.** *AIChE Journal.* Vol.31 p.376-384.

Magnetic EDTA: Coupling Heavy Metal Chelators to Metal Nanomagnets for Rapid Removal of Cadmium, Lead and Copper From Contaminated Water

Fabian M. Koehler, Departement of Chemistry and Applied Biosciences, ETH Zurich, Zurich, Switzerland

Evagelos K. Athanassiou, Department of Chemistry and Applied Biosciences, ETH Zurich, 8093 Zurich, Switzerland

Michael Rossier, Departement of Chemistry and Applied Biosciences, ETH Zurich, Zurich, Switzerland

Markus Waelle, Department of Chemistry and Applied Biosciences, ETH Zurich, 8093 Zurich, Switzerland

Robert N. Grass, Department of Chemistry and Applied Biosciences, ETH Zurich, 8093 Zurich, Switzerland

Detlef Guenther, Department of Chemistry and Applied Biosciences, ETH Zurich, 8093 Zurich, Switzerland

Wendelin J. Stark, Department of Chemistry and Applied Biosciences, ETH Zurich, 8093 Zurich, Switzerland

Nanomagnets combine high specific surface area (small particles) and ease of separation (directed movement of the particles in liquids). Until recently, the use of iron oxide particles had opened fascinating applications of magnetic separation, (Sun, Ding et al. 2006) but limited stability prohibited the use of iron oxide based nanomagnets in combination with complexing agents in acidic media which would immediately dissolve the particles. Carbon-encapsulated metal nanomagnets most recently opened access to chemically stable magnets (Grass and Stark 2006; Grass, Athanassiou et al. 2007) with strongly improved bulk magnetization (stronger magnetic forces result in faster removal). The carbon layers further provide an anchoring point for functional groups on the particles surface, accessible by diazonium chemistry (Tan and Grass 2008). Attachment of EDTA-like chelators to carbon coated metal nanomagnets results in a magnetic reagent for the rapid removal of heavy metals especially lead, cadmium and copper from solutions or contaminated waste water streams by three orders of magnitude to concentrations as low as µg/L. Figure 1. Magnetic separation test in 10 l water with 1 g carbon-coated cobalt particles. Left figure, start of magnetic separation as a permanent magnet was placed in the lower left corner. Middle figure was taken after 1 minute. After 2 minutes the stirrer was stopped and the right figure shows the end of the separation. More than 97% of the particles were removed. Grass, R. N., E. A. Athanassiou, et al. (2007). "Covalently Functionalized Cobalt Nanoparticles as a Platform for Magnetic Separations in Organic Synthesis." Angewandte Chemie Int. Ed. 46(26): 4909-4912. Grass, R. N. and W. J. Stark (2006). "Gas phase synthesis of fcc-cobalt nanoparticles." Journal of Materials Chemistry 16(19): 1825-1830. Sun, Y. B., X. B. Ding, et al. (2006). "Magnetic separation of polymer hybrid iron oxide nanoparticles triggered by temperature." Chemical Communications(26): 2765-2767. Tan, C. G. and R. N. Grass (2008). "Suzuki cross-coupling reactions on the surface of carbon-coated cobalt: expanding the applicability of core-shell nano-magnets." Chemical Communications(36): 4297-4299.

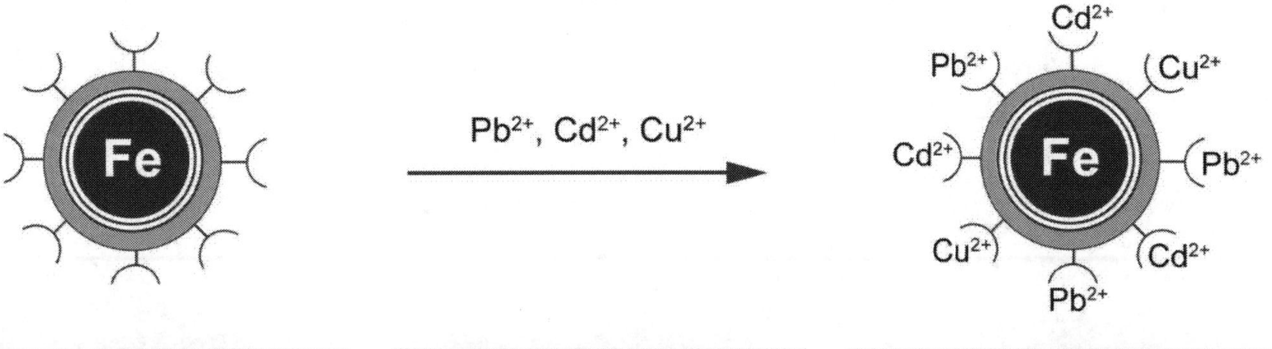

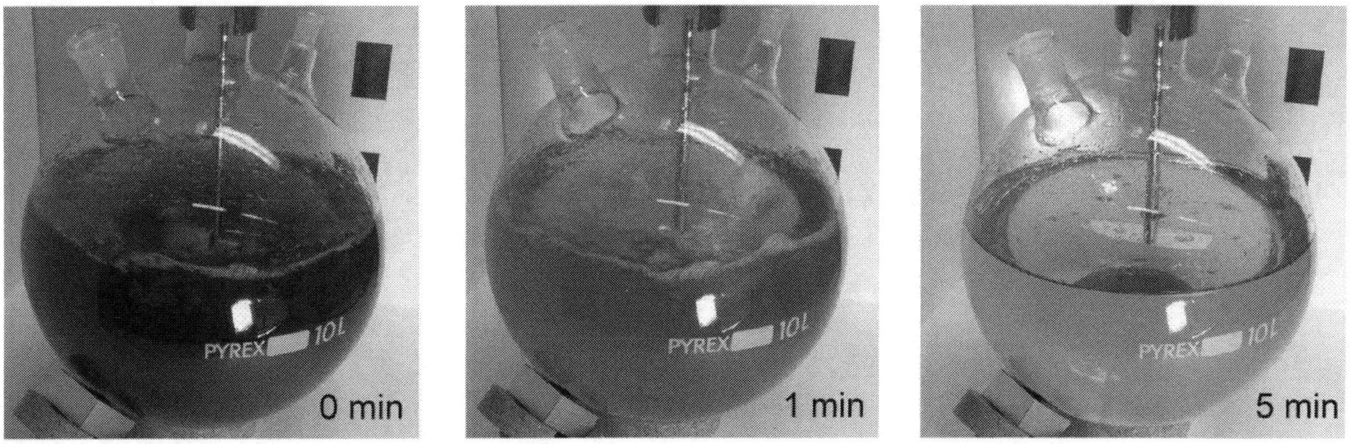

See more of this Session: Trace Contaminants in the Water
See more of this Group/Topical: Environmental Division

Pharmaceuticals in Our Water Supply

Martha J. M. Wells, Center for the Management, Utilization and Protection of Water Resources, Tennessee Technological University, Cookeville, TN

The world's water supply circulates in a cycle in which this precious commodity is constantly reused. If a pharmaceutical is not removed during wastewater treatment, and subsequently, is not removed during drinking water treatment, it has the potential to end up in our drinking water. Excretion of pharmaceuticals by humans and animals and flushing unused pharmaceuticals contribute substantially to the ultimate occurrence of pharmaceuticals in drinking water. No regulations currently require monitoring or public reporting of the presence of pharmaceuticals in our water supply. To date, no known human health effects are attributable to the presence of pharmaceuticals in drinking water. However, most toxicological impacts of chemicals are known for acute exposures at high levels, whereas, environmental exposures to chemicals tend to be chronic, over long periods of time at trace concentrations. The fundamental properties of a chemical, defined by its chemical structure, determine the success of water treatment processes in chemical removal firstly via wastewater treatment and secondly via drinking water treatment. In this presentation, setting priorities for the analysis and treatment of pharmaceuticals in the urban water cycle will be discussed within the context of research completed and underway in our laboratory. A general overview of this issue will be presented including case studies of occurrence of pharmaceuticals in our water supply.

See more of this Session: Trace Contaminants in the Water
See more of this Group/Topical: Environmental Division

Modeling Waste Water Treatment Plant Efficiency: An Integrated Approach

Mrs.A.K.Khambete[1], Dr.R.A.Christian[1]
[1] Department of civil engineering,
Faculty of Environmental Engineering,
S.V. National Institute of Technology
Surat, Gujarat, India.

***E .mail:anjali_khambete3@yahoo.co.in**

ABSTRACT:

Evaluating the efficiency of waste water treatment system often requires decision makers to consider conflicting, vague, and uncertain information. Vagueness in the perception of environmental experts for different techno-scientific parameters in linguistic terms for the specific usage coupled with imprecision in parametric data calls for the application of fuzzy modeling. This study, proposes a systematic approach to evaluate efficiency of waste water treatment system in a fuzzy environment. The approach employs three main concepts: linguistic variables, fuzzy numbers, and an analytic hierarchy process. The linguistic variables represent the degree of appropriateness of decision criteria, which are vague or uncertain. They are then translated into fuzzy numbers to reflect their uncertainties and aggregated into the final fuzzy decision value using a hierarchical structure. Through a case study, the approach is applied to evaluate the efficiency of waste water treatment system consisting of three types of treatment processes, physical, chemical and biological. The results demonstrate that the developed approach can be a useful tool for evaluating the efficiency of waste water treatment system, where criteria are vague or imprecise. The approach detailed herein might be considered as a small step in the formulation of appropriate environmental pollution control strategies

Key words: efficiency, linguistic variables, wastewater treatment plant, fuzzy multi criteria decision making (FMCDM), emission standards.

INTRODUCTION:

The term wastewater treatment has been widely used and it seems that it is well understood, but what does its efficiency mean? It is usually assumed that the criterion by which technological progress should be judged is efficiency. Efficiency is a measure of the closeness to some achievable goal. Efficiency of wastewater treatment plant can be defined as maximization of some desired output to create least impact on environmental component [1]. In general, efficiency is essentially a fraction, having numerator and a denominator. The numerator is the measurement of the quantity one is interested in and the denominator is the standard for comparing it with. The effectiveness of wastewater treatment plant has been judged on the basis of contaminant removal per unit and in terms of net improvement in receiving effluent quality.

Over a period of time depletion of water flows in rivers and the ever increasing pollution load in the natural bodies is a matter of serious concern. In the absence of adequate dilution, the wastewater treatment plants, that has been discharging its wastes within the permissible norms also contributes its share towards the overall environmental degradation. With a view to improve environmental quality and control the pollution levels the regulatory agencies have laid down the pollution norms. However, the discharges from wastewater treatment plants to stream though within the control norms do not stop rivers getting polluted. The surface water bodies have been polluted to such an extent that they have been converted to natural effluent channels. In order to protect and preserve the limited water bodies, the formulation of appropriate pollution abatement strategies is essential. One of the steps in this direction is, to run the treatment plants to its highest efficiency.

The present trend to monitor the treatment plants is on the basis of removal efficiency of the parametric values. For example, the Biochemical oxygen demand (BOD) removal efficiency of treatment plant is 75% or Chemical oxygen demand (COD) removal efficiency is 68% and so on. This is fair and acceptable when dealing with parametric removal studies. However, the parametric values are not individually separated out when discharges are released into the receiving bodies. Also, the parameters do have interrelationship with one another and so cannot be viewed independently in terms of plant efficiency.

Therefore, it is proposed to develop, a methodology for finding integrated efficiency of wastewater treatment plant on the basis of strength of wastewater. The efforts detailed herein could possibly be a first step forward in this direction. An attempt has been made to formulate a fuzzy model employing fuzzy multiple criteria decision making (FMCDM) technique to evaluate integrated efficiency for wastewater treatment plants.

Fuzzy set theory developed by Zadeh [7] is suited to make decisions in complex systems when the context of the problem is often unclear. It has been commonly used for imprecise information in a non- probabilistic sense and allows integration of information of various parameters in to the modeling and evaluation process. Methodology to evaluate efficiency for wastewater treatment plants is also a complex multi criteria integrated evaluation problem in which primary and secondary treatment processes are evaluated on the basis of its integrated pollution parameters and its importance against the removal of the same in the treatment plant.

STUDY AREA:

Data of wastewater samples for one year was collected from wastewater treatment plant situated at Surat city, applying the prescribed methodology for sampling. These samples were analyzed for seven different physical/bio-chemical wastewater quality parameters as per standard procedure (APHA) Decision were made on the basis of expert's opinion. Results of present study were taken for the FMCDM model for finding overall efficiency of wastewater treatment plant using seven parameters, pH, Chlorides, BOD, COD, SS, Temperature and TDS. Oil and grease in raw water was very low hence that parameter was not considered.

Fuzzy Methodology:

➢ The efficiency of WWTP was found by using degree of match and fuzzy rule base system. As stated by GPCB parameters as motioned above were selected. The linguistic terms for all parameters were low, medium, high and very high as shown in fig.1. Perception of experts, about the linguistic description for above selected parameters was obtained through a questionnaire. Average of each selected parameters by expert for particular linguistic variables was taken separately. This average concentration of each parameter was normalized. These normalized values are represented as shown in fig.2

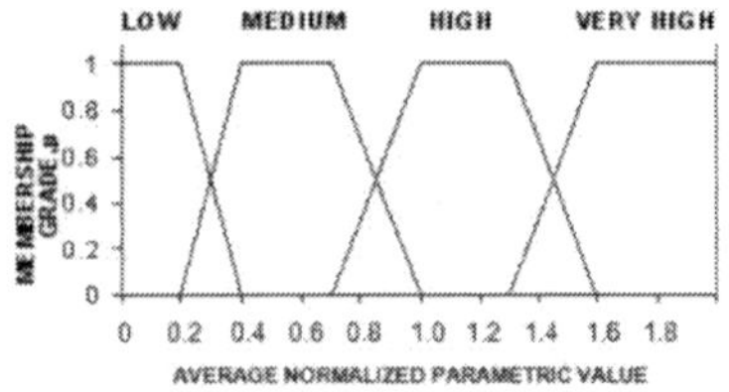

Fig.1 Graphical presentation of Normalized Parametric values for linguistic variables

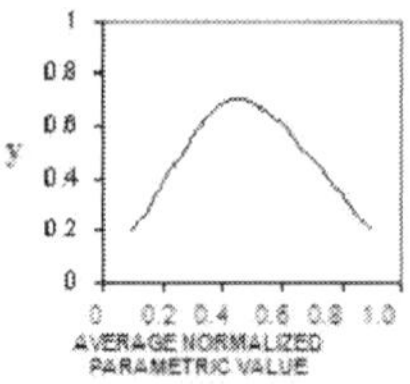

Fig.2 Normal curve

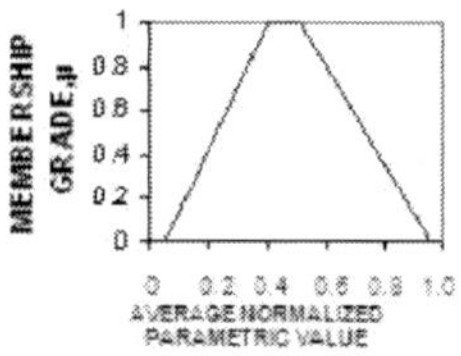

Fig.3 Normalized fuzzy curve

➢ Calculation of the fuzzy numbers for various parameters:

Randomness in the observed data could be ascertained using Gaussian distribution with the probability density function given by the following equations:

$$Y = \frac{1}{\sigma(2\pi)^{\frac{1}{2}}} e^{\frac{-z^2}{2}} \quad \text{--------(1)}$$

$$z = \frac{X - \bar{X}}{\sigma} \quad \text{-------(2)}$$

Where, Y is the frequency
σ is standard deviation of the data
X is the mid value of the class
$\bar{X}$ is mean of the data used

Using Gaussian formulae normal curves were plotted see fig2. The normal values of field data were calculated in MS Excel using 'NORMDIST' inbuilt function. The data was then stretched to make the normalized fuzzy convex curve see fig3. Same data was used for fuzzyfying the data using following equation.

$\mu_A(x) = p(x) / p(x_j)$ ---------(3)

where,

p(x) is the normal frequency obtained ,

$p(x_j)$ is the maximum normal frequency corresponding to the mean

This transforms a random variable into a convex normalized fuzzy number A with membership grade function $\mu_A(x)$.

By same method fuzzy number curves for each linguistic variable were prepared.

- **Matching between fuzzy numbers:**

The fuzzy numbers for field data (A) of parameters and the fuzzy numbers characterizing linguistic terms (A') can be matched together to arrive at a measure called a Degree of Match (DM) defined by

$$\mathbf{DM\ (A, A') = \mu_{AA'}(x)\ /\ \mu_A(x),\ x \in X} \quad \text{---------(4)}$$

Or

$$DM = \frac{\text{Shaded Area}}{\text{Dotted Area}}$$

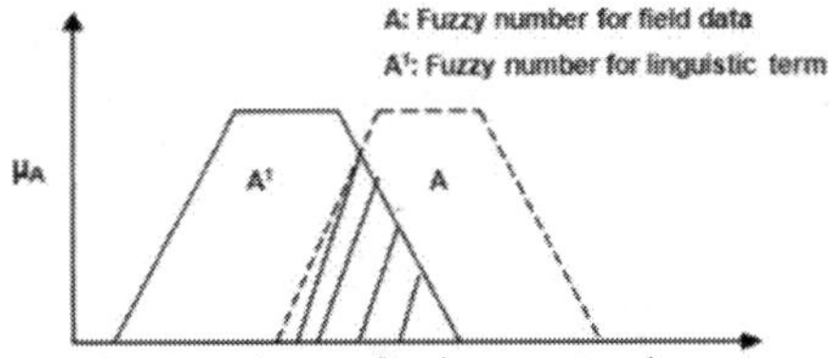

Fig.4 Average normalized parametric value

Thus, for every linguistic variable and the concerned parameters can lead to the corresponding degree of match as per the above procedure.

- **Fuzzy rule-based system:**

After obtaining the fuzzy numbers with their corresponding degree of match we need to define the rule base for the system. A set of rules is constructed for the wastewater under treatment. Each rule has the antecedent proposition connected together using AND operator, resulting in some consequences. The assertions related to its antecedent part are obtained from the experts, which are imprecise or fuzzy. Thus a fuzzy rule based system can be developed for the knowledge representation or reasoning process. Here the partial matching is allowed and the analyst can estimate the extent to which the assertion satisfies the antecedent part of the rule contrary to the rule-base system which examines as to whether the antecedent part is satisfied or not.

The fig.5 presents a hierarchical structure for finding overall efficiency of WWTP. The physical/solids status, chemical/ organic and inorganic status of waste water was judged in the first hierarchical level of knowledge base to arrive at the efficiency at PTS. The second level characterizes efficiency at STP. The third level characterized the overall efficiency of waste water. The rules stored at different hierarchical levels of the knowledge base are shown in the case study.

Conjunction and disjunction operations

If a rule has several antecedent propositions connected through a conjunctive the operator, the degree of match of the rule is given by **Total degree of match of the rule = f [DM_1, DM_2, ...DM_n]**

In which DM_i is the degree of match of the ith antecedent proposition against the corresponding assertion. The function f is **minimum** over DM_1, DM_2, ……..,DM_n.

If more than one rule has the same consequent part, they can be regarded as the rule split from disjunctive rule of which the disjuncts are the antecedent parts of the rule.

DM of disjunction = g (DM of disjunct$_1$, DM of disjunct$_2$,…….,DM of disjunct$_k$)

In which g is **maximum** over all disjuncts. The degree of match of the classification rule indicates the certainty value of classification. The greater the degree of match, the greater is the possibility that the object is classified in that class.

The rules are processed using these operators and as per the hierarchical structure fuzzily describing the efficiency of WWTP. The optimal acceptance strategy is usually that for which the degree of assertion is the maximum.

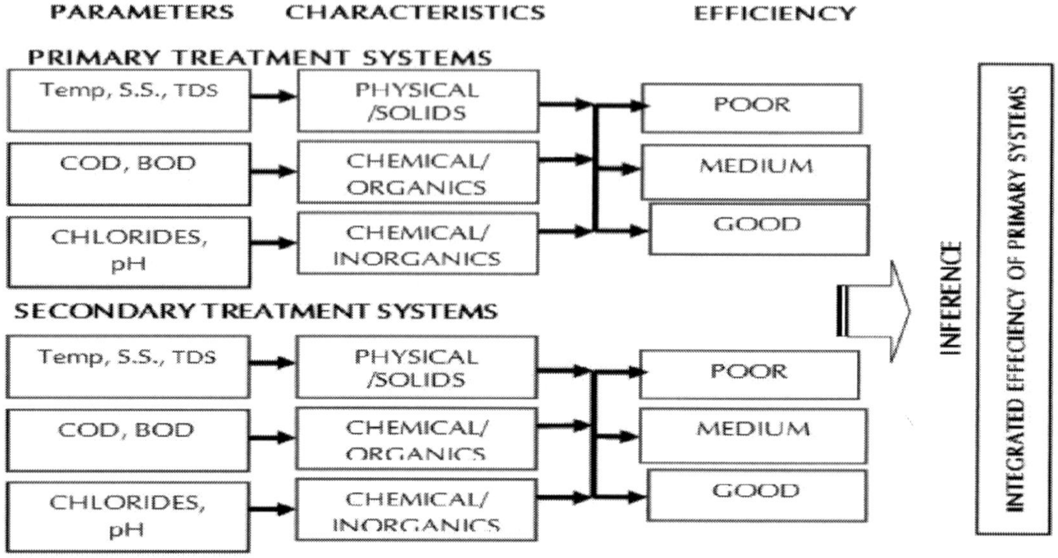

Fig.5 Schematic diagram for finding overall efficency of waste water treatment plant

➢ **Case Study:**

The case study relates to the field data from WWTP in Surat city. Appendix -I shows the permissible limits for effluent discharge. These limits are observed while disposing the treated wastewater into the rivers. The table 1, 2 and 3 overleaf shows the data for BOD for waste after primary treatment, secondary treatment and raw waste respectively. The graphs 1, 2 and 3 give the overall account of the field condition for BOD in waste after primary treatment, secondary treatment and raw water respectively. From the above graphs the degree of match of the field data with linguistic variable was calculated as explained earlier. All the degree of matches pertaining to each linguistic term for all the parameters were calculated and placed in rule base in different tables. The numbers in the bracket denotes the degree of match of that particular parameter with given linguistic variable. For physical/solids, three parameters (SS, TDS and Temp) are considered. The rule base formed for Solids are 64 in numbers while for organic/inorganic rule base formed are 16 because they carry two parameters individually. Table 2 shows the strength of waste water after PTS, STS and raw wastewater in linguistic term L, M, H and VH for solids, organics and in organics. At level 2 final strength of waste water is calculated which is shown in table 3. For determining the final strength, total 64 rules are framed; column 3 in table 3 shows the final strength of treated wastewater in four linguistic terms. The degree of match for the chemical/organics was calculated based on the min-max rule. The rule base was read as (specimen for first rule) If the BOD is low (with the degree of match=0.1992) and COD is low (with the degree of match=0.1954) then the strength of waste was low (with the degree of match=0.1954, minimum of DMs). Min-Max inference pattern is followed which gave the organic strength of the waste water as marked in bold ex. L=0.1954 in column3 in table 1. Finally the strength of waste after PTS and STS was mentioned in table 2. Table 3 shows the final strength of treated waste water with low=0.00, Medium=0.3451 High=0.2131 and Very High=0.00. Same method was followed to find the strength of raw waste water which is presented in table 2. The table 3, column 3 shows removal of the pollution load due to the treatment in treatment plant.

- **Determination of overall efficiency WWTP.**

  From above results reduction in strength of wastewater from raw to treated is for Low 0.008(degree of certainty) to, M 0.1116(degree of certainty), H 0.1981(degree of certainty) and in VH 0.4250(degree of certainty).The overall removal is 0.8155(degree of certainty).The overall efficiency of treatment plant 0.5638 (degree of certainty).

- **Conclusion:**

There is no perfect methodology to evaluate the overall efficiency of the wastewater. From present study the strength of waste after treatment is good with 0.0 (degree of certainty), moderate 0.3451(degree of certainty), poor 0.02131(degree of certainty) and very poor 0.00(degree of certainty). The advantage of FMCDM approach is the transparency of the rules. That is one has a linguistic rule structure that following "translation" in to fuzzy numbers, provides decision makers with explicit quantitative and qualitative insights in judging the strength of waste water, after PTS and STS under variable solids, organic and inorganic strength of wastewater. With this method one can set the fix single number foe discharge which mat help in making decision for efficient reuse of effluent.

- **References:**

1. Arthur Okun. "Equality and efficiency: the big tradeoff". Godkin lectures. Brookings Institute, 1975. Reported by Robert M. Solow, writing in the NYR 24 Mar 1994 p.63.
2. Deshpande, A.W., D.V.Raj, P.Khanna, (1996), Fuzzy description of river water quality. paper for international conference EUFIT-1996.
3. Emery A. Coppola ; Lucien Duckstein; and Donald Davis,2002," Fuzzy Rule-based Methodology for Estimating Monthly Groundwater Recharge in a Temperate Watershed",Journal of hydrologic engineering, ASCE,Vol.7,No.4,326-335.
4. George J. Klir, Bo Yuan, "Fuzzy Sets And Fuzzy Logic – Theory And Applications"2003, aNew Delhi: Prentice Hall of India Private Limited.
5. Seongwon Seo, Toshiya Aramaki, Yongwoo Hwang Keisuke Hanaki, June2003,"Evaluation Of Solid Waste Management System Using Fuzzy Composition", Journal of Environmental Engineering,ASCE,Vol.129,No.6,520-531.
6. Verma A.K., Srividya A. and Prabhu Gaonkar S.,2006, "Selecting optimal condition monitoring technique using fuzzy multi-criteria decision making methods:A case study",Narosa Publishing House,NewDelhi.
7. Zadeh, L.A. (1965).Fuzzy sets and fuzzy logic, theory and applications, New Delhi: Prentice Hall of India Private Limited.

TABLE:1 CHEMICAL/ORGANIC STRENGTH OF WASTE WATER									
	PTS			STS			RAW WASTE WATER		
No	BOD(1)	COD(2)	Organics(3)	BOD(4)	COD(5)	Organics(6)	BOD(7)	COD(8)	Organics(9)
1.	L(0.1992)	L(0.1954)	**L(0.1954)**	L(0.1567)	L(0.3922)	L(0.1567)	L(0.0808)	L(0.0007)	L(0.0007)
2.	M(0.7752	M(0.3451)	**M(0.3451)**	M(0.45820)	M(0.3824)	**M(0.3824)**	M(0.8105)	M(0.084)	**M(0.084)**
3.	H(0.1775	H(0.317)	**H(0.1775)**	H(0.3565)	H(0.2749)	**H(0.2749)**	H(0.9724)	H(0.1300)	**H(0.1300)**
4.	VH(0)	VH(0.2131	VH(0)	VH(0.1395)	VH(0.0824)	VH(0.0824)	VH(0.2735)	VH(0.8725)	VH(0.2735)
5.	L(0.1992)	M(0.3451)	L (0.1992)	L(0.1567)	M(0.3824)	L(0.1567)	L(0.0808)	M(0.084)	**L(0.0808)**
6.	L(0.1992)	H(0.317)	M(0.1992)	L(0.1567)	H(0.2749)	M(0.1567)	L(0.0808)	H(0.1300)	M(0.0808)
7.	L(0.1992)	VH(0.2131	M(0.1992)	L(0.1567)	VH(0.0824)	M(0.0824)	L(0.0808)	VH(0.8725)	H(0.0808)
8.	M(0.7752	L(0.1954)	L(0.1954)	M(0.45820)	L(0.3922)	**L(0.3922)**	M(0.8105)	L(0.0007)	M(0.0007)
9.	M(0.7752	H(0.317)	M(0.317)	M(0.45820)	H(0.2749)	M(0.2749)	M(0.8105)	H(0.1300)	H(0.1300))
10.	M(0.7752	VH(0.2131	**VH(0.2131)**	M(0.45820)	VH(0.0824)	H(0.0824)	M(0.8105)	VH(0.8725)	H(0.8105)
11.	H(0.1775	L(0.1954)	M(0.1775)	H(0.3565)	L(0.3922)	M(0.3565)	H(0.9724)	L(0.0007)	M(0.0007)
12.	H(0.1775	M(0.3451)	M(0.1775)	H(0.3565)	M(0.3824)	H(0.3565)	H(0.9724)	M(0.084)	H(0.084)
13.	H(0.1775	VH(0.2131	H(0.1775)	H(0.3565)	VH(0.0824)	H(0.0824)	H(0.9724)	VH(0.8725)	**VH(0.8725)**
14.	VH(0)	L(0.1954)	H(0)	VH(0.1395)	L(0.3922)	**VH(0.1395)**	VH(0.2735)	L(0.0007)	M(0.0007)
15.	VH(0)	M(0.3451)	H(0)	VH(0.1395)	M(0.3824)	H(0.1395)	VH(0.2735)	M(0.084)	H(0.084)
16.	VH(0)	H(0.317)	VH(0)	VH(0.1395)	H(0.2749)	VH(0.1395)	VH(0.2735)	H(0.1300)	VH(0.1300)

TABLE .2 FINAL STRENGTH OF SOLIDS, ORGANICS AND INORGANICS AFTER PTS, STS AND RAW WASTE WATER						
	PTS			STS		
L.T	Solids (1)	Organics (2)	Inorganic (3)	Solids (4)	Organics (5)	Inorganic (6)
L	0.432	0.1954	0.0	0.0	0.3922	0.0
M	0.5787	0.3451	0.2964	0.5793	0.3824	0.3262
H	0.1587	0.1775	0.3827	0.0187	0.2749	0.6757
VH	0.0002	0.2131	0.7059	0.0	0.1394	0.0

TABLE.2 RAW WASTE WATER		
Solids (7)	Organic (8)	Inorganic (9)
0.4567	0.0808	0.2297
0.4975	0.0840	0.0990
0.425	0.1300	0.8950
0.4975	0.8725	0.0990

TABLE:3 FINAL STRENGTH OF RAW AND TREATED WASTE WATER				
No	LV (1)	Strength of raw (2)	Strength of treated wastewater (3)	Removal of pollution load (4)
1	L	0.0808	0.00	GOOD(0.0808)
2	M	0.4567	0.3451	MEDIUM(0.1116)
3	H	0.4112	0.2131	POOR(0.1981)
4	VH	0.4250	0.00	VERY POOR(0.4250)

- Appendix:1

Table: 4 GPCB LIMITS FOR DISPOSAL OF WASTEWATER IN TO SURFACE STREAM:

No	Parameters	GPCB Norms
1	pH	6.5-8.5
2	Temperature(^{0}c.c)	40
3	Total Suspended Solids (mg/L)	100
4	Total Dissolved Solids (mg/L)	2100
5	Chemical Oxygen Demand (mg/L)	100
6	Biochemical Oxygen Demand (mg/L)	30
7	Chlorides (Cl) (mg/L)	600
8	Oil & grease (mg/L)	10

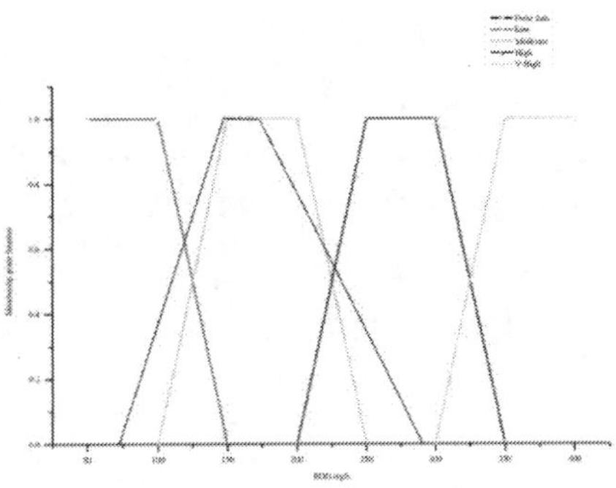

Graph.1BOD strength after PTS
DM L is 0.1992,M is 0.7752.H is 0.1775 and VH is 0.0

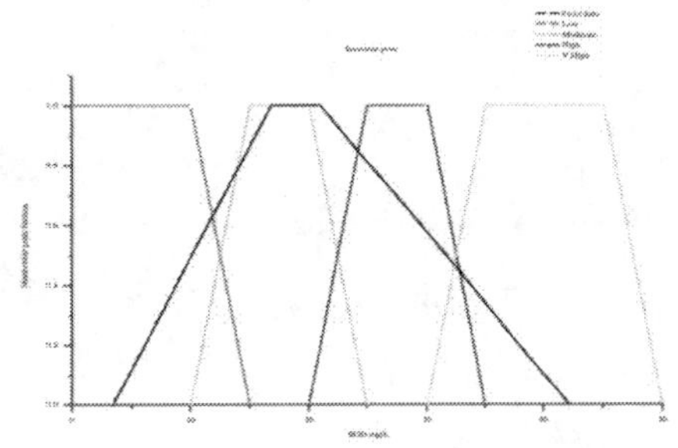

Graph.2 BOD strength after STS
DM for L is 0.1567, M is 0.4820, H is 0.3565 and VH is 0.1395

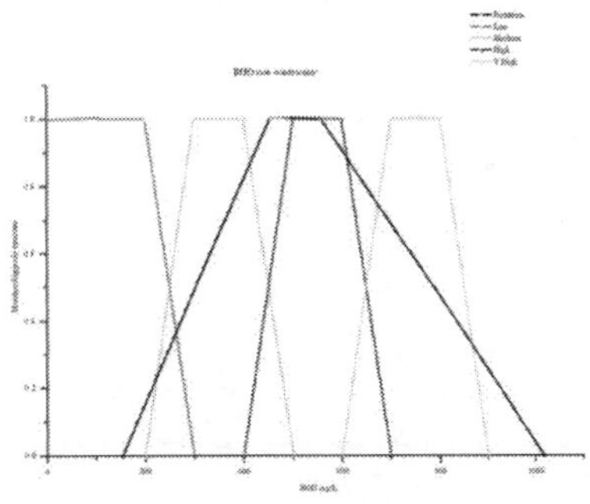

Graph.3 BOD strength in raw waste water
DM:L is 0.0808,M is 0.8105 H is 0.9724,VH is 0.2735

Electrochemical and Photochemical Wastewater Treatment

Christoph Hoislbauer, University of Technology Graz, Department of Chemical Engineering and Environmental Technology, Graz, Austria

Matthäus Siebenhofer, Department of Chemical Engineering and Environmental Technology, Graz University of Technology, Graz, Austria

Photochemical and electrochemical Advanced Oxidation Processes (AOP/EAOP) are efficient oxidation operations for degradation of persistent pollutants such as pharmaceuticals, pesticides, substituted aromatic hydrocarbons, dyes and complexing agents. The UV/H2O2 -, UV/O3- processes and the Photo-Fenton process are representative photochemical AOPs. Anodic oxidation is an electrochemical AOP. Both technologies generate reactive hydroxyl radicals which mineralize contaminants or at least convert pollutants into less toxic or biodegradable products.

The aim of the project is the comparison and documentation of approved photochemical and electrochemical processes for different substance classes. This catalogue should help to evaluate a proper technology for specific application. Criterion for comparison is the potential of degradation and mineralization, as well as the specific energy consumption.

Several substances, the selection was mainly based on the properties of substituents, were investigated. The experiments were carried out with a start concentration of 500 ppm of the investigated substances. Photochemical degradation was carried out in a Heareus Noblelight UV-reactor system (UV-RS-2), equipped with a 15 W mercury low-pressure lamp. Electrochemical degradation was performed in open cell electrolysis with diamond coated electrodes and 0.05 M Na2SO4 for conductivity reasons. The results demonstrate different rate of mineralization and degradation related to the substituent of the model substance and the absorption properties in the UV Range. In general electrochemical experiments show a simultaneous degradation and mineralization whereas photochemical experiments show degradation of substances in a first step followed by mineralization.

See more of this Session: Advanced Oxidation/Reduction Applications: Liquid/Gas Phase I
See more of this Group/Topical: Environmental Division

Mechanistic Study On the Electrochemical Decolorisation of Rhodamine 6G Using RuO2 Coated Ti DSA

Yu-Ming Zheng, Division of Environmental Science and Engineering, Natioanl University of Singapore, Singapore, Singapore

Rita Farida Yunus, Environmental Science and Engineering, National University of Singapore, Singapore, Singapore

J. Paul Chen, Division of Environmental Science and Engineering, National University of Singapore, Singapore, Singapore

In this study, the performance of electrochemical treatment of Rhodamine 6G containing water by using RuO_2 coated Ti mesh DSA is evaluated, as well as the mechanism of decolorisation process is explored. The effects of solution pH, temperature and addition of auxiliaries on the electrochemical process were examined. Results show solution pH and temperature play import roles during the electrochemical treatment of Rhodamine 6G containing water. Lower solution pH and lower temperature facilitate the decolorisation of the dye-containing water. The addition of auxiliaries dose not significantly effect the electrochemical decolorisation of the dye-containing water. Under optimal experimental condition, complete decolorisation of the dye-containing water can be obtained within 5 min. The electrochemical process can achieve 40.6 and 37.3% mineralization of Rhodamine 6G solution without and with auxiliaries within 5 min, respectively. It was found that water-insoluble foam product formation and mineralization of the Rhodamine 6G were involved during the electrochemical decolorisation of the dye-containing water. Total organic carbon, total nitrogen mass balance analyses, UV-Vis, FTIR and XPS spectroscopy were used to explore the mechanism of the electrochemical decolorisation process. UV-Vis spectra analysis shows the foam product has different molecular structure from Rhodamine 6G. FT-IR, XPS, carbon and nitrogen analyses indicate: amine and methyl groups are first detached from the dye molecule, which changes the dye from polar into non-polar and results in formation of water-insoluble substance; then partial of the water-insoluble substance are floated by the hydrogen bubble generated on cathode, and the others in the solution are further mineralized via aromatic opening and oxidizing. A conceptual model for the electrochemical decolorisation is proposed to illustrate the mechanism.

See more of this Session: Advanced Oxidation/Reduction Applications: Liquid/Gas Phase I
See more of this Group/Topical: Environmental Division

Eliminating Contaminants From Drinking Water or Fluids Using Pulsed Corona Discharges

Chinyere P. Mbachu, Chemical Engineering, Tennessee Technological University, Cookeville, TN

Pedro E. Arce, Chemical Engineering, Tennessee Technological University, Cookeville, TN

Dennis George, Center for the Management, Utilization and Protection of Water Resources, Tennessee Tech University, Cookeville, TN

The use of high oxidation methods to eliminate contaminants from drinking water or fluids is an attractive and efficient approach. Earlier work (Sharma et al., 1993) has shown the elimination of highly toxic contaminants such as phenol and more recently fumic traces (Smith, 2004) and proteins from water fluents (Ibañez, 2004). Pulsed corona high oxidation can be enhanced by adding either oxygen, in order to produce high active radicals, or iron to increase the effectiveness by the presence of the photo-Fenton reaction.

In this research proposal, we are interesting in determining the efficiency of the method in application such as sanitation of drinking water or other related fluids such as fruit juices. In particular, the technique has a low heat generation and leads to negligible or moderated temperature increases. This option is highly desirable as an alternative to high temperature methods such as pasteurization. Both organic and biological contaminants will be tested experimentally to determinate the ability of pulsed corona in eliminating them and leading to a safe drinking product.

See more of this Session: Advanced Oxidation/Reduction Applications: Liquid/Gas Phase I
See more of this Group/Topical: Environmental Division

Investigating the Photocatalytic Performance of Rapid Settling Pnipam-Titania Microcomposites

Cecil Coutinho, Chemical & Biomedical Engineering, University of South Florida, Tampa, FL

Vinay K. Gupta, Chemical & Biomedical Engineering, University of South Florida, Tampa, FL

Novel composites were prepared that were composed of titania (TiO2) nanoparticles embedded within cross-linked microgels of poly(N-isopropylacrylamide). Interpenetrating linear chains of poly(acrylic acid) were used to functionalize the nanoparticles of TiO2 for dispersal within the porous framework of the stimuli responsive microgels. The microcomposite particles showed rapid sedimentation, which is useful for the gravity separation of these particles in applications such as environmental remediation via photocatalytic degradation. The settling behavior of microcomposites was studied optically using turbidity measurenets. The settling of the polymer-titania composites occurred over minutes and was much faster than solid, impermeable spheres. As the content of TiO2 increased within the particles from 10wt% to 75wt%, the increased effective particle density led to significant decrease in the settling time. A simple mathematical model was used to interpret the optical measurements in terms of the distribution of settling velocities of the particles. These microcomposites were also very photocatalytically active and the degradation kinetics in aqueous solutions was investigated using methyl orange as a model contaminant. Kinetics of the photodegradation was evaluated by monitoring the decline in the methyl orange concentration using UV-Vis spectroscopy. Degradation of methyl orange using freely suspended titania (DegussaTM P25) was also conducted for comparison with the composites. The results reported here are a necessary input for the development of processes that could utilize composites for photocatalysis.

See more of this Session: Advanced Oxidation/Reduction Applications: Liquid/Gas Phase I
See more of this Group/Topical: Environmental Division

Kinetic Modeling of Phenol Oxidation by Fenton's Reagent in Culture Media

Pardeep Kumar, Chemical Engineering, University of Saskatchewan, Saskatoon, SK, Canada

Hossein Nikakhtari, Chemical Engineering, University of Saskatchewan, Saskatoon, SK, Canada

Mehdi Nemati, Chemical Engineering, University of Saskatchewan, Saskatoon, SK, Canada

Gordon Hill, Chemical Engineering, University of Saskatchewan, Saskatoon, SK, Canada

Phenols are commonly found in the wastewater effluents of a variety of chemical industries. Higher concentrations of phenols are toxic and resistant to biological degradation. Advanced oxidation processes (AOPs) can be combined with biological processes to effectively remove higher concentrations of phenols. In this study, chemical oxidation of phenol has been investigated by Fenton's reagent in both pure water and culture media required for the purpose of microbial growth and bioremediation. The oxidation of phenol by Fenton's reagent shows that extent of conversion and the oxidation rate of phenol in the culture media are quite different from that reported for the same reaction in pure water.

Intermediate compounds, catechol, hydroquinone and maleic acid, have been experimentally identified and quantified during phenol oxidation by Fenton's reagent. A kinetic model for phenol degradation that includes the dynamic formation and destruction of these intermediate compounds has been developed. The final oxidized product is carbon dioxide. Experiments were conducted at different initial concentrations of Fenton's reagents in both pure water and culture media. The phenol removal increased from 62% to 100% with the increase in hydrogen peroxide dosage from 1000 mg/L to 5000 mg/L at an initial phenol concentration of 2000 mg/L. The kinetic model developed fits the experimental data reasonably well and is capable of predicting the dynamic concentrations of each intermediate compound. The comparison of kinetic constants in pure water and culture media will be shown.

See more of this Session: Advanced Oxidation/Reduction Applications: Liquid/Gas Phase I
See more of this Group/Topical: Environmental Division

Formation of Humic-Like Secondary Organic Material in Aqueous Aerosol Mimics

V. Faye McNeill, Chemical Engineering, Columbia University, New York, NY

Erica L. Shapiro, Chemical Engineering, Columbia University, New York, NY

Allison N. Schwier, Chemical Engineering, Columbia University, New York, NY

Neha Sareen, Chemical Engineering, Columbia University, New York, NY

We have observed that methylglyoxal forms surface-active and light-absorbing oligomeric products in aqueous inorganic salt solutions mimicking tropospheric aerosol particles. The solutions studied were saturated in the salt of interest ((NH4)2SO4, NaCl, NH4NO3) and the concentration of organics in the mixture was maintained at 1-10% of the solute mass, consistent with ambient aerosol composition. The product mixtures were characterized using UV-Vis spectrophotometry, MALDI-MS, pendant drop tensiometry, Aerosol-CIMS, and HPLC with UV detection. The products formed in aqueous solutions containing (NH4)2SO4 or NH4NO3 absorb light at UV and visible wavelengths. The absorption spectrum consists of a nearly featureless broad absorption band that extends to longer wavelengths with increasing reaction time. Significant absorption at 550 nm is exhibited after 12 hours, with absorption at up to 700 nm developing within 2-3 days. The results suggest a mechanism involving the participation of the ammonium ion. Solutions initially containing > 0.05 M methylglyoxal exhibit significant surface tension depression compared to ammonium sulfate solutions without organics. The surface tension depression follows a Langmuir-like dependence on initial methylglyoxal concentration, with a minimum (saturation) surface tension of 39.9(+/-0.6) dynes/cm. Mixtures of methylglyoxal and glyoxal also form surface-active organic products. The formation of surface-active secondary organic material in aqueous tropospheric aerosols, besides affecting CCN activity, could limit SOA formation via heterogeneous chemistry by inhibiting the uptake of glyoxal, methylglyoxal, and other gas-phase SOA precursors.

See more of this Session: Atmospheric Chemistry and Physics - I
See more of this Group/Topical: Environmental Division

A Semi-Empirical Correlation Between Saturation Concentration and Heat of Vaporization: Application to a Thermodenuder Model

Scott A. Epstein, Department of Chemical Engineering, Carnegie Mellon University, Pittsburgh, PA

Ilona Riipinen, Department of Chemical Engineering, Carnegie Mellon University, Pittsburgh, PA

Neil M. Donahue, Department of Chemical Engineering, Carnegie Mellon University, Pittsburgh, PA

Current laboratory and atmospheric models that deal with temperature induced re-partitioning of semi-volatile mixtures often assume a constant heat of vaporization for lumped species, or else use specific values for a small set of surrogate compounds. To improve on this shortcoming, we have developed an easy to implement semi-empirical model to correlate the heat of vaporization (ΔH_{VAP}) with saturation concentration (C*). A database of Antoine coefficients was separated into compounds containing only carbon and hydrogen which were used to model primary organic aerosol (POA) and compounds containing only carbon, hydrogen, and oxygen to model secondary organic aerosol (SOA). These data at relevant temperatures were fit to a thermodynamically derived function relating ΔH_{VAP} and C* using a non-linear least squares minimization. These fits allow us to extrapolate the ΔH_{VAP} vs. C* relationship to compounds that have vapor pressures that are difficult to measure in the laboratory but are commonly present in secondary organic aerosol. Statistical analysis shows that these relationships can be used with enough certainty to improve model accuracy over the current ΔH_{VAP} treatment. We have applied the SOA fit to a thermodenuder model to reduce the model uncertainty and to improve model agreement with experimental results.

See more of this Session: Atmospheric Chemistry and Physics - I
See more of this Group/Topical: Environmental Division

Evaluating the Mixing of Organic Aerosol Components Using High-Resolution Aerosol Mass Spectrometry

Lea Hildebrandt, Chemical Engineering, Carnegie Mellon University, Pittsburgh, PA

Kaytlin Henry, Chemical Engineering, Carnegie Mellon University, Pittsburgh, PA

Jesse H. Kroll, Civil and Environmental Engineering, Massachusetts Institute of Technology, Cambridge, MA

Marissa A. Miracolo, Mechanical Engineering, Carnegie Mellon University, Pittsburgh, PA

Allen L. Robinson, Mechanical Engineering, Carnegie Mellon University, Pittsburgh, PA

Spyros N. Pandis, Chemical Engineering, Carnegie Mellon University, Pittsburgh, PA

Neil M. Donahue, Chemical Engineering, Carnegie Mellon University, Pittsburgh, PA

Small particles in the atmosphere can scatter or absorb radiation, as well as influence cloud formation and lifetime, and therefore affect climate (IPCC, 2007). The aerosol particles also affect human health by, among others, damaging the respiratory and cardiovascular systems (Dockery *et al.*, 1993; Davidson *et al.*, 2005; Pope and Dockery, 2006). Even though organic material represents about half of the fine-particle mass on average (Kanakidou *et al.*, 2005; Zhang *et al.*, 2007), the formation and properties of organic aerosol are not well understood.

Traditionally, the organic particulate matter is classified as primary or secondary organic aerosol (POA or SOA). In this classification, POA refers to compounds that are emitted as particles. SOA is formed when volatile organic compounds (VOCs) undergo one or more chemical transformations in the gas phase, forming less volatile compounds that then partition to the particle phase. SOA is further separated into anthropogenic SOA (ASOA) and biogenic SOA (BSOA). The formation of SOA is modeled as absorptive partitioning of semivolatile compounds between the gas and particle phases (Pankow, 1994; Odum *et al.*, 1996; Donahue *et al.*, 2006). According to these models, the addition of (anthropogenic) organic aerosol will enhance the formation of biogenic organic aerosol, if anthropogenic and biogenic components mix. Practically all air-quality models employ a pseudo-ideal-mixing assumption, by which organic-aerosol components from different sources mix ideally. Recent studies measuring SOA yields in the presence of POA have challenged this assumption (Song *et al.*, 2007).

Experimentally, it is very difficult to distinguish between different types of organic aerosol such as ASOA versus BSOA. Organic aerosol looks increasingly similar as it is processed in the atmosphere, and even different types of fresh organic aerosol formed in the laboratory have very similar organic mass spectra when analyzed with a unit-resolution aerosol mass spectrometer (AMS). This is partly because the aerosol in the AMS is vaporized at 600°C. The high temperature is necessary for vaporization, but it also enhances fragmentation of the organic species after electron-impact ionization, making the species lose most of their chemical distinction. In our mixing experiments, we want to measure how much aerosol is formed from each source, and we therefore need to be able to distinguish the aerosol from different sources. Our approach is to use isotopically labeled compounds and a High Resolution Time-of-Flight Aerosol Mass Spectrometer (HR-AMS) from Aerodyne, Inc. (DeCarlo *et al.*, 2006). The HR-AMS is able to distinguish between mass spectral peaks at the same *m/z*. We use this capability to distinguish ^{13}C-containing and ^{12}C-containing mass fragments.

We performed experiments in Carnegie Mellon's environmental chamber to form ASOA from isotopically labeled toluene (^{13}C-toluene), and non-labeled BSOA from a-pinene or limonene, or non-labeled POA from diesel exhaust. These experiments thus allow us to investigate the mixing of ASOA/BSOA and ASOA/POA by comparing SOA yields in the presence of organic precursor seed to the yields in the presence of inorganic, non-reactive seed (Hildebrandt *et al.*, 2009).

We use, compare and discuss three approaches to extract the labeled and non-labeled organic mass spectra. The first method identifies all of the peaks corresponding to organic fragments containing ^{13}C, and those containing only ^{12}C. The sum of these fragments should correspond to the ^{13}C-toluene aerosol versus the non-labeled organic aerosol, respectively. The second method uses positive matrix factorization (PMF) on the mass spectra, an increasingly popular method used to extract the mass spectra of different types of organic aerosol (Ulbrich *et al.*, 2008; Zhang *et al.*, 2005; Zhang *et al.*, 2007). By applying PMF to the unit mass resolution (UMR) data and the high resolution (HR) data, we can infer how much additional information the HR data provide and whether these experiments could also be conducted with a UMR instrument. Finally, the third method separates the organic mass spectra by using the known organic mass spectra of the different types of organic aerosol (e.g. the ^{13}C-toluene spectrum and the a-pinene spectrum) and reconstructing the observed, mixed organic mass spectrum as a linear combination of the two known mass spectra.

Separation of the organic mass spectra is possible and the results from the three methods agree reasonably well. We can thus calculate the mass yield of one type of aerosol in the presence of the other and compare these yields to the yields in the presence of inorganic seed. This allows us to infer the extent of mixing of the different types of organic aerosol. Understanding the mixing behavior is necessary to understand the formation of organic aerosol, and how organic aerosol concentrations will respond to emission reductions.

This work was supported by the National Science Foundation and the Environmental Protection Agency.
References:

Davidson, C. I., Phalen, R. F., and Solomon, P. A.: Airborne particulate matter and human health: A review, Aerosol Science and Technology, 39, 737-749, 2005.

DeCarlo, P. F. *et al.*: Field-Deployable, High-Resolution, Time-of-Flight Aerosol Mass Spectrometer, Analytical Chemistry, 78, 8281-8289, 2006.

Dockery, D. W., Pope, C. A., Xu, X. P., Spengler, J. D., Ware, J. H., Fay, M. E., Ferris, B. G., and Speizer, F. E.: An Association Between Air-Pollution And Mortality In 6 United-States Cities, New England Journal Of Medicine, 329, 1753-1759, 1993.

Donahue, N. M., Robinson, A. L., Stanier, C. O., and Pandis, S. N.: Coupled partitioning, dilution, and chemical aging of semivolatile organics, Environmental Science & Technology, 40, 2635-2643, 2006.

Hildebrandt, L., Donahue, N.M., and Pandis, S.N.: High formation of secondary organic aerosol from the photo-oxidation of toluene, Atmospheric Chemistry and Physics, 9, 2973-2986, 2009.

IPCC: Climate Change 2007 - The Physical Science Basis. Contribution of Working Group I to the Fourth Assessment Report of the IPCC, 2007.

Kanakidou, M. et al.: Organic aerosol and global climate modelling: a review, Atmos. Chem. Phys., 5, 1053-1123, 2005.

Odum, J. R. *et al.*: Gas/particle partitioning and secondary organic aerosol yields, Environmental Science & Technology, 30, 2580-2585, 1996.

Pankow, J. F.: An Absorption-Model Of The Gas Aerosol Partitioning Involved In The Formation Of Secondary Organic Aerosol, Atmospheric Environment, 28, 189-193, 1994.

Pope, C. A., and Dockery, D. W.: Health Effects of Fine Particulate Air Pollution: Lines that Connect, Journal of the Air and Waste Management Association, 56, 709-742, 2006.

Song, C. *et al.*: Effect of hydrophobic primary organic aerosol on secondary organic aerosol formation from ozonolysis and a-pinene, Geophysical Research Letters, 34, 2007.

Ulbrich, I. M., Canagaratna, M. R., Zhang, Q., Worsnop, D. R., and Jimenez, J. L.: Interpretation of organic components from positive matrix factorization of aerosol mass spectrometric data, Atmospheric Chemistry and Physics, 9, 2891-2918, 2009.

Zhang, Q., Worsnop, D. R., Canagaratna, M. R., and Jimenez, J. L.: Hydrocarbon-like and oxygenated organic aerosols in Pittsburgh: insights into sources and processes of organic aerosols, Atmospheric Chemistry and Physics, 5, 3289-3311, 2005.

Zhang, Q. *et al.*: Ubiquity and dominance of oxygenated species in organic aerosols in anthropogenically-influenced Northern Hemisphere midlatitudes, Geophysical Research Letters, 34, 2007.

See more of this Session: Atmospheric Chemistry and Physics - I
See more of this Group/Topical: Environmental Division

The Impact of Elemental Composition On Volatility Profiles in Organic Aerosol Particles

Sean H. Kessler, Chemical Engineering, Massachusetts Institute of Technology, Cambridge, MA

Jesse H. Kroll, Civil and Environmental Engineering, Massachusetts Institute of Technology, Cambridge, MA

Although the presence of organic aerosols in the atmosphere has been implicated in matters pertaining both to human health and to the global climate, the actual role played by these aerosols is not yet fully understood, due to the immense number of components that may be found in the atmosphere. The development of a method for separating components of an aerosol particle allows us to readily analyze the correlation between elemental composition and the volatility of the organic components. A flow-tube reactor is employed to generate particles at varying degrees of oxidation. This stream of particles is then passed through a thermal denuder (TD) at a set of temperatures such that we are able to determine via high-resolution time-of-flight aerosol mass spectrometry (HR-ToF-AMS) the elemental composition of organic components that are transferred to the vapor phase over a focused temperature range. In this paper, a set of submicron (~160 nm) particles of squalane (a highly branched C_{30} alkane) is subjected to oxidation and sent to the TD/AMS system. Results will include comparison with model predictions of the relative breakdown of molecules by degree of oxidation, as well as discussion of implications for future modeling applications in which a range of volatility sets is used to describe the time-evolution of both primary and secondary organic aerosols.

See more of this Session: Atmospheric Chemistry and Physics - I
See more of this Group/Topical: Environmental Division

New Insights Into the Ozonolysis of Surface-Adsorbed Polycyclic Aromatic Hydrocarbons

V. Faye McNeill, Department of Chemical Engineering, Columbia University, New York City, NY

Sophie Chu, Department of Chemical Engineering, Columbia University, New York City, NY

Sophia Sands, Department of Chemical Engineering, Columbia University, New York City, NY

Michelle Tomasik, Department of Chemical Engineering, Columbia University, New York City, NY

Neha Sareen, Department of Chemical Engineering, Columbia University, New York City, NY

Paul Lee, Department of Chemical Engineering, Columbia University, New York City, NY

The environmental fate of polycyclic aromatic hydrocarbons (PAHs), from small species consisting of a few fused aromatic rings to man-made graphene-based nanomaterials, is an issue of pressing importance in ecology and public health. We have used a combination of laboratory experiments and ab initio simulations to gain insight into the mechanisms of ozone oxidation of surface-bound molecules in the PAH family, including anthracene, benzo[a]pyrene (BaP), and graphene sheets. We performed geometry optimizations and vibrational frequency calculations using density functional theory (B3LYP/6-31g**) for anthracene/O3 and BaP/O3 oxidation products and intermediates that have been suggested in the literature. We have identified oxidation pathways that require significant distortion of the PAH molecule from the native planar geometry in order to be energetically favorable, and others that do not. A pathway requiring distortion may be unfavorable for species interacting strongly with a rigid substrate. Our results provide new insight into the role of substrate in PAH oxidation. We will discuss these results in the context of the experimental data from our laboratory and in the literature.

See more of this Session: Atmospheric Chemistry and Physics - I
See more of this Group/Topical: Environmental Division

Kinetic and Product Study of the Gas-Phase Reactions of OH Radicals with Trans-2-Hexenal, Trans-2-Octenal and Trans-2-Nonenal

Tingting Gao, Civil, Environmental and Sustainable Engineering, Arizona State University, Tempe, AZ

Jean M. Andino, Chemical Engineering, Arizona State University, Tempe, AZ

Carlos Rivera, Chemistry Department, Universidad Metropolitana, San Juan, PR

Misaela Francisco Márquez, Chemistry Department, Universidad Autónoma Metropolitana, Mexico D. F, Mexico

The rate constants of the gas-phase reaction of OH radicals with *trans*-2-hexenal, *trans*-2-octenal and *trans*-2-nonenal were determined at 298±2 K and atmospheric pressure using the relative rate technique. Two reference compounds were selected for each rate constant determination. The relative rates of OH + *trans*-2-hexenal versus OH + 2-methyl-2-butene and b-pinene were 0.452±0.054 and 0.530±0.036, respectively. These results yielded an average rate constant for OH + *trans*-2-hexenal of (39.3±1.7) $*10^{-12}$ $cm^3 molecule^{-1} s^{-1}$. The relative rates of OH+ *trans*-2-octenal versus the OH reaction with butanal and b-pinene were 1.65±0.08 and 0.527±0.032, yielding an average rate constant for OH + *trans*-2-octenal of (40.5±2.5) $*10^{-12}$ $cm^3 molecule^{-1} s^{-1}$. The relative rates of OH+ *trans*-2-nonenal versus OH+ butanal and OH + *trans*-2-hexenal were 1.77±0.08 and 1.09±0.06, resulting in an average rate constant for OH + *trans*-2-nonenal of (43.5±3.0) $*10^{-12}$ $cm^3 molecule^{-1} s^{-1}$. In all cases, the errors represent 2 sigma (95% confidential level) and the calculated rate constants do not include the error associated with the rate constant of the OH reaction with the reference compounds. The rate constants for the hydroxyl radical reactions of a series of *trans*-2-aldehydes were compared with the values estimated using the structure activity relationship. The mechanism of hydroxyl radial reaction with *trans*-2-hexenal was also investigated experimentally. The major product of the OH+ *trans*-2-hexenal reaction is butanal.

See more of this Session: Atmospheric Chemistry and Physics - I
See more of this Group/Topical: Environmental Division

A Coating Based Approach to Zeolite Membranes: Hydrogen Separation by MCM-22/Silica Composite Membranes Made Using Layer-by-Layer Deposition

Jungkyu Choi, Department of Chemical Engineering and Materials Science, University of Minnesota, Minneapolis, MN

Michael Tsapatsis, Department of Chemical Engineering and Materials Science, University of Minnesota, Minneapolis, MN

The realization of zeolite membranes in industry has been limited largely because the reproduction of ultra-thin and defect-free zeolite membranes on a large scale is quite challenging. Moreover, any probable methodology such as in-situ or secondary (seeded growth) growth for membrane manufacturing requires considerable time-consuming procedures, especially because of conventional batch-wise hydrothermal growth. Recently, composite zeolite/silica inorganic films with hierarchical structures have been reported .[1-4] These films were physically formed via a one or multiple time spin coatings or castings of nanocrystal suspension and silica precursor mixture. No conventional hydrothermal growth was involved during film manufacturing.

We have also introduced a coating-based strategy, i.e., layer-by-layer deposition, for the preparation of c-oriented MCM-22-zeolite/silica composite films that showed a potential for gas separations and corrosion prevention.[5] MCM-22 layers were deposited in the composite film in such way that they form an array of transport limiting 6 member ring (MR) apertures which allowed the passage of a small molecule like hydrogen across the membrane, while the transport of larger molecules was hindered.

Here we report improvements in the layer-by-layer deposition method which result in membranes with high (larger than 100) H2/N2 selectivity.

References

1. R.A. Farrell, N. Petkov, H. Amenitsch, J.D. Holmes, and M.A. Morris, J. Mater. Chem., 2008. 18(19): p. 2213-2220.

2. N. Petkov, M. Hoelzl, T.H. Metzger, S. Mintova, and T. Bein, J. Phys. Chem. B, 2005. 109(10): p. 4485-4491.

3. N. Petkov, S. Mintova, B. Jean, T.H. Metzger, and T. Bein, Chem. Mater., 2003. 15(11): p. 2240-2246.

4. T. Seo, T. Yoshino, Y. Cho, N. Hata, and T. Kikkawa, Jpn. J. Appl. Phys., Part 1 2007. 46(9A): p. 5742-5746.

5. J. Choi, Z. Lai, S. Ghosh, D.E. Beving, Y. Yan, and M. Tsapatsis, Ind. Eng. Chem. Res., 2007. 46(22): p. 7096-7106.

See more of this Session: Hydrogen Separation/Storage
See more of this Group/Topical: Environmental Division

Hydrogen Storage Using Mg-Mixed Metal Hydrides

Canan Acar, Department of Chemical and Biological Engineering, Illinois Institute of Technology, Chicago, IL

Alan Zdunek, Department of Chemical and Biological Engineering, Illinois Institute of Technology, Chicago, IL

Hamid Arastoopour, Department of Chemical and Biological Engineering, Illinois Institute of Technology, Chicago, IL

Persistent urban air pollution, demand for zero or low emission vehicles and the need to reduce foreign oil imports increase the need for renewable energy concept. Hydrogen is a promising material since it is an environmentally harmless energy carrier. The lack of satisfactory hydrogen storage systems which are safe, cheap and simple is one of the main problems for the transition to a hydrogen-based energy system. Mg-based alloys show potential as hydrogen storage materials because of the high gravimetric density of MgH_2 (7.6 wt. %), as well as its abundant supply and low cost as a raw material. However, they exhibit high enthalpies of formation, poor hydrogenation/dehydrogenation kinetics, poor charge/discharge cycling stability and high temperature requirements. Also, MgO formation and a high dissociation barrier of clean magnesium surfaces are other major problems with Mg-based alloys. These problems could be solved by mixing Mg with another chemical component, adding a small amount of a catalytically active metal on the surface of the Mg clusters, or by using small particle sizes [1]. The addition of $LaNi_5$ to MgH_2 reduced the temperature of hydrogen absorption, accelerated the kinetics at room temperature and yielded higher hydrogen storage capacity at elevated temperatures [2]. Furthermore, Barkhordarian et al showed that Nb_2O_5 is the most effective catalyst for the hydrogen sorption reaction of Mg known so far [3].

In this work, Mg-$LaNi_5$ and Mg-Nb_2O_5 composite systems were studied and compared to obtain a better understanding of hydrogen storage capacity, adsorption/desorption rates and to observe the effect of different transition metals on hydrogenation kinetics. In order to quantify the effect of particle size on hydrogen storage capacity, different ball-milling times were utilized to prepare the samples before testing. Hydrogen storage capacities of the materials are measured by performing adsorption/desorption experiments at different temperatures and generating pressure-composition isotherm data. The results of the experiments are described and compared with previous data from the literature.

References

1 David, E., "*An overview of advanced materials for hydrogen storage*", Journal of Material Processing Technology 162-163 (2005) 169-177.

2 Au, M., "*Hydrogen storage properties of magnesium based nanostructured composite materials*", Materials Science and Engineering B 117 (2005) 37-44

3 Barkhordarian G. et al, "*Fast hydrogen sorption kinetics of nanocrystalline Mg using Nb_2O_5 as catalyst*", Scripta Materialia 49 (2003) 213-217

See more of this Session: Hydrogen Separation/Storage
See more of this Group/Topical: Environmental Division

Complex Borohydride for Reversible Hydrogen Storage

Pabitra Choudhury, Chemical and Biomedical Engineering, University of South Florida, Tampa, FL

Nianthrini Balakrishnan, Chemical and Biomedical Engineering, University of South Florida, Tampa, FL

Venkat R. Bhethanabotla, Chemical and Biomedical Engineering, University of South Florida, Tampa, FL

Elias Stefanakos, Electrical Engineering, University of South Florida, Tampa, FL

Materials that are light weight, low cost and have high hydrogen storage capacity are essential for on-board vehicular applications. Some reversible complex hydrides are alanates and amides but have low reversible capacity which is inadequate for onboard application. High capacity, light weight, reversibility and fast kinetics at lower temperature are the primary desirable aspects for any type of hydrogen storage material. Borohydride complexes as hydrogen storage materials have recently attracted great interest. Understanding the above parameters for designing efficient complex borohydride materials requires modeling across different length and time scale. In this study, a density functional theory study is performed to find the hydrogen storage properties for the reversible complex borohydride material $(LiNH_2)_2$ - $LiBH_4$ - (MgH_2) system. The structural stability of the reactants and products involved in the reaction steps are confirmed and thermodynamic properties such as enthalpy of reactions and Gibbs energy are also calculated of these reactions. These theoretically predicted reaction steps compare well with the experimental proposed reaction pathways. Hydrogen storage capacity predicted for this complex borohydride will also be presented from the de-sorption behavior observed at different temperatures.

See more of this Session: Hydrogen Separation/Storage
See more of this Group/Topical: Environmental Division

Methylcyclohexane Dehydrogenation — A Convenient Way for Hydrogen Storage

Muhammad R. Usman, Faisal S. Alhumaidan, David L. Cresswell and Arthur A. Garforth*
*arthur.garforth@manchester.ac.uk

School of Chemical Engineering and Analytical Science, The University of Manchester, PO BOX 88, M60 1QD, United Kingdom

Introduction:

Over the years, many alternate solutions have been proposed to replace or to assist the fossil fuels in order to curtail the environmental damage and future fuel shortage. One such solution is to use hydrogen as fuel in an internal combustion engine or a fuel cell. Hydrogen is a very clean fuel producing nothing other than water. However, hydrogen is not available naturally and must be produced by refining some hydrogen source. Also gaseous in nature and with extremely low critical temperature (−253 °C), it has problems with storage and transportation. A successful transportation system based on the hydrogen economy is thus dependent entirely upon the practical, safe and economical production, storage and transportation of hydrogen and its appropriate utilization to produce the required power. Organic hydrides, such as methylcyclohexane (MCH) are promising as recyclable hydrogen precursors. MCH is a stable, relatively involatile, contains high hydrogen mass (6.5% by wt.) and its product when dehydrogenated, toluene, is capable of undergoing easy hydrogenation to return back to MCH. The Methylcyclohexane-Toluene-Hydrogen system is frequently known as MTH system. The concept of the MTH system is depicted in the Fig.1. Vehicles are filled with MCH fuel instead of conventional fuel like gasoline. The dehydrogenation reactor installed within the vehicle itself produces hydrogen and toluene when fed with reactant MCH. The hydrogen so produced can be used in an internal combustion (IC) engine or in a stack of fuel cells to generate enough power that can be effectively utilized to move the vehicle around. The by-product toluene is sent to a storage tank within the vehicle. On the filling station, the toluene is pumped out while the MCH is pumped in to recharge the vehicle for further travel. From the filling station, the tanker that brings MCH to the filling station takes the toluene away to the hydrogenation plant where it is hydrogenated back to MCH. The hydrogen required for the hydrogenation is produced by water electrolysis using any available, economical and sustainable source like Sun. It is important to note that the existing motor fuel infrastructure for transportation and dispensation of liquid fuel (MCH) can effectively be utilized. The heart of the MTH system is the on-board catalytic dehydrogenation of MCH that requires an efficient dehydrogenating catalyst. The acceptable conversion, high selectivity and long life are the challenges to the most suitable catalyst required for the MTH system. In this study, an effort is made to find a suitable enough catalyst and a number of commercial and in-house catalysts were analysed for their activity, selectivity, and stability. Table 1 shows the list of the catalysts, with their relevant physical properties, that are used in our study.

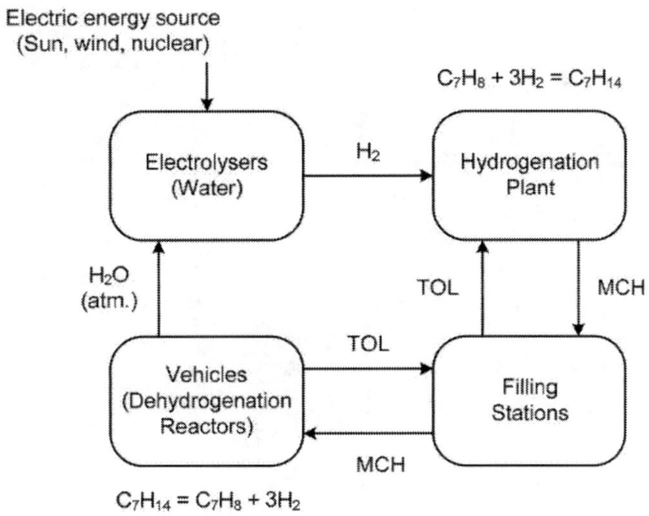

Fig. 1 Concept of MTH System

Experimental:

Experiments were performed in a conventional fixed bed tubular reactor shown in Fig. 2. All the in-house catalysts were made by wet impregnation of a γ-Al_2O_3 support in the form of extrudates of the size 3.2 mm (from Alfa-Aesar). The support was crushed to the required particle size and was impregnated by the right concentration of the relevant metal salt. Hexachloroplatinic acid was used for Pt-catalysts, while $PdCl_2$ for Pd metal and ammonium perrhenate was employed for Re metal loading. Methylcyclohexane was obtained from Sigma-Aldrich and had a purity of 99 wt%. The dehydrogenation experiments were performed under integral conditions at varying conditions of reactor wall temperature, pressure and feed compositions. The major products were analyzed by GC technique while by-products were identified and analyzed by GC-MS. In the case of important by-products the identification was confirmed by injecting the known standards.

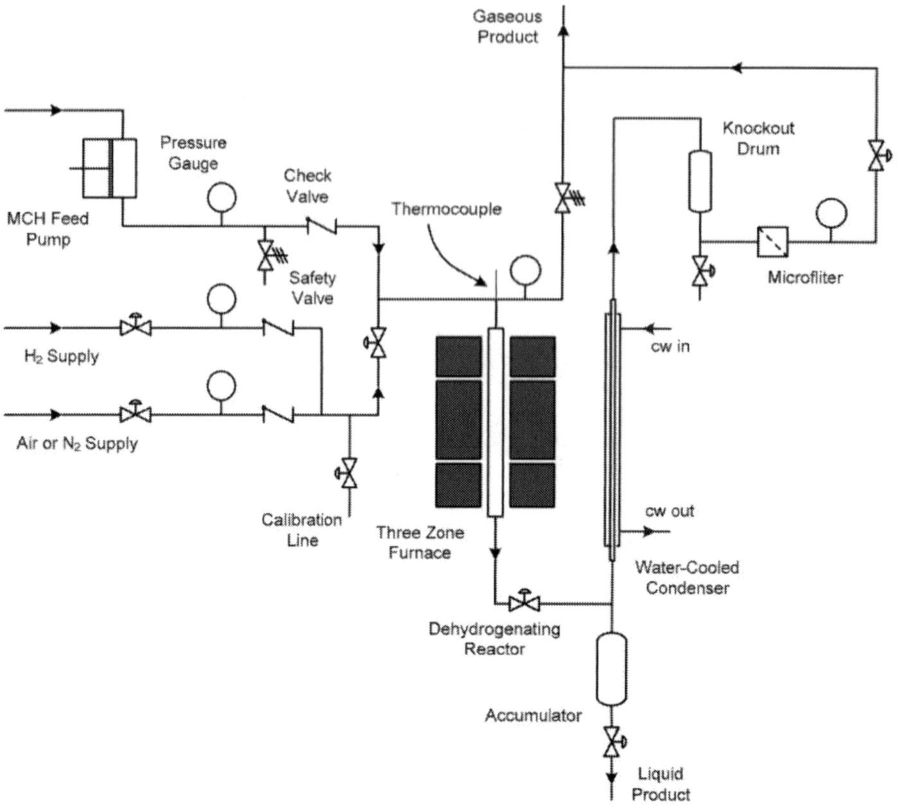

Fig. 2 Experimental setup

Table 1 Catalysts used in the study

Catalyst	Assay	Supplier	BET ($m^2 \cdot g^{-1}$)	Pore Volume	Pore Diameter (Å)
CAT-A	1.0 wt% Pt/γ-Al_2O_3	Commercial	302.3	0.52	67.2
CAT-B	0.3 wt% Pd/γ-Al_2O_3	In-house	220	0.62	70
CAT-C	0.3 wt% Pt/γ-Al_2O_3	In-house	220	0.62	70
CAT-D	1.0 wt% Pt/γ-Al_2O_3	In-house	220	0.62	70
CAT-E	0.3 wt% Pt-Pd/γ-Al_2O_3	In-house	220	0.62	70
CAT-F	0.3 wt% Pt-Re/γ-Al_2O_3	In-house	220	0.62	70

Results and Discussion:

Activity:

Fig. 3 shows the effect of the W/F_{A0} on the conversion of the methylcyclohexane, where W is the mass of the catalyst in g and F_{A0} is the initial molar flow rate of MCH in mol per h. All the catalysts showed comparable activities, however, 0.3wt% Pd/γ-Al_2O_3 showed very low activities compared to the others under the same conditions. The results reveal out that the Pt metal is much better catalyst for the dehydrogenation of MCH than the Pd metal. Equal amounts of the addition of Re or Pd to the Pt catalyst almost had no effect on the activity of the catalyst. Increasing the loading of Pt on the support from 0.3wt% to 1.0wt% had a slight but negative effect on the activity of the catalyst. This may, however, be explained when considering metal dispersion and the orientation of the metal atoms exposed to the surface. Although not shown here, co-impregnation and sequential impregnation of the bimetallic Pt-Re showed comparable activities.

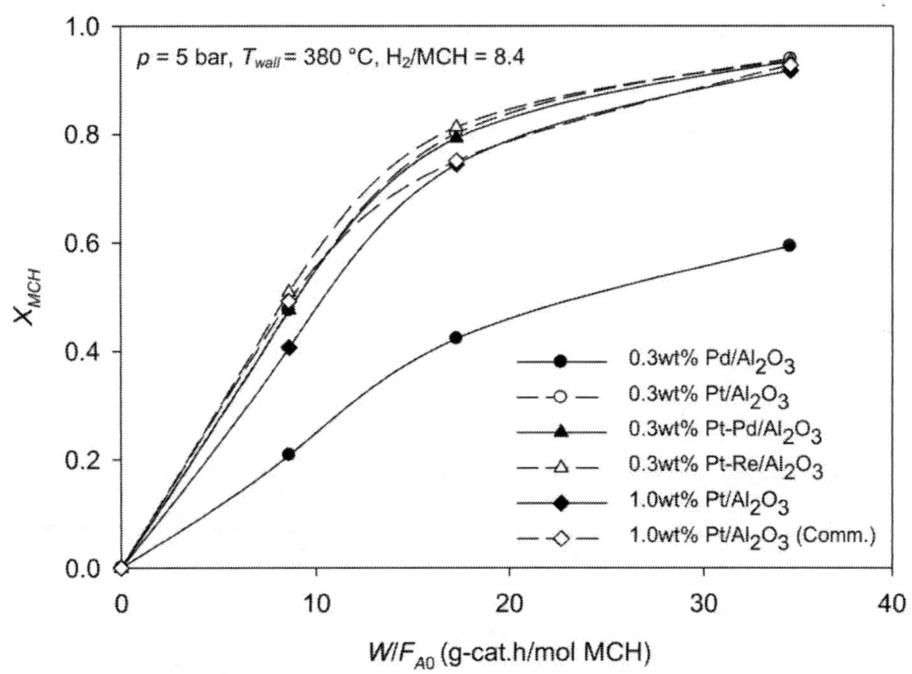

Fig.3 Comparative study of the effect of space time on the conversion of MCH for five in-house and one commercial catalyst, W = weight of the catalyst, g; F_{A0} = initial molar flowrate of MCH, g/h; X_{MCH} is the total conversion of MCH

Selectivity:

The dehydrogenation reaction under all conditions and with all the catalytic samples was found very clean with toluene as the only major by-product. Although the by-products were found in very small quantities, but it was considered important to analyse them qualitatively and quantitatively. Upon analyzing with GC-MS a number of by-products were identified. The somehow important by-products that are detected were benzene, heptane, dimethylcyclopentanes, 4-methylcyclohexene, naphthalenes, fluorenes and biphenyls. Benzene, however, found to be the most important by-product. Fig. 4 describes the graphical relationship between the total selectivity of the by-products and the W/F_{A0}. Total by-product selectivity is defined as the difference between the moles of MCH consumed per mole of MCH fed and moles of the major by-product, toluene, formed per mole of MCH fed. So the smaller the number the more selective is the catalyst. In-house 1.0wt% Pt/Al_2O_3 was found to be the least selective following just after $Pt\text{-}Re/Al_2O_3$ which had an intermediate selectivity. The remaining catalysts had comparable selectivity.

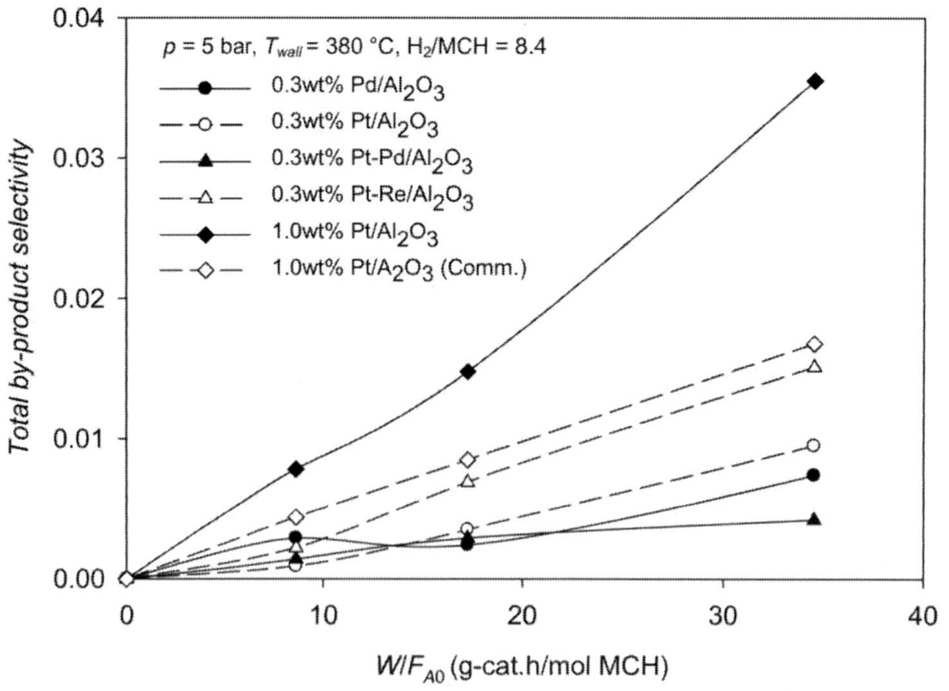

Fig.4 The effect of space time on the total selectivity of the by-products for the catalyst used in this study. Total selectivity of by-products is defined as the conversion of MCH less the total toluene formed per mole of MCH fed, W = weight of the catalyst, g; F_{A0} = initial molar flowrate of MCH, g/h; X_{MCH} is the total conversion of MCH

Stability:

A highly active and highly selective catalyst with active and/or selective life of only few hours or days is of no use for the MTH-system, where the motorist expect the catalyst to work for months without any regeneration. In the long term deactivation tests, catalysts was subject to pure methylcyclohexane at 1 bar and 380 °C (severe conditions of experimentations as no hydrogen in the feed and the highest conversion was expected), reactor wall temperature with W/F_{A0} equal to 34.55 g-cat·h/mol MCH. The long term deactivation was performed for 600 h and the results of deactivation for the selected catalysts are plotted in Fig. 5. Poor stability was acquired by most of the catalysts at the end of 600 h period. Commercial 1.0wt% Pt/Al_2O_3 catalyst showed a rapid decline in activity in the initial 200 h and then showed a rather stable character. Addition of Pd or Re to Pt-loaded catalyst was not found beneficial rather Pt-Pd catalyst showed a trend that could lead the catalyst to the complete death in the very next 50 h. Only the in-house monometallic Pt metal catalysts showed acceptable endurance. Out of the two, 1.0wt% Pt/Al_2O_3 was found more stable with more life. Although 0.3wt% Pt/Al_2O_3 might have better dispersion of Platinum metal clusters, but due to less loading it might have fewer active sites remaining after initial deactivation. In-house 1.0wt% Pt/Al_2O_3 showed extraordinary stability and the drop in conversion after 600 h that is 25 days was only 7.6 % of the conversion after 1 h period.

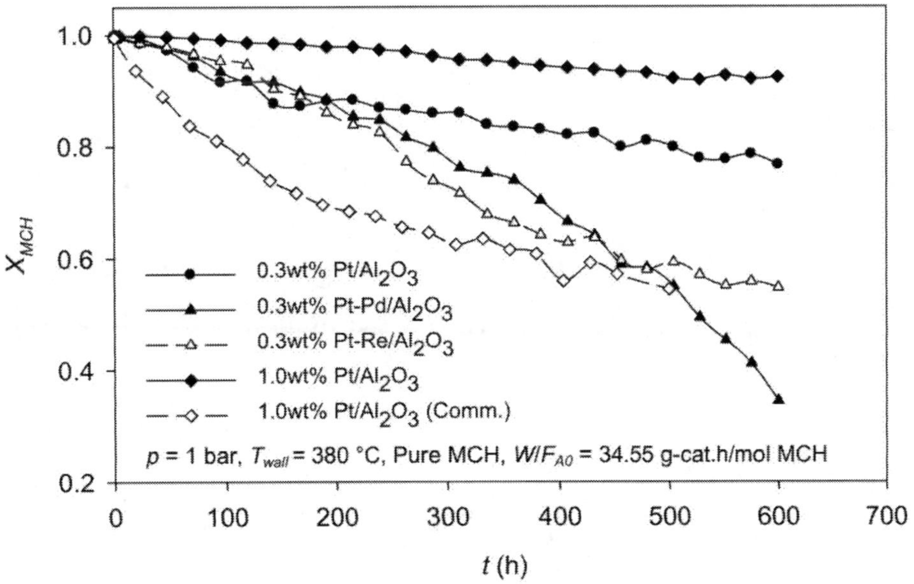

Fig.5 Comparative study of the life of the catalyst for four in-house and one commercial catalyst, t = time on stream, h; X_{MCH} is the total conversion of MCH

Conclusion and Further Work:

We have reported on some highly active and selective precious metal catalysts. Work is on going on the development of catalysts with improved activity, while maintaining high selectivity and long lifetime. Keeping in view the best overall performance of the in-house 1.0wt% Pt/Al_2O_3, the experimental results on the catalyst are being rigorously analysed to elucidate the reaction mechanism of the dehydrogenation. One dimensional pseudo-homogeneous plug flow reactor model is assumed and the regression of the data is carried out for a number of kinetic mechanisms. FORTRAN routine based on the Marquardt algorithm is employed for the regression of the kinetic data and to find the parameters associated with the respective kinetic model. Dual site Langmuir-Hinshelwood-Hougen-Watson (LHHW) model, when the loss of first hydrogen is the rate controlling step, is so far the best at correlating the data. Over 200 data sets, at 1 to 9 bar, have been correlated within ±20%. The energy of activation comes out be 69.90 kJ/mol. Finally, we are targeting the one optimal catalyst to model mathematically and simulate a novel chemical reactor design for the MTH system.

Acknowledgements:

One of the authors, Muhammad R. Usman would like to acknowledge Higher Education Commission of Pakistan for funding his PhD education.

Extending First-Principles Calculations of Complex Metal Hydrides to Predict Impurity Gases During Hydrogen Evolution

Ki Chul Kim, Chemical and Biomolecular Engineering, Georgia Institute of Technology, Atlanta, GA

Mark D. Allendorf, Sandia National Laboratories, Livermore, CA

David S. Sholl, Georgia Institute of Technology, Atlanta, GA

Complex metal hydrides have a potential as candidates for reversible hydrogen storage. First-principles calculations have played a useful role in examining large numbers of potential reaction mixtures to find materials with high H2 capacity and favorable thermodynamics. A significant disadvantage of most calculations of this type is that they have assumed that the only gaseous phase present is H2. We have developed a first-principles methodology for predicting not only the reaction thermodynamics of hydrogen evolution, but also the formation of other gaseous species that may act as impurities, including ammonia, methane, and borazine. Results from the application of this method to a range of "destabilized" metal hydride mixtures will be presented. Calculations of this type will play an important role in future screening of metal hydrides for hydrogen storage applications.

See more of this Session: Hydrogen Separation/Storage
See more of this Group/Topical: Environmental Division

Diffusion of Hydrogen through Layered Polymer Composite Pipes

Simon L. Marshall, Mathematical and Information Sciences, CSIRO, Clayton, Victoria 3168, Australia

James G. Blencoe, Hydrogen Discoveries, Inc., Oak Ridge, TN

The diffusion of hydrogen through polymeric pipeline materials is associated, on the one hand, with potentially significant losses, and on the other, with safety hazards arising from accumulation of hydrogen in confined spaces. The purpose of this paper is to consider the possibility of mitigating the diffusion of hydrogen through pipelines by use of layered polymeric composite materials. For cylindrical pipes constructed from layers of polymeric material, analytical solutions of the steady-state concentration profile are derived, together with the limiting forms that are appropriate for large numbers of extremely thin layers. These analytical results can be used to construct a figure of merit expressing the ratio of the multi-wall diffusion flux to that through a single-walled pipe of the same overall thickness, in terms of the (usually unknown) layer-to-layer diffusional contact resistance or the corresponding dimensionless mass-transfer coefficient. Conservative estimates of this parameter lead to the conclusion that the reduction achievable in this way is rather limited. An analysis of an alternative design concept in which a thin metallic foil is sandwiched between two coaxial layers of polymer is also given. Application of equilibrium and flux-matching conditions at the interfaces results in a nonlinear boundary value problem for the Laplace equation defining the steady concentration profile. Making use of published solubilities and diffusivities of hydrogen in polymers and metals, calculations indicate that reductions of several orders of magnitude could be achieved with foils as thin as 0.01 mm.

See more of this Session: Hydrogen Separation/Storage
See more of this Group/Topical: Environmental Division

Sonochemical Removal of Nitric Oxide Using a Novel Ultrasonic CSTR

Ibraheem R. Muhammad, Chemical & Bioengineering Department, North Carolina Agricultural and Technical State University, Greensboro, NC

Nymul E. Khan, Chemical & Bioengineering Department, North Carolina Agricultural and Technical State University, Greensboro, NC

Naresh N. Mahamuni, Chemical & Bioengineering Department, North Carolina Agricultural and Technical State University, Greensboro, NC

Yusuf G. (Debo) Adewuyi, Chemical & Bioengineering Department, North Carolina Agricultural and Technical State University, Greensboro, NC

The combustion of fossil fuel releases a large amount of NOx and sulfur dioxide (SO2) into the atmosphere which is a major threat to human health and environment. NOx, together with SO2 is the major contributer to acid rain that harms forest crops and buildings, as well as aquatic life. NOx also constitutes one of the main ingredients involved in the formation of ground-level ozone and urban smog through photochemical reactions, which can trigger serious respiratory problems. The efficient, cost effective and environmentally friendlier removal of NOx and SO2 from flue gases poses a considerable industrial problem. Available control technologies such as Selective Catalytic Reduction (SCR) used for high NOX removal has very high capital cost and suffers from various problems including disposal of spent toxic catalysts. Selective noncatalytic reduction (SNCR, urea injection) is highly temperature dependent and SNCR, ammonia injection (Thermal DeNOX) has moderately high capital cost and ammonia handling problems. Low-NOX burners (LNB) also has moderately high capital cost. Commercial applications of methods involving the chemical conversion of NO into soluble NO2 involving numerous liquid adsorbents in various gas-liquid contactors have been slow due to their complexities and additional costs of chemicals. Thus, there is need in the aqueous scrubbing arena for environmentally friendly and cost-effective alternatives for comprehensive treatment of NOx from industrial flue gases. Our research involves the aqueous removal of NOx (NO and NO2) from flue gases induced by ultrasound or hydrodynamic cavitation. Cavitation is the formation, growth and implosive collapse of gas- or vapor-filled microbubbles and can be induced acoustically or hydrodynamically in a body of liquid. Acoustic or ultrasonic cavitation involves the formation and subsequent collapse of microbubbles from acoustical wave-induced compression/rarefaction. On the microscale, the collapse of the bubbles leads to surprisingly high estimated local temperatures and pressures up to and above 5000 C and 500 atmospheres, respectively. These rather extreme conditions are very short lived, but have been shown to produce several highly reactive species in the aqueous systems including hydroxyl (•OH), hydrogen (H•) and hydroperoxyl (HO2•) radicals, and hydrogen peroxide.

In previous studies, we demonstrated that the absorption of NO (50-1040 ppm) into water with simultaneous oxidation to nitrite and nitrate induced by ultrasonic irradiation at a fixed frequency of 20 kHz could be accomplished using a sonochemical bubble column reactor in the absence and presence of SO2 (52-4930 ppm) and NaCl at room temperature (23 ± 2oC). The fractional conversions of NO were found to range from 60 to 85% while complete removal of SO2 was observed for all the inlet gas concentrations studied. The fractional conversions of NO were found to range from 60 to 85% while complete removal of SO2 was observed for all the inlet gas concentrations studied. It was shown that the presence of about 2520 ppm SO2 in combination with 0.01 M NaCl further enhanced NO removal. Combinative effect of sonication and chemical oxidation using 0.005-0.05M oxone was also studied. It was demonstrated that lower concentrations of HSO5- (0.005 M) enhanced NO removal efficiency. This paper will present new results obtained using our one-of it-kind sonochemical gas-liquid continuous stirred tank reactor,newly designed and built to provide drastic kinetic rate enhancement for demanding operations.

References

(1) Adewuyi, Y.G., Owusu, S.O. Ultrasound-induced Aqueous Removal of Nitric Oxide from Flue Gases. Effects of Sulfur Dioxide, Chloride and Chemical Oxidant. Journal of Physical Chemistry A. 2006. 110, 11098-11107.

(2) Owusu, S.O; Adewuyi, Y.G. Sonochemical Removal of Nitric Oxide from Flue Gases. Industrial & Engineering Chemistry Research. 2006, 45, 4475-4485.

See more of this Session: Advanced Oxidation/Reduction Applications: Liquid/Gas Phase II
See more of this Group/Topical: Environmental Division

MODELING THE SONOCHEMICAL REMOVAL OF NITRIC OXIDES USING THE RAYLEIGH-PLESSET-NOLTINGK-NAPPIRAS-PORITSKY (RPNNP) BUBBLE DYNAMICS EQUATION

Nymul E. Khan and Yusuf G. Adewuyi, Chemical and Bioengineering Department, North Carolina A&T State University, Greensboro, NC

Introduction

NO_x (NO & NO_2) and SO_2 – mainly generated from combustion of fossil fuel in power plants and automobiles – have severe detrimental effects on plants, animals and humans. Numerous environmental and health effects such as acid rain, formation of ground-level ozone, eutrophication, respiratory and heart diseases in humans are caused by NO_x. Although emissions of NO_2 and SO_2 can be controlled easily by scrubbing, nitric oxide (NO) is not easily removed by aqueous scrubbing due to its very low solubility in such solutions. Ultrasound has been known to enhance the rate of many chemical reactions. Owusu and Adewuyi previously reported the effectiveness of sonochemical methods for the removal of NO.[1-2]

Not many studies in the field of ultrasound have focused on gas absorption. The few that have are mostly mass-transfer studies [3]. There has been no investigation studying the applicability of ultrasound in the absorption of a commercially important gas. In this perspective our previous studies have been solely about the removal nitric oxide (NO) from flue gases using a sonochemical bubble column reactor [1-2]. Absorption and conversion in the range of 65-80% have been shown to be possible.

In the current paper the development of a mathematical model for this system has been undertaken for the understanding of the different mechanisms involved. We propose a model to relate the cavitation bubble dynamics phenomena with the reaction in liquid phase and the simultaneous transfer of mass from the gas to the liquid phase to predict the rate of absorption in presence of ultrasound. We have developed a comprehensive model which takes into consideration (i) dynamics of cavitation, (ii) generation of oxidizing species from bubble collapse through reaction kinetics, (iii) transfer of NO from from gas to liquid, (iv) chemical reactions of oxidizing species with dissolved NO.

Sonochemical techniques utilize ultrasound to produce an oxidative environment via acoustic cavitation due to the formation and subsequent collapse of microbubbles from acoustical wave-induced compression/rarefaction. Ultrasound creates alternating compression and rarefaction pressure fields in a liquid. This causes tiny gas bubbles trapped in the liquid or in the crevices of surfaces to expand and then violently collapse. During the expansion phase water evaporates into the bubble and condenses during the collapse [4-7]. However, since the collapse is extremely fast and almost adiabatic in nature most of the vapor trapped inside the bubble cannot escape and is disintegrated due to the very high temperature reached inside the bubble (≈ 3000-5000 K) [8-10] to form reactive species such as hydroxyl (·OH), hydroperoxyl (·HO_2) radicals or oxygen (O) and hydrogen (H) atoms and hydrogen peroxide (H_2O_2) [11]. These species diffuse out of the bubble into a thin interfacial layer where they either recombine or react with compounds present in the liquid.

Majority of the studies for the application of sonochemistry are in the field of wastewater treatment [12-21]. The mechanisms related to the reaction-enhancement by cavitation have been elucidated in the many studies done on the sonochemical degradation of hazardous organic compounds found in wastewater [12-19]. This system is different from ours in terms of the number and types of phases involved but we can certainly expect that the mechanisms that lead to the chemical effects of cavitation to remain the same. Three different mechanisms have been proposed to explain the degradation of organic pollutants in wastewater. The first is the pyrolysis in the cavitation bubble interior. Volatile solutes evaporate into the bubble during the expansion phase and during the collapse the extremely high temperatures reached break them down into other species and thus causing their removal. The second mechanism is the hydroxylation of the species in the interfacial region by the ·OH radicals generated in the bubble. The H atom does not penetrate very far outside the bubble [20] and therefore the ·OH radical is the more prominent species and it is also one with the most oxidative potential of any known oxidative species except fluorine. The third possible mechanism, proposed by Hua et. al. [21], is the formation of super-critical water in a small shell around the bubble which promotes the reactions.

Earlier, studies were done on the degradation of gases in aqueous solutions [22-29] and important conclusions about the mechanism of sonochemistry were drawn from these studies. The primary mechanism for the decomposition of gases by ultrasound appears to be pyrolysis in the cavitation bubbles with hydroxylation in the interfacial sheath providing secondary pathways. But in their study using acetylene Hart et. al. [28] found out that the dominant mechanism shifts between that of pyrolysis in the bubble core and that of hydroxylation by ·OH radicals at the interface. They concluded that the lower concentration favors reaction pathways in the interfacial zone. Kotronarou and Hoffmann [29] proposed a model for the aqueous phase oxidation of H_2S by ultrasound. A liquid phase reaction mechanism involving ·OH radicals generated by cavitational collapse was used in their model. By fitting with experimental data they have estimated the uniform release of ·OH radicals into the bulk solution to be of the order of 3.5 μM/min and a steady state concentration ≤ 0.1 μM. In view of these studies it is assumed that NO is destroyed mainly through reaction in the interfacial zone of the cavitation bubble with the OH radicals generated from the bubble interior. Hence, it would be assumed that the NO is removed primarily by attack of OH radicals in the interfacial zone of the cavitation bubble.

Cavitation model

Bubble dynamics

The motion of the bubble wall is described by the Keller equation [30], which is a modification of the well known Rayleigh-Plesset-Noltingk-Nappiras-Poritsky (RPNNP) equation and takes into account the liquid compressibility.

$$\left(1-\frac{\dot{R}}{c_\infty}\right)R\ddot{R}+\frac{3}{2}\dot{R}^2\left(1-\frac{\dot{R}}{3c_\infty}\right)=\frac{1}{\rho_L}\left(1+\frac{\dot{R}}{c_\infty}\right)\left(P_g-P_s(t)-P_\infty\right)+\frac{R}{c_\infty\rho_L}\frac{dP_g}{dt}-4\eta\frac{\dot{R}}{R}-\frac{2\sigma}{R} \quad (1)$$

where · denotes the time derivative (d/dt), R is the radius of the bubble, c_∞ is the speed of sound in the liquid, P_g is the pressure of the gas in the bubble, P_s is the time varying pressure field, and P_∞ is the hydrostatic pressure, σ, μ, ρ_L denotes the surface tension, viscosity and density of the liquid respectively.

Differential mole balance of water vapor (transport and chemical reactions)

The total rate of change of water vapor in the bubble is given by the following expression.

$$\frac{dn_w}{dt}=4\pi R^2\frac{10^3N_A}{M_{H2O}}\dot{m}+\frac{4}{3}\pi R^3\sum_j \alpha_{H_2O,j}r_j \quad (2)$$

where n_w is the number of molecules of water in the bubble, M_{H_2O} is the molar mass of water, N_A is the Avogadro's constant. The second term represents the consumption by chemical reaction where $\alpha_{H_2O,j}$ is the stoichiometric coefficient of water in and r_j is the rate expression of jth reaction.

$\dot{m}$ is the flux of water vapor across the bubble boundary and is given by:

$$\dot{m}=\dot{m}_{eva}-\dot{m}_{con} \quad (3)$$

$$\dot{m}_{eva}=\frac{\alpha}{\sqrt{2\pi R_g}}\left(\frac{p^*(T_L)}{\sqrt{T_L}}\right) \quad (4)$$

$$\dot{m}_{con}=\frac{\alpha}{\sqrt{2\pi R_g}}\left(\frac{\Gamma p_v}{\sqrt{T}}\right) \quad (5)$$

where α is the mass accommodation coefficient (literature suggests a value between 0.17 to 0.32 [31] so we used a conservative value of 0.17), p^* is the saturated vapor pressure and p_v is the partial pressure of water in the bubble. The correction factor, Γ, is given by eq. 6.

$$\Gamma=\exp(-\Omega^2)-\Omega\sqrt{\pi}\left(1-\frac{2}{\sqrt{\pi}}\int_0^\Omega \exp(-x^2)\,dx\right) \quad (6)$$

$$\Omega=\frac{\dot{m}}{p_v}\left(\frac{R_vT}{2}\right)^{\frac{1}{2}} \quad (7)$$

Differential mole balance of other species (chemical reactions)

The chemical rate of change of species i is given by the sum of generation/consumption from all elementary reactions involving that species with their proper stoichiometric weight $\alpha_{i,j}$

$$\left(\frac{dn_i}{dt}\right)_c = V \times \sum_j \alpha_{i,j} r_j \quad (8)$$

A total of 49 reactions involving 19 species have been taken into account [32]. Each of these elementary reactions has a forward and backward rate per unit volume, $r_{f,j}$, $r_{b,j}$, which is described by means of modified Arrhenius laws.

$$r_{f,j} = A_{f,j} T^{B_{f,j}} \exp\left(-\frac{C_{f,j}}{kT}\right) \prod_i N_i^{\upsilon_{j,i}} \quad (9)$$

$$r_{b,j} = A_{b,j} T^{B_{b,j}} \exp\left(-\frac{C_{b,j}}{kT}\right) \prod_i N_i^{\upsilon_{j,i}} \quad (10)$$

The constants A_j, B_j, C_j are constants corresponding to jth reaction. The reactions are taken from ref. [32]. The overall rate is given by the difference of forward and backward rate

$$r_j = r_{f,j} - r_{b,j} \quad (11)$$

Energy balance

The temperature change inside the bubble can be obtained using the following equation.

$$\frac{dT}{dt} \sum_i \frac{\partial e_{th,i}}{\partial T} n_i = 4\pi R^2 \kappa \left.\frac{\partial T}{\partial r}\right|_{r=R} + -4\pi R^2 P_g \frac{dR}{dt} + \left(h_w - e_{th,w}\right)\frac{dn_w}{dt} + \sum_i \left(\frac{dn_i}{dt}\right)_c \Delta H_i^f \quad (12)$$

where T represents the temperature inside the bubble, $e_{th,i}$ denotes the thermal energy and $e_{form,i}$ denotes the enthalpy of formation of species i, n_i is the number of moles of species i, κ is the thermal conductivity of the gas, h_w is the molecular enthalpy of water vapor at the cold bubble wall, $e_{form,w}$ is the enthalpy of formation of water vapor.

The value of $e_{th,i}$ is given by the expression [33]:

$$e_{th,i} = \left(\frac{f_i}{2} kT + \sum_l \frac{k\theta_{i,l}}{\exp\left(\frac{\theta_{i,l}}{T}\right) - 1}\right) \quad (13)$$

and f_i, $\theta_{i,l}$, $l = 1, \ldots\ldots, l_{max}$, denotes the number of translational and rotational degrees of freedom and the characteristic vibrational temperatures of species i, the values are obtained from ref. [33].

Extrapolation of cavitation model to a sonochemical reactor

The above model for a single bubble furnishes information about the amount of OH radicals generated at the collapse and the change of bubble volume over one acoustic cycle. Using these two quantities we can determine on an order of magnitude the values of the range of the number of cavitation bubbles in a sonochemical reactor and subsequently the rate of generation of the OH radical. The mass transfer of NO from gas bubbles entering the reactor together with the following reaction scheme (Table 1) in the liquid phase is then used to predict the rate of removal of NO.

Table 1. Reaction scheme for the consumption of NO in the liquid phase

No.	Reaction	k (L mol^{-1} s^{-1})	Ref.
1	$NO + OH \rightarrow H^+ + NO_2^-$	$2x10^{10}$	[34]
2	$OH + OH \rightarrow H_2O_2$	$5.5x10^9$	[34]
3	$OH + H_2O_2 \rightarrow O_2^- + H_2O$	$2.4x10^7$	[34]
4	$O_2^- + OH \rightarrow OH^- + O_2$	$7x10^9$	[34]
5	$OH + OH^- \rightarrow O.^- + H_2O$	$1.3x10^{10}$	[34]
6	$NO_2^- + OH \rightarrow NO_2 + OH^-$	$1x10^{10}$	[34]
7	$NO_2 + OH \rightarrow HNO_3$	$1x10^{10}$	[34]

8	$O_2^{\cdot-} + NO \rightarrow ONOO^-$	$4.3 \pm 0.5x10^9$	[34]
9	$O.^{-} + H_2O \rightarrow OH^- + OH$	$1.7x10^6$	[34]
10	$O.^{-} + OH \rightarrow HO_2^-$	$\leq 2x10^{10}$	[34]
11	$H_2O_2 \leftrightarrow H^+ + HO_2^-$	pK = 11.65	[29]
12	$HNO_2 \leftrightarrow H^+ + NO_2^-$	pK = 3.15	[29]
13	$HO_2 \leftrightarrow H^+ + O_2^{\cdot-}$	pK = 4.46	[29]

Yasui [35] has suggested an expression to calculate the rate of dissolution of the different species in the liquid using uptake coefficient (Θ).

$$r_{d,i} = \Theta \sqrt{\frac{kT_B}{2\pi m_i} \frac{n_i}{V}} \times 4\pi R^2 \quad (14)$$

where m_i is the molecular mass of the species.

The time-averaged rate of dissolution of OH radicals over one acoustic cycle from a single bubble is,

$$\langle r_{d,OH}(R_0) \rangle_t = \frac{\int_0^{t_{cycle}} r_{d,OH}(R_0,t)dt}{t_{cycle}} \quad (15)$$

where $\langle\ \rangle_t$ represents averaged quantity over t.

The average volume of a bubble over one acoustic cycle is,

$$\langle V(R_0) \rangle_t = \frac{\int_0^{t_{cycle}} V(R_0,t)dt}{t_{cycle}} \quad (16)$$

The number of bubbles has been estimated from the void fractions calculated in Tsochatzidis et. al. [36].

$$N = \frac{\epsilon}{\langle V(R_0) \rangle_t} \quad (17)$$

where ϵ is the void fraction and N is the number of bubbles per unit volume.

Once this is calculated the average rate of generation of OH radicals in the reactor can be given by the expression,

$$S_{OH} = \langle r_{d,OH}(R_0) \rangle_t N \quad (18)$$

Results and Discussion

Simulation of cavitation dynamics

The bubble dynamics equations have been used to generate information such as bubble volume vs. time and no. of reactive radicals vs. time. These information allow us to calculate the number of bubbles and the rate of generation of OH radicals per unit time per unit volume. The illustrative results from the bubble dynamic simulations are shown in Figs. 1-4. The conditions used for these simulations are taken from Didenko and Suslick [37] in order to be able to validate our results against theirs. For the validation, we have compared our results against those found experimentally by Didenko and Suslick [37] and those from a more detailed model by Yasui et. al. [35].

The following are the bubble dynamics simulation results.

Conditions: f = 52 kHz, P_A = 1.52 bar, T_{amb} = 276 K, P_h = 1 atm, R_0 = 4 μm.

Table 2. Comparison of our results with other models and experiment

	Experiment [37]	Yasui et. al. [35]	Diffusion limited [35]	Present 4 95% Ar, 2.5% N2 and 2.5% O2
T_{max}	...	10900 K	9400 K	10200 K
Photon	$7.5x10^4$	$8.0x10^4$	$4.9x10^4$	...
OH	$8.2x10^5$	$6.6x10^5$	$9.2x10^5$	$1.1841x10^5$
O	...	$1.3x10^7$	$1.1x10^7$	$1.9562x10^7$

H_2O_2	...	$6.3x10^6$	$9.2x10^6$	$2.6197x10^6$
O_3	...	$3.4x10^4$	$4.0x10^4$	$3.1830x10^4$
HNO_2	$9.9x10^6$	$2.3x10^6$	$2.0x10^6$	$6.3503x10^1$
HNO_3	...	$8.4x10^5$	...	2.107710^4

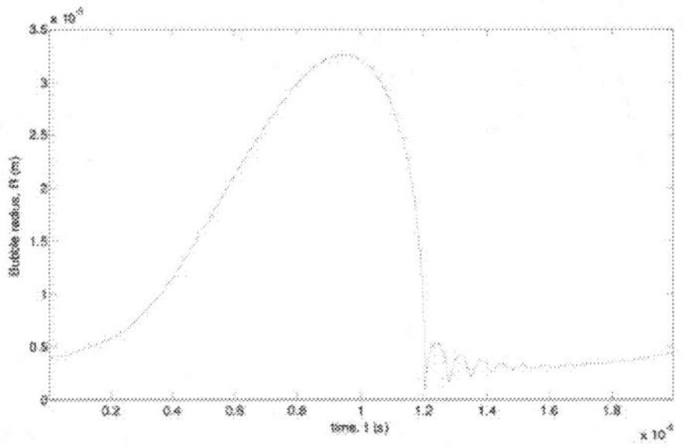

Figure 1. Cavitation bubble radius vs. time (1 acoustic cycle)

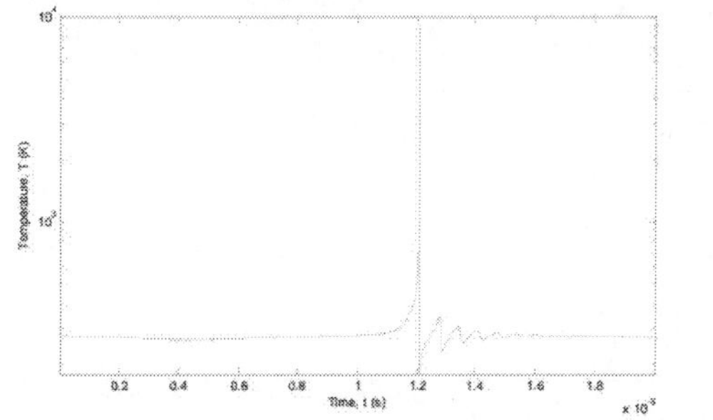

Figure 2. Temperature inside the cavitation bubble vs. Time (1 acoustic cycle)

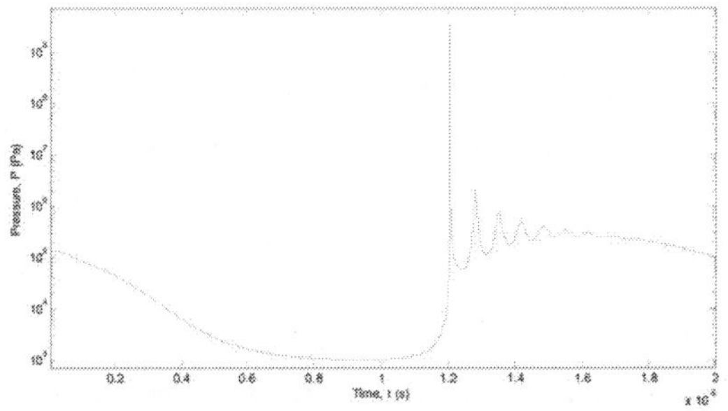

Figure 3. Pressure inside a cavitating bubble vs. Time (1 acoustic cycle)

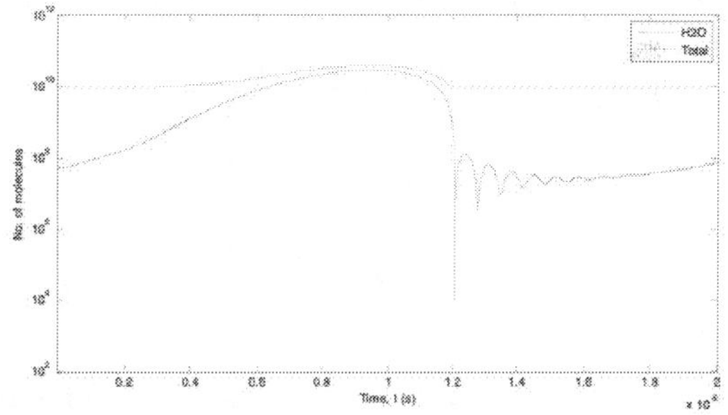

Figure 4. No. of molecules inside a cavitating bubble vs. Time (1 acoustic cycle)

This cavitation model applied to the conditions used in our experiments (f = 20 kHz, P_A = 2 bar, T_{amb} = 293 K, P_h = 1 atm, R_0 = 5 μm) yields the following values:
The average volume of a cavitation bubble,

$$\langle V(R_0)\rangle_t = 3.51 \times 10^{-12}\ m^3$$

and rate of dissolution of OH radicals,

$$\langle r_{d,OH}(R_0)\rangle_t = 8.33 \times 10^{10}\ molecules/s$$

Therefore, the range of number of bubbles:

$$N = 2.85 \times 10^7\ to\ 1.42 \times 10^9\ bubbles/m^3$$

and the corresponding rate of generation of OH radicals:

$$S_{OH} = 8.14 \times 10^{-9}\ to\ 4.07 \times 10^{-7} \frac{mol}{L.s}$$

Simulations for NO absorption

Simulation conditions:

We are simulating a semibatch system for the absorption of NO. The gas phase is flowing continuously and the liquid phase is constant, with the ultrasonic probe dipped into it. The following are the global variables that are used for the simulations corresponding to the experiments. They remain constant for all the simulations.

Gas flow rate, $Q^g = 0.1\ slpm = 1.667\frac{cm^3}{s}$

Liquid volume, $V^l = \frac{\pi}{4}D^2h = \frac{\pi}{4}x10^2x16 = 1256.64\ cm^3$

Gas holdup, $V^g = \frac{\pi}{4}D^2h = \frac{\pi}{4}x5^2x1 = 19.63\ cm^3$ (assuming a height increment of 1 cm)

Henry's law constant for NO, $H_{NO} = 5.1546\ x\ 10^5\ atm/(mol/cm^3)$

Physical mass transfer coefficient, $k_La = 1.57x10^{-2}\ cm/s$ (calculated following ref. [38])

Following is an experimental data from NOx absorption studies (Fig. 5). The steady state conversion of NO is about 78%.

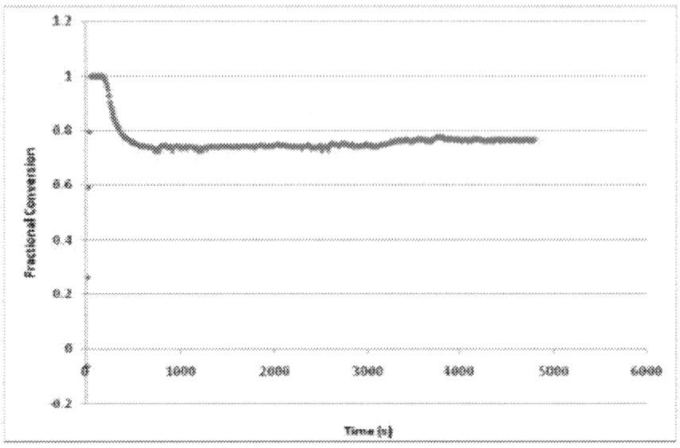

Figure 5. Experimental data for a NOx absorption study

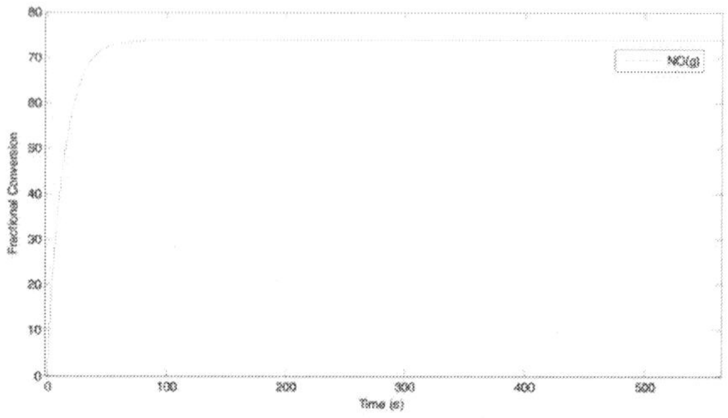

Figure 6. Simulation result of NOx absorption

In order to fit the experimental data, the simulation has been carried out at different mass transfer coefficient keeping the other parameters constant, since the mass transfer coefficient is the only parameter that is not known under ultrasonic irradiation. Following is the result of such a simulation (Fig. 6).

It was seen that around 75% conversion of NO is obtained at a mass transfer coefficient that is 5 times the physical mass-transfer coefficient calculated before. Hence, it is concluded that ultrasonic irradiation, apart from generating OH radicals which react with the dissolved NO to remove it, also increases the mass transfer at least 5 fold.

Conclusion

A sonochemical bubble column for the absorption of NO was modeled. Bubble dynamics simulations of acoustic cavitation, incorporating liquid compressibility, mass and heat transport and chemical reaction, was performed. Coupling the chemistry generated from these microscopic phenomena to the reaction occurring in the liquid phase and mass transfer from the gas phase, the fractional conversion of NO was obtained. The results of the simulations indicate that only sonochemical oxidation cannot explain the previously reported results of NO removal by ultrasonic cavitation. Further simulations, with higher mass-transfer coefficients (equivalent to those found by other investigators in presence of ultrasound), show that only a combination of chemical activity and mass transfer enhancement by ultrasound can explain the results obtained.

References

1. Adewuyi, Y. and S. Owusu, *Ultrasound-induced aqueous removal of nitric oxide from flue gases: Effects of sulfur dioxide, chloride, and chemical oxidant.* J. Phys. Chem. A, 2006. **110**(38): p. 11098-11107.
2. Owusu, S. and Y. Adewuyi, *Sonochemical removal of nitric oxide from flue gases.* Ind. Eng. Chem. Res, 2006. **45**(13): p. 4475-4485.
3. Laugier, F., et al., *Ultrasound in gas–liquid systems: Effects on solubility and mass transfer.* Ultrasonics-Sonochemistry, 2008.
4. Suslick, K., *Sonochemistry.* Science, 1990. **247**(4949): p. 1439-1445.
5. Adewuyi, Y., *Sonochemistry: environmental science and engineering applications.* Ind. Eng. Chem. Res, 2001. **40**(22): p. 4681-4715.
6. Mason, T. and J. Lorimer, *Applied sonochemistry.* 2002: Wiley-VCH Weinheim.
7. Thompson, L. and L. Doraiswamy, *Sonochemistry: science and engineering.* Ind. Eng. Chem. Res, 1999. **38**(4): p. 1215-1249.
8. Suslick, K., D. Hammerton, and R. Cline, *Sonochemical hot spot.* Journal of the American Chemical Society, 1986. **108**(18): p. 5641-5642.
9. Ciawi, E., et al., *Determination of temperatures within acoustically generated bubbles in aqueous solutions at different ultrasound frequencies.* J. Phys. Chem. B, 2006. **110**(27): p. 13656-13660.
10. Kanthale, P., M. Ashokkumar, and F. Grieser, *Estimation of Cavitation Bubble Temperatures in an Ionic Liquid.* 2007.
11. Makino, K., M. Mossoba, and P. Riesz, *Chemical effects of ultrasound on aqueous solutions. Formation of hydroxyl radicals and hydrogen atoms.* The Journal of physical chemistry, 1983. **87**(8): p. 1369-1377.
12. Hoffmann, M., I. Hua, and R. Höchemer, *Application of ultrasonic irradiation for the degradation of chemical contaminants in water.* Ultrasonics-Sonochemistry, 1996. **3**(3): p. 163-172.
13. Weavers, L., F. Ling, and M. Hoffmann, *Aromatic compound degradation in water using a combination of sonolysis and ozonolysis.* Environ. Sci. Technol, 1998. **32**(18): p. 2727-2733.
14. Hung, H., F. Ling, and M. Hoffmann, *Kinetics and mechanism of the enhanced reductive degradation of nitrobenzene by elemental iron in the presence of ultrasound.* Environ. Sci. Technol, 2000. **34**(9): p. 1758-1763.
15. Jiang, Y., C. Petrier, and T. Waite, *Effect of pH on the ultrasonic degradation of ionic aromatic compounds in aqueous solution.* Ultrasonics-Sonochemistry, 2002. **9**(3): p. 163-168.
16. Hua, I. and M. Hoffmann, *Kinetics and mechanism of the sonolytic degradation of CCl4: intermediates and byproducts.* Environ. Sci. Technol, 1996. **30**(3): p. 864-871.
17. Kotronarou, A., G. Mills, and M. Hoffmann, *Ultrasonic irradiation of p-nitrophenol in aqueous solution.* The Journal of physical chemistry, 1991. **95**(9): p. 3630-3638.
18. Tauber, A., H. Schuchmann, and C. von Sonntag, *Sonolysis of aqueous 4-nitrophenol at low and high pH.* Ultrasonics-Sonochemistry, 2000. **7**(1): p. 45-52.
19. Sivakumar, M., P. Tatake, and A. Pandit, *Kinetics of p-nitrophenol degradation: effect of reaction conditions and cavitational parameters for a multiple frequency system.* Chemical Engineering Journal, 2002. **85**(2-3): p. 327-338.
20. Hart, E. and A. Henglein, *Sonochemistry of aqueous solutions: H2-O2 combustion in cavitation bubbles.* Physical Chem, 1987. **91**(13): p. 3654-6.
21. Hua, I., R. Hoechemer, and M. Hoffmann, *Sonolytic hydrolysis of p-nitrophenyl acetate: the role of supercritical water.* The Journal of physical chemistry, 1995. **99**(8): p. 2335-2342.
22. Suslick, K., et al., *Alkane sonochemistry.* The Journal of physical chemistry, 1983. **87**(13): p. 2299-2301.
23. Fischer, C., E. Hart, and A. Henglein, *Ultrasonic irradiation of water in the presence of oxygen 18, 18O2: isotope exchange and isotopic distribution of hydrogen peroxide.* The Journal of physical chemistry, 1986. **90**(9): p. 1954-1956.

24. Fischer, C., E. Hart, and A. Henglein, *Hydrogen/deuterium isotope exchange in the molecular deuterium-water system under the influence of ultrasound.* The Journal of physical chemistry, 1986. **90**(2): p. 222-224.
25. Hart, E. and A. Henglein, *Sonolysis of ozone in aqueous solution.* The Journal of physical chemistry, 1986. **90**(14): p. 3061-3062.
26. Hart, E. and A. Henglein, *Sonolytic decomposition of nitrous oxide in aqueous solution.* The Journal of physical chemistry, 1986. **90**(22): p. 5992-5995.
27. Hart, E., C. Fischer, and A. Henglein, *Isotopic exchange in the sonolysis of aqueous solutions containing deuterium and methane.* Journal of Physical Chemistry, 1987. **91**(15): p. 4166-4169.
28. Hart, E., C. Fischer, and A. Henglein, *Pyrolysis of acetylene in sonolytic cavitation bubbles in aqueous solution.* Journal of Physical Chemistry, 1990. **94**(1): p. 284-290.
29. Kotronarou, A. and M.R. Hoffmann, *The chemical effects of collapsing cavitation bubbles: Mathematical modeling*, in *Advances in Chemistry series 244*, C.P. Huang, C.R. O'Melia, and J.J. Morgan, Editors. 1995, American Chemical Society: Washington DC.
30. Prosperetti, A. and A. Lezzi, *Bubble dynamics in a compressible liquid. Part 1. First-order theory.* Journal of Fluid Mechanics Digital Archive, 2006. **168**: p. 457-478.
31. Li, Y., et al., *Mass and thermal accommodation coefficients of H2O (g) on liquid water as a function of temperature.* J. Phys. Chem. A, 2001. **105**(47): p. 10627-10634.
32. Yasui, K., et al., *Theoretical study of the ambient-pressure dependence of sonochemical reactions.* The Journal of Chemical Physics, 2003. **119**: p. 346.
33. Toegel, R. and D. Lohse, *Phase diagrams for sonoluminescing bubbles: A comparison between experiment and theory.* The Journal of Chemical Physics, 2003. **118**: p. 1863.
34. Buxton, G., et al., *Critical review of rate constants for reactions of hydrated electrons, hydrogen atoms and hydroxyl radicals(•OH/•O) in aqueous solution.* Journal of Physical and Chemical Reference Data, 1988. **17**(2): p. 513-886.
35. Yasui, K., et al., *Theoretical study of single-bubble sonochemistry.* The Journal of Chemical Physics, 2005. **122**: p. 224706.
36. Tsochatzidis, N., et al., *Determination of velocity, size and concentration of ultrasonic cavitation bubbles by the phase-Doppler technique.* Chem. Eng. Sci., 2001. **56**(5): p. 1831-1840.
37. Didenko, Y. and K. Suslick, *The energy efficiency of formation of photons, radicals and ions during single-bubble cavitation.* Nature, 2002. **418**(6896): p. 394-397.
38. Dudley, J., *Mass transfer in bubble columns: a comparison of correlations.* Water Research, 1995. **29**(4): p. 1129-1138.

TiO2 Particle Property Investigation Featuring Moving Bed Adsorption and Photo-Assisted Degradation of Azo Dyes

David M. Follansbee, Department of Chemical and Biological Engineering, Rensselaer Polytechnic Institute, Troy, NY

John Paccione, Environmental Health Sciences, SUNY at Albany, School of Public Health, Troy, NY

Lealon L. Martin, Chemical and Biological Engineering, Rensselaer Polytechnic Institute, Troy, NY

The use of photon-driven chemical decomposition of organic pollutants such as benzene is a potentially viable solution for water polishing applications. Various attempts at advanced oxidation process (AOP) have yielded inherent operation and design limitations. To overcome these mass transfer, cross sectional area, and light scattering limitations a novel water treatment process has been developed relying on the characteristics of a circulating fluidized bed (CFB) system. This system allows for adsorption of a target contaminant onto the surface of a photocatalyst composite particle in a moving packed annular bed and photocatalytic degradation/regeneration of laden particles within a photon chamber in a continuous fashion.

The goal of this work is to independently study the adsorption and photocatalytic capabilities of photocatalyst-composite particles by effectively decoupling these two processes within the system. This study will give insight into the bottlenecks and limitations of photocatalyzed reactions by assisting in optimal design of these particles with a high quantum yield and efficient use of photons.

Azo dyes such as Procion Red and orange-II are used as model contaminants. TiO_2 coated silica particles (.5mm, 1mm, and 2mm) are used as photocatalyst composite particles. An initial dye concentration of 500 ppm is used for all experiments. Photon chamber consists of a 2" OD quartz tube surrounded by six 50W UV bulbs (350 nm) in which the particles sediment through in lean phase fashion. Particles are removed prior to entering the UV chamber and upon exiting the UV chamber for analysis. Concentration of the contaminated water is monitored within the annular bed by a Beckman UV-Vis spectrophotometer.

See more of this Session: Advanced Oxidation/Reduction Applications: Liquid/Gas Phase II
See more of this Group/Topical: Environmental Division

TiO2-BASED PHOTOCATALYSTS IN UV and VISIBLE REGION: Study of Activity and Stability

Aaron Ortiz-Gomez, Chemical and Biochemical Engineering, The University of Western Ontario, London, ON, Canada

Jesus Moreira, Chemical Engineering, University of Western Ontario, London, ON, Canada

Benito Serrano, Chemical Engineering, Universidad Autonoma de Zacatecas., Zacatecas Zacatecas., Mexico

Hugo H. De Lasa, University of Western Ontario, London, ON, Canada

Heterogeneous photocatalysis has emerged as a potential alternative for the degradation of organic pollutants in water. The TiO_2 DP25 is the most active catalyst but it is limited to be used in the range of UV light. In order to make the DP25 active in the visible region, ion doping and coupling the DP25 with another photocatalysts have been tested. In this study, various catalysts were prepared using the sol-gel and wet impregnation techniques. Fe (III), Ni (III), and Ag (I) ions were used for the doping of DP25. AgBr was coupled with TiO_2 using the deposition-precipitation method. Catalysts prepared by the sol-gel techniques and wet impregnation are tested in the UV region, while the AgBr/TiO_2 catalysts are tested in the visible region. Both experiments were evaluated for the degradation of maleic acid in water. TiO_2 + 5 ppm Fe catalyst has the higher activity in UV region but showed no activity in the visible region. On the other hand, AgBr/TiO_2 catalyst was the most active in the visible region but showed lack of stability as it decomposed into Ag and Br_2 while being irradiated. Thus, it is concluded that in order to utilize the AgBr/TiO_2 catalyst under Vis radiation, it is necessary to find ways to make it stable under visible light. In this regard, it was found that AgBr/TiO_2 catalyst could be stabilized for longer periods of time when the pH of the reacting solution was maintained at 2. A full characterization of the prepared catalysts is carried out to understand their structure and operational mechanism under the UV and Vis range to develop materials that are stable and active for longer reaction periods. This will allow the use of photocatalysts under both visible light and UV such as that naturally present in the sunlight.

See more of this Session: Advanced Oxidation/Reduction Applications: Liquid/Gas Phase II
See more of this Group/Topical: Environmental Division

Developing the adjoint of ISORROPIA to equip CMAQ-ADJ

Shannon L. Capps[1], Armistead G. Russell[2], and Athanasios Nenes[1,3]

[1]School of Chemical & Biomolecular Engineering, Georgia Institute of Technology
[2]School of Civil & Environmental Engineering, Georgia Institute of Technology
[3]School of Earth & Atmospheric Sciences, Georgia Institute of Technology

Abstract

Data assimilation has been widely applied in weather models to improve forecasts. Efficient inverse modeling techniques are necessary for these large scale, high-temporal resolution applications, and the adjoint method has proved suitable. In view of this success, chemical transport models have been identified as excellent candidates for similar transformations, and one of the few that have been developed is the adjoint of the Community Multiscale Air Quality model, CMAQ-ADJ. Optimizing model parameters, including cell-by-cell emissions inventories, by assimilating spatially and temporally varying data sets is a primary goal of adjoint model application.

This work advances the capabilities of the aerosol module for adjoints of chemical transport models by delivering the same level of specificity for deliquesced aerosol as ISORROPIA, a widely applied inorganic thermodynamic equilibrium model. The adjoint of ISORROPIA captures the behavior of the system of salts containing the species nitrate, sulfate, ammonium, sodium, and chloride. Previously, only limited inorganic components were treated within adjoints of inorganic thermodynamic equilibrium models. Accuracy of the sensitivities produced by the adjoint of ISORROPIA has been evaluated against finite difference sensitivities produced by the model for the range of atmospherically relevant conditions. Such an adjoint will be suitable for application within the adjoint of regional chemical transport models including CMAQ-ADJ.

Introduction

Atmospheric aerosols significantly influence quality of life today and shape the evolution of Earth's climate. These solid or liquid particles suspended in the air alter the radiative heat balance of Earth directly by absorbing and scattering solar radiation and indirectly by enhancing cloud formation and reflectivity (Trenberth et al., 2009). Additionally, their interactions with light reduce visibility, a measure of air quality (Altshuller et al., 1984). Furthermore, extensive epidemiological evidence suggests that particulate air pollution increases cardiovascular deaths as well as pulmonary problems (Dockery, 2001). In an effort to understand mechanisms of climate change and to optimize methods of improving air quality, atmospheric models have been developed to elucidate the effects of natural and anthropogenic emissions of chemicals and particles into the atmosphere.

Across this broad spectrum of climate and health effects, inquiries impossible to make via observation or experiment due to the relevant time scale or spatial range have become feasible with the application of chemical transport models (CTMs) as investigative tools. New policies intended to improve air quality and reduce climate change have been implemented in large part because of the results of atmospheric modeling (Craig et al., 2008). Nevertheless, improvements are still needed in the applicability of regional air quality models and in the accuracy of global climate models, especially with regard to aerosols (Forster et al., 2007).

A promising means of refining atmospheric modeling is using the increasingly large sets of observational data from satellites, monitoring sites, and field campaigns. Inverse modeling is one method of extracting meaningful and specific information out of vast and heterogeneous observational datasets. In the last thirty years, measurements of atmospheric composition including aerosol abundance and type have become much more robust and dense due to increasing satellite technology and coverage (e.g., CALIPSO, MODIS) (Vaughan et al., 2004; Remer et al., 2005). CTMs rely heavily upon input information such as initial or boundary conditions and emissions rates that are often uncertain and unreasonable to assess directly. Therefore, application of inverse modeling of aerosols to vast and heterogeneous data gathered by satellites as well as field missions would assist in refining CTMs and increasing their predictive accuracy (Hartley and Prinn, 1993; Muller and Stavrakou, 2005; Kurokawa et al., 2009). An efficient method of inverse modeling with atmospheric aerosol observations is necessary for optimizing these parameters (Henze and Seinfeld, 2007). Such tools can also be employed to calculate efficiently the extent to which various sources contribute to a given location's atmospheric composition. This type of information provides a means of answering a long-standing inquiry that has developed into the field of source apportionment with the most quantitative, physically-based methods available.

Methodology

Background

CTMs solve the atmospheric diffusion equation (eq. 1) based on supplied boundary and initial conditions (Seinfeld and Pandis, 2006).

$$\frac{\partial C_i}{\partial t} = -\mathbf{u}\cdot\nabla C_i + \frac{1}{\rho}\nabla\cdot(\rho\mathbf{K}\nabla C_i) + R_i + E_i \qquad (1)$$

The concentration of species *i*, C_i, at a given time, *t*, and location is determined by the advection of the species due to wind with speeds, ***u***, from a meteorological model; the diffusion through air given its density, ρ, and the diffusivity, **K**, of each species *i*; the consumption and production of species *i* through reactions, R_i; and the emissions, E_i, respectively. Given this framework, a formulation appropriate for discussion of the inverse problem is $\mathbf{y} = \mathbf{F}(\mathbf{x}) + \varepsilon$, where observations of atmospheric concentrations, ***y***, equal the sum of the observation and model error, ε and the operation of the CTM, ***F***, on the vector, ***x***, of model parameters influencing predicted concentrations.

For developing an inverse model with atmospheric observations, considering a linearization of the model about a point $\mathbf{x_0}$ is convenient and satisfactory assuming that **F(x)** is linear within the bounds of the observation error, ε (eq. 2) (Rodgers, 2000)

$$\mathbf{y} - \mathbf{F}(\mathbf{x}_o) = \frac{\partial \mathbf{F}(\mathbf{x})}{\partial \mathbf{x}}(\mathbf{x} - \mathbf{x}_o) + \varepsilon = \nabla_x \mathbf{F}(\mathbf{x})(\mathbf{x} - \mathbf{x}_o) + \varepsilon \quad (2)$$

where $\nabla_x \mathbf{F}(\mathbf{x})$ is the kernel or tangent linear model of the function. The aim, then, is to solve an inverse problem with noise, or error, associated with both the ambient observations and the model predictions. Prior knowledge about the state of the atmosphere, **x**, and the degree of error, ε, can be expressed quantitatively with more or less uncertainty. Given this noisy inverse problem, Bayes' theorem provides a formalism by which the model can be inverted to update the *a priori* state (i.e., estimated) of the atmosphere with information from the observation in order to establish the *a posteriori* estimate of the atmosphere if a probability distribution function is assumed for the model state vectors and the observation information. The Gaussian distribution is suitable for describing the error associated with many processes and is amenable to algebraic manipulation; additionally, it retains the minimum amount of information a distribution function can given only mean and variance information (Rodgers, 2000). Therefore, it is frequently applied in the inverse modeling of atmospheric chemistry (Tarantola, 1987; Hartley and Prinn, 1993; Kopacz et al., 2009).

Reducing the error-weighted mismatch between the *a priori* estimate of the state vector to the true state as well as the error-weighted mismatch between the observation and the model predictions requires minimizing the scalar-valued cost function, *J*(**x**) (eq. 3) (Rodgers, 2000).

$$J(\mathbf{x}) = (\mathbf{F}(\mathbf{x}) - \mathbf{y})^T \mathbf{S}_\Sigma^{-1}(\mathbf{F}(\mathbf{x}) - \mathbf{y}) + \gamma(\mathbf{x} - \mathbf{x}_a) S_a^{-1}(\mathbf{x} - \mathbf{x}_a) \quad (3)$$

where $\mathbf{x}_a$ and $\mathbf{S}_a$ denote the *a priori* values of the state vector and the model covariance matrix, respectively; $\mathbf{S}_\Sigma^{-1}$ denotes the observational error covariance matrix; and γ denotes the regularization parameter that controls the extent to which the observations or *a priori* estimate constrains the solution (Kopacz et al., 2009). By minimizing the cost function with respect to the state vector (i.e., $\nabla_x J(\mathbf{x}) = 0$), statistically optimal agreement between observations and predictions of the model as well as the revised *a priori* estimate of the state vector and the true state vector can be obtained (Kopacz et al., 2009). Therefore, solving equation 4 is the means by which inverse modeling can be performed.

$$\nabla_x J(\mathbf{x}) = 2(\nabla_x \mathbf{F}(\mathbf{x}))^T \mathbf{S}_x^{-1}(\mathbf{F}(\mathbf{x}) - \mathbf{y}) + 2\gamma S^{-1}(\mathbf{x} - \mathbf{x}_o) = 0 \quad (4)$$

In the last two decades, minimization of the cost function to determine better estimates of atmospheric composition has been pursued in at least two manners. Analytical solutions to $\nabla_x J(\mathbf{x}) = 0$ have been extensively applied to reconcile emissions of trace gases with observations (Levelt et al., 1998; Kaminski et al., 1999; Mendoza-Dominguez and Russell, 2001; Patra et al., 2003); however, computational limitations on the number of parameters that can be optimized have driven the field beyond this method in an effort to better understand emissions that vary on smaller spatial and temporal scales. Adjoint operators can be applied to minimize the cost function without computational constraints on the number of parameters to be optimized (Marchuk, 1977; Cacuci, 1981; Tarantola, 1987).

Variational data assimilation in the fields of meteorology and oceanography began exploiting the power of adjoints two decades ago (Courtier and Talagrand, 1987; Talagrand and Courtier, 1987; Navon, 1998), and inverse modeling of atmospheric gases has since been accomplished through the application of adjoints (Kaminski et al., 1999; Kaminski et al., 1999; Elbern et al., 2000; Muller and Stavrakou, 2005; Hakami et al., 2006; Martien and Harley, 2006; Hakami et al., 2007; Zhang et al., 2008; Kopacz et al., 2009; Kurokawa et al., 2009; Zhang et al., 2009). Most recently, adjoints have been a focus of developments in the atmospheric modeling field with regard to aerosol emissions which have high spatial and temporal variability (Hakami, 2005; Sandu et al., 2005; Henze et al., 2007; Dubovik et al., 2008; Yumimoto et al., 2008). The primary motivation for development of an aerosol adjoint is that it can overcome the limitations of other approaches to minimizing the cost function.

Recently, implementations of global adjoints of CTMs, including aerosol, have proved successful due to the capability of this method to optimize significantly more model parameters than the analytical method at computational times comparable to forward model executions (e.g., Dubovik et al., 2008; Henze et al., 2009). Optimization of the state vector occurs iteratively in a solution-seeking algorithm such as the steepest descent method wherein

$$\mathbf{x}_{n+1} = \mathbf{x}_n - \delta_n \nabla_{x_n} J \qquad (5)$$

with a step size of δ_n is iterated until the change in the state vector is sufficiently small (Rodgers, 2000). To do so, the observations that have been translated into model observation space are fitted to the model prediction values in accordance with the cost function (eq. 3) in four dimensions across an interval of time, $t = 0 \ldots T$ (Rodgers, 2000). Practically, the steps involved are execution of the forward model (i.e., a typical model run) in which values are saved at particular time points and subsequent integration of the adjoint model backwards in time. In doing so, the model acts as the primary constraint in optimization, which implicitly attributes truth to the model's treatment of atmospheric processes. The primary computational savings arise from the adjoint model operating on the forcing term (i.e.,⍰⍰−⍰) directly rather than a relying on a recalculation of the Jacobian at each time step. Due to this operation, only matrix by vector multiplications are required rather than vector by matrix multiplications. These distinctions provide the necessary efficiency for optimizing such high-resolution model parameters as aerosol emissions and boundary conditions.

ISORROPIA Adjoint

Development of the adjoint of ISORROPIA is carried out within the model framework of CMAQ-ADJ, the adjoint of the Community Multiscale Air Quality (CMAQ) model (Byun and Schere, 2006), an internationally employed regulatory model maintained by the EPA. Hakami et al. developed the adjoint of the gas phase processes captured by the CMAQ model (2007), providing a framework within which an adjoint of aerosol properties would be of profound relevance. The modular nature of the CMAQ forward model allows one to treat it as individual box models (Hakami et al., 2007); in this sense, the box model for the thermodynamics, chemistry, and dynamics of aerosol species has not yet been treated in its adjoint.

The key components of atmospheric aerosol are water, inorganic salts, crustal minerals, organics, and trace metals (Seinfeld and Pandis, 2006). Inorganic species constitute about 25 to 50% of the dry mass of the particle depending on the air mass, wherein ammonium (NH_4^+), sodium (Na^+), sulfate (SO_4^{2-}),, bisulfate (HSO_4^-), nitrate (NO_3^-), and chloride (Cl^-), are the most abundant species (Heitzenberg, 1989). In areas where dust contributes significantly to the aerosol loading, crustal species such as Ca^{2+}, K^+, and Mg^{2+} are also significant constituents, which can perturb equilibria of other species and are important for accurate modeling of aerosol composition and size distributions (Ansari and Pandis, 1999). These species are treated in ISORROPIA II. In the atmosphere and for small aerosol sizes, thermodynamics govern the partitioning of these species which can exist as gases, as solutes in the aerosol water, or as precipitated salts. Modeling the complex phase diagram that describes this behavior requires optimization of nonlinear (local) convex problem (Fountoukis and Nenes, 2007). Of the models developed to describe these states of aerosol precursors and species, CMAQ primarily makes use of ISORROPIA's description of the behavior of deliquesced aerosol; therefore, the adjoint of ISORROPIA has been developed for these aspects of the inorganic thermodynamic equilibrium model.

Results

The adjoint of ISORROPIA consists of the original code supplemented with additional calculations to produce sensitivity values as well as transformations to allow these sensitivities to be traced through the code in reverse order. Due to the complexity of such a task, the ISORROPIA code was manipulated for processing with the automatic differentiation tool TAPENADE (Hascoet and Pascual, 2004), which produced augmented code. In cases where the derivative was not traceable through the original mathematical solution, alternate solution algorithms have been developed and validated against the original solutions. Evaluation of the adjoint consists of comparing its sensitivities against finite difference sensitivities for the range of atmospherically relevant concentrations and conditions. The preliminary results in Figure 1 demonstrate the capability of this approach to treat the complex solution algorithm of the ISORROPIA models.

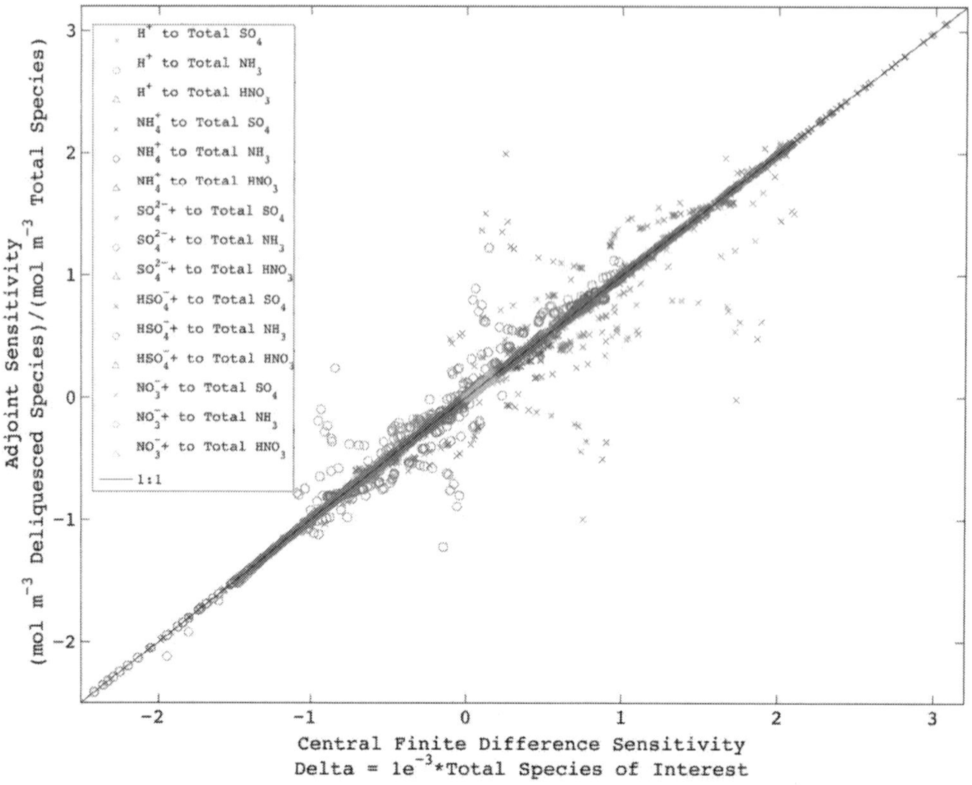

Figure 1. Comparison of the sensitivity results from the brute force method (i.e., central finite difference) and the adjoint method. The brute force results have been filtered based on stoichiometric constraints in order to exclude sensitivities that are thermodynamically unreasonable. The results span the range of atmospherically relevant concentrations of total sulfate, nitrate, and ammonia present in an air mass in ratios such that the aerosol phase is sulfate rich. The range of relevant relative humidities that are treated properly by ISORROPIA (i.e., 5% to 95%).

With the completion of the adjoint of ISORROPIA, coupling with CMAQ-ADJ will provide the adjoint of a regional chemical transport model including its treatment of inorganic aerosol and its gaseous precursors. Such a tool is useful for source apportionment as well as reconciliation of emissions inventories with observations of aerosol from satellites and field campaigns.

References

Altshuller, A. P., R. A. Linthurst, et al. (1984). The acidic deposition phenomenon and its effects : critical assessment review papers. Washington, DC, U.S. Environmental Protection Agency, Office of Research and Development.

Ansari, A. and S. Pandis (1999). "An Analysis of Four Models of Predicting the Partitioning of Semivolatile Inorganic Aerosol Components." Aerosol Sc. & Tech. **31**: 129-153.

Byun, D. and K. Schere (2006). "Review of the Governing Equations, Computational Algorithms, and Other Components of the Models-3 Community Multiscale Air Quality (CMAQ) Modeling System." Applied Mechanics Reviews **59**: 51.

Cacuci, D. (1981). "Sensitivity theory for nonlinear systems. I. Nonlinear functional analysis approach." Journal of Mathematical Physics **22**: 2794.

Courtier, P. and O. Talagrand (1987). "Variational assimilation of meteorological observations with the adjoint vorticity equation. II: Numerical Results." Q J Roy Meteor Soc **113**(478).

Craig, L., J. R. Brook, et al. (2008). "Air pollution and public health: a guidance document for risk managers." J Toxicol Environ Health A **71**(9-10): 588-698.

Dockery, D. W. (2001). "Epidemiologic evidence of cardiovascular effects of particulate air pollution." Environmental Health Perspectives **109**: 483-486.

Dubovik, O., T. Lapyonok, et al. (2008). "Retrieving global aerosol sources from satellites using inverse modeling." Atmos Chem Phys **8**: 209.

Elbern, H., H. Schmidt, et al. (2000). "4D-variational data assimilation with an adjoint air quality model for emission analysis." Environmental Modelling and Software **15**(6-7): 539-548.

Forster, P., V. Ramaswamy, et al. (2007). Changes in Atmospheric Constituents and in Radiative Forcing. Climate Change 2007: The Physical Science Basis. Contribution of Working Group I to the Fourth Assessment Report of the Intergovernmental Panel on Climate Change. S. Solomon, D. Qin, M. Manninget al. Cambridge, UK and New York, NY, Cambridge University Press.

Fountoukis, C. and A. Nenes (2007). "ISORROPIA II: a computationally efficient thermodynamic equilibrium model for K^+–Ca^{2+}–Mg^{2+}–NH_4^+--Na^+--SO_4^{2-}-NO_3^--Cl^--H_2O aerosols." Atmos. Chem. Phys **7**: 4639-4659.

Hakami, A. (2005). "Adjoint inverse modeling of black carbon during the Asian Pacific Regional Aerosol Characterization Experiment." J. Geophys. Res. **110**(D14): 1-17.

Hakami, A., D. K. Henze, et al. (2007). "The Adjoint of CMAQ." Environ. Sci. Technol.

Hakami, A., J. Seinfeld, et al. (2006). "Adjoint sensitivity analysis of ozone nonattainment over the continental United States." Environ Sci Technol **40**(12): 3855-3864.

Hartley, D. and R. Prinn (1993). "Feasibility of determining surface emissions of trace gases using an inverse method in a Three-Dimensional Chemical Transport Model." J Geophys Res-Atmos **98**(D3): 5183-5197.

Hascoet, L. and V. Pascual (2004). TAPENADE 2.1 User's Guide, INRIA.

Heitzenberg, J. (1989). "Fine particles in the global troposphere: a review." Tellus **41B**: 149-160.

Henze, D. and J. Seinfeld (2007). "Development of the adjoint of GEOS-Chem." Atmos Chem Phys **7**(9): 2413-2433.

Henze, D. K., A. Hakami, et al. (2007). "Development of the adjoint of GEOS-Chem." Atmospheric Chemistry & Physics **7**(9): 2413-2433.

Henze, D. K., J. Seinfeld, et al. (2009). "Inverse modeling and mapping US air quality influences of inorganic $PM_{2.5}$ precursor emissions using the adjoint of GEOS-Chem." Atmos. Chem. Phys. **9**: 5877-5903.

Kaminski, T., M. Heimann, et al. (1999). A coarse grid three-dimensional global inverse model of the atmospheric transport. 2. Inversion of the transport of CO2 in the 1980s. J. Geophys. Res. **104:** 18555-18581.

Kaminski, T., M. Heimann, et al. (1999). "A coarse grid three dimensional global inverse model of the atmospheric transport, 2, Inversion of the transport of CO2 in the 1980s." J Geophys Res **104**: 18555-18581.

Kopacz, M., D. Jacob, et al. (2009). "Comparison of adjoint and analytical Bayesian inversion methods for constraining Asian sources of carbon monoxide using satellite (MOPITT) measurements of CO columns." J. Geophys. Res. **114**(D4): 1-10.

Kurokawa, J., K. Yumimoto, et al. (2009). "Adjoint inverse modeling of NOx emissions over eastern China using satellite observations of NO2 vertical column densities." Atmospheric Environment **43**(11): 1878-1887.

Levelt, P. F., B. V. Khattatov, et al. (1998). Assimilation of MLS ozone measurements in the global three-dimensional chemistry three-dimensional chemistry transport model ROSE. Geophys. Res. Lett. **25:** 4493-4496.

Marchuk, G. I. (1977). Methods of numerical mathematics. Applications of Mathematics. **2:** 316.

Martien, P. T. and R. A. Harley (2006). "Adjoint sensitivity analysis for a three-dimensional photochemical model: Application to Southern California." Environ Sci Technol **40**(13): 4200-4210.

Mendoza-Dominguez, A. and A. Russell (2001). Estimation of emission adjustments from the application of four-dimensional data assimilation to photochemical air quality modeling. Atmospheric Environment. **35:** 2879-2894.

Muller, J. F. and T. Stavrakou (2005). "Inversion of CO and NOx emissions using the adjoint of the IMAGES model." Atmos Chem Phys **5**: 1157-1186.

Navon, I. M. (1998). "Practical and theoretical aspects of adjoint parameter estimation and identifiability in meteorology and oceanography." Dynamics of Atmospheres and Oceans **27**(1-4): 55-79.

Patra, P. K., S. Maksyutov, et al. (2003). "An evaluation of CO2 observations with Solar Occultation FTS for Inclined-Orbit Satellite sensor for surface source inversion." J Geophys Res-Atmos **108**(D24).

Remer, L., Y. J. Kaufman, et al. (2005). The MODIS aerosol algorithm, products, and validation. J Atmosph Sci. **62:** 947-973.

Rodgers, C. D. (2000). Inverse Methods for Atmospheric Sounding: Theory and Practice. River Edge, NJ, World Scientific Publishing Co. Pte. Ltd.

Sandu, A., W. Liao, et al. (2005). "Inverse Modeling of Aerosol Dynamics Using Adjoints: Theoretical and Numerical Considerations." Aerosol Sc. & Tech. **39**(8): 677-694.

Seinfeld, J. H. and S. N. Pandis (2006). Atmospheric Chemistry and Physics: From Air Pollution to Climate Change. Hoboken, New Jersey, John Wiley & Sons, Inc.

Talagrand, O. and P. Courtier (1987). "Variational Assimilation of Meteorological Observations with the Adjoint Vorticity Equation 1. Theory." Q J Roy Meteor Soc **113**(478): 1311-1328.

Tarantola, A. (1987). Inverse problem theory : methods for data fitting and model parameter estimation. Amsterdam ;, Elsevier ;.

Trenberth, K., J. Fasullo, et al. (2009). "Earth's Global Energy Budget." Bull Amer Met Soc **90**(3): 311.

Vaughan, M. A., S. Young, et al. (2004). "Fully automated analysis of space-based lidar data: an overview of the CALIPSO retrieval algorithms and data products." Proc. of SPIE Vol **5575**: 16-30.

Yumimoto, K., I. Uno, et al. (2008). "Adjoint inversion modeling of Asian dust emission using lidar observations." Atmos Chem Phys **8**(11): 2869-2884.

Zhang, L., E. Constantinescu, et al. (2008). "An adjoint sensitivity analysis and 4D-Var data assimilation study of Texas air quality." Atmospheric Environment **42**(23): 5787-5804.

Zhang, L., D. Jacob, et al. (2009). "Intercontinental Source Attribution of Ozone Pollution at Western US Sites Using an Adjoint Method." Geophys. Res. Lett.

Aerosol Production From Solution Atomization and the Important Influences of Solvent, Concentration, Solution Age and Chemistry

Timothy M. Raymond, Bucknell University, Lewisburg, PA

Juan Alberto Lopez Ruiz, Chemical Engineering, Bucknell University, Lewisburg, PA

Solution atomization and subsequent drying is the primary method by which aerosols are generated for laboratory studies such as CCN activity, hygroscopic growth, or smog chamber seeding experiments. The typical assumptions for this atomization process include:

1) The solvent does not affect the aerosol properties

2) The concentration of solution only affects the size distribution

3) The solution produces the same aerosols regardless of its age

4) The solution contains chemicals that do not react in any way

We have conducted experiments using both organic and inorganic compounds and mixtures dissolved in a variety of solvents. We have investigated CCN activity and particle morphology of the resulting atomized particles after increasing aging of the solution. While Koehler theory accurately describes the activation of soluble inorganic salts, it is still not possible to accurately predict the water interaction behavior of organic aerosols. In recent years, several studies of CCN activity for organic aerosols have investigated the effects of solvents, organic coatings, mixed particles, and smog chamber-generated aerosols. The results of these studies are not in agreement and have resulted in conflicting findings. The current study has used a commercially available cloud chamber (DMT CCNC) and the recently reported Scanning Mobility CCN Activity (SMCA) method to measure several different organic species and mixtures of varying complexity.

Additionally, we have investigated the morphology for atmospherically-relevant sized particles using Atomic Force Microscopy (AFM). Results have been analyzed to determine the relationships among chemistry, solvent, aging, and concentration. We have discovered that particle morphology may vary depending on the particle generation methods and conditions and that the morphology may influence properties such as CCN activity.

See more of this Session: Atmospheric Chemistry and Physics - II
See more of this Group/Topical: Environmental Division

Size-Resolved Scanning Flow CCN Analysis (SFCA): A Method for Fast Measurements of CCN Spectra

Richard H. Moore, Chemical and Biomolecular Engineering, Georgia Institute of Technology, Atlanta, GA

Athanasios Nenes, Schools of Chemical & Biomolecular Engineering and Earth & Atmospheric Sciences, Georgia Institute of Technology, Atlanta, GA

The Continuous-Flow Streamwise Thermal-Gradient Cloud Condensation Nuclei Counter (CFSTGC) is rapidly becoming the instrument of choice for both laboratory and field measurements of cloud condensation nuclei (CCN). Typically, the instrument is operated at a constant flow rate, and supersaturation is adjusted by changing the column temperature difference, which is relatively slow (20-40 seconds for the column temperatures to stabilize and as much as 3 minutes for the OPC temperature to stabilize) and may introduce particle volatilization biases[1]. Data collected during transient periods are often discarded, which is particularly problematic for aircraft measurements in heterogeneous environments. We present a new mode of operation, termed "Scanning Flow CCN Analysis" (SFCA), in which the flow rate is changed over time in the flow chamber, while maintaining a constant temperature gradient; this causes supersaturation to continuously change, allowing the measurement of a CCN supersaturation spectrum much more quickly over timescales relevant for airborne measurements (~10-60 seconds and potentially even less) and without loss of data continuity. We present results of size-resolved CCN spectra obtained using SFCA, coupled with an electrostatic classifier to demonstrate the application of the technique to ambient measurements of urban aerosol.

[1]Asa-Awuku, A., et al., Atmos. Chem. Phys., 9, 2009.

See more of this Session: Atmospheric Chemistry and Physics - II
See more of this Group/Topical: Environmental Division

The Importance of Adsorption Activation for Mineral Dust and CCN Activity of Diverse Regional Dust Samples

Prashant Kumar, School of Chemical & Biomolecular Engineering, Georgia Institute of Technology, Atlanta, GA

Irina N. Sokolik, School of Earth & Atmospheric Sciences, Georgia Institute of Technology, Atlanta, GA

Athanasios Nenes, Georgia Institute of Technology, Atlanta, GA

Mineral dust aerosols represent a dominant source of particulate (by mass) in the atmosphere and have been found to play important role in warm clouds due to their ability to serve as effective Cloud Condensation Nuclei (CCN). Current climate models, use traditional Köhler theory to describe supersaturation required for dust activation based on its soluble fraction, and completely ignore interactions of hydrophilic insoluble core with water vapor even in appreciable. In this study, we address the activation of hydrophilic insoluble dust particle using multilayer Frenkel-Halsey-Hill (FHH) adsorption isotherm model modified to account for particle curvature. A comparison of different aerosol-water vapor interaction mechanisms for cloud droplet formation suggests that traditional Köhler theory can lead to substantial differences in parcel supersaturation and cloud droplet number to that predicted by the adsorption activation framework. A comparison based on the CCN spectra of different mineral dusts suggests that the traditional Köhler theory overestimates the CCN number relative to that predicted by multilayer physical adsorption models.

We also investigate the CCN-relevant properties of mineral dust samples representative of major regional dust sources using the Continuous-Flow Streamwise Thermal Gradient CCN Counter (CCNc). The CCN activation behavior of mineral dusts from Northern America, African soils, and East Asian soils are studied experimentally in the laboratory condition to determine their respective affinity to water. In parallel, CCN activities of various individual minerals (such as clays and carbonates) are compared with regional dust samples, to assess the role of the mineralogical composition of dust samples on their CCN activity. This new framework of aerosol-water interaction by the "adsorption activation" has been incorporated into aerosol-cloud interaction parameterizations for mineral dust activation.

See more of this Session: Atmospheric Chemistry and Physics - II
See more of this Group/Topical: Environmental Division

Size-Resolved CCN Activity, Mixing State, and Droplet Growth Kinetics of Atlanta Aerosols

Luz T. Padro, School of Chemical and Biomolecular Engineering, Georgia Institute of Technology, Atlanta, GA

Richard H. Moore, Chemical Engineering, Georgia Institute of Technology, Atlanta, GA

Xialou Zhang, School of Earth and Atmospheric Sciences, Georgia Institute of Technology, Atlanta, GA

Neeraj Rastogi, School of Earth and Atmospheric Sciences, Georgia Institute of Technology, Atlanta, GA

Rodney Weber, Earth and Atmospheric Sciences, Georgia Institute of Technology, Atlanta, GA

Athanasios Nenes, Georgia Institute of Technology, Atlanta, GA

The effects of aerosol composition (especially from organic species) on droplet formation is highly variable and uncertain, but can be largely constrained by size-resolved Cloud Condensation Nuclei (CCN) measurements. A fast method for obtaining size-resolved CCN measurements is Scanning Mobility CCN Analysis (SMCA; Nenes and Medina, in review). In SMCA, ambient monodisperse aerosol from a Differential Mobility Analyzer (operated in scanning voltage mode) is concurrently introduced into a Streamwise Thermal Gradient CCN Chamber (STGC, Roberts and Nenes, 2005) and a Condensation Particle Counter. Inversion of the concentration timeseries from both instruments yields the activation ratio (i.e., fraction of particles that act as CCN) as well as the activated droplet size, as a function of dry mobility diameter and supersaturation. In this study, we performed SMCA for aerosols sampled at Atlanta, GA during August and September of 2008. The daily trend of the mixing state, CCN activity, and droplet size of the size-resolved CCN are investigated and compared to chemical composition measurements. Inferences about the impact of photochemistry and mixing state on CCN activity and droplet growth kinetics are carried out. We also perform CCN closures (i.e., comparison of predicted and measured CCN) in which different compositional assumptions (e.g., size-invariant composition, organic insoluble/soluble with constant hygroscopicity) are addressed.

See more of this Session: Atmospheric Chemistry and Physics - II
See more of this Group/Topical: Environmental Division

Boundary Layer Interactions with Ultrafine Aerosol

Alicia Pettibone, Department of Chemical and Biochemical Engineering, University of Iowa, Iowa City, IA

William Eichinger, Department of Civil and Environmental Engineering, University of Iowa, Iowa City, IA

Charles Stanier, Department of Chemical & Biochemical Engineering, University of Iowa, Iowa City, IA

Field measurements and collocated balloon and LIDAR measurements show pronounced diurnal patterns where new particle formation and growth is often seen at the time of ventilation of the stable nocturnal surface layer. For example, in the March 2006 MILAGRO campaign in Mexico City, observations at the T0 location could often be characterized by morning conditions with high particle mass concentrations, low mixing heights, and good correlation between particle number and carbon dioxide, indicative that particle number is controlled by primary emissions. In the afternoon, the CO2 level drops during ventilation of the daily polluted layer, and the coupling between CO2 and particle number breaks down, with particle number sometimes increasing as CO2 decreases. New particle formation events were observed both simultaneous to and separately from the afternoon ventilation. Simultaneous particle and mixed layer height measurements from Eastern Iowa also show a tight correlation in time between new particle formation events and boundary layer mixing. The use of rapid afternoon ventilation of the polluted layer to infer semivolatility of primary emissions is also discussed. We calculate size-resolved emission factors for Mexico City using a novel analysis method that uses the relative areas of short term elevations in CO2 and particle number to derive emission factors. Using this correlation between particle number and carbon dioxide, an average number based-emission factor of 3.4 x 10^(13) #/vehicle-km has been determined.

See more of this Session: Atmospheric Chemistry and Physics - II
See more of this Group/Topical: Environmental Division

Direct Measurement of Mercury in Simulated Coal Combustion Flue Gas

Bihter Padak, Energy Resources Engineering, Stanford University, Stanford, CA

Jennifer Wilcox, Energy Resources Engineering, Stanford University, Stanford, CA

Coal-fired power plants are the greatest anthropogenic source of mercury emissions in the United States. Mercury exists in coal combustion flue gas in a variety of forms depending on the coal type and combustion conditions, i.e., elemental, oxidized, and particulate. Understanding mercury speciation during combustion and how the transformations occur between different forms of mercury is essential to develop an effective control mechanism for removing mercury from flue gas. Homogeneous oxidation of mercury in the flue gas of coal combustion utility boilers has been studied for many years to understand the speciation of mercury. In spite of a vast amount of experimental studies, supported by modeling efforts, there are still many questions to be answered and the speciation of mercury is not fully understood yet. One needs to be able to make precise mercury measurements to understand the mercury speciation and accurately predict the extents of mercury oxidation. Traditionally "difference" techniques are used which involves the direct measurement of elemental mercury. These techniques do not allow for distinguishing between the two different oxidized forms, Hg+ and Hg+2 which makes it difficult to understand mercury speciation. It has recently been shown that mercury measurements performed with wet chemical conditioning systems are biased, resulting in inaccurate partitioning between oxidized and elemental mercury species. Given the shortcomings of the difference techniques, it is essential to measure oxidized and elemental mercury directly and hence separately to have a complete understanding of mercury speciation. In this study an electron ionization quadrupole mass spectrometer will be used to directly measure mercury species in combustion flue gas. A benefit of employing a mass spectrometer is, unlike traditional impinger methods, the oxidized forms can be isolated and individually identified because it separates the products based on their mass to charge ratio. By directly measuring mercury species accurately, one can determine the actual extent of mercury oxidation in the flue gas, which will aid developing effective mercury control technologies.

See more of this Session: Control of Metal Hap Emissions (Hg, Se, As, Cr, etc.)
See more of this Group/Topical: Environmental Division

Removal of NOx, SO2 and Hg From Flue Gases by Combinative Processes Involving Aqueous Strong Oxidants and Sonochemical Methods

Yusuf G. (Debo) Adewuyi, Chemical & Bioengineering Department, North Carolina Agricultural and Technical State University, Greensboro, NC

Nymul E. Khan, Chemical & Bioengineering Department, North Carolina Agricultural and Technical State University, Greensboro, NC

Recent studies indicate that about 70-85% of total anthropogenic Hg emissions are caused by combustion sources, mainly coal-fired utilities and waste incinerators. The task of Hg pollution abatement and management is arduous due to the complex environmental cycling of Hg compounds. Lime/limestone wet scrubbers widely used for flue gas desulfurization because of their high removal efficiency, reasonable cost and wide applicability, have been shown, in general, to achieve less than 70% mercury removal efficiency. Selective catalytic reduction (SCR) used for high NOx removal has very high capital cost and suffers from various problems including disposal of spent toxic catalysts. Selective noncatalytic reduction (SNCR, urea injection) is highly temperature dependent and SNCR, ammonia injection (Thermal DeNOx) has moderately high capital cost and ammonia handling problems. Low-NOx burners (LNB) also has moderately high capital cost. It is also known that NO can promote or inhibit the homogeneous Hg oxidation, depending on its concentration. Also, oxidized mercury, Hg2+ could also be reduced by aqueous S(IV)(sulfite and/or bisulfite) species resulting in Hg emissions under typical wet FGD scrubber.

We recently demonstrated the effective use of both oxone, NaClO2 and •OH radical via ultrasonic irradiation for the simultaneous absorption-oxidation of NOx and SO2 using a bubble column reactor. If a high removal efficiency of mercury could be achieved from these chemical systems, wet scrubber systems could be considered for cost-effective multi-pollutant control of Hg, NOx and SO2. The persulfate anion is one of the most powerful oxidant of the peroxygen family of compounds. The objective of this study is to develop an innovative and cost-effective aqueous scrubber for the removal of elemental mercury (Hg) and other atmospheric pollutants (e.g., NOx, SO2) simultaneously from flue gases under wet FGD scrubber operating conditions using aqueous solutions of sodium persulfate (Na2S2O8)alone or in combination with ultrasonic irradiations, and to investigate the aqueous chemistry and kinetics of elemental mercury (Hg) absorption in the absence and presence of standard flue gas scrubbing system components like NO/NO2, SO2, HCl, CO2, O2, and salts (NaCl, CaCl2). The results of our studies to date will be presented.

References

(1) Adewuyi, Y.G., Owusu, S.O. Ultrasound-induced Aqueous Removal of Nitric Oxide from Flue Gases. Effects of Sulfur Dioxide, Chloride and Chemical Oxidant. Journal of Physical Chemistry A. 2006. 110, 11098-11107.

(2) Owusu, S.O; Adewuyi, Y.G. Sonochemical Removal of Nitric Oxide from Flue Gases. Industrial & Engineering Chemistry Research. 2006, 45, 4475-4485.

(3) Adewuyi, Y.G.; Owusu, S.O. The Aqueous Absorption and Oxidation of Nitric Oxide with Oxone for the Treatment of Tail Gases. 1. Process Feasibility, Stoichiometry, Reaction Pathways and Absorption Rate. Ind. Eng. Chem. Res., 2003, 42, 4084-4100.

(4) Adewuyi, Y.G.; He, X.; Shaw, H.;Lolertpihop, W. Simultaneous Absorption and Oxidation of NO and SO2 by Aqueous Solutions of Sodium Chlorite. Chemical Engineering Communications, 1999, 174, 21-51.

See more of this Session: Control of Metal Hap Emissions (Hg, Se, As, Cr, etc.)

See more of this Group/Topical: Environmental Division

Study of a Method of Removing Mercury From Coal Power Plant Flue Gases and Its Effects On NOx and SOx Emissions

M. P. Sharma, Chemical and Petroleum Engineering, University of Wyoming, Laramie, WY

This paper will present the results of ongoing study of a method of mercury removal from flue gas of coal-fired power plants. The state-of-the-art of flue gas mercury removal and different methods of removing mercury including calcium chloride injection, activated carbon injection, bromine gas, and photochemical oxidation will be discussed. At the University of Wyoming, we have conducted extensive study of data of Calcium Chloride injection, Pulverized Activated Coal (PAC) injection and Hybrid injection (of combination of Calcium Chloride and PAC) methods for removing mercury from flue gas of Powder River Basin coal. Process mechanism and technology have been studied. Data were collected and analyzed on mercury removal efficiency. Also, to study the multi-component emission effects of the mercury removal technology, data on emissions of sulfur oxide and nitrogen oxide were collected before and after the implementation of mercury removal method. The collected data from the field test study are being analyzed for making conclusions on mercury removal efficiency and other performance parameters (including effects on multi-component emissions). Details of the study including these conclusions will be discussed.

See more of this Session: Control of Metal Hap Emissions (Hg, Se, As, Cr, etc.)
See more of this Group/Topical: Environmental Division

Effect of the Flue Gas Composition On Mercury Speciation Under Homogeneous and Heterogeneous Conditions

Paula A. Buitrago, Chemical Engineering, University of Utah, Salt Lake City, UT

Geoffrey Silcox, Chemical Engineering, University of Utah, Salt Lake City, UT

Constance L. Senior, Reaction Engineering International, Salt Lake City, UT

JoAnn Slama Lighty, Chemical Engineering, University of Utah, Salt Lake City, UT

Mercury is commonly found in the flue gas derivated from the combustion of fossil fuels such as coal and is considered a hazardous pollutant for humans and the environment. In flue gas, mercury can be present in both elemental and oxidized forms. The oxidized form is water soluble and can generally be collected by existing particulate and SO2 control devices. In contrast, elemental mercury is not soluble in water and is highly volatile at typical combustion temperatures. These factors make it difficult to capture by conventional control devices; hence, considerable attention has been given to finding ways to oxidize elemental mercury. Complex mechanisms for heterogeneous oxidation have been proposed. A better understanding of mercury speciation and oxidation mechanisms is essential for the development of air pollution control strategies and devices.

The focus of this work is to study how the composition of the flue gas components (HBr, SO2, and HCl) can affect mercury speciation under homogeneous and heterogeneous conditions. Experimental data were obtained with two reactors: a 300-W, methane-fired, tubular, quartz-lined reactor for studying homogeneous oxidation reactions and a fixed-bed reactor, also of quartz, for studying heterogeneous reactions. The latter was attached to the exit of the former to provide realistic combustion gases. The fixed-bed reactor was packed with one gram of coconut-shell carbon and kept at a temperature of 150oC. All methane, air, SO2, and halogen species were introduced through the burner to produce flue representative of real combustion systems. Speciated mercury measurements were performed using a Tekran 2537A Analyzer coupled with a wet conditioning system.

Gas phase measurements showed that in presence of bromine the observed mercury oxidation ranged from 28% at 18 ppm bromine (as HBr equivalent) to 80% at 55 ppm bromine as (HBr equivalent), while values of mercury oxidation were 2% with 100 ppm chlorine (as HCl equivalent) and 6% with 500 ppm chlorine (as HCl equivalent). In general, the SO2 showed an inhibitory effect on the mercury oxidation obtained by either bromine or chlorine, with mercury oxidation values of 25 % for the mixture 45 ppm bromine (as HBr equivalent), 500 ppm SO2 and 5% for the mixture 500 ppm chlorine (as HCl equivalent) and 500 ppm SO2 . Under heterogeneous conditions, the coconut-shell carbon showed a mercury capture efficiency of about 20% but in presence of bromine or chlorine this uptake increased up to 90%. In the absence of halogens, SO2 increased the mercury adsorption efficiency to up to 30 percent. The extent of adsorption decreased with increasing SO2 concentration when halogens were present.

A model of fixed-bed capture of mercury that incorporates Langmuir adsorption kinetics was developed to predict mercury adsorption and the effect of the flue gas components. The model neglects intraparticle diffusion resistances and is only applicable to pulverized carbon adsorbents. The model roughly describes experimental data from the literature. The current version includes the ability to account for competitive adsorption between mercury, SO2, and NO2. Future work will focus on the development of surface reactions between mercury and halogens.

See more of this Session: Control of Metal Hap Emissions (Hg, Se, As, Cr, etc.)
See more of this Group/Topical: Environmental Division

AUTHOR INDEX

AUTHOR INDEX

AUTHOR INDEX

AUTHOR INDEX

AUTHOR INDEX